D0908584

RIVERS OF CHANGE

RIVERS OF CHANGE

Essays on Early Agriculture in Eastern North America

Bruce D. Smith

with contributions by C. Wesley Cowan and Michael P. Hoffman

Smithsonian Institution Press

Washington and London

Editor: Robin A. Gould
Supervisory Designer: Janice Wheeler
Designer: Chris Hotvedt

Library of Congress Cataloging-in-Publication Data
Smith, Bruce D.
Rivers of change : essays on early agriculture in eastern North America / Bruce D. Smith with contributions by C. Wesley Cowan and Michael P. Hoffman
p. cm.
Includes bibliographical references.
ISBN 1-56098-162-8
1. Paleo-Indians—Agriculture. 2. Paleo-Indians—Food. 3. Agriculture—North American—Origin. 4. Agriculture. Prehistoric—North America. 5. Plant remains (Archaeology)—North America. 6. North America—Antiquities. I. Title.
E59.A35S645 1992
630′ .97—dc20 91-30324
CIP

British Library Cataloguing-in-Publication Data available
Manufactured in the United States of America
10 9 8 7 6 5 4 3 2 1
00 99 98 97 96 95 94 93 92

∞ THE PAPER USED IN THIS PUBLICATION MEETS THE MINIMUM REQUIREMENTS OF THE AMERICAN NATIONAL STANDARD FOR PERFORMANCE OF PAPER FOR PRINTED LIBRARY MATERIALS Z 39.48–1984.

Publisher's Note: Chapters 3, 5, 6, 7, 9, 11, and 12 of this work have appeared in a slightly different version in Southern Illinois University at Carbondale's Center for Archaeological Investigations Occasional Papers series, *Southeastern Archaeology, Ethnobiology,* Monographien des Roeisch-Germanischen Zentralmuseums series, and *Science.* Bibliographic form as published has been adapted to the style of this volume.

*For the
Batemans of
Upper Canada,
Farmers of the
Long Woods of
Caradoc*

The publication of this book was made possible by an award from the Smithsonian Regents Publications Program.

Contents

3

The Independent Domestication of Indigenous Seed-Bearing Plants in Eastern North America 35

4

Is It an Indigene or a Foreigner?

Bruce D. Smith, C. Wesley Cowan, Michael P. Hoffman 67

III
Premaize Farming Economies in Eastern North America

5

The Role of *Chenopodium* as a Domesticate in Premaize Garden Systems of the Eastern United States 103

6

Chenopodium berlandieri ssp. *jonesianum:* Evidence for a Hopewellian Domesticate from Ash Cave, Ohio 133

IV
Synthesis

11

12

List of Figures and Tables

Figures

Tables

I.
Rivers of Change

CHAPTER 1

❑ ❑ ❑

Introduction: Fields of Opportunity, Rivers of Change

Organized into four sections, the twelve chapters of this volume are concerned with prehistoric Native American societies in eastern North America and their transition from a hunting and gathering way of life to a reliance on food production. Written at different times over the past decade, the chapters vary both in length and topical focus. They are joined together, however, by a number of shared "rivers of change."

The most obvious of the rivers of change that flow as common themes through the chapters of this book are the actual watercourses of the Eastern Woodlands of North America. Continually transformed by violent spring floods and the relentless meandering forces of the millennia, joining the past with the present, these rivers shape the floodplain landscapes that were the setting for the initial development and subsequent elaboration of prehistoric food producing economies in the region. The floodwaters of spring constantly created open habitats—habitats that in turn encouraged colonizing plants to develop attributes which preadapted them for manipulation and domestication by Native Americans. These rivers of the East, and the landscapes of open habitats and annually renewed arable soils they create and constantly change, are inexorably linked to the domestication of plants and the development of farming economies in the East. Their currents flow, silently and powerfully, across the pages of this book.

The lengthy transition from foraging to farming that took place in the East can also be seen and charted as rivers of change—a complex developmental network involving a number of indigenous seed plants, each with its own distinct course of transformation. Indications of the interwoven, braided courses of development taken by these plants, and the increasingly important roles that they played in the economies of prehistoric Native Americans, are recorded in the archaeological sites and archaeobotanical sequences long buried in the region's river valley soils.

Many of the crop plants thus recorded began as wild components of floodplain plant communities. In recent years river valleys of the East have witnessed research focused on the modern wild and weedy descendant populations of the species that were important cultivated plants in prehistory. This focus on modern plants as a way of learning more about ancient crops represents one of many currents in yet another, more recent, and quite turbulent, river of change. Like the floodplain landscape of a major river, subject to powerful floodwaters, the discipline of archaeobotany in eastern North America has witnessed dramatic changes during the decade of the 1980s. Support for analysis of archaeological plant assemblages has increased substantially, and like the nutrient-rich waters of spring floods, this increased federal funding has been widely dispersed, fueling larger scale and better quality analysis of archaeobotanical assemblages over broad geographical areas. In addition, a number of new and powerful streams of innovation in methodology and instrumentation have radically reworked the subfields of research in eastern North American archaeobotany. Large and long-established problem areas have literally been swept away by the currents of technological innovation. At the same time, like the open habitats produced by powerful floodwaters, entirely new fields of inquiry have been created by these forces of innovation and change. Flotation recovery has resulted in vastly improved and expanded fields of archaeobotanical information. Scanning electron microscopy has dramatically expanded areas of micromorphological analysis of seed structure. Accelerator mass spectrometer (AMS) radiocarbon dating has provided a means of accurately placing specific plant assemblages or even individual seeds on the temporal-developmental landscape of the region. These interrelated advances of the 1980s—powerful currents of innovation in instrumentation and methodology, and the resultant expansion and reworking of the research landscape of archaeological plant assemblages—have combined to cause, in part, a radically altered picture of plant domestication and the development of farming economies in eastern North America. Richard Yarnell (1991) provides an excellent summary of the growth and expansion of archaeobotany in the 1980s.

The history of research on the question of agricultural origins in eastern North America before the turbulent 1980s also winds quietly through many of the chapters, from Le Page du Pratz's description in the 1720s of sand-bank cultivation of a mystery grain by the Natchez, past Ebenezer Andrews's discovery in 1876 of stored seeds in Ash Cave, Ohio, and the pioneering work of Melvin Gilmore and Volney Jones in the 1930s. This stream of history includes the theories of Edgar Anderson and the floodplain weed research of Jonathan Sauer in the 1950s, the subsequent hypotheses of Melvin Fowler and Stuart Struever and the theories of Jack Harlan, and the landmark research of Richard Yarnell and Patty Jo Watson, leading up to the rapid currents of change in the 1980s.

So while this book is focused on tracing the course of a major process of transformation in eastern North America—the domestication of plants and the initial appearance of farming economies—it also has a number of other streams of change, sometimes obvious, sometimes embedded, meandering as integrating subthemes through its pages.

Drawing on my own research as well as the work of many other researchers, and written during the past decade of rapid advancement in eastern North American archaeobotany, the chapters of this volume also chart a personal course of changing and expanding research interests through the 1980s that was strongly influenced by the surrounding context of debate and emerging opportunity. This process of personal expansion of inquiry initially involved the extensive investigation of a single prehistoric crop plant—*Chenopodium berlandieri*. My initial research on *C. berlandieri* in turn provided a stable vantage point from which to address larger issues of change, leading me to consider a wide range of related questions regarding the transition from foraging to farming economies in the East.

The transition from hunting and gathering to farming was one of the major ecological changes in the history of our planet. This significant turning point in human history occurred at different times in different parts of the world, and involved a wide range of different species of plants and animals (Cowan and Watson 1992). In eastern North America the long transformation occurred way before the arrival of Europeans and played a central role in shaping the long course of Native American cultural development in the region. But the timing and nature of this major developmental transition has been a continuing subject of debate for more than sixty years. A central focus of this lengthy debate has been the role of three Mesoamerican food crops, maize, beans, and squash, in the development of plant husbandry and farming economies in the East. Was Mesoamerica the vital source of initial agricultural concepts and crops for eastern North America? Or did domesticated plants and farming economies develop in eastern North America independently of Mesoamerica and before the introduction of maize or other Mesoamerican crop plants into the region?

This latter possibility was first put forward in 1924 when Ralph Linton suggested that farming economies based on indigenous seed plants may have existed prior to the adoption of maize agriculture in the East. Although Linton may have known of seed caches being recovered from Ozark rock shelters, he may also have drawn his inspiration from Le Page du Pratz's early eighteenth-century account of Natchez Indians sowing the ancient small grain *choupichoul.* Evidence of the indigenous small grain seed crops suggested by Linton had in fact been recovered by Ebenezer Andrews in Ash Cave, Ohio in 1876. But the temporal placement and domesticated status of the large cache of chenopod seeds *(Chenopodium berlandieri* ssp. *jonesianum)* discovered by Andrews would not be established for more than half a century after Linton's conjecture. While this and other museum collections bearing on the question of indigenous plant husbandry systems would not be reanalyzed until the 1980s, far more abundant and varied archaeological evidence regarding possible indigenous plant husbandry came to light soon after Linton's inspired suggestion.

Excavations in the dry rock shelters of the Arkansas Ozarks and the Red River gorge of eastern Kentucky during the 1920s and 1930s provided pioneering archaeobotanists Melvin Gilmore and Volney Jones with abundant evidence of North American crop plants, both cultivated and domesticated, including chenopod *(Chenopodium),* amaranth *(Amaranthus),* ragweed *(Ambrosia),* marshelder *(Iva),* maygrass *(Phalaris caroliniana),* and sunflower *(Helianthus annuus).* But the critically important question of the relative age of these indigenous crop plants relative to introduced tropical crops, including maize, remained unclear. Squash *(Cucurbita pepo)* and maize *(Zea mays)* were often recovered in the same contexts as local cultigens. In 1946 George Quimby succinctly stated this unresolved issue of temporal precedence—an issue that for another forty years would cast a shadow on the possible existence of distinctly indigenous food producing economies in the East: "In the available archaeological evidence. . .there is nothing to show that the eastern plant complex is earlier than the tropical agricultural assemblage" (Quimby 1946:5). So there was good evidence by the 1930s, from two widely separated areas of the East, for the existence of farming economies that included a variety of indigenous small-seed crop plants. But it was not possible to determine whether these local crops were the product of the independent development of domesticated plants and food production prior to the introduction of Mesoamerican crops such as squash and maize, or if, on the other hand, they were peripheral and minor crops brought under domestication on the coattails of introduced tropical domesticates.

Certainly from a common sense approach the latter possibility would seem far more likely. Early European explorers and settlers, after all, provide rich descriptions of Native American economies solidly based on the tropical crop trinity of maize, beans, and squash. With the exception of Le Page du Pratz and a few other narratives, on the other hand, there was no mention of any small-seed crops in the early European descriptions of eastern Native American agriculture. As they cut out their paper turkeys and pilgrim hats, school children across America learn that Indians of the East grew maize, beans, and squash, and college students enrolled in courses on Native American culture learn that southeastern tribes made more than ninety different dishes from corn.

More importantly, maize is an ever-present dietary element in modern America. We consume corn oil and margarine ("some call it corn, we call it maize"), corn on the cob, creamed corn, popcorn, caramel corn, corn nuts, corn flakes, corn fritters, and corn dogs. Corn blankets much of the midsection of America through the growing season, and is a major world crop. Beans and squash are also grown worldwide, and have become important dietary components in many countries. Corn, squash, and beans are clearly important foods in American culture and are universally acknowledged as such. As familiar and commonly grown crop plants they have considerable stature; they command respect. There is no need to describe or explain maize or squash to the average American, or to American archaeologists. It is not necessary to convince anyone of their potential as dependable food crops, capable of sustaining and supporting human societies, present day or precontact, for that is self-evident. After all, we know what we eat.

In contrast, with the exception of sunflower, native North American seed crops are not exactly household words. Sunflower is known primarily as a cooking oil, a bird-feeder filler, a salad bar selection, and a garden ornamental. Little barley *(Hordeum pusillum)* does not stare up at us from the cereal bowl in the morning. You can't find maygrass *(Phalaris caroliniana)* and erect knotweed *(Polygonum erectum)* included in ingredient lists of products on the shelves of your local grocery store, or even in granola bars. Very few archaeologists (and not that many archaeobotanists) have even observed marshelder *(Iva annua)* or *Chenopodium berlandieri* growing in the wild. With the exception of sunflower, indigenous eastern North American crop plants are known to archaeologists, if at all, only as small carbonized seeds counted and sorted by archaeobotanists, perhaps as obscure weeds unnoticed on the modern landscape, or as lists of latin names.

Of only limited interest to most archaeologists, these plants are completely unfamiliar to average Americans. After listening to lengthy and glowing descriptions of these indigenous crop plants, people invariably ask "How were these seeds prepared? How were they eaten? What do they taste like?" Having never seen or eaten any of these seeds, most Americans, and most archaeologists, have a very difficult time conceptualizing them as forming the basis of prehistoric farming economies. A strong cultural bias exists against these largely unknown crop plants. In terms of eastern North American prehistory, this bias has pushed them into the background, into the shadows cast by the familiar, therefore important, tropical crops that have occupied center stage for so long.

Even when considered by archaeologists, these indigenous crop plants are culturally diminished in the presence of the tropical triad of "real crop plants." Although trained to slip free of their own cultural attitudes and perceptions in order to more objectively interpret the societies they study, archaeologists nonetheless are often clearly reluctant to consider these indigenous crop plants as anything but minor players in the ancient fields of eastern North American farming economies. It could well be that this culturally driven tendency to diminish the possible importance of these indigenous crop plants is not so much a matter of lack of knowledge as it is the presence of such a powerful and culturally internalized comparison in the form of the Mesoamerican trinity. For unlike other object-class aspects of the archaeological record that do not extend into our own cultural sphere of direct experience and knowledge, squash, maize, and beans do. Whatever the relative importance of a lack of knowledge of native crops on the one hand, and cultural familiarity with introduced tropical domesticates on the other, these two factors have worked together as blinders over the years, blocking consideration of an alternative scenario—that eastern North America may have been an independent center of plant domestication, and that substantial food production economies might have developed in the East prior to the introduction of Mesoamerican crop plants.

Another important, if rarely acknowledged, obstacle to a consideration of independent eastern origins of plant domestication and farming economies has been the elegant simplicity of explanations based on the introduction of Mesoamerican domesticates. Such explanations are invariably straightforward, easy to outline, and easy to comprehend: squash, maize, and beans were first domesticated in what is now Mexico. Long after initial domestication these crops were introduced into eastern North America, and in a simple cause and effect developmental shift, eastern North American societies embraced the new crop(s) and the concept of agriculture, and were transformed from foragers to farmers. Details of the actual process of transformation in the East are rarely addressed, perhaps because it is, after all, a secondary event—the transplantation of a fully developed socioeconomic innovation. Difficult questions involving the complex processes of initial plant domestication and the first emergence of farming economies need not be addressed, since they occurred earlier and far to the south, over the horizon. While avoiding the messy and complex puzzles of primary (Mesoamerican) and secondary (eastern North American) developmental processes, such explanations at the same time offer a simple and appealing core consisting of the plants themselves and their embedded agricultural concepts. Mesoamerican origin explanations can, as a result, be described in fewer than a hundred words, and fit comfortably into introductory texts or other formats where only brief coverage of the topic is warranted.

The primary questions engendered by such models involve the origin, route, and timing of the northward movement of tropical domesticates, and occasionally the specific mechanisms of diffusion. Such explanations also allow the subsequent derivative cultivation and domestication of eastern seed plants to be viewed comfortably as a spin-off or copy-cat application of newly acquired concepts to locally exploited wild plants, causing their minimal and peripheral accommodation into emerging farming economies. Mesoamerican origin explanations for the eastern North American transition from forager to farmer thus derive much of their strength from their overlapping common sense appeal to a desire for simple and comforting Cartesian solutions to difficult processes of social transformation on the one hand, and their confirmation, on the other, of our strong cultural attitudes toward "real" tropical crops as opposed to minor indigenous "bird seed" plants.

In addition, such explanations fall within a broader "South of the Border Story" category, and thereby also draw support of sorts from a more general core-periphery perspective that places eastern North America on the marginal, passive, and recipient far northern fringe of influence of "nuclear centers" of Middle and even South America. Under a general hyper-diffusionist banner almost every major cultural innovation in eastern North America, from the invention of ceramics up through Poverty Point and Hopewell to Mississippian has been traced at one time or another over the past fifty years, back along lines of diffusion to nuclear America core areas of cultural innovation. Simplistic accounts that look southward in search of the source of eastern North American cultural change and which thrive in limited information situations have been largely rejected over the years, however, as the level of archaeological knowledge in the region has increased beyond being able to accommodate such "big arrow" explanations. The Mesoamerican origin explanation for the initial development of eastern North American food producing economies in fact was the solitary, still viable, "South of the Border Story" to survive into the 1980s. As such it both invited critical scrutiny and encouraged careful consideration of the alternative—that eastern North America was an independent center of plant domestication and food producing economies.

But this intriguing alternative had very little support in the early 1980s, either among archaeologists and archaeobotanists or in the archaeological record. For in addition to the obstacles briefly outlined above, the archaeological evidence in 1980 clearly argued against an independent origin of domesticated plants or farming in the East. The tropical crop plants squash and maize appeared to control the temporal high ground in two important respects. First, at 4,300 B.P. the squash seeds and rind recovered in the late 1970s from the Phillips Spring site in Missouri clearly predated by a few hundred years the earliest evidence for any eastern domesticated plants. The archaeological record thus appeared to indicate that the first domesticated plant in eastern North America was Mesoamerican in origin. Second, carbonized maize had been recovered from a number of archaeological contexts in the East, indicating that it had been introduced into the region by perhaps 1000 B.C. to A.D. 0.

At the same time, Richard Yarnell's landmark analysis of archaeobotanical assemblages from Salts Cave, Kentucky in the late 1970s, along with work by David and Nancy Asch in the Lower Illinois River Valley and by Jefferson Chapman and Andrea Shea in the Little Tennessee Valley, had documented a significant increase in the relative abundance of the seeds of eastern cultigens that was perhaps contemporary with or more likely just after the apparent arrival of maize in the East. Thus in 1980, archaeological evidence seemed to indicate that (1) two North American plants, marshelder and sunflower, had been brought under domestication on the coattails of, and as a result of, the introduction by 4,300 B.P. of the tropical domesticate *Cucurbita pepo* into the region; and (2) that the dramatic increase in utilization of these domesticates, along with a number of eastern cultivated crop plants (chenopod, erect knotweed, little barley, maygrass) at ca. 500 B.C.–A.D. 0 was preceded and precipitated by the arrival of a second major tropical crop plant—maize. The initial domestication of eastern crop plants could be and was explained therefore by the arrival of squash. Subsequently, following a long period of low visibility of crops in the East, the initial emergence by A.D. 0 of true food producing economies was, in a similar manner, attributed to the arrival of maize. So by the early 1980s a number of different cultural and research currents had coalesced to seemingly close off, in a convincing manner, the possibility of the East being an independent center of plant domestication and food producing economies. The stubborn question of temporal precedence of crop plants that had cast a shadow over eastern North America for more than forty years had been resolved in favor of Mesoamerican origins. A Mesoamerican origin explanation in turn provided a comfortingly straightforward cause and effect explanatory framework, and one that conformed both with the then current cultural concepts regarding what was and what was not a crop plant worthy of recognition, and what was the rightful place of eastern North America in the scheme of agricultural origins—on the recipient periphery.

Given that this general question of agricultural origins in the East appeared closed off and largely resolved by the beginning of the 1980s, one could reasonably ask both what initially sparked my interest in the topic at that time and what subsequently structured and sustained my curiosity over the past decade.

In retrospect, looking back over the nine-year span during which the chapters of this book were written, the tandem possibilities of eastern North America being an independent center of plant domestication and the development of food producing economies can now be recognized as having had a near perfect target profile as a worthwhile research problem. As both a specific case study situation of the transition from forager to farmer, and as a major central element in gaining an understanding of the course of cultural development in eastern North America, it qualified as a research problem worthy of consideration. It also offered a large and growing data base of varied and high quality information, along with newly emerging advances in relevant technology, methodology, and theory. It represented an engaging and long unresolved research problem of considerable complexity. Because of improvements in the data base and frameworks of analysis and interpretation, this research area also showed increasing promise through the 1980s of actually yielding answers. Things often appear more obvious in retrospect, of course, and in the early 1980s the origins of agriculture in eastern North America did not seem, on the surface at least, to be an open question, much less one that held much long-term research potential. But since I am one of the archaeologists Lathrup (1987:348) recently accused of harboring ". . .a kind of Late Archaic Monroe doctrine protecting the cultural purity of our native Indians from insidious cultural influences from south of the Rio Grande or Key West. . ."[1] it was easy enough to generate healthy skepticism for the Mesoamerican origin explanations. Armed with a healthy suspicion that prevalent Mesoamerican origin explanations were simply wrong, both the available evidence and the promise of forthcoming information could be viewed from a quite different perspective. It was possible to think about how to go about building a case in support of an alternative explanation having far more appeal to a Monroe Doctrinite—that the Eastern Woodlands witnessed an independent process of plant domestication and the indigenous development of farming economies based on native North American seed crops.

The chapters that follow in this volume present evidence and analysis in support of this alternative theory of an agriculturally independent East. Inherent in the diverse range of topics covered in the following chapters is the complex interlocking nature of the numerous arguments necessary to construct an adequate support structure for such a theory. In their varied and interrelated nature these different support arguments also had a seemingly endless propensity for harboring contingency requirements in the form of other radiating lines of inquiry that required pursuit, often far afield. While sometimes involving considerable frustration, venturing out along these radiating lines of inquiry also invariably brought substantial enjoyment and enlightenment, particularly when I crossed into the problem domains of other scholars specializing in various aspects of the study of plants and plant husbandry in the East. In this regard the following chapters carefully acknowledge and reference the research and publications of numerous archaeologists, archaeobotanists, botanists, evolutionary biologists, and plant taxonomists.

In the following sections of this introductory chapter these research areas and associated support arguments are outlined, along with their myriad contingent lines of inquiry, in order to provide an overall frame of reference for the topics and recurrent themes that tie together the chapters to follow.

The History of Maize in Eastern North America and the Existence of Premaize Farming Economies

In many respects research on maize in eastern North America is just beginning. The application of new approaches and methods developed late in the 1980s promises to rewrite, in considerably greater detail, the complex history of this crop plant in the region (Fritz 1990, 1992). In terms of the present discussion, however, two simple questions regarding maize are of critical importance. When did maize first arrive in eastern North America, and when did it become the major food crop across much of the region? The inability to answer these two questions over a period of fifty years, from the late 1930s well into the 1980s, blocked consideration of the possibility that food production economies centered on native North American seed crops developed prior to the introduction of this major Mesoamerican domesticate. This question of the initial introduction of maize into the East was all the more intriguing in that it impinged not only on the question of the possible existence of earlier indigenous farming economies in the region, but also on the economic basis of Middle Woodland period Hopewellian societies that flourished over a broad area of the East between circa 100 B.C. and A.D. 200. Scholars had argued since the 1950s over the presence and importance of maize in the diet of Hopewellian populations, and by the early 1980s a series of radiocarbon dates on charcoal found associated with maize in apparent Early and Middle Woodland period contexts had yielded age determinations in the 1000 B.C. to A.D. 0 range. Confidence in these dates and even earlier evidence of maize in the East was not strong, however, and pioneering stable carbon isotope analysis of Hopewellian populations failed to show any evidence of corn consumption. At the same time, research in the Lower Illinois Valley and American Bottom regions of Illinois, and the Duck River in Tennessee, when combined with the earlier work in the Little Tennessee Valley and central Kentucky caves, indicated a broad-scale upsurge in the economic importance of native North American seed crops between 500 to 250 B.C. and A.D. 0. Was this a coattail response to the introduction of maize, or evidence of the emergence of indigenous small-seed food producing economies?

Fortunately, a technological solution to the problem of accurately establishing the age of the purported early maize emerged during the first half of the 1980s, as the particle accelerator mass spectrometer (AMS) radiocarbon dating method graduated from experimental status to become a reliable and easily accessible way of directly establishing the age of extremely small samples. As a result of pioneering work by Nicholas Conard and his colleague, Richard Ford, and by Jefferson Chapman and Gary Crites, it was clear by the middle of the decade both that the early maize dates were in error, and that this Mesoamerican crop was not present in the East until A.D. 200. In addition, mounting results of stable carbon isotope research through the 1980s indicated no detectable evidence of maize consumption in the East prior to circa A.D. 800. Although there is still purported early corn from eastern sites, such as the Meadowcroft sample, that await direct dating, the maize arrival obstacle to recognition of substantial premaize food production economies in eastern North America was in effect removed by the AMS research of Richard Ford and others. This AMS based reanalysis of early maize in eastern North America is of interest, and is discussed in a number of contexts in this book, because with the arrival date of maize pushed forward in time to the end of the Middle Woodland period, the now antecedent evidence of increased reliance on native seed crops and their role in Hopewellian economies invited attention. The importance of this temporal repositioning of maize in opening wide a new area of research inquiry is reflected in the "premaize" label often given to these antecedent plant husbandry systems (the term premaize was apparently coined by William Sturdevant in 1965). This premaize label serves to underscore the independent and indigenous status of small-seed economies in the East. No longer consigned to the coattail shadows of maize, and clearly an integral part of Hopewellian economies, native North American seed crop complexes earned general premaize and Hopewellian labels, and invited consideration of a wide range of more specific research questions. A second shadow still remained, however—that cast by the apparent early presence of a second Mesoamerican domesticate—squash.

Early Gourds in the East—Introduced Tropical Domesticate or Indigenous Wild Plants?

As mentioned earlier, the rind fragments and seeds identified as belonging to the genus *Cucurbita* and assigned an age of 4,300 B.P. that had been recovered in the 1970s from the Phillips Spring site in Missouri predated any evidence of native North American plants having been brought under domestication. The first half of the 1980s produced additional unequivocal evidence of the early presence of *Cucurbita* in the East in the form of direct AMS dates in the 7,000–5,000 B.P. range on rind fragments from sites in Tennessee, Kentucky, and Illinois. The distinctive cross-section cell profile of the rind fragments left no doubt that these early specimens belonged to the genus *Cucurbita.* The initial conclusion and rapid consensus drawn from these early rind fragments was that they were evidence of the arrival in eastern North American of a domesticated Mesoamerican squash and associated agricultural practices before any indigenous plants were domesticated. Although the common name "squash" was familiar to many of the individuals forming this consensus, the plant represented by these fragments was not. It was not a true squash in the culinary sense, but rather was a small, hard, thin-walled gourd about the size of a hardball. Judging from these early rind fragments, this small thin-walled *Cucurbita* gourd was morphologically indistinguishable from modern North American wild gourds that have a current geographical range extending across a broad area of the East (Chapter 4). And even as the consensus was building into the middle 1980s that these rind fragments represented a domesticated "squash" from Mesoamerica, an eastern wild gourd alternative explanation was also being suggested by C. Wesley Cowan, Charles Heiser, Patty Jo Watson, Richard Yarnell, and others. Do these small carbonized rind fragments in eastern North America represent the early Mesoamerican gift of a domesticated crop and agricultural concepts to a recipient periphery region? Or, on the other hand, do they simply mark the existence of a wild North American gourd?

In arguing for the East as an independent center of plant domestication (Chapter 3), the history and basis of these alternative interpretations of early eastern cucurbits is discussed, while in Chapter 4 this important topic is expanded and updated with a consideration of recent alternative explanations and a discussion of the range of distribution and habitat of the present day "free-living" gourds of the Arkansas and Missouri Ozarks. Although the Mesoamerican squash scenario still has some lingering proponents (see Chapter 4), a very strong case can now be made in support of the eastern wild gourd hypothesis. Of central importance in this regard has been the research of Deena Decker-Walters, who proposed that while the pumpkin group of *Cucurbita pepo* was domesticated in Mesoamerica, the entire summer squash side of the species *Cucurbita pepo* could well have emerged from wild progenitors in eastern North America. Largely as a result of Decker-Walters's research, the evidence of early *Cucurbita* gourds in eastern North America has been transformed from a Mesoamerican shadow cast on the region's potential status as an independent center of plant domestication, into evidence of the early portion of a trajectory toward domestication of an additional native North American crop plant. With this remarkable transformation—more than any archaeological Monroe Doctrinist might hope for—the last obstacle has been lifted to full recognition of eastern North America as an independent center of both plant domestication and the indigenous development of food producing economies.

When brought out of the coattail shadows of squash and maize, these tandem research questions of independent plant domestication and the development of indigenous farming economies in the Eastern Woodlands of North America can finally be perceived as complex developmental processes, each with its own set of specific and mutually impinging pathways of inquiry.

Plant Domestication in Eastern North America

In order to both document and understand the process of plant domestication in the East, it was necessary to consider five distinct research domains.

1. What different species of indigenous plants were brought under domestication in the East?
2. What is the nature of the evidence of domestication?
3. Over what period of time were these plants initially domesticated?
4. Over what geographical area of eastern North America did this developmental process occur?
5. What was the nature of the developmental process leading up to domestication, and what are the specific, relevant, and necessary components of an adequate framework of explanation of the process of domestication?

By the early 1980s two native North American seed plants, sunflower and marshelder, had been granted domesticated status on the basis of morphological change (an increase in seed size). A third seed plant species, *Chenopodium berlandieri,* had been the focus of consideration as a possible domesticate, based on the thin testa or seed coat of some specimens recovered from Salts Cave, Kentucky and several sites in Illinois. But research on modern stands in the late 1970s had led David and Nancy Asch (1977:20–22) to conclude that such thin seed coat specimens represented phenotypic polymorphism in wild populations, rather than domestication.

My own initial consideration of plant domestication in the East focused on this third species, *C. berlandieri,* and then expanded to include all five of the general research questions listed above. My interest in *Chenopodium* stemmed from its consistent abundant occurrence in archaeobotanical assemblages across the East, along with its present-day status as a domesticated field crop in both Mexico and South America. When combined, these two factors suggested that it was a good candidate for having been domesticated in the East, in spite of the Asches' conclusion to the contrary. In addition, judging from their relative abundance in seed assemblages in comparison with *C. berlandieri,* both sunflower and marshelder appeared to be less important as crop plants, yet both had been brought under domestication while *Chenopodium* seemingly had not. Large caches of *Chenopodium* had been recovered from storage pit contexts at a number of dry caves in the East, and such sizable assemblages, perhaps resulting from single harvests, seemed a good place to begin to look for morphological indications of domestication in this species.

Of the different *Chenopodium* caches mentioned in the literature, the one excavated by Carl Miller from a grass-lined pit in Russell Cave, Alabama in 1956 was an obvious first choice for study since the uncatalogued collections from the site were stored just upstairs from my office in the National Museum of Natural History. After a search of several hours, a portion of the *Chenopodium* excavated by Miller was discovered mixed in with other materials, but safely stored in a cigar box (Tampa Nugget Sublimes). Initial comparison of the Russell Cave *Chenopodium* with modern wild specimens of *Chenopodium* from the collections of the National Herbarium indicated considerable morphological differences. The Russell Cave specimens seemed to have a different, flat sided shape, and to have a much thinner seed coat. What did such differences mean, and what caused them? Were they evidence of domestication? Emerging out of a cigar box of small black *Chenopodium* seeds, these straightforward questions in turn led to a number of intermingled research inquiries that were necessary to build a case for domestication of this species. Consideration of the question of whether the Russell Cave *Chenopodium* represented a domesticated plant in turn provided a case study platform from which to view the larger process of domestication of indigenous crop plants.

What clearly was needed next in the analysis of the Russell Cave *Chenopodium* was a more detailed documentation of the morphology of the cigar box seeds, and a detailed comparison with seeds from both modern eastern North American wild stands and modern Mexican domesticated varieties of the same chenopod species. Hugh Wilson, a botanist at Texas A&M University and an authority on *Chenopodium,* generously provided seeds from a number of Mexican domesticated varieties of chenopod. Wild eastern North American specimens were obtained initially from herbarium sheets and then from a field research project initiated to obtain voucher plant specimens and to harvest modern wild and weedy stands of *C. berlandieri* and marshelder across eastern North America (Chapters 7 and 8). With comparative material in hand, the scanning electron microscopes of the National Museum of Natural History made it possible to document in detail the micromorphology of modern wild and domesticated chenopod seeds, and to demonstrate the marked structural similarity between the Russell Cave and modern domesticated forms (Chapter 5). A sample of Russell Cave *Chenopodium* seeds yielded a direct AMS age determination of 390 B.C.

Following analysis of the Russell Cave, Alabama chenopod assemblage, a second assemblage of *Chenopodium* seeds, this one recovered from a dry rock shelter in Ohio, was located and analyzed. Excavated by Ebenezer Andrews in Ash Cave, Ohio in 1876 and subsequently stored in a Red Pepper jar in the Peabody Museum at Harvard University for over a century before being restudied, this second *Chenopodium* assemblage was very similar to the Russell Cave specimens, providing more evidence that this seed plant had been a domesticated crop plant in prehistory (Chapter 6). A sample of Ash Cave *C. berlandieri* specimens yielded a direct AMS age determination of A.D. 230.

While these two detailed studies of the large chenopod assemblages from Ash Cave and Russell Cave documented its status as a domesticated crop in terms of morphological change, small samples of only a few seeds from two eastern Kentucky rock shelters subsequently provided key evidence regarding the timing of initial domestication of this species. Direct AMS dates on specimens from the Cloudsplitter and Newt Kash shelters produced age determinations of 3,400 B.P. and 3,450 B.P. respectively, in close temporal association with the earliest evidence of domestication of marshelder, sunflower, and more recently, *Cucurbita pepo.*

Research on *Chenopodium* as *a domesticate* thus quickly added another dimension to the overall pattern of plant domestication in the East, and led to a clearer recognition of the general temporal and spatial boundaries of this developmental process. Although the temporal pattern of initial plant domestication in the East is based on few specimens, the available evidence indicates that marshelder, *Chenopodium,* sunflower, and *Cucurbita pepo* were domesticated during the second millennium B.C. (2000–1000 B.C.). As more information becomes available the period of initial domestication will in all likelihood narrow down to the first half of the second millennium B.C. (2000–1500 B.C.). This overall temporal pattern provides additional support for an indigenous *Cucurbita* gourd being domesticated in the East, in that the timing of morphological change indicative of domestication in *Cucurbita* conforms so closely to comparable changes in the three other indigenous domesticates. As it emerged during the 1980s, this general temporal pattern also served to strengthen the case for an independent, indigenous process of domestication, with all four crop plants being brought under domestication within the same developmental process (Chapter 2).

The archaeological evidence for this process of domestication is distributed across a broad interior midlatitude riverine zone, from eastern Kentucky and Tennessee, west to the Ozarks, and north to Illinois. Geographically, the domestication of local plants appears to have been a broadly distributed network of interactive change, without any smaller or more specific core.

While research focusing on establishing *Chenopodium* as a domesticated crop provided new evidence for, and a new perspective on, the broader temporal and spatial dimensions of plant domestication in eastern North America, it also led directly to the deeper general questions of causation. It was not enough to simply document morphological change in *Chenopodium*. It was also necessary, in building a case for domesticated status, to establish a framework of causation that could account for the observed morphological changes within the context of increasing human intervention in the life cycle of this particular plant species. What caused the morphological changes? This consideration of specific causation in the case of *Chenopodium* entailed taking a broader look at the general continuum of human-plant relationships and the constituent escalating levels of human intervention. Such a sequence of escalating levels of intervention provides a logical progression leading to plant domestication, and a number of evolutionary biologists, most notably J. R. Harlan and J. M. de Wet, have developed detailed explanations for how cereals and other seed plants respond, through morphological change, to anthropogenic modifications in their growing environment. The work of Harlan and de Wet was not only directly applicable to *Chenopodium,* but also to the other indigenous North American domesticates. The 2000–1000 B.C. increase in seed size documented for marshelder, sunflower, and *Cucurbita,* as well as the concurrent reduction in seed coat thickness in *Chenopodium,* all represent the automatic adaptive response by plant populations under strong seed bed selective pressure for rapid germination and seedling growth. Such situations of seed bed competition are produced, in turn, by the major step of deliberate human planting of seed. Simply put, the second millennium B.C. morphological changes observed in the seeds of these four plants reflect the initial deliberate planting of crops by human populations in eastern North America.

With the deliberate winter storage and spring planting of seed stock identified as the proximate cause of the set of morphological changes observed in *Chenopodium* and the other three local domesticates, and the spatial and temporal dimensions of this significant escalation in human manipulation of these selected floodplain plants firmly established, attention once again was directed outward, to a consideration of the larger developmental context surrounding the initial domestication of plants in eastern North America. What was the likely progression of increasing human intervention in the life cycle of these plant species that led up to deliberate planting and domestication? Why were these plants involved and not others? What was the locational context of this process on the cultural and natural landscape? These more general processual questions led to the formulation of the Floodplain Weed Theory of plant domestication in eastern North America. The origin and underlying organizational structure of this explanatory framework is considered in Chapter 2, while the relevant information and supporting arguments are presented in detail in Chapter 3.

Premaize Farming Economies in Eastern North America

As is the case with the initial domestication of plants in eastern North America, the subsequent initial emergence of food producing economies in the region was a complex developmental process having a set of specific initial research questions.

1. What is the nature of the evidence for premaize food producing economies in the region?
2. What indigenous crop plants were of economic importance in these economies?
3. Over what period of time did indigenous food producing economies initially develop?
4. Over what area of eastern North America did these new economic systems become established?
5. What was the nature and relative importance of this new food production component of the economic system?
6. What was the broader organizational structure of these emergent Middle Woodland food producing societies and what was the nature of their basic household economic unit?

For a variety of reasons outlined earlier in this introduction, recognition of the existence of, much less the importance of, premaize food producing economies in eastern North America has been slow in coming. As recently as 1986 the circa A.D. 1000 transition to maize agriculture in the East was identified as marking the shift from foraging to farming (Lynott et al. 1986:82). Similarly, Middle Woodland Hopewellian groups were offered by Bender (1985) as an archaeological case study example of complex hunter-gatherer society. This lingering resistance to viewing premaize food production as anything more than a minimally important assortment of little known cultigens is also reflected in the continuing characterization of these plants as minor seasonal buffers against shortages in wild resources, and their relegation to garden plant status. As reflected in the chapters of Section III of this volume, one of my overall research goals during the last nine years or so has been to develop as strong a counter-argument as possible to this minimalist position by documenting in detail the economic potential and archaeological evidence for premaize farming economies in the East.

As with the question of independent plant domestication in the East, my consideration of premaize farming economies began with Russell Cave and *Chenopodium* and rapidly expanded into broader issues encompassing the general economic base of Middle Woodland food producing societies. Even though four indigenous seed plants were deliberately planted and brought under domestication by 1000 B.C., they remain at a very low level of archaeological visibility until the period 500–250 B.C. to A.D. 0. Across this brief span of several centuries these domesticates and three other cultivated plants (maygrass, little barley, and erect knotweed) exhibit a considerable increase in abundance in archaeobotanical seed assemblages. Indigenous crop plants thus did not increase in importance gradually, or in low numbers. Rather they were added to economic systems relatively abruptly, and in groups that appear to combine spring and fall maturing crops, some high in oil content, others high in starch. This dramatic upsurge in the representation (and assumed economic importance) of these crop plants also occurred over a broad geographical area, and is documented in archaeobotanical sequences from the Duck and Little Tennessee River Valleys in Tennessee, the American Bottom and Lower Illinois Valley in Illinois, central and eastern Kentucky, central Ohio, and eastern Missouri (see Chapter 12, Table 12.2), as well as in storage pits such as those of Ash Cave, Ohio and Russell Cave, Alabama. Interestingly, this initial emergence of food producing economies mirrored the spatial and temporal dimensions of the earlier domestication of plants. It took place within the same interior mid-latitude landscape, and occurred over a comparably short period of time.

If it occurred in some other part of the world, and involved grains such as wheats or barleys, such an abrupt, broad scale, and highly visible transition to an increased economic presence of seven domesticated and cultivated plants would quickly be acknowledged as marking a major shift toward farming economies. But in eastern North America, where comparison to maize is unavoidable, and the indigenous crops in question have little name recognition, this transition is still often brushed aside as involving minor crops of little economic import, in all likelihood grown only in small garden plots.

Inherent in this central and still debated issue of the actual importance of these premaize crops is their economic potential. What kind of crops are these plants? Did these indigenous cultigens (chenopod, marshelder, squash, erect knotweed, sunflower, little barley, and maygrass) have certain attributes or limitations that restrict their cultivation to small garden plots? Are there reasons why they were not grown as field crops? Did they require considerable field preparation and attention? Were they somehow deficient nutritionally? Did they have low harvest yield values? Were they difficult to harvest? While none of these questions can be addressed through direct study of these premaize crop plants, considerable information is now available regarding their modern domesticated, wild, and weedy descendants.

My initial interest in this research area of economic potential centered on *Chenopodium berlandieri*. While variation in the relative representation of these seven crop plants across the East indicates the development of regionally variable plant husbandry systems, *Chenopodium* was consistently the single most important crop. Large seed caches such as those in Ash Cave and Russell Cave provided the opportunity to study and describe a portion of a single season's harvest of this major premaize crop plant, and to consider whether the harvests came from small garden plots or larger fields. In looking to modern analog situations, it is certainly not difficult to find ample evidence of *Chenopodium* as a field crop. It is grown today as a food crop in field situations in both South America and Mexico. Similarly, sunflower is a major field crop today. Judging just from a consideration of their current status as New World field crops, neither of these two indigenous cultigens can be confined prehistorically behind small garden fences.

In addition, research in the East on modern wild and weedy populations of chenopod and other premaize crop plants has underscored their considerable potential as field crops. While involving the collection of comparative seed materials and documentation of habitat requirements and stand size, my own field research begun in the mid-1980s on modern wild and weedy populations of *C. berlandieri* and marshelder was also directed toward establishing the harvest yield potential of wild stands of these two species (Chapters 7 and 8). When combined with prior pioneering harvest studies by David and Nancy Asch and others, this field research on modern wild and weedy stands provides solid baseline information on the substantial economic potential of indigenous seed plants as field crops in premaize farming economies.

In order to gain a full appreciation of the field crop potential of both *Iva annua* and *C. berlandieri,* one need only take a few days and walk the alluvial bottomlands of the Lower Mississippi Valley. Minor crop perceptions tend to fall away as these two floodplain plants are repeatedly found growing in profusion both in wild stands and as weedy intruders in the sandy soils of agricultural fields. Although Asch and Asch (1978) characterized marshelder as a plant that grows in the wild only in small linear stands, in 1984 and 1985 numerous extensive stands of solid *Iva annua* were observed along the Mississippi Valley, some covering more than 2,500 sq m (Chapter 8). Similarly, stands of *C. berlandieri* growing along the exposed sand banks that parallel the main channel of the Mississippi River extend literally for hundreds of miles through the Lower Mississippi Valley. Interestingly, this is the exact setting described by Le Page du Pratz in his early eighteenth-century account of the casual sowing of the grain *choupichoul* by the Natchez. In addition to underscoring the likelihood that this ethnohistorically described crop plant was *C. berlandieri* (Chapter 10), these sand bank stands of Chenopod (Chapter 7), like the large lower valley wild stands of marshelder (Chapter 8), and the Ozark populations of wild *Cucurbita* gourds (Chapter 4), document the ease with which these plants could have been grown as field crops. Other premaize cultigens such as erect knotweed, little barley, and maygrass are more elusive on today's landscapes, but limited habitat and harvest studies also confirm their economic potential. Certainly there is no evident reason to limit them to garden plant status. Nutritional studies of many of these premaize crop plants have also been completed (Chapter 8, Table 8.3) indicating that they have impressive nutritional profiles. When drawn together, these various lines of archaeobotanical, nutritional, and modern field study evidence provide a clear picture of the rapid broad-scale development of premaize food production economies based on a variety of high harvest yield and high nutritional profile seed crops.

From this initial interest in the habitats, harvest yield potential, and field crop profiles of premaize seed plants, as well as the likely economic contribution of farming to the overall economies of Hopewellian societies, it was a logical next step to consider the broader social context of premaize food production. This broader consideration of the organizational structure of these emergent food producing societies, with a detailed focus on the basic Hopewellian household economic unit, is presented in Chapter 9. As difficult as it was to pull together and interpret various kinds of information regarding premaize crop plants and food producing economies, it proved even more of a challenge to locate and consolidate information regarding the habitation settlements of Hopewellian societies. Once teased apart from the far better documented Hopewellian sites of the corporate-ceremonial sphere, these habitation sites appeared surprisingly similar across a broad geographical area, and quite different from the traditional expectation of Hopewellian village settlements. In addition, knowledge of these Hopewellian settlements allows reasonable projections to be made regarding the amount of land and energy that would have to be invested in fields of *Chenopodium, Iva,* and other seed plants by the basic household economic units of premaize farming societies relative to the harvest return they might reasonably expect (Chapter 8).

While Sections II and III of this volume focus respectively on topics involving the initial domestication of plants in eastern North America and the nature of premaize crops and food production economies, Section IV considers these transitions within a larger developmental framework, and provides both a short and long format synthesis of the long history of food production economies in eastern North America. Excellent summary discussions of the evolution of agriculture in eastern North America from different perspectives can be found in Watson (1989), and Fritz (1990).

Notes

1. Lathrup appears to have borrowed the disparaging "Monroe Doctrine" label from an unidentified British cultural anthropologist (Anderson 1952).

Literature Cited

Anderson, E.

1952 *Plants, Man and Life.* Little Brown and Co., Boston.

Asch, D., and N. Asch

1977 Chenopod as Cultigen: A Re-evaluation of Some Prehistoric Collections from Eastern North America. *Midcontinental Journal of Archaeology* 2:3–45.

1978 The Economic Potential of *Iva annua* and its Prehistoric Importance in the Lower Illinois Valley. In *The Nature and Status of Ethnobotany,* edited by R. I. Ford, pp. 301–342. Museum of Anthropology, Anthropological Papers No. 67. University of Michigan, Ann Arbor.

Bender, B.

1985 Prehistoric Developments in the American Midcontinent and in Brittany, Northwest France. In *Prehistoric Hunter-Gatherers: The Emergence of Cultural Complexity,* edited by T. D. Price and J. A. Brown. Academic Press, Orlando, Florida.

Cowan, C. W., and P. J. Watson (editors)

1992 *The Origins of Agriculture: An International Perspective.* Smithsonian Institution Press, Washington, D.C.

Fritz, G.

1990 Multiple Pathways to Farming to Precontact Eastern North America. *Journal of World Prehistory* 4:387–435.

1992 "Newer," "Better," Maize and the Mississippian Emergence: A Critique of Prime Mover Explanations. In *Late Prehistoric Agriculture: Observations from the Midwest,* edited by W. Woods, Chapter 2. Illinois Historic Preservation Agency, Studies in Illinois Archaeology 7. Springfield.

Lathrup, Donald

1987 The Introduction of Maize in Prehistoric Eastern North America: The View from Amazonia and the Santa Elena Peninsula. In *Emergent Horticultural Economies of the Eastern Woodland,* edited by W. Keegan, pp. 345–371. Center for Archaeological Investigations, Occasional Paper 7. Southern Illinois University, Carbondale.

Lynott, M., T. Boutton, J. Price, and D. Nelson

1986 Stable Carbon Isotope Evidence for Maize Agriculture in Southeast Missouri and Northeast Arkansas. *American Antiquity* 51:51–65.

Quimby, George

1946 The Possibility of an Independent Agricultural Complex in the Southeastern United States. *Human Origins: An Introductory General Course in Anthropology.* Selected Reading Series No. 31. University of Chicago, Chicago.

Watson, Patty Jo

1989 Early Plant Cultivation in the Eastern Woodlands of North America. In *Foraging and Farming,* edited by D. Harris and G. Hillman, pp. 555–565. Unwin Hyman, London.

Yarnell, R.

1991 Investigations Relevant to the Native Development of Plant Husbandry in Eastern North America: A Brief and Reasonably True Account. In *Agricultural Origins and Development in the Midcontinent*, edited by W. Green. Office of the State Archaeologist, Report 19. Iowa City, Iowa.

II.
An Independent Center of Plant Domestication

CHAPTER 2

❑ ❑ ❑

The Floodplain Weed Theory of Plant Domestication in Eastern North America

Introduction

First presented at a conference in Carbondale, Illinois, in the spring of 1986, Chapter 3 in this volume of essays presents an integrated and detailed set of arguments in support of the idea that eastern North America was an independent center of plant domestication. In this chapter I focus on the theory and framework of explanation that structures the constituent arguments presented in Chapter 3. A primary emphasis is placed both on the origin of the quite varied kinds of information and ideas that are brought together in these arguments, and how they are joined to form a coherent and relatively comprehensive account of the initial domestication of plants in the region. This discussion of how and why such different kinds of evidence are linked together in turn provides an opportunity to consider the important attributes of, and general adequacy criteria for, such explanatory frameworks of domestication. The "Floodplain Weed" theory of plant domestication in eastern North America is comprised of six different interlocking segments or areas of relevant information (Figure 2.1), each of which in turn consists of a set of interrelated component parts.

The first two of these general segments encompass some of the ideas and observations made forty years ago and more by Edgar Anderson and several other botanists regarding the general niche and habitat of different floodplain "weeds," and the reiteration and expansion of their concepts in subsequent decades.

A third major segment of the theory brings in the research and theoretical writing of Jack Harlan and J. M. J. de Wet during the 1970s regarding those sets of human behavior that actually produce the various morphological changes associated with seed-plant domestication.

A fourth segment of the Floodplain Weed theory involves documenting the timing and nature of the morphological changes that were associated with the initial domestication of indigenous North American seed plants. It was not until the early 1980s that two advances in instrumentation (scanning electron microscopy and accelerator mass spectrometer radiocarbon dating) allowed the accurate determination of when domesticated plants were first developed in

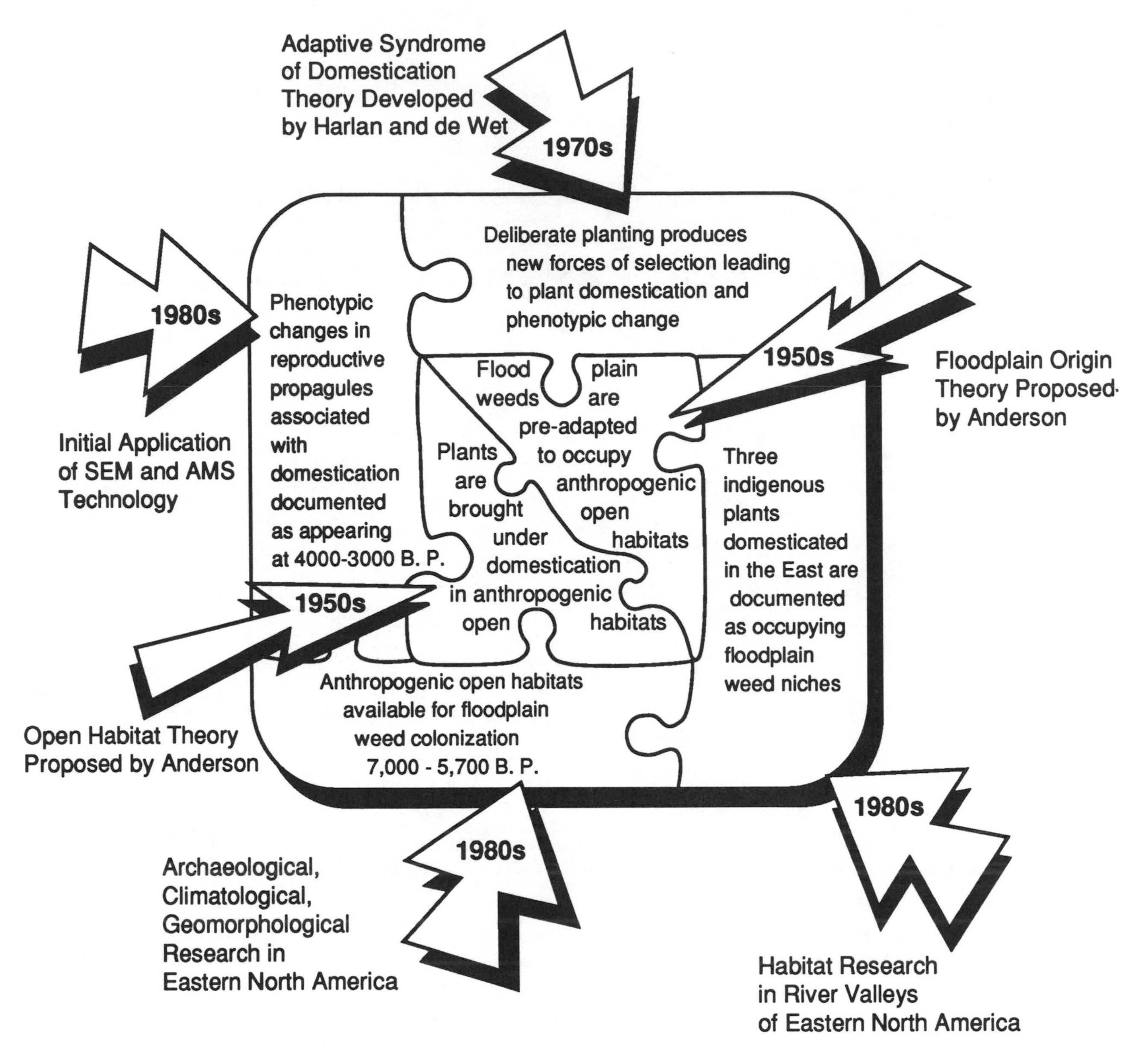

Figure 2.1 Diagrammatic representation of the six interlocking segments of the Floodplain Weed Theory of plant domestication in eastern North America.

eastern North America, and for at least one crop plant, what morphological changes signaled domestication.

A fifth general segment of the theory includes information regarding the nature and first appearance of human-created river valley habitats that were similar to those areas colonized by the floodplain weeds described by Anderson and others. Much of this information relevant to the initial formation of anthropogenic open habitats in eastern North America did not become available until the 1980s.

A sixth and final area of relevant information encompasses field research carried out over the past eight years in the river valleys of eastern North America. Focusing on present-day wild and weedy populations of three species of plants that were brought under domestication 3,000 to 4,000 years ago, field investigations produced information regarding niche and habitat, harvest yield potential, and seed morphology (for comparative baseline values). Coming full circle, these field studies showed that the plants in question fall squarely into the floodplain weed category characterized four decades ago by Edgar Anderson.

When these six interlocking segments of the theory are put into place, they track, in a fairly comprehensive manner, the transition from wild plant to domesticate, from human hunting and gathering to initial food producing economies. The original habitat of the plants involved is identified, their pre- adaptation to domestication is considered, their movement into similar anthropogenic habitats is discussed, and the human activities causing morphological change are identified.

Edgar Anderson and the Plants of Open Habitats

> Many of the plants which follow us about have the look of belonging originally on gravel bars. (Anderson 1952:149)

One of the primary research interests of the botanist Edgar Anderson was the developmental history of domesticated plants. He was particularly interested in the disturbed "open" areas created by humans, and the role played by these anthropogenic habitats in the creation and subsequent survival of new hybrid or "mongrel" plants. These human habitats were "open" in terms of being clear of vegetation and open to the sky, but they were also "open" in comparison to the closed habitats of established plant communities:

> What do we mean by an open habitat? . . . Most plants are very choosy about where they will and will not grow, and some places, like gardens and dump heaps, are relatively open habitats, receptive to a good many kinds of plants, while other places, such as meadows and mountaintops are relatively closed habitats in which aliens will have trouble getting a footing. . . . Here is a strange new kind of habitat. Many of the plants in the native flora do not fit into it, some aliens will, and some hybrids. Plants which can grow in such places will have less interference from other plants. Kitchen middens would be likely places in which fruit pits, seed heads, and the like, brought to the village from some distance, might germinate and survive. (Anderson 1952:145, 146–147)

Anderson observed that such open habitats were rare in nature, and he proposed that as a result, introgression in these areas between closely related plants would produce entire new communities of previously unknown forms—forms that were very competitive within the disturbed open habitats of humans, but unable to survive outside it, in already closed habitats. Anderson and other botanists employed the term "weed" as a general label for these new hybrids or mongrels confined to the open habitats created by human populations:

> I would define a weed as a plant which grows in places in some way disturbed by man or his domesticated animals. . . . Weeds are able to become established and to reproduce on disturbed sites. This difference, a physiological one, between weeds and other plants is of paramount importance. (Heiser 1949:148, 150)

In addition to outlining how humans, by creating anthropogenic open habitats, in turn generated entire new "weed" communities, Anderson also expanded on several speculations of Carl Sauer in proposing that the early habitation sites and refuse heaps of humans, as open habitats, "may have played a key role in the origin of cultivated plants" (Anderson 1952:144). These habitation sites were locations where:

> Seeds and fruits brought back from up the hill or down the river might sometimes sprout and to which even more rarely would be brought seeds from across the lake or from another island. Species which had never intermingled might do so there, and the open habitat of the dump heap would be a more likely niche in which strange new weed mongrels could survive than any which had been there before man came along. Century after century these dump heaps should have bred a strange new weed flora and when man first took to growing plants, these dump heap mongrels would be among the most likely candidates. (Anderson 1952:149)

With what was in retrospect remarkable prescience, given the context of the time and the limited information available to him, Anderson also considered the weed communities of the open habitats created and maintained not by humans but by natural forces, primarily rivers, and their developmental link to human habitats:

> Man's only natural partner was the big rivers. They too make dump heaps of a sort; they too plow up the mantle of vegetation and leave raw scars in it. Rivers are weed breeders; so is man. . .(Anderson 1952:149)

In developmentally linking the pre-existing open habitats of river valleys with those created by humans, Anderson proposed that the "wild habitat" weeds of river valleys were well prepared to rapidly invade the disturbed soil areas created by humans, and that many of our weeds and cultivated plants began as the weeds of open areas that were created by floodwaters:

> Where did these open-soil organisms come from in the first place, these weeds of gardens and fields?. . .well, they must have come mostly from pre-human open-soil sites. River valleys did not supply all of them, but rivers are certainly, next to man, the greatest of weed-breeders. Our large rivers plow their banks at flood times, producing raw-soil areas. Every river system is provided with plants to fill this peculiar niche, all those known to me act as weeds in the uplands. . . . Man and the great rivers are in partnership. Both of them breed weeds and suchlike organisms. The prehuman beginnings of many of our pests and fellow-travelers are to be sought in river valleys. River valleys also must have been the ultimate source of some of the plants by which we live: gourds, squashes, beans, hemp, rice, and maize. (Anderson 1956:773)

> If we now reconsider the kitchen middens of our sedentary fisherfolk, it seems that they would be a natural place where some of the aggressive plants from their riverbanks might find a home. (Anderson 1952:149)

> If we look over our cultivated plants with the dump-heap theory in mind, we find that a goodly number of our oldest crops look as if they might well have come from some such place. Hemp is difficult to keep off of modern dump heaps once it is established in a neighborhood; squashes, pumpkins, beans all have the look of such an origin. (Anderson 1952:150)

Unfortunately, Anderson did not carry his consideration of domesticated plants and their possible floodplain origin any further than the two brief quotations presented above. He does not provide any additional discussion of the floodplain weed characteristics of the domesticates he mentions (gourds, squashes, pumpkins, beans, hemp, rice, maize, and amaranth), nor does he speculate regarding their wild floodplain weed progenitors.

In addition, it is somewhat surprising that there is no suggestion by Anderson that his theory might be specifically applicable to eastern North America, or to the eastern seed plants Melvin Gilmore and Volney Jones had proposed as possible domesticates, based on their respective analysis of Ozark and eastern Kentucky archaeobotanical assemblages in the 1930s (Chapter 3). Anderson mentions Volney Jones's research on the domestication of the sunflower in several places (Anderson 1952), but does not consider Jones's ideas regarding the possible prehistoric domestication of eastern seed plants within the context of his dump heap model. Similarly, even though Anderson characterizes amaranth as a good floodplain weed candidate and a "dump heap plant par excellence. . .common in barnyards, middens, and refuse dumps throughout the world" (1952:150), he does not make any connection with Gilmore's earlier mention of amaranth as a possible eastern domesticate.

Cucurbita provided another potential species link to eastern North America that Anderson might have employed in focusing his general model on the river valleys of the East. On the one hand he identified gourds *(Lagenaria?),* squashes, and pumpkins as domesticates having a likely origin as floodplain weeds, yet did not make any connection with either the wild river valley *Cucurbita texana* populations in Texas, or the free-living populations being collected by his colleague Julian Steyermark in the Missouri Ozarks (Chapter 4, Figure 4.6).

This absence of a specific application of Anderson's theory to eastern North America is all the more ironic because it is clear that his general theory regarding floodplain weed communities and their pre-adaptation to colonizing anthropogenic open habitats was firmly rooted in his observations and research on floodplain landscapes and weed communities in the Missouri Ozarks.

The arboretum of the Missouri Botanical Garden where Anderson was on the staff in the 1940s and 1950s bordered on the Meramac River, and as early as 1948 he had published on the geomorphology of flood episodes and gravel bar formation along the river (Anderson 1948). Anderson also encouraged and supported other more detailed studies of Ozark river flood episodes and their effect on floodplain landscapes and plant communities (for example, Deitz 1952).

At the same time, Anderson's examples of wild floodplain weed species that invade anthropogenic open habitats were all drawn from the Ozarks. He describes winter cress, a kind of wild mustard, as common in areas of the Meramac River Valley where

> . . . .standing water made enough bare spots in the floodplain meadows so that cress got started there . . . cress grew down in the lowlands not just because it wanted more water, but because the water in the lowlands held in check some of the plants, such as bluegrass, which competed with the cress. (Anderson 1952:147)

Anderson also observed this plant of open wet floodplain areas growing in upland settings in the arboretum, but only after plowing provided it with an open habitat: "it required plowing to give cress a start and repeated plowing if it was to persist" (Anderson 1952:147). The sycamore is also offered as an example of river valley species of trees adapted to open floodplain areas that also successfully invade upland disturbed soil areas (Anderson 1952:148), as is pokeweed *(Phytolacca americana)* (Anderson 1956:773). Suggested and guided by Anderson, Jonathan Sauer's classic study of the distribution of pokeweed within a section of the Meramac River Valley and adjacent uplands provides an interesting description of the niche of a floodplain weed and its ability to radiate into open habitat areas in the uplands. Sauer found pokeweed plants growing in a variety of upland settings where human activities had produced open habitats, but their continued presence was dependent on such habitats staying open, and they did not persist in the absence of continual soil disturbance. In the floodplain of the Meramac River, on the other hand:

> Many river-bottom colonies, including abundant seedlings, occupied sites where there was no trace of human activity. They were scattered through the more open river-bank woods, most often among sycamores. Although there was no sign that man had ever disturbed many of the river-bottom sites occupied by poke colonies, another factor causes repeated and violent disturbance of this habitat. During an average spring or early summer, the Meramac leaves its banks at least once. The poke colonies lie in the zone of maximum flood frequency. Piles of river drift, beds of fresh sand and other alluvium, caving banks, and raw cuts give good evidence of the powerful disruption effected by the river. (Sauer 1952:120)

> The microdistribution patterns of poke provide what seems to be a significant clue to the ancient habitat of the plants. Poke seedlings appear to be able to establish themselves only where some external factor has intervened to obliterate the potential competitors and open up a patch of raw soil. Before the coming of man, poke could have found a niche in habitats disrupted by natural agencies. Its stronghold may have been in open stream-bank woods, where new ground was constantly opened by cutting and filling. Poke undoubtedly colonized other natural scars in the mantle of vegetation—gullies, landslides, burns, blowdowns—but away from the constant intervention of the stream such colonies could ordinarily persist only briefly until they were overwhelmed by the slow advance of more aggressive vegetation. (Sauer 1952:123)

The pokeweed habitat research of Sauer, along with the similar modern uniformitarian analog examples offered by Anderson, demonstrate in a convincing manner one of the main axial propositions of Anderson's theory: plants that evolved within disturbed "open" habitats in river valleys are today (as they were in the past) pre-adapted to radiation into, and successful competition within, comparable open habitats of human creation.

The second main axial element of his theory asserts that once established as dominant weeds in the open habitats of human origin (likely by human transport), wild floodplain "weeds" are in a majority, if not exclusive, position to participate in the various levels of human manipulation and intervention along the continuum leading to domestication and food production. Anderson thus situates these activities of intervention on the human landscape, placing them in the disturbed open areas of mankind. He furthermore identifies those plants belonging to the "weed" category as being likely to dominate such open areas and be the subject of manipulation and experimentation, and proposes that the most obvious place of origin for these initial weed candidates for domestication is the flood-maintained open habitats of river valleys.

These two axial elements of Anderson's theory provide a solid central core on which to anchor the other general segments of the explanatory framework presented in this chapter. To a considerable degree, the anchoring of an additional explanatory framework to Anderson's axial theoretical elements is required in order to transform what is essentially a minimally developed broad, brush-stroke theory of universal applicability into a more detailed and comprehensive theory specifically tailored to a particular set of crop plants and a single region of the world—eastern North America. Most of the various blocks of information, methods, and scientific instruments that contribute to the larger explanatory framework to be outlined below had not yet been developed in the 1950s, and were not available to Anderson. I still wonder, however, if he might not have focused his theory more directly on eastern North America, with impressive results, if he had shifted his attention just slightly, and chosen different species of floodplain weeds as his modern analog examples.

The natural river valley habitat of marshelder *(Iva annua)*, for example, appears comparable to that described by Anderson for winter cress—it thrives in open floodplain settings where wet ground inhibits the growth of other plants (Chapter 8). Like winter cress, marshelder is a frequent visitor to human-created open habitats.

Similarly, while not as dominant a feature on the floodplain landscape as the sycamores described by Anderson, small *Cucurbita* gourds grow today in the same flood zone gravel bar and sandy terrace settings of Ozark river valleys as the ubiquitous tree, sometimes trailing their vines up into the branches of younger sycamores (Chapter 4). These gourds also are pre-adapted to, and thrive within, human-created open habitats.

Finally, *Chenopodium berlandieri* grows in the same open stream bank woods habitat as does pokeweed (Chapter 7, Figure 7.7), as well as in other open disturbed soil situations of both natural and human origin, sometimes providing a trellis for *Cucurbita* gourds. For Jonathan Sauer, it would have been an equally appropriate focus of study as pokeweed.

While marshelder, free-living *Cucurbita* gourds, and *C. berlandieri* are similar to the floodplain weed examples cited by Anderson in terms of both their adaptation to the disturbed soil river valley habitats produced by the force of floodwaters, and their propensity for colonizing anthropogenic disturbed soil habitats, they differ from Anderson's examples in one important respect—they were all brought under domestication in eastern North America. It is all the more ironic that *I. annua, Cucurbita,* and *C. berlandieri* did not catch Anderson's attention as modern examples of floodplain weeds, and lead him to focus attention specifically on the domestication of plants in eastern North America, since they constitute, to date, the best documented example of the applicability of at least aspects of his general theory to an actual developmental sequence leading to domestication.

While two elements of Anderson's theory are adopted as a central core for the expanded theory outlined here, another aspect of his explanation that has not held up over the past four decades, and which is discarded as not applicable to the Eastern Woodlands, involves his reliance on hybridization as the major mechanism producing domesticated plants within anthropogenic open habitat settings.

Jack Harlan, J. M. J. de Wet, and the Adaptive Syndrome of Domestication

Edgar Anderson considered the open habitats created by humans as providing excellent opportunities for introgression between closely related plant taxa. He argued that of the many hybrid forms that would result from such introgression (and which could only survive in the open habitats that fostered them) some would prove attractive to human cultivators and would be selected and grown as domesticated plants. While hybridization has certainly produced new and important varieties of cultigens, a far more comprehensive theory of how domesticated seed crops initially develop within open habitats of human creation was presented by Jack Harlan and J. M. J. de Wet in several articles that appeared in the first half of the 1970s (Harlan et al. 1973; de Wet and Harlan 1975).

In addition to fostering introgression, the disturbed soil, open habitat settings of human settlements provided the opportunity for human experimentation with colonizing species of plants, with this experimentation logically progressing along a continuum of increasing intervention by plant managers that extends from simple toleration, through various forms of encouragement, to eventual cultivation and plant husbandry (Chapter 5, Figure 5.1). Along this continuum of increasing intervention in the life cycle of plant species, de Wet and Harlan identify a single major point of transition, represented by a particular set of behaviors, that marks a critically important shift in the way in which plants are manipulated by humans.

The behavior set identified by Harlan and de Wet is the initiation and continuation of an annual cycle of harvesting and storing of seed stock and the planting of the stored seed in prepared areas at the beginning of subsequent growing seasons. It is the deliberate planting of harvested seed, according to de Wet and Harlan, that shifts plant species over into the category of being domesticated. This sowing of seed stock adds a human behavioral component to such open habitats, imposing an entirely new and quite different set of selective pressures on the target species, creating in effect a uniquely human environment that is quite different from any found in nature.

Within this seed bed environment competition under the new set of rules for reproductive success, and associated differential selection, over time produces populations that are well adapted to the new habitat, and as a result, are morphologically changed. Although developed specifically to account for morphological changes occurring in the major cereals during domestication (for example, wheats, barley, oats, rice, sorghum, millet, and maize), the theoretical framework of Harlan and de Wet is clearly applicable to the domestication of seed plants in eastern North America. Two of the phenotypic changes they list (Harlan et al. 1973:314) as occurring in response to increased seedling competition in prepared seedbeds are of specific relevance here—an increase in seed size and a reduction in germination dormancy:

> Sowing increases competition between individuals because of an increase in population density. Seedlings that germinate as soon as conditions are favorable, and grow most vigorously are best adapted for survival in sown fields. Increased seedling vigor is frequently associated with an increase in seed size, since the endosperm supplies the initial food for rapid seedling development. Seeds that germinate as soon as conditions become favorable lack a prolonged dormancy. In wild plants dormancy is advantageous. Seeds that germinate over a period of several years insure offspring even when one season is unfavorable. Domestic seed plants lack dormancy. Seeds that do not germinate soon after planting will contribute little to the harvest. Their seedlings will be crowded out, and the ones that survive will probably mature after the harvest. Dormancy is therefore strongly selected for during domestication. (de Wet and Harlan 1975:104)

Harlan and de Wet's elegant early 1970s Adaptive Syndrome of Domestication theory thus contributes essential explanatory elements to the theory proposed here for eastern North America—elements absent from Anderson's theory of the 1950s. While Anderson's theory identifies disturbed open habitats of human origin as the spatial/behavioral context of domestication, and identifies preadapted floodplain weeds as major open habitat colonizer candidates for human manipulation, Harlan and de Wet focus on the specific cause and effect behavioral sequences that transform colonizer candidates into domesticated plants. They provide a clear definition of seed plant domestication that is framed in a compelling evolutionary context. Specific causal human behavior sets (deliberate human planting of harvested seed stock) are identified as creating an entirely new habitat setting with quite different sets of selective pressures. These selective forces (involving both differential seed capture at harvest and seed bed competition) in turn cause, over time, a set of adaptive syndromes or responses by the plant populations under selection. These adaptive syndromes of domestication are reflected in phenotypic changes in the plant (many of which involve the reproductive propagules) that are directly linked to the selective forces at work (larger seed size reflects an adaptive response to seedling competition).

Fortunately, a number of the morphological changes that de Wet and Harlan's theory link to deliberate planting can be observed in the fragmentary plant parts recovered from archaeological sites. In eastern North America the appearance of several such changes provide direct evidence of the initial domestication of seed plants.

Technological Advance and Documenting Eastern Domesticates

In providing a framework of explanation for particular sets of morphological changes under domestication, the theory of Harlan and de Wet links particular observable results, such as an increase in seed size, to specific causal patterns of human behavior. Inherent in the incorporation of their theory here and its application to the prehistoric East is the recognition that the simple identification of a morphological change of some form does not, in itself, constitute adequate documentation of a plant species having been brought under domestication. An adequate case for domestication must first of all provide linkage, as do de Wet and Harlan, between the observed morphological change and a set of causal behavior patterns, such as deliberate harvesting and sowing that result in a new and artificial environment with distinctive pressures of differential selection. It is not enough to simply document phenotypic change. It is also necessary to explain why such change appears in response to a newly created environment of domestication.

In addition to linking morphological change to causal environments of domestication, it is also necessary, in building an adequate case for domestication, to be able to compare accurate measurements taken on archaeological specimens under study with similar sets of measurements taken on numerous control samples of both wild and known domesticate populations (the larger the sample size the better). When the measurements taken on archaeological specimens of the morphological feature in question fall outside of those recorded in control populations of wild plants, and within the range documented for domesticated comparison groups, *and* the phenotypic change being studied can be linked to the adaptive syndrome of domestication, then a reasonable case for domestication can be said to have been established.

Once a reasonable case for domestication has been established, it is important to temporally place the initial domestication of the seed plant in question as accurately as possible. Accurate placement of the timing of the initial date of domestication of seed plants in eastern North America first occurred in the 1980s, due in large measure to the successful application of a technological advance in radiocarbon dating—accelerator mass spectrometer (AMS) age determination. AMS dating allowed, for the first time, direct age determination of extremely small samples. As discussed in some detail in the following chapter, marshelder *(Iva annua)* was brought under domestication by 4,000 B.P., sunflower *(Helianthus annuus)* and chenopod *(Chenopodium berlandieri)* by 3,500 B.P., and *Cucurbita pepo* sometime between about 4,300 and 3,000 B.P.

In marshelder and sunflower a well-documented increase in seed (achene) size (Figures 3.2, 11.2) associated with increased seedling competition marks the beginning of deliberate planting of these two crop plants and their transition to domestication. A similar increase in seed size along with increasing rind thickness, also marks the domestication of *C. pepo* by 4,300 to 3,000 B.P., but the transition to domestication of this species in eastern North America remains to be documented in detail. As discussed in Chapters 3 and 4, *Cucurbita pepo* is not yet universally accepted as an indigenous plant brought under domestication in the East, rather than an introduced Mesoamerican crop plant.

In the case of *Chenopodium berlandieri* a reduction in seed coat (testa) thickness, associated with reduced germination dormancy, is the morphological marker that signaled deliberate planting, seedbed competition, and the adaptive syndrome of domestication (Chapter 3, Figure 3.2). A reasonable case for domestication of this seed plant in eastern North America was not established until the middle 1980s, however, for several reasons. David and Nancy Asch initially argued that thin-testa *Chenopodium* fruits recovered from archaeological contexts were produced by wild plants (Asch and Asch 1977), delaying the recognition that such a reduction in seed coat thickness was a morphological marker of domestication. More importantly, however, was the initial application in the early 1980s of another technological innovation to the problem of accurately measuring the extremely small scale changes in seed coat thickness that signaled domestication in *Chenopodium berlandieri.* Scanning electron microscopy (SEM) (Wilson 1981) provided accurate measurement of large samples of archaeological seed coats, and their comparison with modern control populations of wild and domesticated *Chenopodium berlandieri,* allowing a strong case to be made for its domesticated status (Smith 1984; Chapters 5 and 6).

In summary, with the application of two technological innovations in the early 1980s (AMS, SEM), it was possible to substantially strengthen this third component of the present theory, as reasonable cases were by then established for the domestication of three indigenous seed plants (marshelder, sunflower, chenopod) in eastern North America. A strong case for a fourth indigenous domesticate *(Cucurbita pepo)* has only been developed since 1985 (Chapters 3 and 4). Containing the archaeological evidence of morphological change in four indigenous seed plants, this third general component of the theory outlined herein is firmly anchored into the overall explanatory framework by Harlan and de Wet, who link such morphological change to causal sets of human behavior.

At the same time that this third general component was being substantially strengthened both by the application of AMS and SEM innovations and by Harlan and de Wet's bridging theory, essential information was also becoming available regarding the initial creation of open habitats by human populations in eastern North America.

The Initial Appearance of Anthropogenic Open Areas in Eastern North America

While underscoring the important role of human-created open habitats in the initial domestication of plants, neither Carl Sauer nor Edgar Anderson addressed the question of when such anthropogenic environments first appeared in different parts of the world. Similarly, while de Wet and Harlan consider the adaptive process of plant domestication that took place within such disturbed soil settings, they do not focus on when such habitats were first created.

As early as 1957, however, Melvin Fowler recognized the applicability of Anderson's "Dump Heap" theory to eastern North America, and with remarkable vision, proposed that cultivation of local plant species earlier mentioned by Gilmore and Jones began in the region during the Late Archaic period in association with the establishment of more permanent settlements:

> In a sense Anderson is proposing that a somewhat sedentary existence is almost a prerequisite for agricultural beginnings. . .dating between 2000 and 3000 B.C. are large middens, built up from accumulated refuse, largely shellfish remains, of successive occupations. The size and extent of the sites indicate a stable occupation and a large number of people dwelling in one place. (Fowler 1971:124)

> The Hypothesis based upon the foregoing discussion is at this point fairly obvious. It can be stated as follows. (1) Most plants domesticated by man were open habitat plants, (2) in the culmination of the Archaic Stage many sites are middens of semi-sedentary peoples, (3) peoples of the Archaic Stage were strongly oriented toward the use of foods, (4) there are indications of plants local to the area and other than maize having been cultivated, therefore (5) plant cultivation in the eastern United States probably had its beginnings in the Archaic Stage. (Fowler 1971:125)

Research in the 1980s indicated that the initial transition from relatively short term camp sites by hunter-gatherer groups to more sedentary settlements in at least some river valley segments of the East took place several thousand years earlier than the Late Archaic settlements mentioned by Fowler, between about 7,000 B.P. and 5,700 B.P. (Brown 1985; Smith 1986:21–27; Chapter 3). The two essential aspects of these new settlements were that they were occupied throughout the growing season, as well as being reoccupied and reused on an annual basis over a long period of time, thus constituting permanent disturbed soil open habitat settings. Combining on a long-term basis the key elements of sunlight, inadvertent human fertilization and disturbance of the soil, and the continual introduction of seeds from harvested wild species, these settlements represented experimental plots for plant manipulation that were established several thousand years prior to the first indications of deliberate planting and plant domestication in the region. As discussed in Chapter 3, these permanent open habitat settings appear to have developed as the result of hunter-gatherer groups targeting rich river valley resource zones which in turn appeared in response to shifting stream flow patterns. This fourth general component of the floodplain weed theory establishes the presence of the cultural and locational context of plant domestication in eastern North America by 7,000 to 5,700 B.P., and accounts for the formation of these open habitats in terms of human adaptation to changing patterns of resource distribution in river valley environments. These changes in river valley landscapes would have also improved and expanded the habitats of three plant species initially brought under domestication in eastern North America, plants that conform to Anderson's floodplain weed profile.

The Floodplain Niche of Indigenous Domesticates

Because of their weedy colonizing adaptations, all four of the seed plants brought under domestication in eastern North American between 4,000 and 3,000 years ago were pre-adapted to becoming components of the plant communities of anthropogenic open habitats. Of the four, the sunflower is perhaps the least well known in terms of its original natural habitat prior to colonizing the sedentary mid-latitude habitation sites described in the preceding section. Recent molecular evidence suggests "a single origin of the domesticated sunflower from a very limited gene pool" (Rieseberg and Seiler 1990:79), and is compatible with Heiser's hypothesis that it was introduced from the west into eastern North America as a "camp following weed" prior to 3,000 B.P., where it was brought under domestication and subsequently diffused into the Southwest and Mexico (Heiser 1985:58–59). The sunflower was apparently a western weed that could be characterized as having been "shirt-tailed" into eastern anthropogenic habitats, and domestication, along with three indigenous floodplain weeds.

Over the past eight years a considerable amount of field research has been carried out in river valley landscapes across twelve eastern states to document the floodplain habitat, specific niche, relative abundance, and potential harvest yield of populations of free-living *Cucurbita* gourds, *Chenopodium berlandieri,* and *Iva annua.* The other three plants that were brought under domestication in the region between 4,000 and 3,000 B.P. With the exception of a single study of marshelder (Asch and Asch 1978), little previous information was available regarding these plant species and their place in floodplain ecosystems.

As described briefly above, and in considerable detail in Chapters 4, 7, and 8, *Cucurbita pepo, Chenopodium berlandieri,* and *Iva annua* exist today as well-adapted, long-established components of river valley vegetational communities. More specifically, the niches of all three plants are tightly tethered to the disruptive ability of floodwaters to modify the landscape and create open habitats. All three plants colonize the natural open areas created annually by floodwaters, while also expanding into the open habitats created by human activities whenever opportunities arise. All three plants thus conform quite closely to the general "floodplain weed" profile outlined by Edgar Anderson and Jonathan Sauer four decades ago.

It is interesting that a period of more than thirty years elapsed from when Anderson first proposed his general ideas regarding the floodplain origin of domesticated plants until field research was initiated in the 1980s that would confirm that the floodplain weed portion of his theory was applicable to three of the four plants brought under domestication in eastern North America. As discussed above, Anderson himself did not pursue any of the plants proposed by Gilmore and Jones as possible eastern domesticates (including *Iva* and *Chenopodium*) even though he was familiar with their work. Jonathan Sauer's classic 1952 study of pokeweed provided an important early case study of a floodplain weed, but it did not lead to any expanded consideration of other plants with similar niches, nor did it address the topic of floodplain origins of domesticated plants. As I was unaware of Sauer's work with pokeweed until quite recently, it did not serve as any sort of impetus for the research described in Chapters 4, 7, and 8. Melvin Fowler embraced Anderson's "Dump Heap" theory of plant domestication and proposed its application to the river valleys of the Midwest as early as 1957. He cited both Anderson's identification of most domesticated species as "open habitat plants" (see above quotation), and the earlier ideas of Gilmore and Jones regarding potential eastern domesticates, including *Iva* and *Chenopodium*. But even Fowler's hypothesis of the late 1950s, which was remarkably accurate given the information available to him at that time, did not stimulate and direct consideration and documentation of eastern domesticates as floodplain weeds. In the early 1960s, Stuart Struever did edge closer in some respects to the theory presented here, putting forward a "Mud Flats" hypothesis of plant cultivation. Citing both Anderson's "Dump Heap" theory and Jonathan Sauer's research on pokeweed, and clearly influenced by Anderson's writing, he characterized *Chenopodium*, amaranth, and marshelder as floodplain weeds:

> *Chenopodium, Amaranthus,* and *Iva* establish themselves only when some external factor has intervened to obliterate the potential competitors and open up a patch of raw soil. These genera do not belong to a stable plant association. Their niche is a habitat disrupted either by man or by natural agencies. They colonize natural scars in the mantle of vegetation. Such disturbed places were created by the annual spring floods which by cutting and filling constantly opened new areas. The most extensive of such scars would be the [mud] flats of shallow backwaters covering large sectors of the flood plain. It is difficult to imagine another situation in which a disturbed habitat is created with a regularity and geographic extent comparable to that of the broad alluvial bottomlands. It is here one might expect the early manipulation of commensal plants to have played an important role prehistorically in the development of higher levels of economic productivity. (Struever 1964:102–103)

While certainly correct in his general characterization of *Chenopodium* and *Iva* as disturbed context floodplain plants that were the subject of human manipulation, Struever's recognition of one of the major components of the theory presented here failed to either gain much recognition or lead to habitat studies of the plants in question. Struever's general niche characterizations do not appear to have been based on any habitat studies, since neither *Chenopodium berlandieri* nor *Iva annua* do well in the mud flats habitat he describes (Munson 1984). It is unfortunate that Struever focused his attention and his terminology so narrowly on "specific [mud flat] locales" (Struever 1964:106), since his niche characterization of these two floodplain weeds was generally accurate, and his incorporation of them into an application of Anderson's theory to eastern North America held considerable promise. Struever's comments were also temporally off target, however, in that they were directed toward the Middle Woodland time frame of intensification of food production economies rather than the initial domestication of seed plants that had taken place several millennia earlier.

Like Fowler's paper of seven years earlier, Struever's study demonstrates both remarkable predictive insight, given the quite limited information available to him, and the important contribution of each of the major components outlined in preceding sections of this chapter in the formation of a cohesive and comprehensive overall framework of explanation. While both Fowler and Struever were aware of and expanded upon the research and ideas of Gilmore, Jones, Anderson, Sauer, and others, and offered impressive insights regarding some aspects of plant domestication in eastern North America, much of the information needed to carry their interpretations any further was not yet available. Harlan and de Wet's Adaptive Syndrome of Domestication theory would not appear until the 1970s. SEM and AMS technology and the morphological and temporal evidence of seed plant domestication, along with habitat studies and information regarding the initial formation of anthropogenic open habitats, would not become available until the 1980s.

Conclusions

The Floodplain Weed theory of plant domestication in eastern North America can be briefly summarized as follows. It begins with pre-7,000 B.P. gatherer-hunter groups in the mid-latitudes utilizing a wide variety of animals and plants as food sources, including three seed plant species that grew in the flood disturbed open habitats of river valley landscapes (*Iva annua, Chenopodium berlandieri, Cucurbita pepo*). These three floodplain weeds were attractive food resources both because of their abundant occurrence in floodplain environments and their substantial harvest yield potential. They were harvested each fall, mostly likely by women, who carried them back to their small, short-term habitation sites, along with other food resources, for preparation and consumption. During processing some of the seeds of these floodplain weeds were invariably dropped into the disturbed soil, open habitat settings of these habitation sites. Some of these seeds would germinate and grow the following summer, with the resultant plants in turn reseeding the location in the fall. These habitation sites do not appear to have been occupied for very long periods of time, however, perhaps six months at most, and they were not reoccupied on a consistent annual basis. As a result, the small disturbed soil pockets of open habitat they represented were ephemeral in nature, representing a dead end for the colonizing floodplain weeds that required a constantly disturbed open habitat setting.

This situation changed in the period 7,000–5,700 B.P. however, as rich resource zones appeared along some mid-latitude river valley corridors and more sedentary habitation sites were established in close proximity to them. Such settlements were occupied for longer periods of the year, perhaps throughout the growing season, and just as importantly, they were apparently reoccupied on an annual basis for thousands of years. These settlements represented both an entirely new type of habitation site for their human occupants (more sedentary, larger, reoccupied on a long-term annual basis) and an entirely new type of habitat for their colonizing plant inhabitants (anthropogenic, continually disturbed, open habitat). The seeds of the three floodplain weeds of interest here continued to be harvested in the fall and returned to such "permanent" open habitat settlements where a small percentage would have been dropped during processing, to germinate and grow the following year. Since these settlements were not quickly abandoned, but stable and consistent, these colonizing floodplain weeds were not closed out after a few seasons, but rather became established as permanent components of the plant communities of an entirely new type of open habitat.

The relative success and abundance of weedy wild forms of *Iva annua, Chenopodium berlandieri,* and *Cucurbita pepo* would likely have varied considerably over time and between the geographically dispersed anthropogenic open habitat settings of the region, depending upon the degree to which human occupants maintained an open habitat, and how and to what extent they encouraged these potential domesticates. Joined by the sunflower *(Helianthus annuus),* an apparent third or late fourth millennium weedy introduction from further west, the three floodplain weeds were pre-adapted for domestication both in terms of economic potential and colonizing abilities. Their seeds had long represented a dependable, high yield river valley food source, and they were probably well represented among the pool of potential candidates for domestication that comprised the plant communities of anthropogenic open habitat settings.

The archaeological record provides few indications of the variety of different ways in which these plants may have been encouraged, managed, and manipulated by the human occupants of such open habitat settings between 7,000 and 4,000 B.P. Clear evidence in the form of phenotypic changes in the reproductive propagules of *I. annua, C. berlandieri, H. annuus,* and *C. pepo,* on the other hand, marks the millennium from 4,000 to 3,000 B.P. as witnessing a significant turning point in the human history of eastern North America. Larger seeds and thinner seed coats indicate the deliberate planting of stored seed stock and the beginning of food production economies, resulting in the independent domestication of four indigenous seed-bearing plants, three of which began their long journey toward domestication as floodplain weeds.

The Floodplain Weed theory of plant domestication in eastern North America, briefly summarized above, consists of the five major interlocking components discussed in the preceding sections of this chapter. Each of the five components is essential to the strength and stability of the overall explanatory framework, with each of the five reinforcing and cross-illuminating the others. Although the following chapter (Chapter 3), which was originally published in 1987, presents the Floodplain Weed theory in considerable detail, it was not quite complete at that time, as indicated by the last minute addition of note 11 to Chapter 3. This end note, which discusses the research of Deena Decker-Walters up to 1987 and the loose end represented by the wild indigenous versus introduced domesticate status of *C. pepo,* serves as a segue from Chapter 3 to Chapter 4. By documenting the wild floodplain weed niche of eastern North American *Cucurbita pepo* gourds, Chapter 4 in turn ties off this loose end and fits the final section of this theory into place. In addition, Pat Watson and Mary Kennedy have recently offered a valuable correction of the theory as originally proposed in 1987. They point out that deliberate planting was down-played as one of a number of steps along the continuum of increasing plant manipulation, rather than highlighted as a major transition point in human history that resulted in the initial domestication of seed plants (Watson and Kennedy 1991). In an interesting and thought provoking potential expansion of the theory, they also suggest that deliberate planting and plant domestication should be acknowledged as a gender specific accomplishment rather than attributed at the population or regional cultural level.

While certainly open to further modification and expansion, the Floodplain Weed theory of plant domestication, as outlined in this chapter, also provides an opportunity to consider the essential attributes of, and adequacy criteria for, theories of domestication in general. It should be fairly obvious, for example, given the above description of the Floodplain Weed theory, that explanatory frameworks of domestication need to be specifically tailored to the environments, plant and animal species, and developmental sequences of the particular areas under study. As a result, it is not surprising that in large measure the development of adequate explanations of plant/animal domestication in different regions of the world will depend upon the gathering of detailed information of quite a diverse nature regarding the landscape, floral and faunal communities, and climatic and cultural history of the regions under study. While more universal theories of domestication are often initially imposed on regions in the absence of much information, adequate regionally specific explanatory frameworks are invariably indigenous, and only emerge as relevant information becomes available.

A second fairly obvious conclusion to be drawn from the Floodplain Weed theory outlined above is that domestication as a process should not be conflated with the initial development of agriculture—substantial human investment in, and reliance on, food production economies. The initial domestication of plants and animals may be temporally and developmentally far removed from the transition to agriculture, and as a research topic domestication is certainly important and complex enough to warrant consideration in its own right.

Another complicating factor highlighted in the Floodplain Weed theory involves the various and often vague definitions assigned to the term "domestication." The Adaptive Syndrome of Domestication theory of Harlan and de Wet resolves this problem, at least for eastern North America, by providing a clear and convincing explanation of the specific set of behaviors (planting) that results in new plant forms that can persist only so long as humans maintain the new and artificial environment, defined in terms of a new set of selective pressures. Since such definitions of domestication will vary across species of root and seed crops and species of animals, it is essential that explanatory frameworks both specify exactly what is meant by the term, and identify the causal pattern of human behavior involved.

De Wet and Harlan's theory also highlights the importance of identifying, to the extent possible, clear and unequivocal effects or results of domestication that can be observed in the archaeological record. Fortunately in regard to seed plants, and to eastern North America, Harlan and de Wet have identified several distinctive phenotypic changes in the reproductive propagules of plants that are associated with the adaptive syndrome of domestication.

The Floodplain Weed theory also underscores the obvious if not always acknowledged necessity of both documenting in detail the archaeological indicators of domestication, and placing them in an accurate temporal context. When morphological change in target species constitute the archaeological indicators of domestication, as is the case in eastern North America, then direct dating of representative specimens (rather than dating by association) should be required, once their domesticated status is established by comparison with wild and domesticated control groups.

It is also essential in any theory of domestication to identify, to the extent possible, the spatial and cultural context of domestication and what factors may have operated in forming the context or milieu of major human manipulation and modification of target species of plants and animals. In eastern North America, for example, the initial establishment of permanent human-disturbed open habitat settings where plant domestication is proposed as having taken place can be linked to changes in the abundance of river valley resources, which in turn resulted from changes in seasonal rainfall and stream flow characteristics. Thus climatic change and modification of river valley landscapes can be identified as important, if indirect, contributing factors in the domestication of seed plants in the region.

Another element of central importance in the Floodplain Weed theory, and any theory of domestication, involves explaining how and why the target species moved or were drawn into the spatial and behavioral context of domestication, and what attributes of the species eventually domesticated contributed to them being singled out for deliberate attention and manipulation. In eastern North America, for example, all four domesticated plants were preadapted to human intervention in terms of their weedy ability to prosper in a variety of disturbed soil contexts. Not only did they establish themselves with little assistance as components of habitation site plant communities, they would also have been relatively flexible and responsive to human management. They were in this regard relatively "user friendly." In addition, three of the four had long histories in the region as abundant, high yield, and dependable wild food sources, while the fourth probably came highly recommended in these regards. The seeds of these four species were also relatively easy to harvest and process, in contrast to other potential domesticates such as giant ragweed, with its hard achenes. In other areas of the world good candidates for domestication are often also as easy to recognize, in retrospect, as they are in the East.

Finally, any consideration of alternative competing theories of domestication in particular regions of the world should take into account their relative overall integrity and coherency—the degree to which their constituent component arguments are well integrated and compatible, without reliance upon assumptions that are either too numerous or too tenuous. In comparison to several alternative theories discussed in the next two chapters, the Floodplain Weed theory exhibits considerable integrity, with its framework of explanation strengthened by two overarching attributes of coherency that are so obvious as to be easily underemphasized. These two attributes involve an integrity of origin of the plants in question, and a coherency of process. Three of the four plants domesticated in the East originated in the open habitats of river floodplains. They shared numerous general adaptations to this habitat and filled very similar ecological niches. In addition, these three floodplain weeds, along with a weedy arrival from the west, all shared the process of domestication, being deliberately planted and transformed into products of the human environment during the same five- to ten-century span. Such basic and obvious attributes of shared origin and shared transition to domestication together indicate a coherent, independent, indigenous process of domestication in eastern North America.

Literature Cited

Anderson, E.

1948 Gravel Bars Evolve Their Own Flood Control. *Bulletin of the Missouri Botanical Garden* 36:54–57.

1952 *Plants, Man and Life.* Little Brown and Co., Boston.

1956 Man as a Maker of New Plants and New Plant Communities, In *Man's Role in Changing the Face of the Earth,* edited by W. L. Thomas, pp. 763–777. University of Chicago Press, Chicago.

Asch, D. L., and N. B. Asch

1977 Chenopod as Cultigen: A Re-evaluation of Some Prehistoric Collections from Eastern North America. *Midcontinental Journal of Archaeology* 2:3–45.

Asch, N. B., and D. L. Asch

1978 The Economic Potential of *Iva annua* and its Prehistoric Importance in the Lower Illinois Valley. In *The Nature and Status of Ethnobotany,* edited by R. Ford, pp. 301–343. Museum of Anthropology, Anthropological Papers No. 67. University of Michigan, Ann Arbor.

Brown, J. A.
1985 Long-Term Trends to Sedentism and the Emergence of Complexity in the American Midwest. In *Prehistoric Hunter-Gatherers: The Emergence of Cultural Complexity,* edited by T. D. Price and J. A. Brown, pp. 201–234. Academic Press, Orlando, Florida.

Deitz, R. A.
1952 The Evolution of a Gravel Bar. *Annals of the Missouri Botanical Garden* 34:249–254.

Fowler, M.
1971 The Origin of Plant Cultivation in the Central Mississippi Valley: A Hypothesis. In *Prehistoric Agriculture,* edited by Stuart Struever, pp. 122–128. Natural History Press, Garden City, New York.

Harlan, J. R., J. M. J. de Wet, and E. G. Price
1973 Comparative Evolution of the Cereals. *Evolution* 27:311–325.

Heiser, C.
1949 Enigma of the Weeds. *Frontiers* 13:148–150.

1985 Some Botanical Considerations of the Early Domesticated Plants North of Mexico. In *Prehistoric Food Production in North America*, edited by R. Ford, pp. 57–72. Museum of Anthropology, Anthropological Papers No. 75. University of Michigan, Ann Arbor.

Munson, P.
1984 Weedy Plant Communities on Mud-Flats and Other Disturbed Habitats in the Central Illinois River Valley. In *Experiments and Observations on Aboriginal Wild Plant Food Utilization in Eastern North America,* edited by P. Munson, pp. 379–385. Prehistoric Research Series 6(2). Indiana Historical Society, Indianapolis.

Rieseberg, L. H., and G. J. Seiler
1990 Molecular Evidence and the Origin and Development of the Domesticated Sunflower (*Helianthus annuus,* Asteraceae). In New Perspectives on the Origin and Evolution of New World Domesticated Plants, edited by P. K. Bretting. *Economic Botany* [Supplement] 44(3):79–91.

Sauer, J.
1952 A Geography of Pokeweed. *Annals of the Missouri Botanical Garden* 39:113–125.

Smith, Bruce D.
1984 *Chenopodium* as a Prehistoric Domesticate in Eastern North America: Evidence from Russell Cave, Alabama. *Science* 226:165–167.

1986 The Archaeology of the Southeastern United States from Dalton to de Soto (10,500 B.P.–500 B.P.). In *Advances in World Archaeology,* vol. 5, edited by F. Wendorf and A. E. Close, pp. 1–92. Academic Press, Orlando, Florida.

Struever, S.
1964 The Hopewell Interaction Sphere in Riverine-Western Great Lakes Culture History. In *Hopewellian Studies,* edited by J. Caldwell and R. Hall, pp. 87–106. Illinois State Museum, Scientific Paper No. 12. Springfield.

Watson, P., and M. Kennedy
1991 The Development of Horticulture in the Eastern Woodlands of North America: Women's Role. In *Engendering Archaeology,* edited by J. Gero and M. Conkey, pp. 255–275. Basil Blackwell, Oxford, England.

de Wet, J. M. J., and J. R. Harlan
1975 Weeds and Domesticates: Evolution in the Man-Made Habitat. *Economic Botany* 29:99–107.

Wilson, H. G.
1981 Domesticated *Chenopodium* of the Ozark Bluff Dwellers. *Economic Botany* 35:233–239.

CHAPTER 3

❑ ❑ ❑

The Independent Domestication of Indigenous Seed-Bearing Plants in Eastern North America

Introduction: An Independent Origin

The possibility that the Eastern Woodlands of North America was an independent center of cultivation and domestication of indigenous seed plants is often discussed under the general heading "Eastern Agricultural Complex."[1] While the specific term "Eastern Agricultural Complex" was apparently not used in print until the early 1960s (Struever 1962:584–586, 1964:98) the independent development of plant husbandry in the eastern United States was first mentioned as a possibility almost sixty years earlier.

The 1920s: Linton and "Various Small Grains in the Southeast"

In an article comparing the nature and technology of maize cultures of Mexico, the Southwest, and the eastern United States (the "Eastern complex"), Ralph Linton briefly mentioned the possibility of an "older food complex" in the region "which did not center around maize":

> The Eastern complex, on the other hand, differs so much from the other two that we must either suppose maize culture to have there undergone a local development along lines quite outside those of the original pattern, or consider it the result of the super-position of maize upon some older food complex which was itself rather elaborate. The latter hypothesis seems much more probable, and I believe that the Indians of the eastern United States were already in possession of the hoe and mortar at the time they acquired maize. The wild foods of the region were sufficient to support a considerable population, and it is not impossible that the eastern tribes had developed at least the beginnings of agriculture, for in historic times there were certain practices, such as the planting of wild rice in the north

This chapter originally appeared in *Emergent Horticultural Economies of the Eastern Woodlands,* edited by W. Keegan. Southern Illinois University at Carbondale, Center for Archaeological Investigations, Occasional Paper 7(1987):3–47.

> and of *various small grains in the southeast,* which bore little resemblance to maize culture and may well have originated independently. . . . In the East, on the other hand, maize probably arrived as a result of gradual diffusion, lost much of its cultural context in route, and was adopted into a preexisting cultural pattern which had grown up around some other food or foods. (Linton 1924:345, 349, emphasis added)[2]

The 1930s: Gilmore and Jones

Linton's brief speculations on the subject foreshadowed the work of Gilmore (1931) and Jones (1936), who provided impressive arguments in support of the prehistoric cultivation, and in several cases domestication, of seed plants indigenous to the eastern United States, based on their respective pioneering analysis of archaeobotanical assemblages recovered from rock shelters in the Ozarks and eastern Kentucky.

While Gilmore did not directly address the issue of whether any eastern plant species had been brought under cultivation independently of, and prior to, the arrival of any tropical crops, he did propose in his 1931 article that a number of indigenous seed plants be assigned cultigen status based on their abundant storage in apparent seed stock situations:

> Like all good and prudent farmers those ancient farmers of the Ozarks put away carefully in woven bags a good supply of specially selected seed stock of all their own crop plants for the next spring's planting.
>
> But besides the well-known staple crops of corn, beans, squashes, pumpkins and sunflowers, which were cultivated by those ancient people and by many other tribes of Indians from that time down to the present, there is evidence that the ancient Ozark Bluff-Dwellers also had certain other species of plants not cultivated at the present time. The ground for this statement lies in the fact that supplies of the seed of these other species of plants were carefully put way together with the selected seed of corn, beans, sunflowers, squashes and pumpkins.
>
> Among the other species of plants which appear to have been cultivated by those ancient farmers were a species of *Chenopodium,* goosefoot or lamb's-quarters, very closely related to *Chenopodium nuttalliae* Stafford, rough pigweed (*Amaranthus* sp.), giant ragweed (*Ambrosia trifida* L.), burweed marshelder (*Iva xanthifolia* [Fresen.] Nutt.) and Carolina canary grass (*Phalaris caroliniana* Walt.). (Gilmore 1931:85–86)

The preference that these species show for disturbed soil situations is also mentioned by Gilmore as evidence for their prehistoric cultivation: "There is another indication in the habits and behavior of all these species which gives the impression that they have been accustomed to domestication. All of them seem to have become attached to man and to prefer to dwell near human habitations, in ground which has undergone disturbance due to human occupation" (Gilmore 1931:86).

In addition to identifying the goosefoot seeds recovered from Ozark Bluff-Dweller sites as *Chenopodium nuttalliae,* the domesticate still grown today in Mexico ("or a species related most closely to it"),[3] Gilmore also mentions both ragweed and sunflower as exhibiting morphological characteristics associated with domestication (sunflower: increase in seed size; ragweed: increase in seed size, pale color).[4]

In his analysis of plant remains from the Newt Kash Hollow rock shelter in eastern Kentucky, Jones (1936:149–152) similarly proposed prehistoric cultigen status for goosefoot, ragweed, sumpweed, and sunflower, based both on large seed size and the abundance of the seeds of these four species in recovered human paleofecal material.[5]

In addition, based on the "surprisingly small amount of material of tropical agricultural origin" (Jones 1936:163) present in the archaeobotanical assemblage from Newt Kash Shelter analyzed by Jones[6] and the apparent abundance of indigenous cultigen seeds in preceramic, premaize contexts, Jones cautiously speculated that eastern seed-bearing plant species might have been brought under cultivation prior to the introduction of tropical cultigens:

> Gilmore is of the opinion that these plants may have been local additions to the tropical agricultural complex. In the Kentucky region they seem earlier than any of the tropical plants except possibly the gourd and the squash. We shall have to await further knowledge of stratigraphy in the bluff shelter cultures to determine priority of the various cultivated plants. If it should develop that these local prairie plants were indeed cultivated and preceded the gourd and squash we could hardly escape the startling conclusion that agriculture had a separate origin in the bluff shelter area. The evidence is yet so meager, however, that further speculation is idle. (Jones 1936:163)

The 1940s: Carter and Quimby

As frequently happens in archaeology, Jones's quite cautious suggestion that the eastern United States might have been an independent center of plant domestication was subsequently recast as fact: "A widespread, rather well developed agriculture in the Southeast prior to the appearance of the typical Middle American plants such as corn, moschata, and kidney beans is thus well established" (Carter 1945:28).

Carter's characterization fortunately spurred George Quimby (1946) to write a quite insightful analysis of the independent origin issue and the group of possible indigenous cultigens identified by Gilmore and Jones, which he variously termed the "eastern plant complex," the "southeastern agricultural complex," the "southeastern plant complex," and the "southeastern complex."[7] While acknowledging that the indigenous plants assigned cultigen status by Gilmore and Jones were probably domesticated, Quimby suggested that the observed increase in seed size might alternatively be explained by deliberate selection during harvesting of wild stands,[8] and that the storage of seeds was not sufficient to confer cultigen status, since wild plant foods were also frequently recovered from storage contexts.

More importantly, while Quimby (1946:1) was willing to assume that "the plants under discussion were domesticated," he also correctly pointed out that "neither Jones nor Gilmore presented evidence which indicated that the southeastern agricultural complex was domesticated prior to the introduction of some (if not all) of the Middle American plants" (1946:1):

> In the available archaeological evidence (admittedly incomplete) from the Ozark Bluff Shelters there is nothing to show that the eastern plant complex is earlier than the tropical agricultural assemblage. The available evidence from the Newt Kash Shelter is not sufficient to determine whether or not the eastern plant complex was domesticated prior to the arrival of any items of the tropical agricultural assemblage.
>
> Consequently, the two best studies of the eastern plant complex (Gilmore 1931 and Jones 1936) only demonstrate the existence of the eastern plant complex in pre-Columbian times, the certainty that the plants were eaten, and the strong probability that any or all of the plants were domesticated. (Quimby 1946:5)[9]

First raised by Jones in 1936 and reiterated by Quimby a decade later, this central question of temporal precedence—whether seed plants indigenous to the Eastern Woodlands of North America were independently brought under cultivation and domesticated prior to the introduction of tropical cultigens—has remained unanswered and actively debated over the past half-century.

The 1950s: Anderson and Fowler

Another decade would pass before Fowler (1957, 1971) would revive the question of independent plant domestication in the Eastern Woodlands, combining Anderson's (1956) "dump heap" theory of agriculture with newly developed interpretations of Late Archaic residential stability in support of an independent Late Archaic beginning of plant cultivation based on indigenous eastern cultigens [see Chapter 2].

The 1960s: Yarnell and Struever

In the 1960s Stuart Struever (1962, 1964) presented interpretive arguments in support of the likely Woodland period economic importance of indigenous cultigens, and the independent eastern development of a "technologically simple horticulture in specific (mudflat) locales" (Struever 1964:106).[10]

At the same time Richard Yarnell (1963, 1964, 1965, 1969, 1972) was providing important critical assessments of the nature and relative strength of the data and arguments presented in support of assigning cultigen or domesticated status to proposed indigenous crop plants.

At the close of the 1960s Watson and Yarnell would provide in the conclusion section of the report on Salts Cave the following assessment of the independent origin question:

> Using these facts (absence of maize and beans, use of squash and gourd primarily as containers, strong dependence on at least two or three cultivated North American plants for food) one could argue that the Salts Cave diet pattern supports the old idea (Gilmore 1931; Jones 1936) of an early eastern United States horticultural complex based primarily on native plants; a complex which preceded cultivation of maize, fleshy squash, and beans. The latter three were presumably added later (it could be suggested that the beginning of this process is evidenced in the Salts paleofeces), and to a large extent superseded the older native cultigens.

> However, the situation described by Quimby in 1946 is not altered by the Salts Cave evidence: there is still no site where the supposedly older species are found as cultigens in the absence of any Mesoamerican species.
>
> The Salts Cave data indicate nothing directly about the nature of the origin of plant cultivation in the Eastern United States: whether this was a case of independent invention (beginning with native plants, and only later adding species traded from Mesoamerica); a result of diffusion of seeds and cultivation techniques from Mesoamerica; or a result of stimulus diffusion involving ideas (and possibly one or two transported species such as squash and gourd as well) from the south. (Watson and Yarnell 1969:76)

Clearly, the nature of the extant 1970 data base did not allow a resolution of the temporal precedence issue in favor of either independent development or Mesoamerican diffusion. While the Salts Cave deposits suggested that indigenous seed plants had been brought under cultivation prior to the introduction of maize, beans, and fleshy squash into the East, the continuing recovery of cucurbit material (bottle gourd and *Cucurbita* gourd) in association with the indigenous premaize cultigens implied the presence, on the same time level, of introduced cultigens.

In the conclusion to their summary of the available archaeological and archaeobotanical data relevant to early plant husbandry in the east, Struever and Vickery (1973) take an interesting approach to this cooccurrence of cucurbit material and early indigenous cultigens in regard to the question of independent development of horticulture in the East. In adopting the argument in support of an "early eastern United States horticultural complex" outlined by Watson and Yarnell (see above quotation), Struever and Vickery, through omission, imply that cucurbits were irrelevant to the independent development of food production systems based on eastern seed plants:

> Out of these varying discussions of an Eastern Agricultural Complex, three common ideas emerge: (1) that certain local plant species were brought under cultivation at an early date in the eastern United States; (2) that this phenomenon occurred independently of experiments with and dependence on cultivation in Mesoamerica; and (3) the cultivation of local plants preceded the diffusion of the tropical maize-beans-squash complex into the east. (Struever and Vickery 1973:1213)

Struever and Vickery's admonition (1973:1213) that "the concept of the 'Eastern Agricultural Complex' is deserving of serious consideration" was swept away, however, by discoveries of the mid-1970s.

The 1970s: Discovery of Middle Holocene Cucurbits

Well-preserved cucurbit remains (bottle gourd and *Cucurbita* sp.) dating circa 4,300 B.P., were recovered from Phillips Spring, Missouri, from 1974 through 1978 (Kay et al. 1980). These Phillips Spring materials, along with *Cucurbita* rind fragments recovered from similar temporal contexts at shell mounds along the Green River, Kentucky, were considered to represent cultivated plants introduced from Mesoamerica. Since these two purported tropical cultigens predated the earliest indications of cultivation or domestication of any indigenous seed plants, Chomko and Crawford (1978:405) concluded that: "the eastern horticultural complex was not an independent development but was a regional adaptation of the concept of horticulture that originated in Mesoamerica."

This confident assertion that eastern plant husbandry did not emerge from an independent developmental process, but rather was a regional adaptation—a secondary, derivative development growing out of a received Mesoamerican horticultural concept—gained further support from west-central Illinois in the early 1980s. *Cucurbita* rind fragments recovered from both the Koster and Napoleon Hollow sites (Figure 3.1) yielded direct accelerator radiocarbon dates of circa 7,000 B.P. (Conard et al. 1984), and were identified as *Cucurbita pepo,* a tropical cultigen introduced from Mexico or perhaps southwest Texas. In addition, the early *Cucurbita* material recovered from Missouri, Kentucky, and Illinois has frequently been referred to as "squash."

These discoveries of cucurbit material in Middle Holocene contexts in the eastern United States are very important for a number of reasons, primarily in focusing attention and research interest on the complex developmental history of the genus *Cucurbita,* and the taxonomic relationships existing between the present day and prehistoric profusion of wild/weedy and domesticated taxa subsumed under the *Cucurbita pepo* label. Since the initial discoveries of the mid-1970s, this early cucurbit material has become

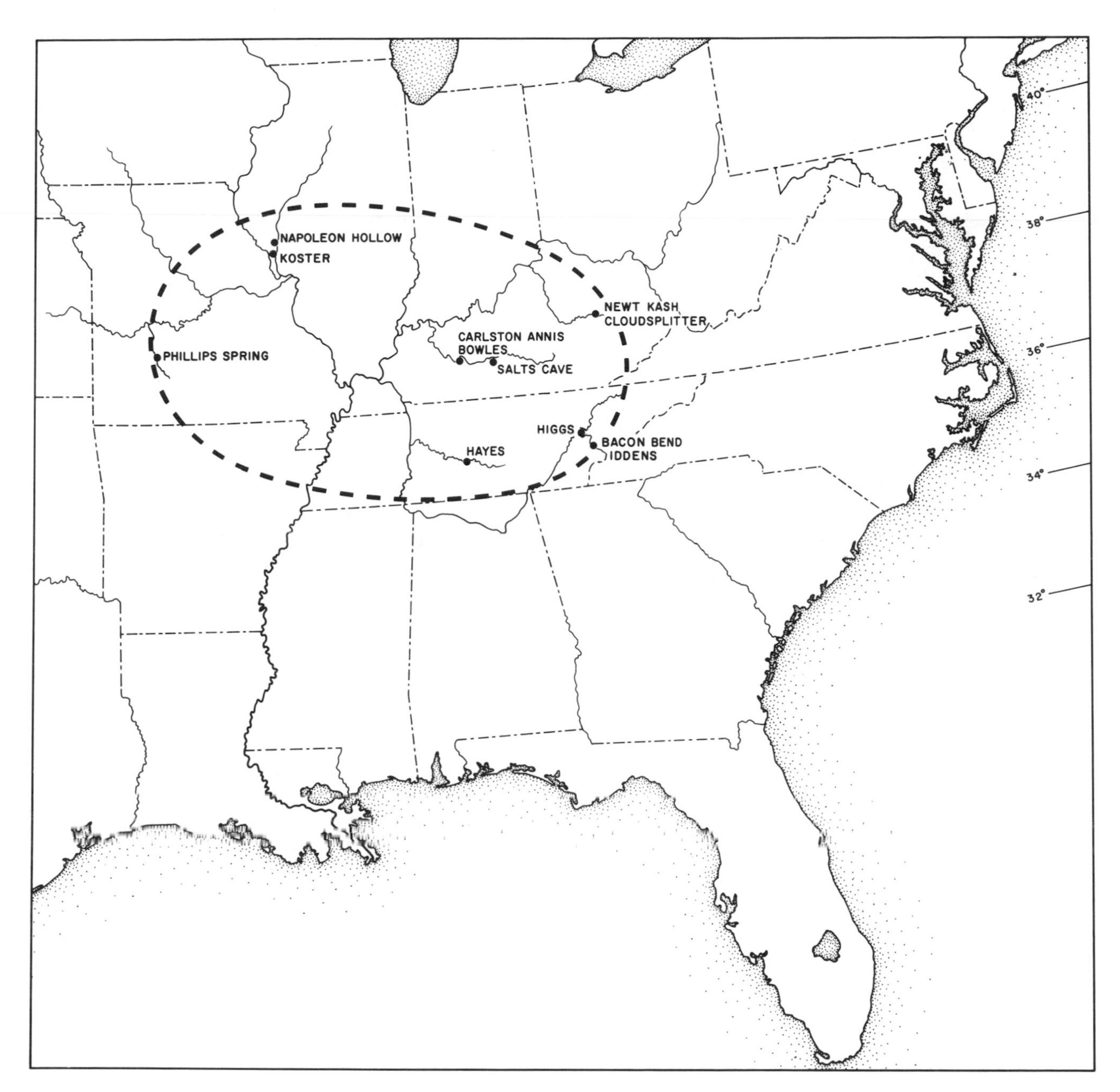

Figure 3.1 The geographic location of selected sites that have provided archaeobotanical information concerning the independent domestication of indigenous seed-bearing plants in the Eastern Woodlands of North America.

the subject of an intense and increasingly more interesting debate regarding its identification as a tropical cultigen, and its significance in regard to the initial cultivation and domestication of seed plants indigenous to eastern North America.

At the same time, however, the discovery of this early cucurbit material and its identification as "*pepo* squash" has deflected research interest over the past decade away from a consideration of the developmental process leading to the cultivation and domestication of local crop plants. The early presence of a purported tropical domesticate has led to the apparently prevalent opinion that plant husbandry was introduced into eastern North America from Mexico during the Middle Holocene (8,000 to 4,000 B.P.), and that this introduction adequately accounts for the subsequent cultivation and domestication of eastern plants: "in the east the idea of agriculture came from Mexico many millennia before even native plants were obviously cultivated, much less domesticated" (Ford 1985:18).

In this chapter an alternative viewpoint will be presented—that the early cucurbit material recovered from Middle Holocene sites does not negate what was an independent developmental process leading to the emergence of distinctly eastern plant husbandry systems centered on indigenous starchy-seeded and oily-seeded food crops. In attempting to build a case for the prehistoric eastern United States as an independent center of food production and plant domestication, the timing and nature of this separate and distinct developmental process will be outlined, following a discussion of alternative perceptions and explanations of the Middle Holocene cucurbit material.

Middle Holocene Cucurbits in the Eastern Woodlands

Based on distinctive differences in both rind structure and seed morphology, two taxa of cucurbits have been identified in Middle Holocene contexts in the eastern United States. The earliest occurrence of the bottle gourd *(Lagenaria siceraria),* is 7,290 B.P. at the Windover site on the east coast of Florida [Doran et al. 1990], while material from west-central Illinois assigned to the species *Cucurbita pepo* has been directly dated to 7,000 B.P.

To attempt an understanding of the kind of plant that is represented by the *Cucurbita* material recovered from eastern Middle Holocene contexts, rather than what labels have been applied to it, a brief consideration of *Cucurbita pepo* terminology and taxonomy is necessary.

Cucurbita Terminology and Taxonomy

The common names "squash" and "summer squash" are frequently used in reference to *C. pepo.* These names are misleading, however, in that the species *Cucurbita pepo* subsumes a wide range of modern cultivar varieties, often differentiated by the culinary terms "pumpkin" (coarse and strongly flavored fruit), "squash" (finer textured, milder flavored fruit), and "gourd" (inedible hard shelled ornamental form) (King 1985:94). While the ornamental gourds comprise the only commonly recognized variety *(C. pepo* var. *ovifera),* Decker's recent analysis of allozyme variation in this morphologically diverse species tends to support the usefulness of Castetter's (1925) division of "edible *C. pepo* cultivars into six horticultural groups according to fruit shape, color, and size" (Decker 1985:300). In addition, the Texas wild gourd *(C. texana)* is morphologically similar to *C. pepo* var. *ovifera,* and a subspecies level grouping of *C. pepo* var. *ovifera* and *C. pepo* var. *texana* has been proposed (Rhodes et al. 1968).

Recent hybridization of *C. texana* and *C. pepo,* with analysis of allozyme differentiation in the progeny, however, indicates that *C. texana* "seems to possess only a subset of alleles present in the *[C. pepo]* species complex" (Kirkpatrick et al. 1985:297). "Low coefficients of genetic identity among groups of *C. pepo* cultivars," however, did not allow for "a clear taxonomic assessment" of *C. texana* (Kirkpatrick et al. 1985:297). The close but unclear taxonomic relationship between *C. texana* and *C. pepo* led Kirkpatrick and others (1985:289) to subsume *C. texana* and the cultivar varieties of *C. pepo* within a "*C. pepo* complex" rather than assigning *C. texana* to the species *C. pepo.* This "boundary blur" is mirrored by the taxonomic treatment given to spontaneous (either wild or escapes from cultivation) "gourds" of the *C. pepo* complex depending upon where they are found. Such spontaneous populations are identified as the Texas wild gourd if found in Texas, while morphologically similar noncultivated plants growing outside of Texas are assumed to be escapes from cultivation and are classified as *C. pepo* var. *ovifera* (Kirkpatrick et al. 1985:94).

The high level of polymorphism documented by Decker (1985:308) also suggests early diffusion and rapid diversification and evolution of various cultivar varieties within the *C. pepo* complex, making comparisons of equality rather than similarity between archaeological specimens and modern materials inappropriate, particularly below the species level and for older materials (King 1985:94).

Two useful points in regard to the use of *Cucurbita* terminology in reference to archaeological plant materials from the eastern United States can be drawn from the foregoing brief discussion (King 1985:94 also provides an excellent discussion of the pitfalls of cucurbit terminology). First, since it carries such strong culinary meaning for most people, the term "squash" should be applied to archaeologically recovered *Cucurbita* material only when the culinary meaning is appropriate—if the material can be determined to represent a fine textured, mild flavored cultivar variety. Secondly, while the Texas wild gourd is at present included within the *C. pepo* complex, a clear taxonomic assessment and assignment to the species *C. pepo* (which at present only includes cultivated forms) is not possible. Identification of archaeological material as belonging to the species *C. pepo* should therefore be done only when it can be demonstrated that the material in question can be distinguished in some convincing manner from morphologically similar noncultivated taxa belonging to the genus *Cucurbita* (for example, *C. texana, C. foetidissima*).

In turning to an assessment of the strength of the case for applying the common term "squash" and the species designation "*Cucurbita pepo*" (both of which carry the embedded meaning "tropical cultigen") to Middle Holocene *Cucurbita* material of the Eastern Woodlands, characteristics of *Cucurbita* rinds and seeds will be considered, along with the present day and prehistoric geographical range of *Cucurbita* species in North America.

Cucurbita Rinds

Nine sites in Illinois (Koster, Napoleon Hollow, Kuhlman, Lagoon), Missouri (Phillips Spring), Kentucky (Carlston Annis, Bowles), and Tennessee (Hayes, Bacon Bend)(Figure 3.1) have yielded *Cucurbita* rind fragments from Middle Holocene (8,000 to 4,000 B.P.) contexts. The six direct accelerator dates on *Cucurbita* rind recovered from Koster (7,100 ± 300, 6,820 ± 240 B.P.), Napoleon Hollow (7,000 ± 250 B.P.), Carlston Annis (5,730 ± 640), Hayes (5,430 ± 120 B.P.), and Bowles (4,060 ± 220) leave no doubt as to the presence of a *Cucurbita* form in the middle latitude riverine area west of the Appalachian wall during the Middle Holocene period.

All of the Middle Holocene *Cucurbita* rind fragments recovered to date are thin, however, with reported maximum thickness values all falling below 2.0 mm (Koster and Napoleon Hollow 1.7 mm, Carlston Annis and Bowles 1.4 mm, Phillips Spring 1.8 mm). The rind thickness values reported for these sites overlap considerably with published rind thickness values for two modern wild *Cucurbita* gourds; the buffalo gourd *(C. foetidissima)* and the Texas wild gourd (*C. texana*)(King 1985:92; Asch and Asch 1985a:156–157), leading King (1985:91) to conclude that *Cucurbita* rind fragments thinner than "perhaps 2.0 mm" can not be conclusively identified as a gourdlike variety of the tropical cultigen *C. pepo* (as opposed to a wild gourd).

If "King's Rule" on rind thickness is applied to the *Cucurbita* rind fragments recovered from Middle Holocene contexts in the Eastern Woodlands, none can be assigned to *C. pepo* cultigen status. It was not until after 3,000 B.P. that rind thickness values exceeded 2.0 mm (at Salts Cave and Cloudsplitter Rock Shelter in Kentucky) (Figures 3.1, 3.2), and that it can be confidently assumed that *C. pepo* was present, based on rind thickness values alone. In their analysis of similar thin rind *Cucurbita* material from the Guila Naquitz site in Oaxaca, Mexico, Whitaker and Cutler (1986:278) state: "the rind fragments of cultivated and noncultivated species of cucurbits are indistinguishable." In an earlier publication Whitaker and Cutler (1965:344) declared that "at present, however, we have no means of differentiating the rinds of species of *Cucurbita.*" Based on the above discussion, and the small sample of rind fragments (n = 8) measured from the Koster and Napoleon Hollow sites, the conclusion reached by Asch and Asch (1985a:157, see also Conard et al. 1984) that the "Archaic squash rind from west-central Illinois is *C. pepo,*" cannot be accepted.

Cucurbita Seeds

To date, *Cucurbita* seeds have been reported from Middle Holocene contexts at only two sites in the Eastern Woodlands of North America. Sixty-five

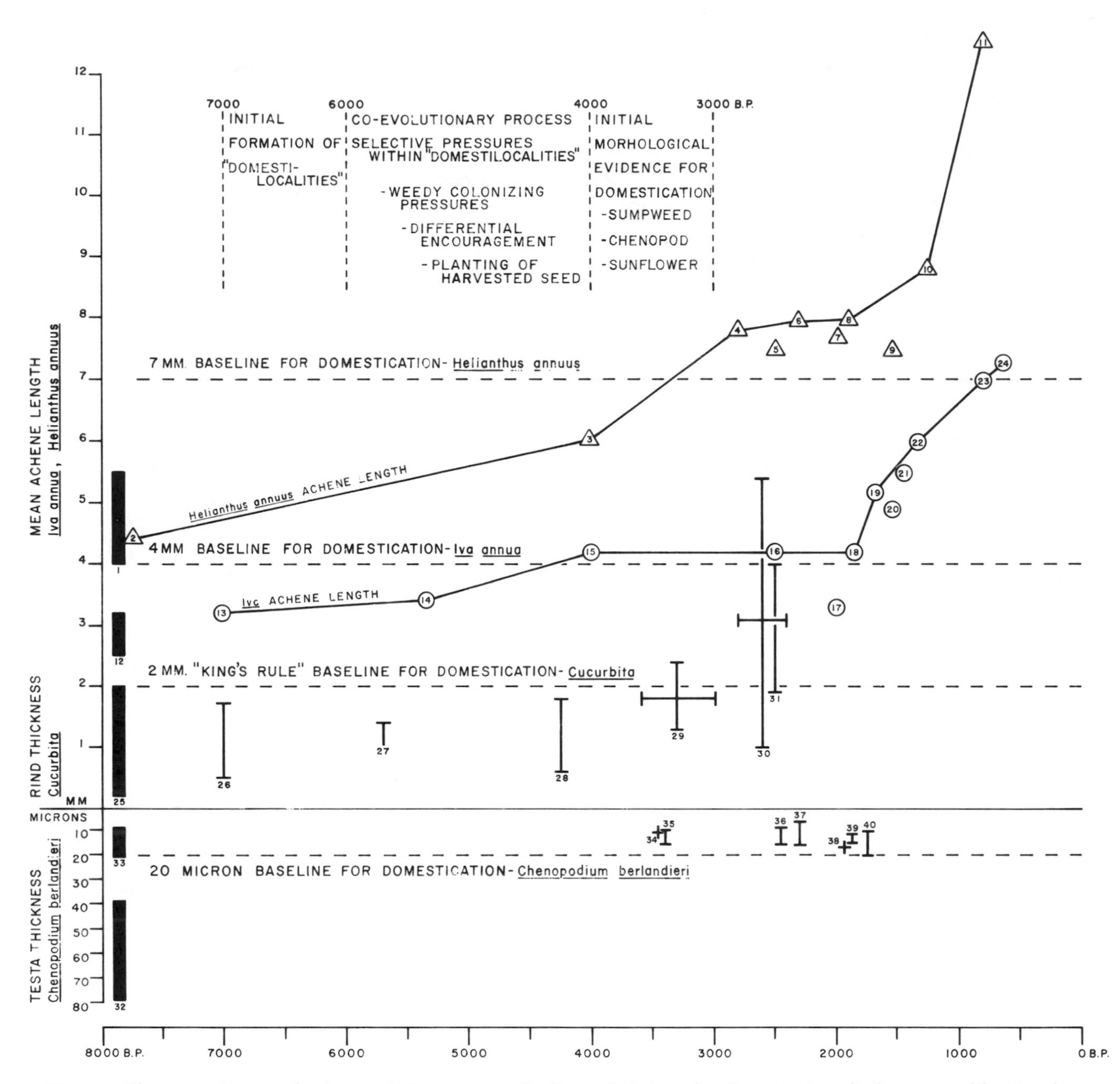

Figure 3.2 The temporal context for the coevolutionary process leading to the independent domestication of indigenous seed-bearing plants in the Eastern Woodlands of North America.[a] Modern or prehistoric assemblage information (sample size), and reference(s).

Sunflower *(Helianthus annuus):* 1. Modern wild size range[b]—Yarnell 1978, Heiser 1954; 2. Koster Horizon 9 (1)—Asch and Asch 1985a; 3. Napoleon Hollow (24)—Asch and Asch 1985a; 4. Higgs (24)—Yarnell 1978; 5. Salts Cave (J4:4) (47)—Yarnell 1978; 6. Westmoreland Barber (1)—Gremillion and Yarnell 1986; 7. McFarland (2)—Crites 1985a; 8. Patrick (4)—Schroedl 1978; 9. Owl Hollow (15)—Crites 1985a; 10. Rogers (11)—Cowan 1985a; 11. McCullough (1000)—Yarnell 1981.

Marshelder *(Iva annua):* 12. Modern wild populations range of means (11)[c]—Asch and Asch 1978; 13. Koster Horizon 9 (3)—Asch and Asch 1985a; 14. Koster Horizon 7–6 (288)—Asch and Asch 1985a; 15. Napoleon Hollow (44)—Asch and Asch 1985a; 16. Salts Cave feces (879)—Yarnell 1978; 17. McFarland (3)—Crites 1985a; 18. Patrick (5)—Schroedl 1978; 19. Smiling Dan (13)—Asch and Asch 1985b; 20. Owl Hollow (1)—Crites 1985a; 21. Newbridge (224)—Asch and Asch 1985a; 22. Haystack (74)—Cowan 1985a; 23. McCollough (19)—Yarnell 1981; 24. Turner-Snodgrass (33)—Yarnell 1978.

seeds securely dated to circa 4,300 B.P. were recovered from the Phillips Spring site in west-central Missouri, and a total of 14 seeds and seed fragments were reported from well-sealed features and midden deposits of Stratum 7 at the Bacon Bend site in east Tennessee (Chapman 1981). Described as representing a short-term campsite occupied on a number of occasions over a period of perhaps 100 to 500 years, Stratum 7 at the Bacon Bend site yielded three radiocarbon dates on samples of wood charcoal recovered from the fill of two adjacent rock-filled fire pits. Feature 16, which contained three *Cucurbita* seed fragments, produced a date of 4,390 ± 155 B.P., while a split sample from Feature 17, a similar pit located 1 m away, produced dates of 4,070 ± 70 and 3,580 ± 225 B.P. [The Bacon Bend *Cucurbita* seeds have been reanalyzed and are not *Cucurbita*.]

Although no *Cucurbita* seeds have yet been recovered from secure Middle Holocene contexts earlier than 4,300 B.P., the presence of rind fragments promises the eventual recovery of such seed specimens dating to the period 7,000 to 4,300 B.P. [*Cucurbita* seeds

Gourd *(Cucurbita)*: 25. Modern and archaeological buffalo gourd *(C. foetidissima)* range (44)[d]—King 1985; 26. Koster Horizon 9, Napoleon Hollow (8)—Asch and Asch 1985a; 27. Carlston Annis, Bowles (11)—Watson 1985; 28. Phillips Spring (10)—King 1985; 29. Cloudsplitter (6)—Cowan 1985b; 30. Cloudsplitter (37)—Cowan 1985b; 31. Salts Cave (10)—King 1985, Yarnell 1969.

Goosefoot *(Chenopodium berlandieri)*: 32. Modern wild range (58)[e]—Smith 1985b; 33. Modern *C. berlandieri* ssp. *nuttalliae* cv. "chia" (2)—Smith 1985b; 34. Cloudsplitter (F.S. 1361) (1)—Smith and Cowan 1987; 35. Newt Kash (4)—Smith and Cowan 1987; 36. Cloudsplitter (F.S. 2316) (6)—Smith and Cowan 1987; 37. Russell Cave (20)—Smith 1985a; 38. White Bluff (1)—Fritz 1986; 39. Edens Bluff (1)—Fritz 1986; 40. Ash Cave (20)—Smith 1985b.

[a]The lines representing temporal trends of morphological change in sunflower, marshelder, and goosefoot represent the leading edge of the domestication process as reflected in selected archaeobotanical samples. As these plant species came under domestication, wild varieties of each of these species continued to be utilized, and if plotted on this figure, would fill in areas below the "domestication baselines" all along the temporal sequence.

[b]Yarnell (1978:291) estimates the mean achene length values for "original wild ancestral sunflowers" to probably have ranged from 4.5 to 5.0 mm, based on the modern achene length range of 4.0–5.5 mm given by Heiser (1954:299) for *H. annuus lenticularis* (no sample size given). The 7 mm "baseline for domestication" in sunflower is based on the position taken by Heiser (1985:60) that "It is likely that any achene over 7 mm long should be interpreted as coming from a domesticated plant." It should be kept in mind that this 7 mm baseline refers to individual seeds. Yarnell (personal communication, 1986) considers 6 mm a conservative baseline mean achene length value for assigning archaeobotanical assemblages of sunflower achenes domesticated status.

[c]Asch and Asch (1978:323) provide achene size information for 11 modern wild *Iva annua* populations, with population mean values for achene length ranging from 2.5 to 3.2 mm (population samples ranged from 97 to 204 achenes). Achenes as large as 4.6 mm were occasionally observed in these modern wild stands. In a more recent publication, Asch and Asch (1985a:162) state that "In any wild population of *Iva* there are occasional achenes longer than 4.0 mm, but these are a very small percentage of the total. We have found no technique of selective collecting from different regions or different habitats, from large or small plants, or from different parts of a plant that could raise the mean size of achenes in a wild harvest sample significantly above the natural mean (2.5–3.2 mm)." The 4 mm "baseline for domestication" in marshelder is based both on the documented very rare occurrence of achenes larger than 4 mm in modern wild populations (Asch and Asch 1978, 1985a), and the consensus acknowledgment that the mean achene length value of 4.2 mm documented for the Napoleon Hollow *Iva* assemblage represents a domesticated crop (see text discussion). Yarnell (personal communication, 1986) considers 3.5 mm as a tentative baseline mean achene length value for assigning archaeobotanical assemblages of marshelder achenes domesticated status.

[d]King (1985:92) provides rind thickness values for a total of 44 measured specimens of *Cucurbita foetidissima*. Twenty-four modern specimens (four specimens from each of six collections from Kansas, Colorado, and Arizona) ranged in thickness from .2–1.5 mm (average .7 mm). Twenty archaeological specimens from Tularosa Cave ranged in thickness from .7 to 2.0 mm (average 1.0 mm). King's 2 mm "baseline for domestication" is discussed in the text.

[e]Smith (1985b:128) provides testa thickness values for eight modern wild populations of *Chenopodium berlandieri* from the Eastern Woodlands, with thickness values ranging from 39 to 78 microns (population samples ranged from five to nine fruits). Based on the testa thickness values documented for the modern Mexican cultivar *C. berlandieri* ssp. *nuttalliae* (Figure 3.2, line 33) a 20 micron baseline for domestication can be proposed for this species.

predating those recorded from Phillips Spring have now been reported from Cloudsplitter Rock Shelter, Kentucky, see Chapter 12]. Likely to be single carbonized seeds or seed fragments when eventually identified, *Cucurbita* seed material from pre-4,300 B.P. contexts will present a difficult dilemma for researchers. On the one hand, a recent direct accelerator date on a carbonized *Cucurbita* seed recovered from an apparently well sealed pre-4,300 B.P. cultural context in eastern North America demonstrated it to be a modern contaminant, underscoring the necessity of obtaining direct accelerator age determinations on apparent Middle Holocene *Cucurbita* material. On the other hand, direct dating of pre-4,300 B.P. *Cucurbita* seed material in order to unequivocally establish its age will necessitate sacrificing rare and potentially critically important specimens relevant to the taxonomic assignment of Middle Holocene *Cucurbita,* and the larger question of independent plant domestication in eastern North America.

Seed morphology is useful in distinguishing among modern species of cucurbits (Whitaker and Cutler 1965), allowing, for example, the easy identification of seeds as belonging to the *C. pepo* complex rather than to *C. foetidissima.* The value of seed characters, however, has yet to be convincingly demonstrated in regard to differentiating between closely related wild and cultivated varieties of *Cucurbita,* and in marking the transition from wild to cultivated and domesticated status along a developmental trajectory of incipient plant husbandry. As King (1985) points out in an excellent discussion of the topic, morphological characteristics of *Cucurbita* seeds exhibit little variability relative to other fruit characters and are genetically independent to a large degree from them. In addition, within the *Cucurbita pepo* complex, there is "considerable overlap in length and width measurements for various modern cultivars. Therefore, very few can distinctly be separated on the basis of seed characters. In a letter to the author (23 January 1980) Bemis says that he does 'not think that one can separate certain cultivars of *C. pepo, C. pepo* var. *ovifera, C. texana* or other species of the Sororia group based entirely on seed characteristics'" (King 1985:93). [There now appear to be morphological characteristics that can be used to distinguish at least some wild and cultivated forms of *C. pepo,* see note 11.]

In their analysis of Early and Middle Holocene cucurbits from the Guila Naquitz site in Oaxaca, Mexico, Whitaker and Cutler (1986), however, distinguish seeds of a cultivar variety of *C. pepo* from morphologically quite similar seeds of a wild species of *Cucurbita.* Judging from the illustrations provided of *Cucurbita* seeds from Guila Naquitz, neither seed size nor shape (width/length) criteria were employed in distinguishing wild from cultivated *Cucurbita.*

Until Whitaker and Cutler, or others, describe and validate those specific morphological characteristics of seeds that can be employed in differentiating between closely related wild and cultivated varieties of *Cucurbita,* it would be premature to identify the Middle Holocene *Cucurbita* seeds from eastern North America (i.e., the circa 4,300 B.P. materials from Phillips Spring and Bacon Bend) as representing, on the basis of seed characters alone, the tropical cultigen *C. pepo* as opposed to a prehistoric wild/weedy variety of *Cucurbita.*

The difficulties inherent in attempting to assign cultigen *C. pepo* status to the Phillips Spring *Cucurbita* seeds on the basis of seed morphology can be illustrated by comparing them with present day and prehistoric wild and domesticated varieties of the *C. pepo* complex. The Middle Holocene *Cucurbita* seeds from Bacon Bend have yet to be described [now withdrawn as *Cucurbita*], and have not been directly dated by the accelerator method. King provides a detailed description of the 65 measurable *Cucurbita* seeds from the Phillips Spring site, which ranged in length from 8.3 to 12.2 mm (x = 10.5) and from 5.4 to 8.5 mm in width (x = 7.03), while shape (width/length) ratios varied from 0.54 to 0.82.

In comparison, the nine measurable seeds from the Guila Naquitz site identified by Whitaker and Cutler (1986) as representing the earliest known cultivar variety of *C. pepo* (circa 9,800 B.P.) are closely comparable to the Phillips Spring seeds in width (range 6 to 8.5 mm, x = 7.4) and shape (width/length ratio values range from 0.5 to 0.72), with only the largest of the Guila Naquitz seeds falling outside of the Phillips Spring range of variation in length (range 10 to 17 mm, x = 11.6). In addition, Whitaker and Cutler illustrate two seeds identified as representing a wild species of *Cucurbita,* and these appear to measure 11.6 by 7.2 mm and 9.3 by 7.6 mm, with respective width/length ratios of 0.62 and 0.81, placing them within the size range of the Phillips Spring seeds.

In addition, samples of 25 seeds from each of two modern Texas wild gourds *(C. texana)* (King 1985:86) yielded size and shape statistics that overlap considerably with the Phillips Spring *Cucurbita* assemblage (mean seed length = 9 mm, mean seed width = 5.8 mm, width/length ratio = 0.64).

While both prehistoric and present-day seeds of wild species of the *Cucurbita pepo* complex fall comfortably within the range of size variation exhibited by the Phillips Spring seed assemblage, modern cultivar varieties of *C. pepo,* as well as seed assemblages from more recent (Late Holocene) contexts, also contain seeds that fall within the Phillips Spring size range curve (King 1985, figs. 4.3, 4.4).

Although the Phillips Spring *Cucurbita* seed assemblage shows considerable overlap in size with both wild and cultivated varieties of the *Cucurbita pepo* complex, it is interesting to note that of the mean seed length values given by King (1985, figs. 4.3, 4.4) for 39 populations of *Cucurbita* (32 populations of modern cultivar varieties of *C. pepo,* along with 7 archaeological *Cucurbita* seed assemblages from both the eastern United States and Mexico), only three assemblages have mean seed length values of less than 11 mm: Phillips Spring (10.5 mm), the Texas wild gourd (9 mm), and the three "smaller seeded type" specimens from the Cloudsplitter Rock Shelter in eastern Kentucky (8.7 mm). Because an increase in seed size is an automatic result of deliberate planting, and, as will be discussed in detail later in this chapter, is often used as both a morphological indicator of domesticated status in seed plants (see King 1985:90), and as a marker for the transition from wild harvesting of a plant to its cultivation and initial domestication, an 11 mm mean seed length boundary line would appear to perhaps be of value in separating seed assemblages of wild and domesticated varieties of the *C. pepo* complex. While this 11 mm mean seed length dividing line certainly serves to separate present day wild and cultivated taxa within the *C. pepo* complex, based on the measurement of large samples of seeds, archaeological assemblages consisting of only a few seeds would be difficult to assess with any reasonable level of confidence. This 11 mm line does, however, separate the admittedly small samples of wild and domesticated *Cucurbita* seeds from Guila Naquitz described by Whitaker and Cutler (1986).

When applied to the prehistoric Eastern Woodlands of North America, this seed length dividing line suggests the same developmental trajectory as that indicated by *Cucurbita* rind fragments (Figure 3.2): it is not until after 3,000 B.P. that average seed length values for *Cucurbita* exceed 11 mm (at Salts Cave and Cloudsplitter Rock Shelter in Kentucky).

Because there is thus a complete absence of any morphological indicators of domesticated status in Middle Holocene *Cucurbita* assemblages, and both rind thickness and mean seed length values suggest that a domesticated variety of *C. pepo* was not present in the eastern United States until after 3,000 B.P., what then is the basis for assigning Middle Holocene *Cucurbita* materials cultivar *Cucurbita pepo* status?

Geographical Range Arguments

Setting aside the morphological evidence, the case for assigning cultigen status to Middle Holocene *Cucurbita* material from eastern North America rests entirely on the apparent present-day absence of any indigenous wild *Cucurbita* forms in the East. Such geographical range arguments, sometimes employed in support of assigning a cultigen label to prehistoric plant material (for example, sumpweed and maygrass in the eastern United States), involves the recovery of prehistoric specimens of a plant species in locations considerably outside of its documented present-day geographical range. Such prehistoric range extensions relative to the present-day distribution of a plant species are explained in terms of human transportation and cultivation of the plant in question beyond its "normal" range. While range extension arguments are used in support of assigning cultigen status to prehistoric plant species, rarely if ever have they been considered of much value in establishing a claim for domestication in the absence of any demonstrated morphological change which allows differentiation from wild populations. Even when used in support of cultigen status, rather than domestication, such arguments are relatively weak. In the absence of other types of evidence, they invariably engender debates concerning both the accuracy with which the present day geographical range of the plant in question has been established, and the possibility of prehistoric fluctuations in a plant's distribution that are unrelated to human activities.

This is certainly the case for the Middle Holocene *Cucurbita* material of the Eastern Woodlands, with alternative explanations to it being a domesticated cultivar variety of *C. pepo* introduced from Mexico centering on two wild species of North American *Cucurbita*.

Alternative Explanations

THE HYPSITHERMAL HYPOTHESIS. The buffalo gourd *(C. foetidissima)*, which prefers dry, often rather barren habitats, has a present-day geographical range that extends north from Mexico into the Southwest and as far east as Texas, southwest Missouri, Nebraska, and Kansas, and is considered as recently adventive east of the Mississippi River in Illinois and Indiana (King 1985:91; Asch and Asch 1985a:156).

Before eventually rejecting the possibility, Asch and Asch (Conard et al. 1984; Asch and Asch 1985a:156) speculated that the *Cucurbita* rind fragments recovered from 7,000 B.P. contexts in west-central Illinois might reflect a Hypsithermal related eastward range extension of the ancestor populations of the present-day buffalo gourd: "No wild *Cucurbita* is native to Illinois, but *Cucurbita foetidissima* extends eastward into Missouri (perhaps even farther east during the Hypsithermal climate of 7,000 B.P.?)" (Conard et al. 1984:443). "Speculatively, the Hypsithermal climatic conditions of 7,000 years ago might have favored a temporary eastward extension of this species' natural range" (Asch and Asch 1985a:157).

The Hypsithermal climatic episode (see Smith 1986 for a recent brief summary) witnessed the eastward expansion of the geographical ranges of a variety of species of plants and animals into a middle latitude "direct impact zone." The buffalo gourd may also have expanded its range eastward in response to changing climatic conditions. It is interesting to note in this regard that the temporal sequence of available accelerator dates on Middle Holocene *Cucurbita* rind (west-central Illinois by 7,000 B.P., west-central Kentucky by 5,700 B.P., and west-central Tennessee by 5,400 B.P.) parallels the time-transgressive eastward expansion of the Hypsithermal climatic peak.

A variation on this Hypsithermal hypothesis would not require a range extension of the buffalo gourd as far east as central Kentucky and Tennessee, but only far enough east within the expanding prairie peninsula to bring it within the range of developing prehistoric exchange catchments of the Eastern Woodlands. A number of different exchange mechanism models have been offered in explanation of how the tropical cultivar *C. pepo* could initially have been introduced into the Eastern Woodlands from Mexico. These same models of exchange could also be invoked, over much smaller distances, to explain the rapid and widespread exchange across the Midwest and Southeast of what would have been highly prized utilitarian items—buffalo gourd containers and rattles.

While certainly speculative, some combination of these two Hypsithermal-related buffalo gourd explanations, on the basis of available evidence, are as plausible as the "introduced tropical domesticate" hypothesis in accounting for Middle Holocene *Cucurbita* rind fragments recovered from pre-4,300 B.P. contexts. The eventual recovery of *Cucurbita* seeds from pre-4,300 B.P. contexts should shed considerable light on the relative strength of these alternative explanations, since buffalo gourd seeds can be morphologically distinguished from those of the *C. pepo* complex. A continued absence of Middle Holocene *Cucurbita* seeds east of western Missouri (Phillips Spring) and prior to 4,300 B.P. would strengthen the likelihood that early *Cucurbita* containers were exchanged beyond the geographical range of the species as finished utilitarian objects, containing, if anything, other trade materials rather than seeds.

THE INDIGENOUS WILD GOURD HYPOTHESIS. The Texas wild gourd *(C. texana)* has a geographical range limited to central Texas where it is "rare but abundant where found, occurring in debris and piles of driftwood, often climbing into trees, along several rivers, especially the Guadalupe, that drains the Edwards Plateau" (Heiser 1985:65, quoting Correll and Johnston 1970). As mentioned earlier in this chapter, the present geographical range of *C. texana* may be taxonomically masked.

When plants morphologically similar to *C. texana* are found growing outside of cultivation in the Southeast, they are classified as escaped feral examples of the domesticated yellow-flowered egg gourd (*C. pepo* var. *ovifera*), rather than as the closely related *C. texana*. A comprehensive biosystematic study of these spontaneous populations of southeastern *Cucurbita* gourd, which have been reported growing along a number of southeastern rivers, particularly in Alabama, is both essential and long overdue. If such southeastern floodplain populations turn out to be indigenous rather than naturalized *C. pepo* var. *ovifera* (an escape from cultivation that has reverted to the wild state and become firmly established) they would provide considerable support for the proposition that the *C. texana* populations of central Texas represent a modern remnant of a population system/species complex of wild *Cucurbita* that had a much broader prehistoric distribution. [Chapter 4 addresses this issue.]

Charles Heiser, among others, has suggested the possible prehistoric presence of a wild *Cucurbita* gourd similar to *C. texana* east of the Mississippi River as a plausible alternative to the "introduced tropical domesticate" hypothesis: "This *wild* gourd [from Koster], probably similar to *C. texana*, at one time could have grown wild as far as Illinois so that calling for an introduction from outside may be unnecessary" (Heiser 1985:71, emphasis added).

Two variations on this indigenous wild gourd hypothesis also warrant consideration. The first involves an indigenous wild gourd with a relatively limited prehistoric range, perhaps not much beyond the modern range of *C. texana*. These gourds could have been traded, as highly prized containers, far beyond the plant's natural range. It is of course possible than wild gourd containers of different taxa were being exchanged into the Eastern Woodlands from both the west (*C. foetidissima*) and from the south (*C. texana*).

A second variation involves the possibility of an indigenous gourd belonging to the *C. pepo* complex being independently brought under domestication in the Eastern Woodlands (Cowan et al. 1981:71; Heiser 1985:64–65): "A plausible argument can be made for its domestication in the eastern half of North America, for it has a candidate for its progenitor in central Texas" (Heiser 1985:70). "Until further research is conducted, the prospect that cucurbit domestication took place in Texas or the East and spread southward remains an intriguing hypothesis" (Ford 1985:346). While it may seem questionable to consider the possible independent domestication of *Cucurbita pepo* in the East, in light of the apparent absence of a modern indigenous wild *Cucurbita* gourd in the region, the absence of an indigenous wild progenitor candidate for *C. pepo* is not a problem limited to the eastern United States. At present it appears that *C. pepo* was first brought under domestication in southwest Mexico, outside of the modern range of any proposed progenitors, including *C. lundelliana* to the east in Guatemala and Honduras and *C. texana* far to the north in Texas (Heiser 1979:312) [see Chapter 4 for a discussion of the recent discovery of *Cucurbita fraterna*, which provides a possible but unlikely progenitor for *C. pepo* in Mexico]. As to the possibility that the Middle Holocene *Cucurbita* material of the eastern United States represents a species of indigenous wild *Cucurbita* that no longer grows in the region, Whitaker and Cutler (1986:275) appear comfortable in recognizing, on the basis of thin rind fragments and small seeds recovered from Guila Naquitz Cave in Oaxaca, the presence of "a wild species of *Cucurbita* growing in the area and probably collected for the edible seeds," even though there are no similar wild cucurbits present today in southwest Mexico.

A number of alternative explanations have been outlined above that appear just as plausible as the "introduced tropical domesticate" hypothesis in accounting for the Middle Holocene *Cucurbita* of the Eastern Woodlands. A final and most plausible alternative explanation involves focusing on the likely habitat and niche of these early gourds and the relationship existing between them and human populations, rather than on the original source or "homeland" of these small container plants.

THE UNHUSBANDED GOURD HYPOTHESIS. Until recently, the Phillips Spring (4,300 B.P.) bottle gourd (*Lagenaria siceraria*) represented the earliest record of this species in the Eastern Woodlands. The two fragmented bottle gourds recovered in 1986 from the Windover site in Florida and directly dated to circa 7,300 B.P. [Doran et al. 1990], however, document the presence of this species in the East much earlier in the Middle Holocene. The bottle gourd could have been carried to eastern North America by ocean currents, either from Middle or South America (Heiser 1985:67, 72), or directly from Africa. Alternatively, it may have been brought north by humans. It is extremely rare in pre-2,000 B.P. contexts, having been reported at only four sites other than Phillips Spring (Iddens, Jernigan II, Cloudsplitter, and Salts Cave).

Judging from the bottle gourd rind fragments and seeds recovered from Phillips Spring, this Middle Holocene bottle gourd was quite similar in size to the Middle Holocene *Cucurbita* gourd of the East. Both were small, perhaps weighing up to a pound. Both produced numerous small and perhaps bitter tasting seeds, and both had a thin, hard, and inedible rind or shell, making them excellent containers and very unlikely food crops. Like the Middle Holocene *Cucurbita* rind and seed material, the Phillips Spring bottle gourd specimens cannot clearly be assigned domesticated status: "It should be pointed out that the size of the seed and the thickness of the rind of the bottle gourd from Phillips Spring. . .fall within the range of the smallest known for this species. This might indicate that they come either from a wild gourd or a primitive domesticated sort" (Heiser 1985:72). In the same publication Heiser also states that "the probability of the arrival of an entirely wild bottle gourd to the eastern area appears remote, but it did not necessarily come as a domesticate" (Heiser 1985:70).

In a number of publications, Richard Yarnell (Yarnell 1983, 1985; Yarnell and Black 1985:99) has expanded this nondomesticated, noncultivated, nonlocal gourd category to include the Middle Holocene *Cucurbita* material as well: "At present we probably should not assume that cultigen crops were involved. The older archaeological remains may be derived from weedy *Cucurbita pepo* gourds that did not require significant or intentional husbandry" (Yarnell 1983:4–5). "Currently it cannot be assumed with confidence that squash had achieved cultigen status before the Early Woodland period" (Yarnell 1983).

The impressive ability of both the bottle gourd and the yellow-flowered egg gourd to thrive unassisted in disturbed soil situations is well documented. "Cucurbits are vigorous growers and most are competitive enough for growth on the 'trash heap' without much human aid" (King 1985:78). This ability to thrive without human encouragement is underscored by the fact that the bottlegourd is naturalized (growing wild) today over much of the East, from Texas to Florida, and as far north as Illinois and Missouri (King 1985:79). Similarly, Asch and Asch remark on the pronounced self-propagating ability of *C. pepo* var. *ovifera* (Conard et al. 1984; Asch and Asch 1985a:157), but they argue that, unlike the bottle gourd, it could not have persisted prehistorically without "at least occasional replanting by Indians" (1985a:158). While this need for occasional replanting may be true for more northern latitudes, the poorly documented *C. pepo* complex gourd populations of southeastern floodplains, if not *C. texana,* could very well be naturalized stands of *C. pepo* var. *ovifera* [see Chapter 4 for an expanded discussion of the wild vs. escaped nature of these modern *Cucurbita* gourds].

The unhusbanded gourd hypothesis, while allowing the initial introduction of cucurbits (both bottle gourd and *Cucurbita* gourd) from Mesoamerica into the Eastern Woodlands during the Middle Holocene, at the same time denies them cultigen or domesticated status. Rather, they are viewed as "camp followers." Once introduced, they would have required very little if any intervention on the part of humans to survive. They could have thrived within the context of anthropogenic habitats as little more than tolerated or minimally encouraged weeds.

Yarnell's perceptive characterization of these Middle Holocene cucurbit remains directs attention beyond the question of their initial point of origin (indigenous, Texas, Mexico) to the more important issue of where these early container plants should be placed along the continuum of human-plant interaction, from wild to domesticated (Smith 1985a [Chapter 4]). There is no morphological basis on which to assign them domesticated status, and the ability of modern cucurbit analogues to survive in the East with no human attention strongly supports Yarnell's position that they do not warrant a "domesticate" or even a "cultigen" label, but rather should be considered as tolerated, perhaps minimally encouraged, weeds. It should also be noted that it is of course possible that once introduced from Mesoamerica as "camp follower" weeds, these early cucurbits were eventually, and independently, brought under domestication in the East, by around 3,000 B.P.[11]

The Archaeological Evidence for Initial Domestication of Seed-Bearing Plants

The ten-century span from 4,000 to 3,000 B.P. brackets the earliest evidence of morphological changes reflecting domesticated status in all three of the annual seed crops brought under domestication before the first appearance of maize in the East at about A.D. 200–300.[12]

For sumpweed *(Iva annua)* and sunflower *(Helianthus annuus)* the morphological change indicating domestication is an increase in achene size. Figure 3.2 illustrates the temporal trend of increase in the size of sunflower and sumpweed achenes over the past 4,000 years, so well documented by Richard Yarnell (1972, 1978, 1981) and Nancy and David Asch (1978, 1985a).

Iva annua

Based on the mean length values of a sample of 44 *Iva* achenes recovered from a Titterington phase context (circa 4,000 B.P.) at the Napoleon Hollow site (Asch and Asch 1985a:160), Yarnell (1983) characterizes sumpweed as "clearly on the road to domestication in western Illinois by about 2000 B.C.," while Ford (1985:347) places the Late Archaic domestication of *Iva* "in Kentucky and Illinois about 4,000 B.P.," and the Asches (1985a:159) indicate that "the process of *Iva's* domestication evidently was underway in the lower Illinois Valley at least by 2000 B.C."

With a mean achene length value of 4.2 mm, the circa 4,000 B.P. Napoleon Hollow *Iva* assemblage, generally acknowledged by Yarnell, Ford, and the Asches to represent a domesticated crop, reflects a substantial (30 percent) size increase over both modern wild *Iva* populations (with mean achene length values of 2.5 to 3.2 mm) and the unique Middle Holocene Koster site *Iva* assemblages (Figure 3.2) having mean achene length values of 3.2 mm (Horizon 8c: circa 7,000 B.P.) and 3.4 mm (Horizons 7 and 6: circa 5,800 to 4,800 B.P.), which are comparable to modern wild populations (Asch and Asch 1985a:164).

It thus seems reasonable to place the initial domestication of sumpweed at about 4,000 B.P., and to explicitly recognize a mean achene length value of 4.0 to 4.2 mm as the baseline value of domestication for *Iva*. Establishing a baseline value for domestication in *Iva* greater than 4.0 to 4.2 mm would necessitate pushing the initial transition to domesticated status forward more than 2,000 years into the Middle Woodland period (Figure 3.2).

Helianthus annuus

While there is general agreement that the circa 4,000 B.P. Titterington phase *Iva* assemblage from the Napoleon Hollow site marks the earliest archaeological evidence for the domestication of this species, there is no such agreement in regard to the domesticated status of the *Helianthus* achenes recovered from the same site. Asch and Asch (1985a:169) suggest that "it is plausible that the 4,000 year old achenes from Napoleon Hollow represent an early stage in the domestication of common sunflower," but they also mention (1985a:169) the possibility that the four achenes in question, having a mean length value of 6.1 mm, might be either from uncultivated *H. annuus* plants (achene length values for modern wild populations ranged from 4.0 to 5.5 mm; Heiser 1954), or from the indigenous Jerusalem artichoke *(H. tuberosus)*.

Yarnell (1983) and Ford (1985:348) both identify the circa 2,850 B.P. Higgs site sunflower achenes, which have a mean length value of 7.8 mm, as representing the earliest evidence for domestication of this species in the East. In addition, Yarnell (1978:291) suggests that the Higgs site sunflowers were "in early, but not initial stages of domestication." Ford (1985:348) suggests that such initial stages of domestication may have taken place at about 3,500 B.P. Finally, Heiser (1985:60) identifies a baseline value for identifying individual seeds as coming from a domesticated variety of sunflower: "It is likely that any achene over 7 mm long should be interpreted as coming from a domesticated plant." The mean achene length value for the Higgs site sunflower assemblage falls considerably above this 7 mm baseline, while that of the Napoleon Hollow assemblage falls considerably below it. It is interesting to note that this 7 mm baseline value for identifying individual sunflower seeds as coming from a domesticated plant transects the mean achene length curve for sunflower at about 3,400 B.P. (Figure 3.2), conforming to the suggested temporal framework proposed by Yarnell and Ford.

This proposed 3,500 to 3,400 B.P. transition to domesticated status for *Helianthus annuus* also corresponds, interestingly enough, with the earliest available evidence for the domestication of *Chenopodium,* the third seed crop to be brought under domestication prior to the introduction of maize at circa A.D. 200–300.

Chenopodium berlandieri

In contrast to both sunflower and sumpweed, an increase in seed size is not a reliable indicator of domesticated status in *Chenopodium.* The relative thickness of the seed coat or testa can, however, be employed as a morphological indicator of domestication in *Chenopodium* in the prehistoric Eastern Woodlands of North America (Smith 1984, 1985a, 1985b [Chapters 5, 6]), in that it represents a strong selective pressure for reduced germination dormancy. Eight modern wild populations of *Chenopodium* (subsection *Cellulata*) in the eastern United States were found to have mean testa thickness values ranging from 39 to 78 microns (Smith 1985b:128).

In comparison, prehistoric chenopod assemblages (subsection *Cellulata*) from Russell Cave, Alabama (1,975 ± 55 B.P., 2,340 ± 120 B.P.) and Ash Cave, Ohio (1,720 ± 100 B.P.) were found to have mean thickness values ranging from 11 to 15 microns, comparable to the modern Mexican thin-testa cultivar *C. berlandieri* ssp. *nuttalliae* cv. "chia" (16 microns). Although it has not yet been established whether this prehistoric thin-testa domesticated variety of chenopod *(C. berlandieri* ssp. *jonesianum)* was domesticated independently in the East or introduced from Mesoamerica, it seems reasonable at present to consider it an indigenous domesticate in the complete absence of any archaeological record of a thin-testa chenopod from Mexico.

This prehistoric thin-testa domesticated variety of chenopod has also been recognized in Middle Woodland contexts in the White Bluff and Edens Bluff shelters in the northwest Arkansas Ozarks (Fritz 1986), in west-central Illinois (Asch and Asch 1985a), and in central Tennessee, and it was present in central Kentucky (Salts Cave) by 2,500 B.P. (Figures 3.1, 3.2).

In addition, thin-testa fruits recovered from the Newt Kash and Cloudsplitter rock shelters in eastern Kentucky and clearly assignable to *C. berlandieri* ssp. *jonesianum* have recently been accelerator dated to 3,400 ± 150 B.P. and 3,450 ± 150 B.P., respectively (Smith and Cowan 1987). These two dates effectively push the earliest evidence for domesticated *Chenopodium* back a thousand years in the Eastern Woodlands, into close temporal proximity with the transition to domestication documented for both sumpweed and sunflower.

The Fourth Millennium Transition

With sumpweed domesticated by circa 4,000 B.P., *Chenopodium* by circa 3,500 B.P., and sunflower projected to have come under domestication at circa 3,500 B.P., based on the large size of the clearly domesticated circa 2,850 B.P. Higgs sunflower achenes, the ten-century span from 4,000 to 3,000 B.P. effectively brackets the earliest evidence of domestication for these three seed crops.

This time frame for the initial evidence for domesticated plants also coincides closely to that identified by both Yarnell (1985) and by Watson (1985) in her recent excellent review of the consequences of early horticulture in the Midwest and Midsouth. As Watson (1985:99) so aptly puts it, to "trace the antecedents of horticulture," it is necessary to extend the temporal frame of reference back a full 3,000 years prior to the initial appearance of morphological indicators of domestication.

Before turning to a consideration of the Middle Holocene coevolutionary trajectory leading to the initial domestication of indigenous plants, however, it should be emphasized that with three domesticates in place by 3,500 B.P., subsequent cultural events and developmental processes are after the fact in terms of the initial development of eastern plant husbandry systems. This is not to say that the subsequent intensification of indigenous plant husbandry systems during the Early, Middle, and particularly during the Late Woodland periods is not an interesting set of related research problems. On the contrary, the Woodland period developmental trajectory of plant husbandry systems is a central theme in understanding Woodland and subsequent patterns of cultural development in the eastern United States. But these challenging and important research topics are separate from, and subsequent to, the question of the initial development of incipient domesticates.

The Domestication of Indigenous Seed Crops

Early Holocene Foragers

Before the seventh millennium B.P., Early Holocene and initial Middle Holocene forest foragers of the mid-latitude Eastern Woodlands appear to have rather uniformly exploited fairly evenly distributed riverine resources by periodically shifting the location of their late spring to early winter base camps along river valley segment support areas at least several times during each seasonal period of reduced stream flow. Although plant remains have not been recovered in any abundance from Early Holocene cultural contexts (Smith 1986:11) these forager populations were clearly opportunistic dispersal agents for a variety of different plant species, and their apparent substantial seasonal reliance on hickory nuts and acorns would in fact appear to qualify them as specialized dispersal agents for these arboreal species (Rindos 1984:115; Munson 1986).

Judging from the short-term occupational lenses of Early Holocene groups, which are widely scattered along floodplain levees and terraces, their relatively high level of residential mobility resulted in a series of small and ephemeral disturbance "patch" episodes within the local floodplain environment. Although any human disturbance of the environment represents a habitat situation which both plants and animals could potentially successfully occupy within a coevolutionary framework, the spotty and short-lived human disturbance episodes of the Early Holocene and initial Middle Holocene in my opinion would not have represented an adequate window of opportunity for plants dispersed from their natural environment (harvested and brought to such camps for processing and consumption) to evolve disturbed habitat colonizer forms. These small and scattered disturbed habitat patches were closed out too quickly by vegetational succession to represent an opportunity for the evolution of pioneering weedy adaptions. As a result, while human populations were certainly filling the role of both opportunistic and specialized dispersal for a variety of plants including sumpweed, chenopod, and sunflower, this dispersal did not result in any expansion of the niche of the plants in question. The plants continued to be limited by "intrinsic environmental parameters" (Rindos 1984:159). This situation would change dramatically, however, during the seventh millennium B.P.

The Hypsithermal

The general postglacial warming trend had its most dramatic impact west of the Appalachians and north of the southern lower Mississippi Valley during the 8,000 to 4,000 B.P. Middle Holocene period. During this 4,000-year time-transgressive Hypsithermal climatic interval, zonal atmospheric flow across the Midwest (Knox 1983) resulted in a shift of mid-latitude fluvial systems from an Early Holocene pattern of episodic pulses of sediment removal and river incision, to a Middle Holocene phase of river aggradation and stabilization.

This change in stream flow characteristics in turn resulted in the initial formation, during the seventh millennium B.P., of both backwater swamp and open water (oxbow), and active stream shallow water and shoal habitats along segments of some mid-latitude river systems.

These developing slackwater and shoal area aquatic habitats in turn supported abundant and easily accessible aquatic resources. The increasingly abundant aquatic fauna (fish, snails, mollusks) of these slackwater and shoal area habitats represented "fixed-place" resources in that specific habitat locations (particular shoal areas, backswamps, oxbows) remained unchanged for relatively long periods of time (Smith 1986:21–27).

This "enrichment" of floodplain environments, and the intrinsic increase in the environmental gradient between river valley and upland settings, does not appear to have precipitated a dramatic or substantial modification of the pre-existing dichotomous seasonal scheduling of exploitation of riverine versus upland resources. There was a continuation of the basic Early Holocene and initial Middle Holocene pattern of riverine base camps combined with short-term limited activity sites in both the uplands and floodplain. The duration of occupation of dry season residential base camps lengthened, however.

Sedentism and the Emergence of Domestilocalities

With the seventh millennium initial appearance or enhancement of both backswamp-oxbow lake and active-channel, shallow water-shoal area habitats, the biomass level of exploitable aquatic species in certain portions of floodplain corridors increased. In response to this shift away from relative uniformity in riverine resource distribution and within the context of mediating internal social and organizational changes (Brown 1985), the widely scattered short-term occupational lenses of the antecedent Early Holocene and initial Middle Holocene were replaced by deep, rich shell mounds and midden mounds. While earlier seasonal occupational episodes tended to be more widely scattered along floodplain corridors, they began to be stacked one on top of another on a hill or slight rise during the seventh millennium B.P. This narrowing of site preference reflects the selection of suitable topographic highs in the floodplain close to spatially limited aquatic resources, with the size, shape, and composition of floodplain shell mound and midden mound sites being dictated by the existing floodplain topography adjacent to backswamps and the spawning grounds and shellfish beds of shoal areas.

This localization of occupational episodes quite likely reflects both an increasing permanence of occupation on a seasonal basis, and long-term annual reoccupation and re-use. While it is very difficult to quantify accurately this shift toward seasonal sedentism in terms of weeks or months, a number of such Middle Holocene midden mounds have been interpreted as reflecting permanent year-round residential base settlements and a shift from a high mobility forager organization to a fully sedentary harvester strategy (Smith 1986:27).

Even if such sites were not occupied on a permanent year-round basis, but only during the low-water, late spring to early winter (growing season) period of the year, it is their long-term annual reoccupation and re-use that is of importance to the present discussion. This annual reoccupation and re-use of specific floodplain locations through the growing season represented the initial emergence, in the interior river valleys of the eastern United States, of continually disturbed "anthropogenic" habitats. No longer were human disturbance patches within the floodplain environment small, scattered, and ephemeral. They were now larger, and more importantly, they were in all likelihood permanently sustained over long periods of time. Because the coevolutionary trajectory leading to domestication originates and progresses within such continually disturbed anthropogenic habitat patches, they will be termed "domestilocalities."[13]

These floodplain domestilocalities that first emerge in the mid-latitude Eastern Woodlands during the seventh millennium B.P. have four salient attributes in terms of representing a new habitat for plant colonization and coevolution:

1. Sunlight. These areas could be expected to have been cleared, at least to some extent, of an overstory canopy, with this removal of later successional stage competitors providing for increased sunlight.

2. Soil Fertility. In addition to receiving an annual influx of nutrients from floodwaters, domestilocalities also received abundant decaying plant and animal residue as a result of the on-site processing and discard of floral and faunal resources. In addition, human defecation in peripheral areas of such sites would also have improved the soil chemistry of such midden mounds (Rindos 1984:136). Resident earthworm populations, whose presence is frequently indicated by substantial evidence of bioturbation (Stein 1983), provided aeration of the midden deposits.

3. Soil Disturbance. A wide range of human activities would have resulted in a continual disturbance of the soil in domestilocalities. The most obvious types of disturbance activities would have been the construction of facilities such as houses, wind breaks, storage/refuse pits, drying racks, earth ovens, hearths, and so on. In addition, the accumulation of plant and animal debris, particularly in secondary disposal con-

texts at site edges, would have altered the ground surface, as would a wide range of everyday processing and manufacturing activities.

4. Continual Introduction of Seeds. A final essential ingredient of such anthropogenic domestilocality situations would have been the frequent introduction of the seeds of potential colonizing plant species. This introduction is an unintended and automatic result of the harvesting of plants and transporting ("dispersing") them back to the domestilocality for processing and consumption. As a result of loss during processing, loss during storage, and defecation subsequent to consumption, a small percentage of the seeds of harvested wild plants would enter the soil.

It is within the context of such Eastern Woodlands river floodplain domestilocalities that the gradual coevolutionary trajectory of initial domestication of indigenous seed plants occurred during the span from 6,500 to 3,500 B.P. Before turning to a consideration of the likely selective pressures that defined this 3,000-year-long coevolutionary process, it is appropriate to consider the original "natural" niche occupied by two of the three plant species that would be involved: sumpweed *(Iva annua)* and goosefoot *(Chenopodium berlandieri)*. It is not possible to describe a pristine "natural" Eastern Woodlands niche for *Helianthus annuus,* since it is generally considered to have arrived in the East during the Middle Holocene as an adventive weed that rapidly colonized domestilocalities. Spontaneous sunflower stands do occur sporadically today east of the Mississippi, but generally as a weedy colonizer of highly disturbed areas near cities (Heiser 1985:58).

The "Natural" Floodplain Habitat Situations of Initial Indigenous Domesticates

MARSHELDER *(IVA ANNUA)*. One of the most obvious benefactors of the seventh millennium B.P. stabilization of floodplain habitats and the establishment of slackwater zones in mid-latitude river valleys would have been marshelder (sumpweed), which primarily occupies habitats that are "flooded in the spring and often wet throughout the year" (Asch and Asch 1978:308–309). In their excellent study of sumpweed, Nancy and David Asch state that *Iva* is "mainly an edge species occurring between permanently wet and somewhat better drained soils and its extension in either direction is limited by better adapted plant competitors" (1978:309). Their description of the restricted linear nature of *Iva's* primary habitat, however, may be somewhat misleading: "Thus its distribution is linear, following microtopographic drainage contours. Stands are frequently only a few plants wide, and a 5-meter-wide stand would be exceptionally broad" (1978:309). While *Iva* often does occur in such narrow linear bands, it also, not uncommonly, occurs in quite large stands when microtopographic drainage contours allow. During two brief harvesting trips in the fall of 1984 and 1985, seven large stands (greater than 1,000 sq m) of *Iva* were located within the floodplains of mid-latitude river valleys of the Eastern Woodlands [Chapter 8]. With floodwaters dispersing seeds each spring, *Iva* is also a very vigorous and successful colonizer of newly formed wet-area habitats in backswamp and standing-water zones.

CHENOPODIUM BERLANDIERI. The "natural" habitat of this present-day weedy species is relatively poorly documented. It has only rarely been located in "natural" floodplain vegetation associations (Munson 1984; Seeman and Wilson 1984; Asch and Asch 1985a:175) growing along exposed mud flats of the lower Illinois River (Asch and Asch 1977:20), and sometimes along the edge of eroding river banks. During the fall of 1985, however, *Chenopodium berlandieri* was found to be a ubiquitous understory constituent of a black willow *(Salix nigra)* vegetation community (Shelford 1954) situated on the sandy banks of the main channel of the Mississippi River in Arkansas. Growing in a light shade understory setting, these *Chenopodium* plants were tall, delicate, with few branches and quite small and diffuse infructescences. While these understory plants would be rather disappointing in terms of economic potential, they were in striking contrast to plants of the same species growing not 2 m away in the full sun of river margin sand banks. These full sun chenopod plants were robust, with multiple branches and abundant large terminal infructescences [Chapter 7].

This discovery of *Chenopodium berlandieri* growing as an understory constituent of an undisturbed sandy soil Mississippi River margin black willow vegetation community goes a long way toward confirming its identity as Le Page du Pratz's *"belle dame sauvage"*: a plant minimally cultivated by the Natchez along the exposed sand banks of the Mississippi River up into the 1700s (Smith 1987a [Chapter 10]). The almost casual way in which the Natchez scattered the seeds of *"belle dame sauvage"* on exposed sand banks, kicked sand over them, and then largely ignored them until the fall harvest, in turn provides an excellent example of a low energy effort to increase yield which might well have been practiced by forager populations even in advance of the establishment of domestilocalities. This type of planting, if present, certainly represented an important preadaptation to the deliberate planting of seed stock within domestilocality settings.

WEEDY PREADAPTATION TO ANTHROPOGENIC HABITATS. The likely "natural" habitat situation occupied by these two plant species during the Early and Middle Holocene was thus within an active and ever-changing floodplain environment. As Anderson (1967) and Struever (1964) have noted, such changing floodplain environments are quite similar to the disturbed habitats created by human populations: "They too make dump heaps of a sort; they too plow up the mantle of vegetation and leave raw scars in it. Rivers are weed breeders; so is man, and many of the plants which follow us about have the look of belonging originally on gravel bars or mudbanks" (Anderson 1967:148–149). "Such disturbed places were created by the annual spring floods which by cutting and filling constantly opened new areas. . . . It is difficult to imagine another situation in which a disturbed habitat is created with a regularity and geographic extent comparable to that of the broad alluvial bottomlands" (Struever 1964:102–103). While both *Iva* and *Chenopodium berlandieri* are components of established, if early successional, floodplain vegetational communities today, they also exhibit pronounced adaptations to disturbed habitat situations in general. It would be difficult to establish to what extent such adaptation syndromes were present in the gene pools of populations of these plants that occupied the primary "natural" habitat situations just described. Hawkes (1969:19) has argued that such "weedy" tendencies were necessary prerequisites to the colonization of anthropogenically created disturbed habitats, and certainly the floodplain environment occupied by *Iva annua* and *C. berlandieri* would have provided opportunities for "weedy" adaptations to evolve. On the other hand, it is also possible that such disturbed habitat adaptations evolved in great measure within domestilocalities that were initially established in the seventh millennium B.P.

Middle Holocene *Cucurbita* gourd taxa, whether introduced as a weed and naturalized or indigenous, could be expected to have occupied a habitat similar to that of the present-day Texas wild gourd and the spontaneous gourd populations that occur in riverine settings of the Southeast. Apparently dispersed by floodwaters, the small gourds of *C. texana* and morphologically similar spontaneous gourd populations outside of Texas are observed most frequently floating or stranded in river edge piles of driftwood and debris. Colonizing such river-edge habitats, *C. texana* is found growing "in debris and piles of driftwood, often climbing into trees" (Heiser 1985:65). Employing such modern gourd populations as an analog, it seems reasonable to suggest that the Middle Holocene gourd populations of the Eastern Woodlands occupied a similar habitat, colonizing river-edge debris situations and being dispersed along river valley corridors by annual floodwaters [see Chapter 4 for a detailed discussion of the habitat and dispersal of modern free-living gourds].

What then were the selective pressures within domestilocalities over a 3,000-year span of time (6,500 to 3,500 B.P.) that resulted in the evolution of weedy colonizing-domestic cultigen adaptive syndromes in these plant species?

Selective Pressures and the Coevolution of Domesticates within Domestilocalities

Human activities of two general kinds were essential to setting the stage for the coevolution of initial indigenous domesticates within domestilocality settings. Human populations both created and sustained a disturbed habitat setting, and transported (dispersed) the propagules (seeds) of a wide variety of plants from their "natural" floodplain and upland niches to the newly established anthropogenic habitat. Because human activity both created the domestilocality and introduced plant propagules into its disturbed and enriched soil, these activities could in one sense be considered as having played a "causal" role in the subsequent evolutionary process of domestication. But these human activities, while certainly prerequisite and perhaps causal in a sense, were unintentionally causal. Middle Holocene collector/harvester populations neither established domestilocalities nor introduced seeds with any perception or intention of opening up an opportunity for plant evolution in a newly created habitat.

A brief view back along a Middle Holocene floodplain "causal chain" serves to underscore the opportunistic, coevolutionary nature of this process. The emergence of anthropogenically sustained disturbed habitat settings might be considered as having been "caused" by sedentism, which in turn was "caused" by the emergence of enriched aquatic habitats. Pursuing this causal chain, the initial Middle Holocene formation of enriched aquatic habitats might be viewed as having been "caused" by a change in stream flow characteristics, which in turn resulted from the Hypsithermal shift in zonal atmospheric flow across the mid-latitudes. In some tenuous sense, then, it might be argued that Middle Holocene climatic change was the initial independent variable "causing," some 3,000 years later, the appearance of morphological changes in indigenous seed plants reflecting their being brought under cultivation and domestication.

Any such long-term "explanatory framework," however, must necessarily emphasize that the dominant developmental theme that forms the links in this particular causal chain is coevolution. Various plant and animal populations (including human hunter-gatherer groups) were tied together in a sequence of opportunistic adaptations to changes in the structure and composition of floodplain ecosystems. The various plant and animal components of floodplain biotic communities opportunistically responded to the initial formation and stabilization of aggrading stream landforms by expanding to fill the developing habitats. The resultant biotic enrichment of floodplain ecosystems allowed the opportunistic adjustment by human populations to the emergence of abundant localized resources. A major aspect of this opportunistic adjustment by human populations to a change in floodplain ecosystems involved the establishment of, at minimum, seasonally permanent base camps. This establishment of growing season base camps in turn produced a significant change in floodplain ecosystems: the creation of anthropogenically sustained permanently disturbed habitats.

This ecosystem change in turn provided an opportunity for some of the floodplain vegetation community constituent species that had participated in the enrichment of floodplain habitats to occupy and evolve within the newly created disturbed habitat patches. These were an intrinsic, if incidental, aspect of the opportunistic adaptation by human populations to that floodplain habitat enrichment.

The obviously coevolutionary nature of this developmental process continues within the context of the selective pressures present within the domestilocality. While a number of the initial and ongoing selective pressures acting on plants within such disturbed habitats were clearly related to human activities, these activities were unintentional and "automatic" rather than the result of predetermined and deliberate human action toward the plant species in question.

While the basic creation and sustainment of such disturbed habitat situations constituted, in and of itself, a considerable array of selective pressures for the evolution of introduced potential colonizing plant species, it is impossible to accurately establish to what extent sumpweed, *Chenopodium berlandieri,* and sunflower may already have evolved along the path of weedy colonizer of disturbed habitats prior to the emergence of domestilocalities. Certainly the present-day wild/spontaneous descendants of sumpweed and *C. berlandieri* exhibit successful weedy adaptations to man-disturbed habitats, and the "naturally" changing floodplain landscape to some degree parallels human cultivation activities (Anderson 1967:148–149; Rindos 1984:132). But determining the relative extent to which these adaptation syndromes to disturbed habitats evolved within domestilocalities, as opposed to prior to their formation, would be difficult, since many of these adaptations would not be reflected in the fragmentary plant materials comprising Early and Middle Holocene archaeobotanical assemblages.

Introduced into domestilocalities by either fluvial or human dispersal actions, gourd plants could have been established in much the same manner that their present-day spontaneous analogues colonize river bank debris piles. Once established, domestilocality stands of Middle Holocene gourds would have thrived without any subsequent human encouragement or protection. The simple inadvertent sustainment of a disturbed habitat (particularly the site periphery "dump heap" areas) would have ensured the continued presence of this "container" crop and its utilitarian fruit (King 1985:76–78; Asch and Asch 1985a:157–158). Present-day spontaneous stands of *Cucurbita* gourds are self-propagating and thrive in waste ground dump-heap situations.

Such gourds, along with the other colonizers of domestilocalities, would also have been subjected to a selective pressure for increased seed production, which, while usually associated with cultivation, is also present and acting on weedy colonizers of disturbed habitat situations.

Present-day weedy colonizing stands of *Chenopodium berlandieri* impressively demonstrate this strong selective pressure for increased seed production. A single plant occupying 1.5 sq m of an overgrown vegetable garden in McPike County, Ohio, harvested in the fall of 1985, yielded 428 g of fruits/seeds, a weight almost equal to the total dry weight of the stripped plant (497 g) (Smith 1987b [Chapter 7]).

A second disturbed-ground selective pressure acting on the early colonizers of domestilocalities would have encouraged some loss in germination dormancy, ensuring the rapid occupation of disturbed ground. Such an adaptation is reflected in the low frequency (1 to 3 percent) production of "red morph"/"brown morph" thin-testa reduced-dormancy fruits by present-day weedy chenopod plants (Smith 1985b:122 [Chapter 5]). The low-level production of thin-testa reduced-dormancy fruits in weed chenopods foreshadows domestilocality selection for "thin-testa" and "no testa" cultivar varieties of *Chenopodium*.

Thus within the domestilocality, selective pressures that increased seed production and reduced seed dormancy would have directed the initial evolutionary pathway of colonizing species in a direction that would have enhanced their economic potential and their attractiveness to human populations as a food source. This evolutionary increase in economic potential would have occurred with no human intervention in the life cycle of colonizing weeds other than sustaining the disturbed habitat and tolerating those species with economic value.

The next level of human intervention in the life cycle of these weedy colonizers was a transition from simple toleration to inadvertent and then active encouragement. While impossible to document in the archaeobotanical record, this transition from toleration to encouragement was critical in the coevolutionary trajectory leading to domestication. At that point where the human occupants of domestilocalities begin to partition or dichotomize the colonizing plants into valuable versus "weed" categories and to actively discourage the "weeds," then the element of deliberate management or husbandry was introduced, to whatever minimal degree, and domestilocalities were transformed from inadvertent or incidental gardens into managed or "true gardens." The weeding by humans of undesirable plant populations would have reduced interspecies competition within the domestilocality and inadvertently encouraged those colonizers considered to be of economic value. The active encouragement of what might at this point be termed "quasi-cultigen" plant species would have involved the active reduction in "weeds" with the deliberate intent of encouraging economically important colonizers. Such active encouragement could have also included expansion, however small, of the domestilocality at the expense of the surrounding "natural" ecosystem, in order to increase the stand size of quasi-cultigens.

It would have been a small and seemingly minor escalation of the level of human encouragement and intervention in the life cycle of such quasi-cultigens at this point for their stands to be expanded by planting within the domestilocality. It is this simple step of planting harvested seeds, even on a very small scale, that if sustained over the long-term marks both the beginning of cultivation and the onset of automatic selection within affected domestilocality plant populations for interrelated adaptation syndromes associated with domestication. It is important to emphasize that from this point on "the initial establishment of a domesticated plant can proceed through automatic selection alone" (Harlan et al. 1973:314).

This continuing coevolutionary process did not require any deliberate selection efforts on the part of Middle Holocene inhabitants of domestilocalities in the Eastern Woodlands. All that was needed was a sustained opportunistic exploitation and minimal encouragement of what were still rather unimportant plant food sources. No real cultural retooling was required for this process, and none immediately resulted from it. It would not have involved any changes in subsistence patterns or scheduling, seasonal round, settlement systems, or world view. Nor did it result initially in any substantial increase in the dietary contribution of these plant species. They would become, however, with each passing year, a more dependable, more easily monitored, abundant localized resource.

Because the general character of the adaptation syndromes associated with domestication are presented in some detail in other publications, most notably in Harlan et al. (1973) and de Wet and Harlan (1975), and have been discussed in specific reference to *Chenopodium berlandieri* (Smith 1984, 1985a [Chapter 4]), they will be only briefly outlined here.

1. Selection Pressures Associated with Harvesting. Since those plants making the greatest contribution to a harvest will have an increased representation in the next year's gardens, the selective pressure for increased seed production is joined by related selective pressures for a more convenient "packaging" of seeds for the harvester in increasingly compacted and terminalized seed heads. Such compacted and terminalized seed heads would be more likely to be seen and perceived as being worth the time to harvest. In addition, selective pressure for both increased seed retention (loss of "natural" dispersal/"shatter" mechanisms) and uniform maturation of seed are directly related to a plant's relative contribution of viable seed to the next year's garden.
2. Selection Pressures Associated with Seedling Competition. Within Middle Holocene domestilocality spring seed beds there would have been extremely intense competition between young plants: "The first seeds to sprout and the most vigorous seedlings are more likely to contribute to the next generation than the slow or weak seedlings" (Harlan et al. 1973:318). As a result, strong selective pressure within seed beds would favor those seeds that would both sprout quickly because of a reduced germination dormancy, and grow quickly because of greater endosperm food reserves, as reflected by an increase in seed size. By 4,000 to 3,000 B.P. the coevolutionary process of domestilocality cultivation leading to domestication had resulted in morphological changes related to seed bed competition in marshelder (increased seed size), sunflower (increased seed size), and *Chenopodium berlandieri* (reduction in testa thickness) such that they could be recognized as having come under domestication. Although they have not been directly observed in the archaeobotanical record, it is reasonable to assume that the other interrelated adaptation syndromes relating to harvest pressures were, by 4,000 B.P., also present to a greater or lesser degree within domestilocality cultigen populations (i.e., increasing seed production, increasing compaction and terminalization of seed heads, increasing loss of natural shatter mechanisms, greater uniform maturation of seeds).

In comparison, the cucurbits present in the Eastern Woodlands do not show any morphological changes in either rind or seeds, which suggest that they were not brought under domestication until 3,000 B.P.

Discussion: A Coevolutionary Explanation

The independent domestication of indigenous seed plants in the Eastern Woodlands of North America constitutes a unique prehistoric developmental trajectory, with a distinct array of relevant environmental, archaeobotanical, and archaeological data sets. Any description, analysis, or "explanation" of this developmental process therefore should be tailor-made to fit these particular bodies of data, rather than simply being an "off the rack" generic processual garment of one fashion or another.

The particular coevolutionary explanation offered here is specifically tailored to comfortably accommodate Eastern Woodlands bodies of data, and it therefore may not be applicable to other developmental situations. The general outline or structure of the explanatory framework, however, can be seen to reflect the compatible combination of aspects of several different, more general developmental models.

It certainly could be characterized, to some degree, as being a "rubbish heap" or "dump heap" explanation (Anderson 1956, 1960, 1967), and as a supplement to the explanatory model first proposed by Fowler (1957, 1971). As in Fowler's explanation, the domestication process is identified as having taken place within the context of the anthropogenic disturbed habitats created in the vicinity of human habitation sites. Beyond this basic similarity, however, the explanation presented here diverges markedly from Anderson's "Dump Heap" model in that it does not rely upon polyploid hybridization as a major factor in the development of domesticated plants, but rather employs an explicitly coevolutionary approach that relies upon the well-known work of Harlan and his co-workers (Harlan et al. 1973; de Wet and Harlan 1975).

Although not directly observable in the archaeological record in any detail, this general developmental process is clearly applicable to the Eastern Woodlands, with human populations providing a sequence of three critically important, if largely "undirected" actions (i.e., domestication was not consciously perceived as an eventual goal or consequence).

The Initial Establishment of Domestilocalities

While sedentism is often identified as a general requirement for the domestication of plants, it was more specifically the establishment of anthropogenically sustained disturbed habitat situations that was a prerequisite for the coevolutionary processes leading to domestication in the prehistoric Eastern Woodlands (Struever 1964:97). There may be a caveat to the domestilocality as prerequisite premise, however, involving the possible planting of *Chenopodium berlandieri* along exposed river margin sand banks and point bars. As such banks emerged from spring floodwaters, *C. berlandieri* seeds harvested the preceding fall could have been sowed with minimal effort, much as the Natchez did into the 1700s. If deliberate planting took place in such uncolonized "naturally disturbed" habitat contexts, either before or subsequent to the establishment of domestilocalities, it would have represented an opportunity for the domestication process to have been either initiated in advance of domestilocality development, or carried on in tandem with similar planting efforts within domestilocality contexts. In either case, the sowing of harvested seeds within such uncolonized, "naturally prepared" sand-bank seed beds would have brought selective pressures associated with deliberate planting into play.

Selective Encouragement

The transition of domestilocalities from unmanaged or accidental garden settings to deliberately managed or "true gardens" occurred during the time period 6,500 to 3,500 B.P. Marking the initial development of plant husbandry or management, this transition occurred when human populations inhabiting domestilocalities began to deliberately encourage some plants while discouraging their competitors through selective weeding and perhaps expansion of domestilocality perimeters.

Deliberate Planting of Harvested Seeds

While the initial and continued sowing of harvested seeds within domestilocalities to enlarge the stand size of indigenous cultigens would have been a deliberate action directed toward increasing yield, it would not have required or likely involved any perception of either the selective pressures that would be brought to bear on the plants involved, or the adaptive syndromes of domestication that would result from such pressures. This third action by the human groups occupying domestilocalities was the most important in directing the coevolutionary process toward domestication.

Given the largely automatic and unconscious nature of the coevolutionary trajectory leading to the initial domestication of seed-bearing plants in the prehistoric Eastern Woodlands of North America, as outlined in the explanatory framework offered here, it makes little difference whether the Middle Holocene *Cucurbita* material represents an indigenous wild/weedy gourd or one introduced from Texas or Mesoamerica. In either case they would have experienced benign neglect and would have flourished around the periphery of a largely unconscious and undirected coevolutionary process. It is quite improbable that either the idea of domestication or actual domesticated plants, necessitating a substantial degree of life-cycle intervention on the part of human populations for their continued survival, would have found a very receptive audience within the context of pre-4,500 B.P. domestilocalities. To perceive the presence of a small wild/weedy gourd of whatever nationality as explaining or containing the secret or solution to the process of domestication in the East avoids and ignores the far more complex independent developmental process that did take place.

The explanatory framework presented here identifies river valley domestilocalities of the mid-latitude Eastern Woodlands as the disturbed habitat patches within which the coevolutionary process leading to domestication took place, but the question of whether there was a particular localized heartland where domestication initially occurred has not been addressed. On the basis of information currently available, it is possible to identify a broad "Midwest-Riverine" area of the interior mid-latitude Eastern Woodlands which quite likely encompassed the coevolutionary trajectory to domestication (Figure 3.1), but any identification of a more specific "heartland" area within this broad geographical area is highly unlikely.

River valley domestilocalities of the various drainage systems of the Midwest-Riverine area supported a series of semi-isolated populations of quasi-cultivated and cultivated plants, each under a similar set of selective pressures. Different drainage systems and specific domestilocalities within drainage systems would have exhibited variation in terms of both the relative level of selective pressure and the timing of the developmental process. Against this backdrop of spatial and developmental variability the "dispersal" of seed stock, information, and individuals between domestilocalities would have resulted in a complex mosaic of occasionally linked, generally parallel, but distinct coevolutionary histories for different areas of the mid-latitude Eastern Woodlands. So while it is certainly worthwhile to recover archaeobotanical assemblages and information relevant to this developmental process from a wide range of temporal and spatial contexts across the eastern United States in order to obtain a clearer picture of this complex developmental mosaic, it is quite doubtful if a single "heartland" ever existed where indigenous seed-bearing plants were first brought under domestication and from which they subsequently radiated.

By considering the long-term (6,500 to 3,500 B.P.) developmental trajectory leading to plant domestication within human disturbed habitat patches of the mid-latitude Eastern Woodlands as a slow sequence of interlinked opportunistic responses by both plant and human populations to emerging enriched habitat situations, the explanation offered here might also be termed both "stress free" and "minimalist." It is "stress free" in that it does not rely on any "external" prime movers or forcing functions such as population growth or territory reduction to precipitate or drive the developmental process. This is not to imply that such variables are inappropriate in general, only that they are not required to account for the developmental process leading to domestication. Localized population growth and territorial packing/cultural circumscription may well have come into play later, however, during the post 2,500 B.P. intensification of food production systems in the mid-latitude Eastern Woodlands. The explanation is "minimalist" in that it requires only a quite limited degree of conscious and directed innovation on the part of prehistoric human populations, and a quite limited leap of faith on the part of the reader. The seventh millennium B.P. development of anthropogenically disturbed habitat patches is identified as a starting point of the coevolutionary process leading to domestication, while the fourth millennium B.P. appearance of morphological changes indicating domestication in sumpweed, chenopod, and sunflower marks an end point. In between, given the context of domestilocalities, the interrelated and largely automatic set of selective pressures outlined above define the developmental process.

Such a minimalist explanation, which casts human groups in a largely unconscious and automatic role, does not preclude attempts at explanatory expansion through the addition of supplemental layers of interpretation, including transformational or social supplementation, as recently outlined by Brown (1985) in regard to the origins of sedentism in the Eastern Woodlands.

But during the millennium following their domestication, sumpweed, sunflower and goosefoot (a) showed little morphological change, (b) appear to have been of limited dietary importance, and (c) are almost invisible archaeologically. It is not until 2,500 to 2,000 B.P. that these indigenous domesticates, along with a whole host of other quasi-cultigens and cultigens, became economically important and archaeologically abundant. This initial low profile of marshelder, sunflower, and *Chenopodium* in turn suggests that their domestication, which represents the first step of a long and complex developmental trajectory of prehistoric plant husbandry in the mid-latitude Eastern Woodlands (Smith 1985a:51), does not require much explanatory supplementation beyond the minimalist framework offered here.

Even a brief consideration of the 3,000-year-long developmental sequence from the initial cultivation of indigenous seed plants at 4,000 to 3,500 B.P. to the eventual emergence of maize-centered field agriculture after A.D. 1000 indicates that this process of coevolution can not be adequately covered by any single "origins of agriculture" explanatory model, but rather requires a whole set of interlocking explanations to deal with different parts of a unique developmental process. Different types of explanatory models are appropriate for different portions of the developmental trajectory, and while a minimalist explanation appears appropriate for the initial domestication of indigenous seed plants, far more complex models are required in order to approach an understanding of the later stages of this long coevolutionary relationship between prehistoric human and plant populations in the Eastern Woodlands.

Notes

1. The phrase "Eastern Agricultural Complex" should be dropped from usage. Richard Yarnell has noted that it is quite misleading in that implies an integrated and coherent group of indigenous crop plants and related technology. The archaeobotanical record strongly suggests, on the other hand, that "each crop plant is largely an independent entity in its space and time distributions and overall not part of a crop complex, at least not on the basis of our current archaeological evidence" (Yarnell 1983).

In addition to not involving a "complex" in terms of either a set of crops being consistently grown together or an ever-present distinctive set of associated technologies, the initial process leading to the independent domestication of seed-bearing plants in the Eastern Woodlands was not "agricultural" in the usual sense of the word. While a regionally variable set of indigenous plants was clearly brought under cultivation, and some of these exhibited morphological changes associated with domestication, they were not grown initially within an "agricultural" field cropping system that dominated the economic base of populations. Thus the "Eastern Agricultural Complex" was neither agricultural nor a complex.

2. Linton's reference to the planting of "various small grains in the southeast" during the historic period refers to the early descriptions by Hariot of the coastal Carolinas (1586) and of Le Page in the Lower Mississippi Valley (early 1700s) (Smith 1987a [Chapter 9]).

3. The cultivar variety of chenopod referred to by Gilmore as *Chenopodium nuttalliae* is *Chenopodium berlandieri* ssp. *nuttalliae* (see Wilson and Heiser 1979).

4. Although proposed as a possible domesticate based on a large achene size, ragweed is today not thought to have been brought under cultivation in the East.

5. The large chenopod type in the Newt Kash archaeobotanical assemblage identified by Jones as a possible domesticate was in fact not a chenopod (Asch and Asch 1977). Other morphological indicators of domesticated status are exhibited by Newt Kash *Chenopodium* specimens, however (Smith and Cowan 1987).

6. Quimby (1946) has pointed out that Jones was apparently not given all of the maize recovered during excavation of the Newt Kash shelter (Webb and Funkhouser 1936:130), but rather was sent only a sample of materials consisting mostly of materials that Webb and Funkhouser could not readily identify. Webb and Funkhouser state that "corn cobs were numerous" (1936:130).

7. Quimby's various descriptive terms may be the source of the "Eastern Agricultural Complex" label (Struever 1962, 1964).

8. Yarnell addressed this issue of wild stands yielding larger seeds as early as 1965, and subsequent research has upheld the value of increasing seed size as an indicator of domestication.

9. See Martin et al. 1947:234, 519, for other comments by Quimby on this issue.

10. Floodplain plant surveys (for example, Munson 1984) have failed to provide support for Struever's "Mud-Flat Horticulture" hypothesis. The plants in question grow only infrequently in mud-flat habitats.

11. The foregoing discussion of Middle Holocene cucurbits in the east, and the proposal that the wild gourd *C. texana* may have had a much larger geographical range in the prehistoric Eastern Woodlands of North America, was certainly initially stimulated in part by a number of the statements made by Deena Decker in a 1985 article (Decker 1985). But it was only subsequent to having written this chapter and seen it go into production in late 1986, that I learned of the recent research of Dr. Decker regarding the origins, evolution, and systematics of *Cucurbita pepo*. Decker's entirely independent botanical research of the past several years, now reported in a number of landmark articles and papers, provides substantial, and in my opinion convincing, support for the position presented here. While I will attempt to summarize a number of Decker's major conclusions below, I urge the reader to consult her remarkable publications, which have been added to the bibliography (Decker 1986, 1987; Decker and Wilson 1986, 1987; Decker and Newsom 1987).

In reference to a recognized need to "describe and validate those specific morphological characteristics of seeds which can be employed in differentiating between closely related wild and cultivated varieties of *Cucurbita*" Decker in fact now has documented the value of seed morphology in distinguishing different subspecies of *Cucurbita pepo* (Decker and Wilson 1986; Decker and Newsom 1987). On the basis of Decker's morphological criteria, it now appears possible to assess the nondomesticated versus domesticated status of *Cucurbita* seeds recovered from archaeological contexts in the East with a much higher level of accuracy and confidence. I would predict that as the *Cucurbita* seeds recovered from pre-3,000 B.P. contexts in the east are analyzed within the context of Decker's shape analysis, it will not be possible to assign any of them to domesticated taxa of *Cucurbita pepo*. Rather they will likely resemble the seeds of the modern wild gourd *Cucurbita pepo* ssp. *ovifera* var. *texana* (see Decker 1987 for a proposed new classification of *C. pepo*) closely enough to leave the question of their domesticated status unresolved. Additionally, in regard to Middle Holocene rind fragments, Decker (1987) states: "In particular, the rind fragments from the Illinois sites, dated around 5000 B.C., cannot be distinguished from modern samples of *C. texana*." Decker also points out (personal communication, 1987) that since there "have been trends both towards smaller and larger seeds (as well as a lack of change in size) since initial domestication," it is not possible to establish a simple size boundary for distinguishing the seeds of domesticated vs. nondomesticated *C. pepo* taxa: "While large seeds can reasonably be assumed to represent domestication, it cannot be assumed that small seeds came from wild populations."

In regard to the present-day and likely prehistoric range of the eastern North American wild gourd *C. pepo* ssp. *ovifera* var. *texana,* Decker has concluded, on the basis of both seed morphology and allozyme studies, that modern spontaneous gourd populations of *texana*-like plants from Alabama, Arkansas, and Illinois "share various attributes with Texas populations, suggesting that *C. texana* once had a more widespread distribution to the Northeast," and that "it is possible that these populations are relicts of *C. texana*" (Decker 1987).

Decker's allozyme (starch gel electrophoresis) analysis of modern wild and cultivar members of the species *C. pepo* also indicated considerable genetic diversity. She establishes two distinct subspecies categories, *Cucurbita pepo* ssp. *pepo* and *Cucurbita pepo* ssp. *ovifera,* which reflect the "dichotomy separating Mexican accessions, pumpkins, marrows, and a few ornamental gourds from acorn squashes, scallop squashes, fordhooks, crooknecks, and the bulk of ornamental gourds" (Decker 1987). The second subspecies (*C. pepo* ssp. *ovifera*) includes many cultivars that "apparently have their origins in eastern U.S." (Decker 1987), along with the eastern wild gourd *C. pepo* ssp. *ovifera* var. *texana,* which "includes all populations sustaining themselves in the natural environment" (Decker 1987). Decker concludes that this clear separation of the Mexican cultivars from eastern forms could be explained, on the one hand, in terms of a diffusion of a domesticated form into eastern North America shortly after initial domestication, with subsequent genetic divergence in the East. She also discusses, and appears to favor, an alternative multiple domestications hypothesis (Decker 1987), which involves the independent domestication of the non-Mexican *Cucurbita pepo* forms in the eastern United States from an indigenous *texana*-like wild gourd. Based on current evidence, I would speculate that this independent process of domestication in the East, which resulted in many of the cultivar varieties of *Cucurbita pepo* ssp. *ovifera,* occurred between 4,000 and 3,000 B.P., within the same time frame and developmental context as the initial domestication of two other indigenous plants of early successional floodplain habitats, *Chenopodium berlandieri* and *Iva annua*.

In summary, Decker's recent botanical research provides substantial support for the Middle Holocene presence of an indigenous wild *texana*-like gourd in the eastern United

States, and the subsequent independent development from it (at circa 4,000–3,000 B.P.?) of domesticated varieties of *Cucurbita pepo* ssp. *ovifera*.

12. It may be inappropriate to refer to *Helianthus annuus* as an indigenous eastern plant. At present it is thought to have arrived in the East as an adventive weed during the Middle Holocene (perhaps not much earlier than 4,000 B.P. [Richard Yarnell, personal communication, 1986]). While reported today in noncultivated contexts (Heiser 1985), it is not clear to what degree it has become naturalized in the East.

13. The new term "domestilocality," as employed here, refers to human habitation site areas occupied throughout the growing season, which because of substantial human disturbance of the soil, represent a new type of river valley habitat locale within which strong selection pressures favoring plant domestication occur. While I am not particularly comfortable with the term domestilocality, it is certainly preferable to the repeated use of either the descriptive phrase "continually disturbed anthropogenic habitat patch" or any derivative acronym. I also deliberately avoided the term "agrilocality" as coined by Rindos (1985:176–178), even though he employs it in a similar way, because it carries a number of inherent meanings distinctive to his view of agricultural origins.

Acknowledgments

An earlier version of this chapter was read by Gayle Fritz, James B. Griffin, William Keegan, Frances B. King, George Milner, Hugh Wilson, and Richard A. Yarnell. While none of these individuals should be held responsible for any remaining errors of fact or interpretation, their comments and corrections are greatly appreciated. Support for the author's *Iva* and *Chenopodium* research reported in this chapter was provided by Smithsonian Institution Research Opportunity Fund Grants.

Literature Cited

Anderson, E.

1956 Man as a Maker of New Plants and Plant Communities. In *Man's Role in Changing the Face of the Earth,* vol. 2, edited by W. L. Thomas, pp. 763–777. University of Chicago Press, Chicago.

1960 The Evolution of Domestication. In *The Evolution of Man, Culture, and Society,* vol. 2, *Evolution after Darwin,* edited by Sol Tax, pp. 67–84. University of Chicago Press, Chicago.

1967 *Plants, Man, and Life.* University of California Press, Berkeley and Los Angeles.

Asch, D. L., and N. B. Asch

1977 Chenopod as Cultigen: A Re-evaluation of Some Prehistoric Collections from Eastern North America. *Midcontinental Journal of Archaeology* 2:3–45.

1978 The Economic Potential of *Iva annua* and its Prehistoric Importance in the Lower Illinois Valley. In *The Nature and Status of Ethnobotany,* edited by R. I. Ford, pp. 300–341. Museum of Anthropology, Anthropological Paper No. 67. University of Michigan, Ann Arbor.

1985a Prehistoric Plant Cultivation in West-Central Illinois. In *Prehistoric Food Production in North America,* edited by R. I. Ford, pp. 149–203. Museum of Anthropology, Anthropological Paper No. 75. University of Michigan, Ann Arbor.

1985b Archaeobotany. In *Smiling Dan Structure and Function at a Middle Woodland Settlement in the Lower Illinois Valley,* edited by B. D. Stafford and M. B. Sant, pp. 322–401. Center for American Archaeology, Research Series No. 2. Kampsville, Illinois.

Brown, James A.

1985 Long Term Trends to Sedentism and the Emergence of Complexity in the American Midwest. In *Prehistoric Hunter-Gatherers: The Emergence of Cultural Complexity,* edited by T. Douglas Price and James A. Brown, pp. 201–234. Academic Press, Orlando, Florida.

Carter, George

1945 *Plant Geography and Culture History in the American Southwest.* Viking Fund Publications in Anthropology No. 5. New York.

Castetter, E. F.

1925 Horticultural Groups of Cucurbits. *Proceedings of the American Society of Horticultural Science* 22:338–340.

Chapman, Jefferson

1981 *The Bacon Bend and Iddins Sites.* Report of Investigations No. 31. Department of Anthropology, University of Tennessee, Knoxville.

Chomko, S. A., and G. W. Crawford
1978 Plant Husbandry in Prehistoric Eastern North America: New Evidence for its Development. *American Antiquity* 43:405–408.

Conard, N., D. Asch, N. Asch, D. Elmore, H. Gove, M. Rubin, J. Brown, M. Wiant, K. Farnsworth, and T. Cook
1984 Accelerator Radiocarbon Dating of Evidence for Prehistoric Horticulture in Illinois. *Nature* 308:443–446.

Correll, Donovan S., and M. C. Johnston
1970 *Manual of Vascular Plants of Texas.* Contributions of the Texas Research Foundation, Renner, Texas.

Cowan, C. Wesley
1985a *From Foraging to Incipient Food Production: Subsistence Change and Continuity on the Cumberland Plateau of Eastern Kentucky.* Ph.D. dissertation, Department of Anthropology, University of Michigan. University Microfilms, Ann Arbor.

1985b Understanding the Evolution of Plant Husbandry in Eastern North America: Lessons from Botany, Ethnography and Archaeology. In *Prehistoric Food Production in North America,* edited by R. I. Ford, pp. 205–244. Museum of Anthropology, Anthropological Paper No. 75. University of Michigan, Ann Arbor.

Cowan, C. W., Homer E. Jackson, K. Moore, A. Nickelhoff, and T. Smart
1981 The Cloudsplitter Rockshelter, Menifee County, Kentucky: A Preliminary Report. *Southeastern Archaeological Conference Bulletin* 24:60–75.

Crites, Gary D.
1985 Middle Woodland Paleoethnobotany of the Eastern Highland Rim of Tennessee: An Evolutionary Perspective on Change in Human-Plant Interaction. Ph.D. dissertation, Department of Anthropology, University of Tennessee, Knoxville.

Decker, Deena
1985 Numerical Analysis of Allozyme Variation in *Cucurbita pepo. Economic Botany* 39:300–309.

1986 A Biosystematic Study of *Cucurbita pepo.* Ph.D. dissertation, Department of Biology, Texas A&M University, College Station.

1987 Origin(s), Evolution, and Systematics of *Cucurbita pepo* (Cucurbitaceae). *Economic Botany* 42:3–15.

Decker, Deena, and Hugh Wilson
1986 Numerical Analysis of Seed Morphology in *Cucurbita pepo. Systematic Botany* 11:595–607.

1987 Allozyme Variation in the *Cucurbita pepo* Complex: *C. pepo* var *ovifera* vs. *C. texana. Systematic Botany* 12:263–273.

Decker, Deena, and Lee A. Newsom
1987 Numerical Analysis of Archaeological *Cucurbita* Seeds from Hontoon Island, Florida. Paper presented at the 10th annual conference of the Society for Ethnobiology, Gainesville, Florida.

Doran, Glen H., David Dickel, and Lee A. Newsom
1990 A 7,290-Year-Old Bottle Gourd From the Windover Site, Florida. *American Antiquity* 55:354–360.

Ford, Richard I.
1985 Patterns of Prehistoric Food Production in North America. In *Prehistoric Food Production in North America,* edited by R. I. Ford, pp. 341–364. Museum of Anthropology, Anthropological Paper No. 75. University of Michigan, Ann Arbor.

Fowler, M. L.
1957 The Origin of Plant Cultivation in the Central Mississippi Valley: A Hypothesis. Paper presented at the annual meeting of the American Anthropological Association, Chicago.

1971 The Origin of Plant Cultivation in the Central Mississippi Valley: A Hypothesis. In *Prehistoric Agriculture,* edited by Stuart Struever, pp. 122–128. Natural History Press, Garden City, New York.

Fritz, Gayle L.
1986 Starchy Grain Crops in the Eastern U.S.: Evidence from the Desiccated Ozark Plant Remains. Paper presented at the 51st annual meeting of the Society for American Archaeology, New Orleans.

Gilmore, Melvin R.
1931 Vegetal Remains of the Ozark Bluff-Dweller Culture. *Papers of the Michigan Academy of Science, Arts, and Letters* 14:83–105.

Gremillion, Kristen, and Richard A. Yarnell
1986 Plant Remains from the Westmoreland-Barber and Pittman-Alder Sites, Marion County, Tennessee. *Tennessee Anthropologist* 11:1–20.

Harlan, J. R., J. M. J. de Wet, and E. G. Price
1973 Comparative Evolution of Cereals. *Evolution* 27:311–325.

Hawkes, J. G.
1969 The Ecological Background of Plant Domestication. In *The Domestication and Exploitation of Plants and Animals,* edited by Peter Ucko and G. W. Dimbleby, pp. 17–31. Aldine, Chicago.

Heiser, Charles B., Jr.
1954 Variation and Subspeciation in the Common Sunflower *Helianthus annuus. American Midland Naturalist* 51:287–305.

1979 Origins of Some Cultivated New World Plants. *Annual Review of Ecology and Systematics* 10:309–326.

1985 Some Botanical Considerations of the Early Domesticated Plants North of Mexico. In *Prehistoric Food Production in North America,* edited by Richard I. Ford, pp. 57–72. Museum of Anthropology, Anthropological Paper No. 75. University of Michigan, Ann Arbor.

Jones, Volney
1936 The Vegetal Remains of Newt Kash Hollow Shelter. In *Rock Shelters in Menifee County, Kentucky,* edited by W. S. Webb and W. D. Funkhouser, pp. 147–165. Reports in Archaeology and Anthropology 3(4). University of Kentucky, Lexington.

Kay, Marvin, Frances B. King, and C. K. Robinson
1980 Cucurbits from Phillips Spring: New Evidence and Interpretations. *American Antiquity* 45(4):806–822.

King, Frances B.
1985 Early Cultivated Cucurbits in Eastern North America. In *Prehistoric Food Production in North America,* edited by R. I. Ford, pp. 73–98. Museum of Anthropology, Anthropological Paper No. 75. University of Michigan, Ann Arbor.

Kirkpatrick, K. J., Deena S. Decker, and Hugh D. Wilson
1985 Allozyme Differentiation in the *Cucurbita pepo* Complex: *C. pepo* var. *medullosa* vs. *C. texana. Economic Botany* 39:289–299.

Knox, J. C.
1983 Responses of River Systems to Holocene Climates. In *The Holocene,* vol. 2, *Late-Quaternary Environments of the United States,* edited by H. E. Wright, Jr., pp. 26–41. University of Minnesota Press, Minneapolis.

Linton, Ralph
1924 The Significance of Certain Traits in North American Maize Culture. *American Anthropologist* 26:345–349.

Martin, Paul S., George I. Quimby, and Donald Collier
1947 *Indians before Columbus.* University of Chicago Press, Chicago.

Munson, Patrick
1984 Weedy Plant Communities on Mud-Flats and other Disturbed Habitats in the Central Illinois River Valley. In *Experiments and Observations on Aboriginal Wild Plant Food Utilization in Eastern North America,* edited by Patrick Munson. Prehistoric Research Series 6(2). Indiana Historical Society, Indianapolis.

1986 Hickory Silvaculture: A Subsistence Revolution in the Prehistory of Eastern North America. Paper presented at the conference on Emergent Horticultural Economies of the Eastern Woodlands, Carbondale, Illinois.

Quimby, G. I.
1946 The Possibility of an Independent Agricultural Complex in the Southeastern United States. *Human Origins: An Introductory General Course in Anthropology.* Selected Reading Series 31. University of Chicago, Chicago.

Rhodes, A. M., W. P. Bemis, T. W. Whitaker, and S. G. Cramer
1968 A Numerical Taxonomic Study of *Cucurbita. Brittonia* 20:251–266.

Rindos, D.
1984 *The Origins of Agriculture.* Academic Press, Orlando, Florida.

Schroedl, G. F.
1978 *The Patrick Site* (40MR40), Tellico Reservoir, Tennessee. Report of Investigations No. 25. Department of Anthropology, University of Tennessee, Knoxville.

Seeman, M. F., and H. D. Wilson
1984 The Food Potential of *Chenopodium* for the Prehistoric Midwest. In *Experiments and Observations on Aboriginal Wild Plant Food Utilization in Eastern North America,* edited by P. Munson. Prehistoric Research Series 6(2). Indiana Historical Society, Bloomington.

Shelford, V. E.
1954 Some Lower Mississippi Valley Floodplain Biotic Communities; Their Age and Elevation. *Ecology* 35:126–142.

Smith, Bruce D.
1984 *Chenopodium* as a Prehistoric Domesticate in Eastern North America: Evidence from Russell Cave, Alabama. *Science* 226:165–167.

1985a The Role of *Chenopodium* as a Domesticate in Pre-Maize Garden Systems of the Eastern United States. *Southeastern Archaeology* 4:51–72.

1985b *Chenopodium berlandieri* ssp. *jonesianum:* Evidence for a Hopewellian Domesticate from Ash Cave, Ohio. *Southeastern Archaeology* 4:107–133.

1986 The Archaeology of the Southeastern United States: From Dalton to de Soto, 10,500 B.P.–500 B.P. In *Advances in World Archaeology,* vol. 5, edited by F. Wendorf and A. Close, pp. 1–92. Academic Press, Orlando, Florida.

1987a In Search of *Choupichoul,* the Mystical Grain of the Natchez. Keynote address, 10th annual conference of the Society of Ethnobiology, Gainesville, Florida.

1987b The Economic Potential of *Chenopodium berlandieri* in Prehistoric Eastern North America. *Ethnobiology* 7:29–54.

Smith, Bruce D., and C. W. Cowan
1987 The Age of Domesticated *Chenopodium* in Prehistoric Eastern North America: New Accelerator Dates From Eastern Kentucky. *American Antiquity* 52:353–357.

Stein, J.
1983 Earthworm Activity: A Source of Potential Disturbance of Archaeological Sediments. *American Antiquity* 48:277–289.

Struever, S.
1962 Implications of Vegetal Remains from an Illinois Hopewell Site. *American Antiquity* 27:584–587.

1964 The Hopewell Interaction Sphere in Riverine-Western Great Lakes Culture History. In *Hopewellian Studies,* edited by J. Caldwell and R. Hall. Scientific Paper No. 12. Illinois State Museum, Springfield.

Struever, S., and K. D. Vickery
1973 The Beginnings of Cultivation in the Midwest-Riverine Area of the United States. *American Anthropologist* 75:1197–1220.

Watson, P. J.
1985 The Impact of Early Horticulture in the Upland Drainages of the Midwest and Midsouth. In *Prehistoric Food Production in North America,* edited by R. I. Ford, pp. 73–98. Museum of Anthropology, Anthropological Paper No. 75. University of Michigan, Ann Arbor.

Watson, P. J., and R. Yarnell
1969 Conclusions—The Prehistoric Utilization of Salts Cave. In *The Prehistory of Salts Cave, Kentucky,* edited by Patty Jo Watson, pp. 71–78. Reports of Investigations No. 16. Illinois State Museum, Springfield.

Webb, W. S., and W. P. Funkhouser
1936 *Rock Shelters in Menifee County, Kentucky.* Reports in Archaeology and Anthropology 3(4). University of Kentucky, Lexington.

de Wet, J. M. J., and Jack R. Harlan
1975 Weeds and Domesticates: Evolution in the Man-Made Habitat. *Economic Botany* 29:99–107.

Whitaker, T., and H. Cutler
1965 Cucurbits and Culture in the Americas. *Economic Botany* 19:344–349.

1986 Cucurbits from Preceramic Levels at Guila Naquitz. In *Guila Naquitz,* edited by K. Flannery, pp. 275–280. Academic Press, Orlando, Florida.

Wilson, H., and C. Heiser
1979 The Origin and Evolutionary Relationship of "Huauzontle" (*Chenopodium nuttalliae* Stafford), Domesticated Chenopod of Mexico. *American Journal of Botany* 66:198–206.

Yarnell, R.
1963 Comments on Struever's Discussion of an Early Eastern Agricultural Complex. *American Antiquity* 28:547–548.

1964 *Aboriginal Relationships between Culture and Plant Life in the Upper Great Lakes Region.* Museum of Anthropology, Anthropological Paper No. 23. University of Michigan, Ann Arbor.

1965 Early Woodland Plant Remains and the Question of Cultivation. *Florida Anthropologist* 18:77–82.

1969 Contents of Human Paleofeces. In *The Prehistory of Salts Cave, Kentucky,* edited by P. J. Watson. Reports of Investigations No. 16. Illinois State Museum, Springfield.

1972 *Iva annua* var. *macrocarpa:* Extinct American Cultigen? *American Anthropologist* 74:335–341.

1978 Domestication of Sunflower and Sumpweed in Eastern North America. In *The Nature and Status of Ethnobotany,* edited by R. I. Ford, pp. 289–299. Museum of Anthropology, Anthropological Papers No. 67. University of Michigan, Ann Arbor.

1981 Inferred Dating of Ozark Bluff Dweller Occupations Based on Achene Size of Sunflower and Sumpweed. *Ethnobiology* 1:55–60.

1983 Prehistory of Plant Foods and Husbandry in North America. Paper presented at the annual meeting of the Society for American Archaeology, Pittsburgh.

1985 A Survey of Prehistoric Crop Plants in Eastern North America. Ms. in possession of the author.

Yarnell, Richard A., and M. Jean Black

1985 Temporal Trends Indicated by a Survey of Archaic and Woodland Plant Food Remains from Southeastern North America. *Southeastern Archaeology* 4:93–106.

CHAPTER 4

❑ ❑ ❑

Is It an Indigene or a Foreigner?

BRUCE D. SMITH, C. WESLEY COWAN, AND MICHAEL P. HOFFMAN

> Is this Texas cucurbit a garden escape as suggested by Gray, or is it the proto-type from whence came the numerous cultivated forms of *Cucurbita pepo*? Is it an indigene or a foreigner? (Erwin 1938:253)

Introduction

In a little recognized and rarely referenced article published over fifty years ago, A. T. Erwin first posed a question that today remains of central importance in documenting eastern North America as an independent center of plant domestication. As might be expected, the question has become more complex over the intervening years and has expanded in scope, but it still involves the status of the Texas wild gourd and other free-living *Cucurbita* gourd populations of eastern North America. Are these widely distributed plants wild, or are they recent escapes from cultivation? What can be established regarding their possible role as a progenitor in the independent domestication of *Cucurbita pepo* in eastern North America?

The Texas wild gourd grows today along rivers and streams in eastern Texas, and related "free-living" populations of gourds (able to survive outside of cultivation) are being documented in increasing numbers in other areas of eastern North America. In addition, since the 1980s, rind fragments and seeds of a similar gourd have been recovered from archaeological deposits in eastern North America that predate the earliest evidence of domestication of indigenous seed plants by 3,000 years (see preceding chapter). Over the past decade both these present-day and prehistoric gourds have been the topic of considerable discussion. Do the early *Cucurbita* gourds of the East that predate the circa 4,000–3,500 B.P. domestication of local plants represent a wild plant indigenous to the region, or an early domesticate introduced from Mexico? If these pre-4,000 B.P. gourds can be traced back to Mesoamerican agroecosystems, then agricultural developments in the East might be considered derivative rather than independent, since a tropical crop plant would have arrived in advance of domestication of local species. On the other hand, if this early material represents an indigenous wild gourd, then it offers no obstacle to eastern North America being accepted as an independent center of plant domestication. The debate regarding these early gourds (are they indigenous and wild or foreign and domesticated?) has added new significance to a long-standing discussion of present-day free-living gourd populations in eastern North America: do they represent an indigenous wild gourd, or the recent escape from cultivation of a domesticate that ultimately originated in Mesoamerica?

Intertwined with and overlaying these questions regarding the origin of both present-day and pre-4,000 B.P. gourd populations of the East is the possibility that *C. pepo* was independently domesticated in eastern North America a full 6,000 years after it was first domesticated in Mexico (Flannery 1986). Based on available archaeological evidence for the earliest clearly domesticated *C. pepo* in the East (circa 3,000 B.P.), this second episode of domestication would have occurred slightly after other eastern seed crops were brought under domestication in the region.

Without duplicating archaeological information presented elsewhere in the volume, this chapter expands on topics raised in Chapter 3. It summarizes and evaluates the arguments advanced to date regarding both the indigenous versus introduced nature of prehistoric gourd populations, and the wild versus escaped nature of present-day gourds. In addition, single and multiple origin explanations for *C. pepo* are discussed, and additional relevant information regarding the geographical range and habitat of present-day free-living gourd populations of the region is presented.

Single Origin Explanations of the Late 1970s and 1980s

Until the last half of the 1980s, free-living gourd populations of eastern North America, particularly those that existed outside of Texas, were little studied. When gourd populations beyond Texas surviving in the wild were observed, they were assumed to have resulted from the escape of domesticated ornamental gourds from cultivation:

> In sum, the present frequency of *C. pepo* var. *ovifera* apparently represents a recent naturalization of the plant in extensive, highly disturbed floodplain fields. (Asch and Asch 1985:158)

Given the apparent absence of a native wild gourd east or north of Texas, and only sporadic reports of recent "escapes" of domesticated ornamental gourds, it is not surprising that when small carbonized rind fragments having the distinctive cellular structure of the genus *Cucurbita* were recovered in archaeological contexts that clearly predated the domestication of local seed crops, it was assumed that they represented a Mesoamerican domesticate. Based on these finds, a single-origin consensus formed through the first half of the 1980s that derived both the early *C. pepo* specimens and the concept of agriculture from Mesoamerica (Figure 4.1):

> Squash remains from these three Late Archaic archaeological sites constitute the earliest evidence for cultigens in eastern North America. The new data indicate the tropical cultigen, squash, was introduced into the area prior to the domestication of native plant resources. . . .The new data indicate that the eastern horticultural complex was not an independent development but was a regional adaptation of the concept of plant husbandry which originated in Mesoamerica. (Chomko and Crawford 1978:405, 407)

> *C. pepo* was evidently brought into eastern North America near the end of the Middle Holocene. . . . Diffusion of a viable subsistence technology involving at least small-scale gardening appears to have led to later domestication of native plants in the East. (Kay et al. 1980:817, 818)

> Present evidence suggests that the first domesticated plants in the United States originated in eastern Mexico, probably diffused across Texas, and into the southwest and the major river systems of the Midwest. . . . The Early Eastern Mexican Agricultural Complex was the basis of plant husbandry in the East. (Ford 1981:7, 9)

> The first cultivated species, *Cucurbita* sp. (probably pepo squash) was introduced by 5,000 B.C. (Asch and Asch 1985:202)

> In the East the idea of agriculture came from Mexico many millennia before even native plants were obvi-

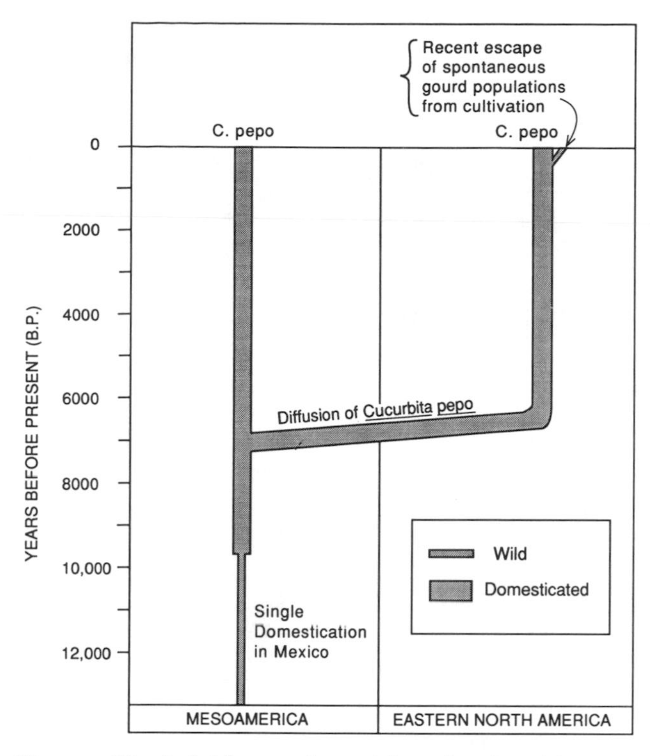

Figure 4.1 The single Mesoamerican origin explanation of the domestication of *C. pepo*.

> ously cultivated, much less domesticated. . . .Only within the last decade has the priority of tropical cultivation been resolved. . . .The gourd and squash preceded the domestication and perhaps even the cultivation of indigenous disturbed habitat plant species in the East. (Ford 1985:18, 10)

> Based on our present knowledge of the archaeological record, it may be argued that plant cultivation was an idea introduced into the eastern U.S. from Mesoamerica, and that *Cucurbita pepo* and *Lagenaria siceraria*, the first plants raised in eastern North America, were also introduced from Mesoamerica. . . .It is hypothesized that shamans were the first to adopt cucurbit gardening as a means of providing ritualistic paraphernalia. (Prentice 1986:103)

This consensus was short-lived, however, and a strong opposing position had developed by the middle of the last decade.

The Emergence of a Multiple Origins Explanation for the Domestication of Cucurbita pepo

The earliest published opposition to this single-origin, introduced-domesticate explanation for the pre-4,000 B.P. *Cucurbita* material in the East appeared in 1981:

> The North American *pepo* squash need not have diffused into the East or Southwest via some Mexican connection; they might have evolved in situ from some distinctive North American stock. (Cowan et al. 1981:71)

Cowan's suggestion was stimulated both by his excavation of very small, pre-4,000 B.P. *Cucurbita* seeds at the Cloudsplitter rock shelter in eastern Kentucky in 1978 (Cowan et al. 1981) and by a paper presented by Charles Heiser at a seminar in 1980 (and not published until 1985) in which he outlined a similar indigenous progenitor-multiple origins position in some detail:

> I would suggest that this wild gourd, probably similar to *C. texana*, at one time could have grown wild as far north as Illinois so that calling for an introduction from the outside may be unnecessary. (Heiser 1985a:71)

The publication quoted above also contains a lengthy discussion of the likely role of a *C. texana*-like gourd as a progenitor of domesticated *C. pepo*, as well as the proposal that *C. pepo* could well have been independently brought under domestication in both Mexico and eastern North America. Heiser also covered these topics in another publication the same year (Heiser 1985b:13–18), and he addressed the issue again in a conference paper presented in 1986 (Heiser 1986, 1989):

> It is well known that *C. pepo* was domesticated in Mexico by 5000 B.C. or earlier. . . .But could it have been independently domesticated in the east as well? (Heiser 1985a:64)

> *Cucurbita pepo* is an old domesticated plant in Mexico, but a plausible argument can be made for its domestication in the eastern half of North America, for it has a candidate for its progenitor in central Texas. (Heiser 1985a:70)

On the basis of our present knowledge separate domestications of C. *pepo* in Mexico and the United States can not be ruled out. In fact, I think they are much more likely than a single domestication. The old idea that a domesticate had but a single origin can no longer be accepted for a number of domesticated species. . . (Heiser 1986:3)

Although a single domestication in Mexico cannot be ruled out, separate domestications in the United States and Mexico appear far more likely. (Heiser 1989:474)

Other researchers, including Richard Yarnell, voiced similar positions in the early 1980s:

At present we probably should not assume that cultigen crops were involved. The older archaeological remains may be derived from weedy *Cucurbita pepo* gourds that did not require significant or intentional husbandry. (Yarnell 1983:4–5)

Currently it cannot be assumed with confidence that squash had achieved cultigen status before the Early Woodland Period. (Yarnell and Black 1985:99)

First presented at a conference in 1986, the preceding chapter in this volume was also in sharp disagreement with the single-origin consensus of the first half of the decade, and offered a detailed argument in support of the early rind fragments representing an indigenous wild gourd rather than an introduced Mesoamerican domesticate.

By the spring of 1986, single-origin (Figure 4.1) and multiple-origin (Figure 4.2) explanations for the domestication of C. *pepo* had been outlined, each offering distinct developmental scenarios for the East, and each casting present-day and pre-4,000 B.P. eastern gourds in quite different roles. In the multiple-origins explanation, C. *pepo* was brought under domestication independently in eastern North America at circa 3,000 B.P. from an indigenous wild gourd. The single-origin explanation, on the other hand, identified the early introduction of a domesticated Mesoamerican variety of C. *pepo* as occurring by 7,000 B.P., and considered any modern free-living gourds in the East to be recent escapes from cultivation of domesticated ornamental gourds.

Considerable botanical research relevant to this question has been reported since 1986, and while it has engendered a second generation of single-origin proposals, it has also produced considerable new evidence in support of a multiple-origins explanation. Of principal importance in this regard has been the research of Deena Decker-Walters (see Chapter 3, note 11).

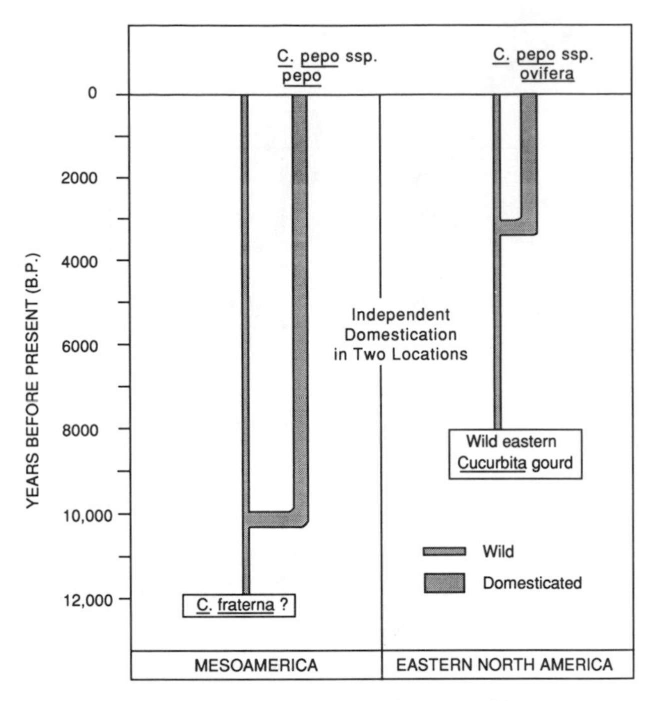

Figure 4.2 The dual multiple origins explanation of the domestication of C. *pepo*.

Documenting a Developmental Dichotomy

In the mid-1980s Decker-Walters carried out an initial study of allozyme variation within a limited sample of cultivar varieties of *Cucurbita pepo,* revealing for the first time a major dichotomy within the species. This separation:

May reflect an early divergence within the species accompanied by rapid diversification under drift and differential selection pressures, or possible multiple origins of domestication. (Decker 1985:308)

Appearing the following year, Decker-Walters's dissertation (Decker 1986, summarized in Decker 1988) presented an expanded allozyme analysis of the species C. *pepo,* along with parallel comprehensive consideration of seed, fruit, and flower morphology and flowering patterns:

> All analyses indicated that a major subdivision exists within *C. pepo*. This may reflect an early geographical divergence within the species, or multiple origins of domestication. *Cucurbita texana* is closely allied to the subdivision that apparently experienced much of its early diversification in eastern U.S. The other subdivision, which exhibits closer ties to Mexico, is exclusively represented in that country today. (Decker 1986:iii–iv)

> Populations of *texana*-like plants from beyond Texas share various attributes with Texas populations, suggesting that *C. texana* once had a more widespread distribution to the northeast. The possibility exists that *C. pepo* was domesticated independently in eastern U.S., as well as in Mexico. . . . (Decker 1988:4)

The separation between Mexican *(C. pepo* ssp. *pepo)* and eastern North American *(C. pepo* ssp. *ovifera)* cultivar subdivisions of the species, and the close association of *C. texana* and related free-living gourds with the eastern subdivision was sustained by subsequent research through the 1980s (Decker and Wilson 1986, 1987; Wilson 1990); leading up to Decker-Walters's recent summary article "Evidence for Multiple Domestications of *Cucurbita pepo*" (Decker-Walters 1990):

> In summary, evidence from various sources indicates a significant divergence within *C. pepo*. Furthermore, this divergence appears to separate a group of cultivars rooted in Mexico from another group whose origins are apparently in the eastern United States. . . .Recent morphological and allozyme analyses strongly suggest that *C. texana* is ancestral to the eastern United States lineage of the domesticate. Exploratory studies on *C. fraterna* indicate that it may be the Mexican counterpoint to *C. texana,* having given rise to another lineage of *C. pepo*. . . .The evidence suggests independent domestications of *C. pepo* in the eastern United States and in Mexico, with *C. texana* and *C. fraterna* as the respective wild progenitors. (Decker-Walters 1990:99, 96)

In an article published the same year, Heiser reached a similar conclusion regarding the respective roles of *C. texana* and *C. fraterna* as likely wild progenitors in the multiple domestications of *Cucurbita pepo*:

> I find most interesting that she [Decker-Walters] recognizes two major groups of squashes today, one of which could have been domesticated from the United States and the other from Mexico. The most recent development in the *C. pepo* story. . .is the discovery of *C. fraterna* Bailey which now provides a progenitor for *C. pepo* in Mexico. (Heiser 1990:114)

While Decker-Walters's discovery of a major developmental dichotomy within *C. pepo* provides strong support for a multiple-origins explanation and the independent domestication of *C. pepo* in eastern North America, more focused allozyme analysis has also considerably clarified the genetic and developmental relationships that exist between present-day domesticated and free-living gourd populations of the eastern United States.

The Cophyletic Model: Recasting the Question of "Wild" versus "Escape"

Since it was first collected in 1835, the wild versus escaped status of the Texas wild gourd *(C. pepo* ssp. *ovifera* var. *texana)* has been a recurring subject of botanical interest (Heiser 1985a:14–17, 1985b:64–65; Decker 1988:6). At the same time, however, similar free-living gourds in other areas of eastern North America have almost always been assumed to be escapes from cultivation, and when infrequently collected, have been identified as feral *C. pepo* var. *ovifera,* a domesticated ornamental gourd (Asch and Asch 1985:157–158, 1991:44–49).

The triangle of interesting taxonomic and genetic questions formed by these three groups of *Cucurbita* gourds (the free-living gourds of Texas, the free-living gourds outside Texas, and the domesticated *C. pepo* var. *ovifera*) has been the subject of a number of recent papers (Decker 1988; Decker and Wilson 1986, 1987; Kirkpatrick and Wilson 1988; Wilson 1989, 1990). In the process of illuminating the nature of the genetic relationships among the three categories of gourds in question these studies have also served to recast the wild versus escape issue, in effect replacing a simplistic either-or question with a more complex and more accurately framed question of developmental history.

The allozyme analysis of Decker-Walters and Wilson (Decker 1985, 1986, 1988; Decker and Wilson 1987) resulted in a proposed new classification for *C.*

pepo (Decker 1988) that divided the eastern North American subspecies *(C. pepo* ssp. *ovifera)* into two varieties. Existing only in cultivation, members of the first variety *(C. pepo* ssp. *ovifera* var. *ovifera)* include acorn, crookneck, and fordhook squashes, along with many of the ornamental gourds previously labeled as *C. pepo* var. *ovifera.* The second variety, *C. pepo* ssp. *ovifera* var. *texana* includes only those populations of gourds sustaining themselves in the natural environment of the central-eastern United States, both in Texas (previously labeled *C. texana*) and beyond. Although placed in a separate variety on the basis of both allozyme differences and seed morphology, the free-living gourds of variety *C. pepo* ssp. *ovifera* var. *texana* were still found to be closely related to the domesticated ornamental gourds of *C. pepo* ssp. *ovifera* var. *ovifera,* with the non-Texas gourds included in the study (from Arkansas, Alabama, and Illinois) falling "between *C. texana* and *C. pepo* in their expression of allozyme frequencies" (Decker 1988:8).

In a study published the same year, Kirkpatrick and Wilson provided further information regarding the close genetic relationship between the free-living gourds of var. *texana* and the cultivated members of var. *ovifera.* They were able to demonstrate the movement of pollen by bees and the exchange of genetic material between experimental populations of var. *texana* and cultivated populations representing both subspecies of *C. pepo* over distances of more than 1,300 m, producing hybrids in 5 percent of the over 1,500 progeny produced during the study:

> The observed lack of interspecific reproductive isolation supports treatment of cultivars and wild types as a single species and, in conjunction with available data concerning temporal/geographical relationships among bees, squash, gourds, and humans in eastern North America, suggests the possibility of long-term genetic interaction between wild types and domesticates. (Kirkpatrick and Wilson 1988:519)

In another recent study that further clarifies the nature of the relationship between modern free-ranging and domesticated taxa within the genus *Cucurbita,* Wilson (1989) examines how such populations of free-ranging *Cucurbita* gourds can be classified relative to domestic varieties in terms of fruit and seed morphology on the one hand, and allozyme frequencies, on the other.

Among domesticated taxa, Wilson confirms the earlier research results of Decker-Walters: that seed and fruit characteristics and allozyme frequencies clearly agree in partitioning the different species (e.g., *C. pepo, C. moschata, C. mixta*) of the genus, and even, in the case of *C. pepo,* the subspecies (ssp. *pepo* and *ovifera*).

In contrast, when free-ranging *Cucurbita* gourds are considered, the results of morphological and allozyme analysis do not agree, but rather differ markedly. Wilson recognizes a considerable degree of morphological similarity between different Mexican varieties of free-ranging *Cucurbita* gourds *(C. sororia, C. fraterna, C. galeotti)* and argues that in terms of general seed and fruit characteristics they form a distinct and internally homogenous grouping quite distinct from domesticated taxa.

The seeds of the free-ranging gourds included in Wilson's study are all considerably smaller than those of domesticated taxa. Similarly, the small, hard-walled fruits of the different free-ranging gourds considered by Wilson, as well as those of eastern North America, comprised a separate homogeneous grouping quite distinct from the domesticated taxa, with the exception of some ornamental gourds.

Genetically, however, members of the morphologically homogeneous grouping of free-ranging *Cucurbita* gourds were different from each other, and also varied considerably in terms of their degree of affinity to various domesticated taxa, ranging from autonomy *(C. galeotti)* through partial association to a domesticate *(C. fraterna/C. pepo),* to full identity with a domesticate *(C. sororia/C. mixta)*:

> Free-living samples, conspicuously distinct from domesticated samples in essentially every traditional key character, show a graded allozymic affinity to domesticated taxa. (Wilson 1989:617)

Wilson (1989:617) accounts for this "unusual opposition of structural and genetic variation" involving a combination of bimodal structural differentiation (crop versus free-living) with genetic affinity, within a framework of coevolution or cophyletic development:

> The cophyletic model places both weed [free-ranging population] and crop as a single monophyletic unit that is unified by sporadic introgressive hybridization and polarized by bimodal (human vs. natural) selection. . . .This explanation demands a bimodal, unlinked, genomic response to disruptive (human vs. natural) selection. Allozyme variation, possibly sheltered from the immediate selective environment, appears to be conservative and basal. Morphological variation, on the other hand, is more exposed to selective forces and therefore, more malleable. (Wilson 1989:618)

When combined with related research on pollen vector mechanisms of introgression between free-ranging and crop populations of *Cucurbita pepo* (Kirkpatrick and Wilson 1988), Wilson's cophyletic explanation offers a potential answer to the question of whether the present-day populations of free-ranging *Cucurbita* gourds of eastern North America are wild or escapes, and does it in a way that imbues these poorly known plants with considerably greater genetic complexity and developmental depth.

The present-day free-living gourds are wild in terms of both their morphology and the niche they occupy. They conform to Wilson's free-ranging morphotype, exhibiting the full range of attributes required to survive outside the agroecosystem, and as will be detailed later in this chapter, they demonstrate these adaptations by growing in remote, undisturbed natural habitats as well as in anthropogenic settings.

Furthermore, Wilson argues forcefully against the possibility that present-day free-living gourd populations, with their wild-morphotype adaptations to natural habitats, could have originated from domesticated cultigen varieties of *C. pepo*:

> If these populations represent secondary weeds, then features that distinguish them from the putative parental cultivars present a problem. With the exception of some ornamental gourds, cultivars breed true for many characteristics that are simply not present in free-living populations. Direct descent of free-living populations via escaped cultigens requires multiple mutation or recombination events that, given the number of genes involved. . ., seem unlikely. (Wilson 1990:452)

While rejecting the idea of direct escape of free-living populations from parental cultivars, Wilson does consider these present-day wild morphotype gourds to be, in a sense, escapes from cultivation. While the gourds themselves have not escaped the agroeconomy, bees have repeatedly freed the pollen of cultivated plants over thousands of years, and the two-way bridge formed by these pollen vectors, capable of extending 1,300 m and more, has developmentally linked the ecologically distinct populations of the wild and human habitats.

In summary, the allozyme and pollen vector research described above, as well as Wilson's cophyletic model, all support a long developmental history in the East of sporadic introgressive hybridization between domesticated cultivars of the agroeconomy and free-living wild morphotype gourds, with present-day populations of free-living gourds representing the descendants of wild ancestral populations:

> Free-living *C. pepo* (including *C. texana* and *Cucurbita fraterna*; Decker 1988) are associated with the most diverse element of the species. . ., which includes hard-fruited cultivars that are morphologically similar (in some cases identical) to free-living populations. Thus a phyletically basal position for this free-living element is supported. (Wilson 1990:452)

While Wilson's cophyletic model and associated allozyme and pollen vector studies also provide additional support for a multiple-origins explanation and the independent domestication of *C. pepo* in eastern North America, Wilson has taken a different direction, proposing a set of three alternative single Mesoamerican origin explanations for the domestication of *C. pepo* that are briefly outlined in Kirkpatrick and Wilson (1988) and Wilson (1990). In addition, Asch and Asch (1991) have recently outlined a single Mesoamerican origin hypothesis that shares common elements with those proposed by Wilson and Kirkpatrick. Interestingly enough, these variations by Wilson, Kirkpatrick, and the Asches on a single Mesoamerican origin explanation serve to strengthen the case for multiple origins by highlighting the difficulties inherent in attempting to establish a developmental link between Mexico and eastern North America.

Recent Single Mesoamerican Origin Models

Each of the four variations on a single Mesoamerican origin explanation to be discussed below acknowledge the long developmental separation of the Mexican *(C. pepo* ssp. *pepo)* and eastern *(C. pepo* ssp. *ovifera)* subspecies groupings, but they view this long separation as occurring after the early introduction of a *Cucurbita* gourd from Mesoamerica rather than as a result of independent domestication from an indigenous eastern North American gourd.

Wilson's 1990 Explanation

Wilson suggests in his 1990 model that *C. pepo* was domesticated once, in southern Mexico at about 10,000 B.P., and with the subsequent northern introduction of a domesticated variety of *C. pepo* into eastern North America, the region became a secondary center of development (Figure 4.3):

> If *C. pepo* originated under human selection in Mesoamerica, then subsequent diffusion to the eastern United States could have produced a secondary center of diversity in that agriculturally active area. (Wilson 1990:452)

Wilson discounts the possibility of *C. pepo* being brought under domestication in eastern North America on the basis of evidence of its earlier domestication in Mexico:

> The archaeological record indicates that domesticated *C. pepo* was present in southern Mexico (Oaxaca) 10,000 years ago. . . .This documented antiquity at the southern portion of the range is not consistent with a phyletic scenario involving a northern origin from northern progenitors. (Wilson 1990:452)

Wilson's argument contains the embedded presumption of a single origin for domesticated *C. pepo.* Early evidence of domestication of the species in Oaxaca, however, does not preclude subsequent, developmentally independent domestication in eastern North America.

Wilson's rejection of a northern origin from northern progenitors for *C. pepo,* along with his opposition to any possibility of escape from cultivation, sets the stage for a third hypothesis, which he favors, regarding the source of present-day free-living gourds in eastern North America:

> Another hypothesis places the free-living populations as extant manifestations of an ancient companion weed that accompanied *C. pepo* as the species dispersed from a center of domestication in Mesoamerica. Companion weeds, as defined by Harlan. . .are genetically interactive, conspecific, cophyletic elements of a single lineage that includes both free-living (companion weed) and domesticated (crop) elements as a crop/weed complex. (Wilson 1990:452)

Although Wilson does not specify when a domesticated form of *C. pepo,* accompanied by its companion weed, was carried from Mesoamerica into eastern North America, two points in time would seem reasonable, based on available archaeological information. The earliest evidence of a clearly domesticated variety of *C. pepo* (exhibiting morphology different from that of the wild morphotype) dates to about 3,000 B.P., while the earliest evidence of a wild morphotype *Cucurbita* gourd in the region dates to 7,000 B.P. (Figure 4.3). The 4,000-year separation between the documented presence of a wild morphotype at 7,000 B.P. and the first clearly domesticated variety of *C. pepo* at 3,000 B.P. poses an interesting logical dilemma for Wilson's scenario of a simultaneous weed/crop introduction. A still growing archaeobotanical data base of *Cucurbita* rind fragments and seeds from pre-4,000 B.P. contexts in Illinois, Missouri, Kentucky, and Tennessee attests to the early presence of a small hard-walled gourd over a broad interior riverine region of the East. Ongoing morphological analyses of both fruits and seeds (rind thickness, anatomy, and fruit shape, peduncle and blossom scar size, seed shape) indicates that all of these early gourds fall comfortably within Wilson's wild adapted or free-living morphotype. At the same time, there is a complete absence in the archaeobotanical record of any *Cucurbita* material that falls outside of the wild adapted or free-living morphotype before about 3,000 B.P. Within the context of Wilson's argument, if this early material represents the companion weed component of an introduced Mesoamerican crop/weed complex, rather than an indigenous wild gourd, why the glaring absence, for 4,000 years, of the domesticated crop, which is a necessary prerequisite for the development of a companion weed gourd?

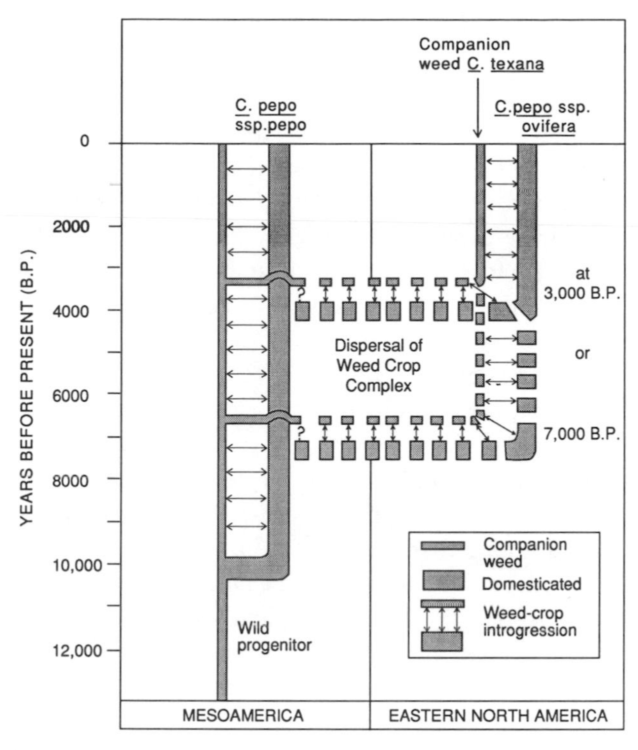

Figure 4.3 Wilson's cophyletic single-origin explanation of the domestication of *C. pepo*.

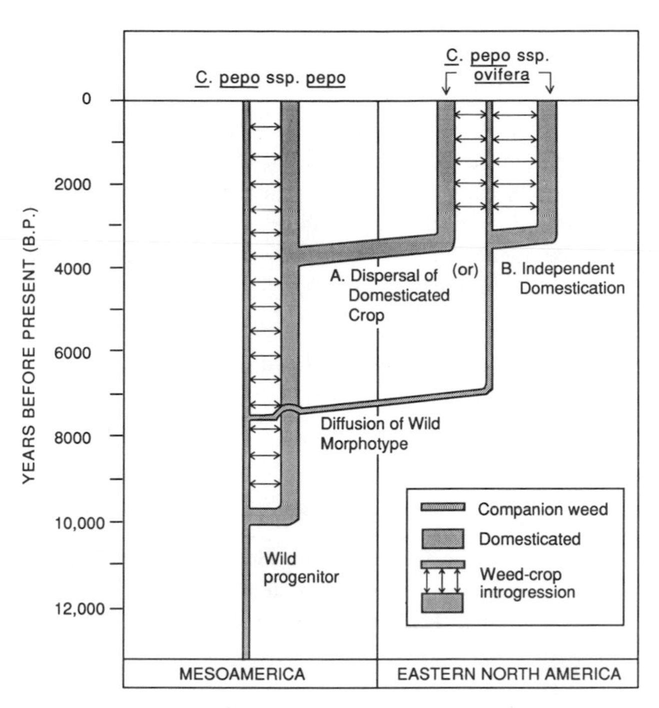

Figure 4.4 Kirkpatrick and Wilson's single-origin explanation of the domestication of *C. pepo*.

If the introduction from Mesoamerica occurs at 3,000 B.P., on the other hand, a similar logical dilemma remains unresolved. While the Mesoamerican weed/crop complex might be invoked to explain present-day free-living gourds of eastern North America, it cannot account for the pre-3,000 B.P. wild morphotype *Cucurbita* gourds of the region. A 3,000 B.P. arrival date would also place the tropical weed/crop complex in the region subsequent to the domestication of local seed crops.

Kirkpatrick and Wilson's 1988 Explanations

The two variations on the single Mesoamerican origin model presented by Kirkpatrick and Wilson (1988) acknowledge the dilemma of having a wild morphotype companion weed gourd present in the East 4,000 years earlier than the domestic crop it is supposed to accompany. In attempting to resolve this problem and explain the early arrival, these variations provide further support for a multiple-origins explanation.

Both are very similar in most respects (Figure 4.4). Each acknowledges the early presence of a wild morphotype *Cucurbita* gourd and the appearance of a domesticated form by 4,000 B.P. (the earliest evidence of clearly domesticated *C. pepo* in the region dates to about 3,000 B.P.):

> Archaeological samples from several sites in the upper Mississippi Drainage indicate that *C. pepo* was a basal element of premaize agriculture, with domesticates dated at 4,000 B.P. . .and remains of a hard-fruited wild type from Illinois dated at 7,000 B.P. . . .As indicated above, the archaeological record indicates the presence of gourd-producing *C. pepo* in Illinois at 7,000 B.P. . .and large-seeded (domesticated, fleshy fruited) types under cultivation by 4,000 B.P. (Kirkpatrick and Wilson 1988:520, 526)

In both the Kirkpatrick and Wilson models the early presence of wild gourds in eastern North America is explained as the result of a human-facilitated northward range extension from Mesoamerica of the wild morphotype, because it was unique and extremely useful as a container:

> Gourd-like fruits of both *Lagenaria* and *Cucurbita* provided a unique and extremely useful tool to preceramic cultural groups. This could account for the early distribution of tropical Cucurbitaceous "container crops" in temperate North America, either through trade or "camp following" dispersal. (Kirkpatrick and Wilson 1988:520)

The two variations differ in how they account for the appearance of domesticated *C. pepo* at 4,000 B.P. [3,000 B.P.]—as either the result of independent development in eastern North America, or a second diffusion from Mexico (Figure 4.4):

> Subsequent selection of edible cultivars could have occurred within the extended range of the basal "container" plants. (Kirkpatrick and Wilson 1988:520)

> It appears that the archaeological material represents early preagricultural and subsequent post agricultural diffusion from Mexico. (Kirkpatrick and Wilson 1988:526)

Once the in situ development or diffusion of edible cultivars takes place, both variations on the single-origin model then link the early arriving "wild type" gourds and the edible cultivars in a weed/crop complex ancestral to present-day free-living gourds:

> Localized weed-crop hybridization/differentiation events involving wild-adapted gourd types and edible cultivars could have generated a companion weed that survives today in relictual populations. (Kirkpatrick and Wilson 1988:520)

As detailed in the above quotations, Kirkpatrick and Wilson consistently make a clear distinction between two different types of *Cucurbita pepo,* with these two types conforming to Wilson's cophyletic model of bimodal structural differentiation:

1. A domesticated, large-seeded fleshy-fruited edible cultivar that either arrives or is locally developed at 4,000 B.P. [3,000 B.P.], after local plants are domesticated ("post-agricultural diffusion");
2. A wild morphotype hard-fruited type that is present in the East by 7,000 B.P. as the result of "early pre-agricultural. . .diffusion from Mexico."

It is the early introduction of this basal container crop that provides the clearest difficulties for Kirkpatrick and Wilson. For even if this "wild adapted" gourd did have its range expanded northward from Mexico into eastern North America by 7,000 B.P. as a result of human action, it could reasonably be expected to subsequently have adapted to temperate forest environments and assumedly filled a niche similar to the one it occupied in its Mesoamerican homeland. In terms of fruit and seed morphology these early gourds conform to Wilson's free-living or wild morphotype, exhibiting a number of attributes associated with the ability to survive outside of cultivation and natural, as opposed to human, selection. Thus even if the hypothesized wild adapted hard-fruited gourd that left Mexico in time for it to become adapted to, and established in, Illinois by 7,000 B.P. was, before it left, a companion weed in a weed/crop complex, and therefore carried some components of the Mesoamerican domesticated *C. pepo* gene pool, it would seem highly unlikely that such a "Mesoamerican genetic core" would have survived intact across 20 degrees of latitude and 3,000 generations of developmental isolation and differential selection within wild temperate forest floodplain habitats. While it is far more likely that the wild hard-fruited gourds present in eastern North America by 7,000 B.P. were indigenous rather than introductions from Mesoamerica, it is also clear that even if originally diffused northward, these early gourds would have over time lost any claim to a Mesoamerican identity as they were exposed over the long term to the selective pressures of river valley floodplains and become part of the native flora.

If the seemingly inevitable transformation of this early arriving gourd into a component of temperate forest ecosystems is considered, the first of Kirkpatrick and Wilson's developmental scenarios becomes, in effect, a multiple-origins explanation for the domestication of *C. pepo*. Acknowledging the long developmental partitioning between Mexico and eastern North America, Kirkpatrick and Wilson propose the separate eastern North American derivation of edible fleshy-fruited, large seeded varieties of *C. pepo* from a wild adapted, wild morphotype gourd that had survived and evolved in eastern floodplain ecosystems for more than 4,000 years, with present-day free-living gourds of the region reflecting subsequent long-term genetic interaction between wild morphotypes and domesticated crop populations.

The Asches' 1992 Explanation

Picking up where Kirkpatrick and Wilson's model leaves off, however, Asch and Asch (1992) addresses its inherent problem—the inevitable "easternization" of a wild adapted gourd introduced from Mexico by 7,000 B.P. They block this process of "easternization" and maintain the Mesoamerican genetic core of this proposed early arrival by identifying it as a domesticated plant; that is, a plant unable to survive in the wild, outside of cultivation, without human management.

They follow Kirkpatrick and Wilson in proposing that this basal container crop was "introduced through human agency" (Asch and Asch 1992:51), and propose that:

> The gourds' technological uses as containers, ladles, floats, and rattles, and the food value of the non-bitter, oily, protein-rich seeds would have provided a sufficient motive for its early spread and protection. (Asch and Asch 1992:51)

But while Wilson does not provide much discussion of the niche occupied by this basal gourd, other than characterizing it as a "wild type," "wild adapted," and "free-living" (able to survive outside of cultivation), the Asches specifically confine it to a cultivated "campsite niche," and in effect reiterate the models of the early 1980s (Figure 4.1) (Asch and Asch 1992:51):

> Cultivation of a few individual plants at or near a campsite would confer a small but obvious economic benefit by assuring an adequate and convenient supply of the specialty product. (Asch and Asch 1992:51)

By restricting the introduced gourd to cultivated contexts and denying it any ability to survive unaided in the wild, the Asches clearly reject its initial introduction into the East and subsequent development as representing a range extension of essentially a wild plant, followed by its in situ development into a component of local floodplain plant communities.

The Asches' scenario represents the strongest argument that currently can be marshalled in support of a single origin for *C. pepo* in Mesoamerica, followed by the diffusion of a small domesticated gourd (and the concept of agriculture) into eastern North America. Rather than strengthening the single-origin explanation, however, the Asches' arguments highlight a number of its prima facie flaws, and in the process of demonstrating the logical extremes one must go to in order to confine the early gourds of eastern North America to contexts of cultivation, they provide further support for the presence of an indigenous wild gourd in the region—a gourd that was eventually brought under domestication, along with three other eastern seed plants, at 4,000 to 3,000 B.P. (Figure 4.5).

A variety of problems surrounding the initial movement of this domesticated Mesoamerican gourd into eastern North America, for example, lend credence to the far more plausible alternative—that it was not introduced as a domesticate but already present

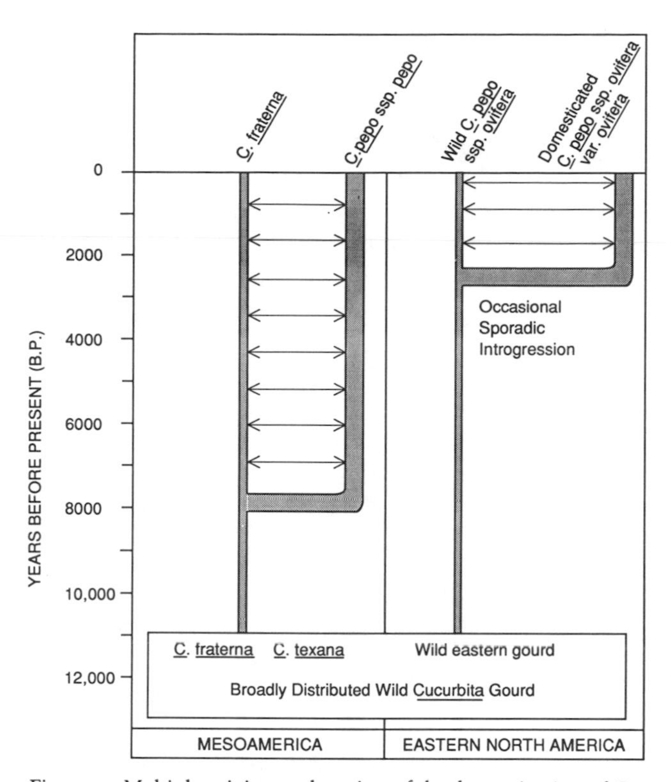

Figure 4.5 Multiple origins explanation of the domestication of *C. pepo*, indicating sporadic introgression between cultivars and free-living wild morphotype populations.

as an indigenous wild plant. In a temporal context showing little evidence of interregional exchange in the East, much less any contact with Mexico, the Mesoamerican domesticated gourd is transported north by humans (it can't survive on its own), reaching Illinois by 7,000 B.P. In the process, it crosses environmental and cultural barriers in northeast Mexico that proved very effective throughout prehistory in blocking the movement of Mexican material culture and crop plants northward. Once past this border it moves northward across 1,400 km of territory inhabited by groups having no previous experience in plant husbandry. While the domesticated *Cucurbita* gourd rapidly transversed this barrier by 8,000–7,000 B.P., later Mesoamerican domesticates (e.g., maize, beans, *C. moschata*) were long delayed in their diffusion out of Mexico, even though interregional trade and contact within eastern North America had by then increased dramatically. In addition, while the strong need for a container crop in Illinois apparently facilitated the rapid transport of the small Mesoamerican *Cucurbita* gourd northward, the bottle gourd *(Lagenaria siceraria),* a larger and far superior container crop, failed to make the trip. Although a companion of *Cucurbita* gourds in Mesoamerica by 9,000 B.P., and documented in Florida at 7,300 B.P. (Doran et al. 1990), the bottle gourd does not in fact reach Illinois until about 2,000 B.P. (Asch and Asch 1985:158), long after local plants have been domesticated. With the exception of the Phillips Spring site in Missouri, which has yielded bottle gourd rind fragments dating to 4,300 B.P. (and perhaps somewhat earlier, Kay 1986), this container crop does not appear north of Florida before 4,000 B.P. The almost total absence of the bottle gourd north of peninsular Florida during the period 7,000–4,000 B.P. further supports the conclusion that the wild morphotype *Cucurbita* gourd that had a broad mid-latitude distribution during this 3,000-year span was an indigenous wild plant.

In addition to the difficulties inherent in moving this Mesoamerican *Cucurbita* gourd, unaccompanied by *Lagenaria siceraria,* north to Illinois by 7,000 B.P., there are clear implausabilities inherent in its cultivation once it arrives. Given that it cannot survive on its own, and must adjust within cultivation to the mid-latitude growing season of Illinois, it is worth noting that this *Cucurbita* gourd is not delivered into the hands of waiting plant managers who add it into pre-existing plant husbandry systems. Rather it becomes the first and only plant cultivated by hunting and gathering groups in the region. Apparently reluctant to expand this newly acquired skill in cultivation to include any local plants, these early cultivators maintain this *pepo* gourd as their only domesticated plant for about 3,000 years.

Finally, in preserving the Mesoamerican genetic identity and single origin of domesticated *C. pepo* by denying these early gourds the ability to survive outside of cultivation, either as an indigenous wild gourd or an introduced wild-adapted "easternized" Mesoamerican gourd, the Asches are also faced with accounting for the absence, for 3,000–4,000 years, of any morphological changes in this plant associated with human manipulation and contexts of cultivation.

The Asches' cultivated Mesoamerican *Cucurbita* gourd is remarkable in this regard. In terms of everything known regarding fruit and seed morphology and size of this early gourd, it conforms to Wilson's wild morphotype, closely resembling a product of natural selection in wild environments. Yet in the Asches' view it could not survive in the wild, and these structural adaptations were unnecessary. At the same time, confined to contexts of cultivation and dependent on human manipulation, this "domesticated" (i.e., unable to survive outside cultivation) gourd shows none of the structural shifts toward Wilson's crop plant morphotype one might reasonably expect under human selection.

In contrast, the initial human manipulation of similar wild morphotype gourds by Mesoamerican groups of mobile foragers having a minimal dependence on agricultural products and apparently using gourds in much the same way as did the Asches' early eastern cultivators (Asch and Asch 1992:52), rapidly produced an increase in the size of *Cucurbita* seeds:

> Our measurement of seed size of *C. pepo*. . .suggest[s] there is a trend toward selection, perhaps largely unconscious, by the early cultivators for larger seed size and presumably larger fruit. (Whitaker and Cutler 1986:276–277)

In eastern North America, however, cultivation of the basal Mesoamerican gourd did not produce a corresponding increase in seed size for 3,000–4,000 years (Asch and Asch 1991:54–55). Similarly, although grown for use as containers, this early domesticate remains small (about the size of a hardball), thin walled (<2.0 mm), and brittle for 4,000 years, with a clear increase in fruit size and wall thickness not occurring until about 3,000 B.P. (Cowan 1990). In direct opposition to what might be expected in terms of either automatic selection or deliberate selection under cultivation, the Asches' broadly dispersed small bands of early eastern North American cultivators uniformly avoid any improvements on the wild morphotype Mesoamerican gourd through 3,000 to 4,000 seasonal cycles of planting, managing, harvesting, and storing.

In summary, over this long period of human husbandry, the Asches' early cultivators steadfastly manage to maintain in this domesticated gourd that is unable to survive in the wild, a set of structural characteristics that reflect adaptation to wild environments, and are unnecessary in cultivated contexts. These cultivators mimicked, to a remarkable degree, natural as opposed to human forces of selection, and in the process avoided any beneficial morphological changes (e.g., larger seeds, better containers) that could have been expected under human selection, and which do, in fact, eventually appear four thousand years later in time.

Given all of these problems, assumptions, and implausabilities associated with the single-origin, Mesoamerican-domesticate diffusion explanation, what then are the weak points of the alternative, far more simple explanation—the existence of an indigenous wild gourd in eastern North America?

In building a case in support of this early Mesoamerican introduction being confined to contexts of human management, the Asches present a variety of arguments regarding why *C. pepo* in general, and present-day free-living gourds of the East in particular, were poorly adapted to survive in eastern North America outside of contexts of deliberate cultivation. They particularly focus on the lack of adaptation by these free-living gourds to disturbed habitat settings, either natural or anthropogenic.

These modern wild-morphotype gourds thus play an essential role in the defense of this single-origin explanation. The perceived inability of modern free-living *pepo* gourds to exist today outside of Texas in natural, nonagricultural habitats is projected back into prehistory as indicative of any comparable wild morphotype gourd's similar inability to survive outside of cultivation.

Even though the Texas wild gourd is documented as having flourished in naturally disturbed floodplain settings in eastern Texas since the 1830s, the Asches argue that it lacks many of the essential characteristics of a weedy pioneer plant capable of surviving in disturbed habitats, man-made or otherwise. The presumed self-sterility of individual *C. pepo* plants, for example, is cited as a barrier to survival outside of cultivation (Asch and Asch 1992:49, 53), as is their nonbitter seeds and the observed absence of populations of var. *texana* in natural disturbance settings adequate to provide the necessary seed reservoir for dispersal into anthropogenic habitats (Asch and Asch 1992:53). Similarly, it is suggested that var. *texana* does not produce the large number of seeds (per unit area) characteristic of weedy pioneer plants, is "poorly adapted to extreme disturbed habitats," and does not grow in places "where few other plants thrive" (Asch and Asch 1992:54).

Based on their field observations of wild morphotype gourds in the Lower Illinois Valley and elsewhere in the East, they consider the plant to be a narrowly adapted agricultural weed incapable of surviving in the wild:

> In Illinois today, the wild *pepo* gourd occurs basically as a soybean-field weed in floodplains. . . .In Illinois, we find the plant as a field-edge weed and in fallow fields but see no evidence that it is moving into more natural floodplain settings. . . .We have never seen it in Illinois growing in a floodplain forest, on drift, or on a mud flat. (Asch and Asch 1992:48)
>
> Smith (1987:30) [Chapter 3] proposed the riverine habitat of *C. texana* in Texas as a midwestern model "colonizing river-edge debris situations and being dispersed along river valley corridors by alluvial floodwaters." This may be appropriate in the Ozarks, but elsewhere the applicability is questionable. Where we are familiar with the plant in the large floodplains of Illinois, it has

> not colonized nonfield settings. (Asch and Asch 1992:50–51)

In addition, the Asches propose that modern free-living gourds in eastern North America only escaped from cultivation after the Second World War, and even then did not go far, moving into a restricted narrow niche that has existed for only a short period of time within modern agroeconomies, and that certainly was not present prehistorically:

> It is more likely that the Midwestern spontaneous *pepo* gourds have originated as escapes from some of the cultivated ornamental forms. . . .In the Midwest, the plowed, cultivated, herbicide-treated fields of row crops where *pepo* gourds now grow spontaneously constitute a habitat that is unlikely to have had a near-analog at prehistoric habitation sites where no agriculture was practiced. . . .It appears to be a new weed which is occupying a niche opening up as farming practices changed in the Midwest. (Asch and Asch 1992:49, 53, 48)

Just as the prehistoric *Cucurbita* gourds are confined to cultivation and denied access to natural habitats by the Asches, so too are present-day *Cucurbita* gourds restricted to a narrow niche within the argoeconomy. The logical alternative to this position is that present-day free-living gourds extant beyond Texas are not a new, narrow-niche agricultural weed, but rather that they are, like the prehistoric pre-4,000 B.P. *pepo* gourds of the region, wild plants existing in natural habitats, while also having the capability of invading cultivated fields in a manner similar to that of other indigenous pioneer plants. Judging from the documented existence of the Texas wild gourd in natural floodplain environments of east Texas since 1835, this alternative position would not seem unreasonable.

While the continuous existence of the Texas wild gourd since it was first observed a century and a half ago calls into question the strength of the Asches' arguments regarding why comparable present-day or prehistoric wild-morphotype gourds couldn't exist outside of human management in areas to the east and north of the Lone Star state, it also underscores the need for field research on modern *pepo* gourds beyond Texas. If present-day populations of *Cucurbita* gourds could be found growing north and east of Texas in natural floodplain settings similar to that of the Texas wild gourd, then the "new narrow-niche agricultural weed" profile of the Asches would be seriously compromised, and the major supporting arguments for the single-origin, introduced Mesoamerican domesticate explanation would in turn be substantially weakened. At the same time, the existence of naturally occurring populations of *pepo* gourd growing beyond Texas would provide additional support for the existence of an indigenous wild gourd in the East prior to 4,000 B.P., and would offer an excellent modern analog opportunity to consider a wide range of questions regarding the likely life cycle and habitat of such ancestral gourds of eastern North America.

The Geographical Range of Free-Living Gourds in Eastern North America

Until the mid-1980s relatively little attention was focused on *Cucurbita* gourd populations outside of Texas, probably because they were assumed to be ephemeral escapes from cultivation (Asch and Asch 1985:157; Heiser 1985a:16–17).

Julian Steyermark appears to have played an important role in establishing free-living *Cucurbita* gourds outside of Texas as a legitimate target for collection in this century. He collected specimens from a number of counties in southwest Missouri in the 1950s, distributed herbarium sheets of the plants to other institutions, and included a distribution map and plant illustration (Figure 4.6) in his landmark publication *Flora of Missouri* (Steyermark 1963). Steyermark's willingness to consider this gourd as worthy of inclusion in his widely consulted reference publication encouraged other botanists to collect it and led to its addition to other state floras. In Illinois, for example, it is absent from a 1955 survey (Jones and Fuller 1955) yet is documented as present in seven counties in the next published atlas (Mohlenbrock and Ladd 1977). Similarly, while it goes unmentioned in earlier surveys, *C. pepo* var. *ovifera* is recorded in ten Arkansas counties in Edwin Smith's 1978 state atlas (E. Smith 1978, 1988), with many of the relevant herbarium specimens collected in the mid-1970s, apparently in anticipation of the forthcoming atlas.

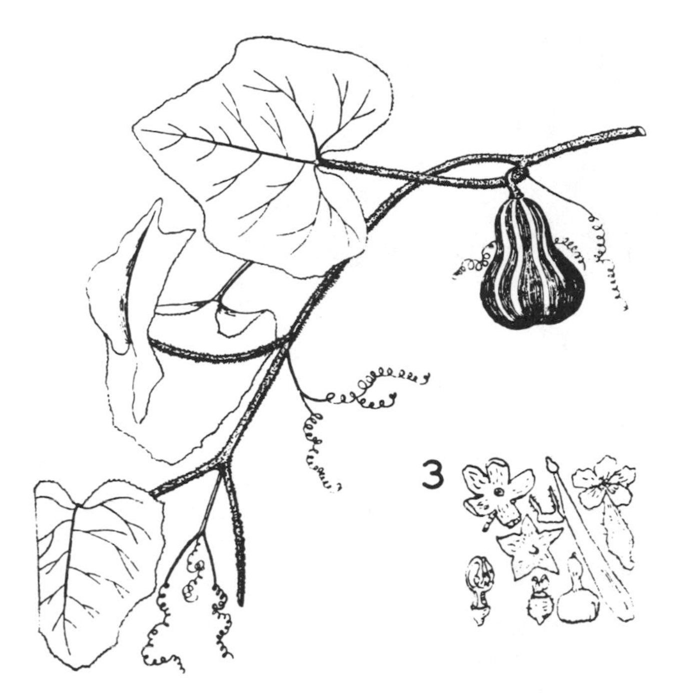

Figure 4.6 Steyermark's illustration of a free-living *Cucurbita* gourd from Missouri (from Steyermark 1963:1425).

Another important impetus to the collecting and documentation of free-living gourd populations outside of Texas came in the mid-1980s when Decker-Walters included gourds from Illinois, Arkansas, and Alabama in her dissertation research on the taxonomy and evolution of *C. pepo* and proposed that these geographical outliers might represent relict populations of a wild indigenous gourd. As a result of Decker-Walters inclusion of these outlier specimens in her research, Heiser added them to his distribution map for *C. texana* (compare Heiser 1986 and Heiser 1989), expanding it to encompass the "distribution of *C. texana* and plants approaching *C. texana*" (Heiser 1989:473). Similarly, Michael Nee includes eastern free-living outliers recorded for Missouri (Steyermark 1963), Illinois (Mohlenbrock and Ladd 1977), Arkansas (E. Smith 1988), and Alabama on his geographical range map for *C. texana,* with the caption note that "some dots outside of Texas may represent feral *C. pepo*" (Nee 1990:61).

Decker-Walters's research, and the subsequent mapping efforts of Heiser and Nee, marked an important shift in that they considered both Texas and "outlier" populations of free-living *Cucurbita* gourds and expanded beyond the state level focus of previous studies to begin to chart the plant's full geographical distribution in the United States.

Building on these studies, a more concerted effort to establish the range of free-living *Cucurbita* gourds in the region was begun in the fall of 1990. Herbariums were canvased regarding accessions of *C. pepo* var. *ovifera,* colleagues were questioned regarding recent sightings of free-living gourds, and an initial survey of drainages in the eastern Missouri and Arkansas Ozarks was conducted in November of 1990, complementing the surveys carried out by Michael P. Hoffman and colleagues in the western Ozarks over the past several years.

The initial results of this distributional study is shown in Figure 4.7 and summarized in Table 4.1. Outside of Texas, a total of 26 county records were added to the 29 previously documented, and fruits were collected by more than ten different individuals from over 30 locations in seven states (Table 4.1). Within Texas, gourd populations are now documented in 20 counties.

Based on Figure 4.7, the primary present-day geographical range of free-living *Cucurbita* gourds west of the Mississippi River extends in a broad north-south band from south-central Texas to central Illinois. Stretching for more than 1,400 km (900 miles) this north-south band appears to be divided into four subareas.

In the large Texas subarea, populations are documented throughout much of the drainage catchment of the eastern part of the state, occurring in all of the southeastern flowing rivers that enter the Gulf of Mexico between Corpus Christi and Galveston (the Nueces, San Antonio, Guadalupe, Colorado, Brazos, and Trinity rivers).

An absence of populations to date from the Neches and Sabine drainages separates the Texas subarea from the populations documented along the Red River and lower Ouachita River in Arkansas. A similar absence of recorded populations within the Ouachita Mountains of west-central Arkansas separates the as yet little surveyed Red River subarea from a major subarea located in the Ozark Plateau of southern Missouri and northern Arkansas.

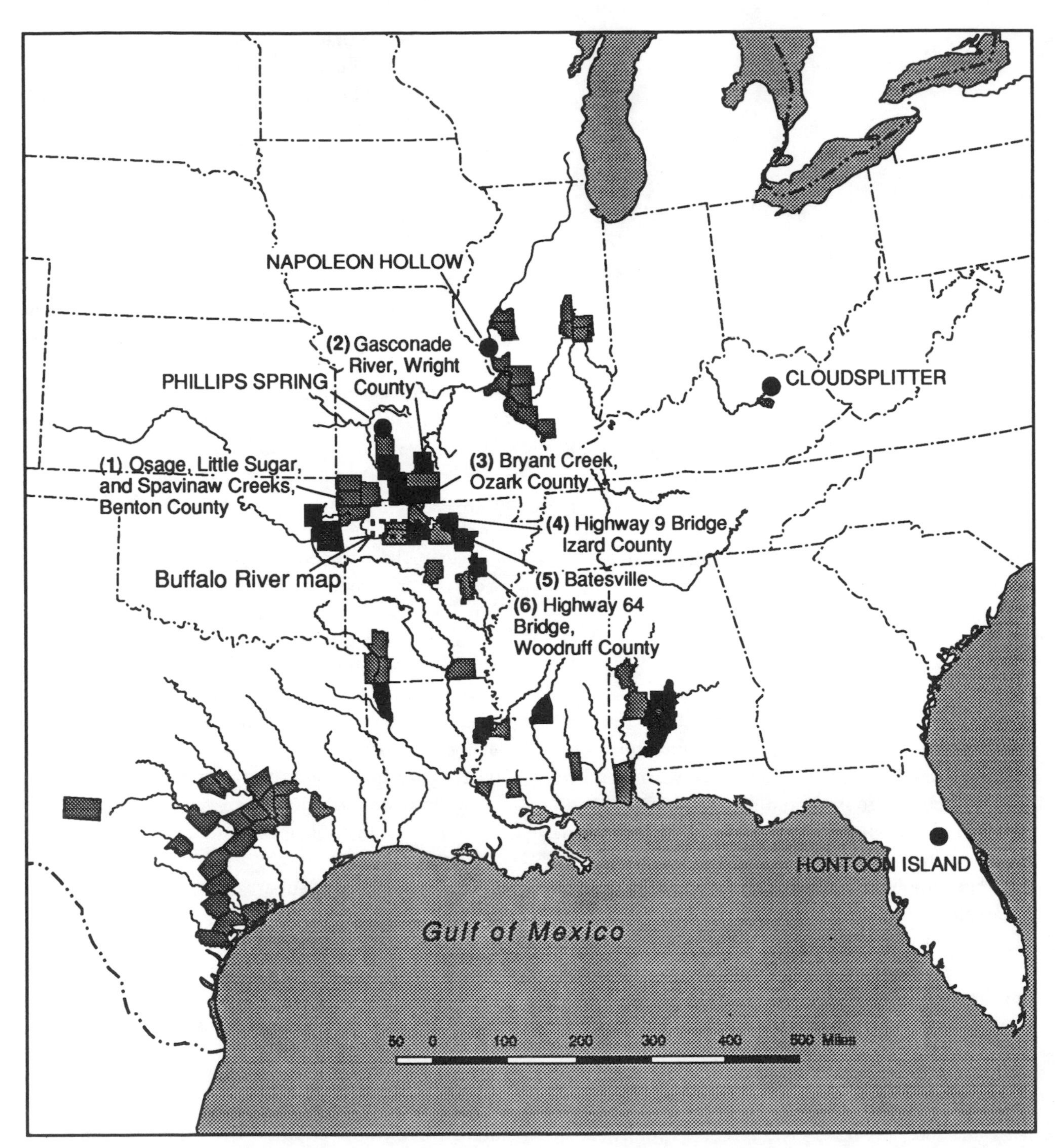

Figure 4.7 The present-day distribution of populations of free-living *Cucurbita* gourds in the Midwestern and Southeastern United States. Herbarium sheet and published county records indicated by light shaded counties, recent field collected specimens by dark shaded counties (Table 4.1 provides county information).

Within the Ozark Plateau, populations have been documented in small western trending streams and larger southern flowing rivers of the drainage system of the Arkansas River, as well as in north-flowing streams of the Osage and Gasconade systems, and along the east-southeastern flowing Buffalo-White River drainage system. About 150 km separates the Ozarks subarea from the northernmost, Illinois subarea, which encompasses the lower Illinois River, the upper reaches of the Kaskaskia River, a 250 km (150 mile) portion of the Mississippi River, and a small section of the lower Ohio River.

A fifth subarea can be recognized along the central coastal plain of the Gulf of Mexico, with populations recorded along a number of major and minor drainage systems emptying into the Gulf, including those of the lower Mississippi, Pascagoula, Pearl, Mobile-Tombigbee, and Alabama rivers.

While it is not possible at the present time to ascertain the degree to which these apparent internal subdivisions are real as opposed to simply reflecting an absence of active survey and collection, planned field research in these intervening areas, along with ongoing comparative genetic analysis of the populations of different subareas, will provide a clearer picture of the degree of extant geographic and genetic separation that does exist.

It would be fairly safe to predict, however, even given this nonuniform distribution, that populations of eastern free-living *Cucurbita* gourds today have a largely unbroken geographical distribution in river systems from south Texas as far north as central Illinois, and along the Gulf coastal plain from Corpus Christi to Mobile.

The time depth of the populations in different areas of this geographical range, however, which extends across portions of nine states and ten degrees of latitude and longitude, is difficult to determine with any high degree of certainty. Although *C. texana* was observed in 1835 in Texas, free-living *Cucurbita* gourds were not noted in the Ozarks until the mid-1950s, seemingly went unrecorded in Illinois and along the Gulf coastal Plain until the mid-1960s, and were only documented along the Red River in the mid-1970s.

This broad geographical distribution of *Cucurbita* gourd populations beyond Texas, and the temporal sequence of their initial collection and documentation (Table 4.1), is considered by the Asches to be indicative of the rapid spread of a specialized agricultural weed associated with post-World War II increasing importance of soybeans as a crop plant (Asch and Asch 1992:48). The Asches identify cultivated ornamental gourd populations as the source for this new narrow-niche soybean field weed.

In our view a much more likely explanation for the seeming shallow time depth of this distribution outside of Texas is twofold, including an increased interest in documenting a long ignored indigenous plant, once Steyermark "legitimized" it in the 1950s, combined with its progressive and ongoing expansion and re-establishment in drainage systems that it had previously occupied, particularly east of the Mississippi River.

Before rejecting it in favor of the escape from cultivation explanation, the Asches briefly discuss the possibility of the Texas wild gourd expanding its range to the east and north over the past 40 years as the result of agricultural commerce (traveling from Texas to Arkansas in a load of hay), dispersal by gourd hobbyists, or in packets of ornamental gourd seeds (Asch and Asch 1992:48–49). These potential human vectors of dispersal, along with any number of other possibilities, could well be involved in the continuing expansion of the range of free-living *pepo* gourds in eastern North America.

The suggestion that seeds of wild gourds, either *C. texana* or other eastern free-living forms, may have been packaged and sold as ornamental gourds does in fact provide a context for the subsequent "escape" of an "ornamental gourd" across the East, but such events, if they occurred, would clearly have to be considered human mediated range extension of a packaged but still wild plant, rather than the escape of a domesticated plant from cultivation.

There are two fairly obvious and straightforward tests of whether this ongoing range expansion of *Cucurbita* gourds involves the radiation and re-establishment of wild gourds in previously occupied territory, or the spread of a new weed into a specific, narrow, and recently created habitat of the agroeconomy.

The first of these tests focuses on how tightly these modern *Cucurbita* gourds are restricted to agri-

Table 4.1 The Geographical Distribution, by County, of Free-living *Cucurbita* Gourds in Eastern North America

County	Collector	Year	Herbarium[a]	Reference
ALABAMA				
Dallas	C. Sheldon	1988		This study
Greene	M.L. Roberts	1982	Alabama	
Greene				Decker 1986
Marengo				Decker 1986
Mobile	R. Deramus	1966	Alabama	
Monroe	C. Sheldon	1988		This study
Wilcox	R. Haynes	1978	Alabama	
Wilcox	C. Sheldon	1988		This study
ARKANSAS				
Ashley				Smith 1988
Benton	M. Hoffman	1990		This study
Benton	D. Dickson	1990		This study
Benton	E. McCollum	1990		This study
Faulkner	D. Oliver	1975	Arkansas	Smith 1988
Hempstead	S. Harrison	1976	Arkansas	Smith 1988
Independence				Smith 1988
Independence	Smith and Cowan	1990		This study
Izard				Smith 1988
Izard	Smith and Cowan	1990		This study
Lafayette				Smith 1988
Marion	B. L. Lipscomb	1975	Arkansas	Smith 1988
Miller				Smith 1988
Newton				Decker 1986
Prairie	D. Oliver	1975	Arkansas	Smith 1988
Searcy	B. Hinterthuer	1977	Arkansas	Smith 1988
Searcy	Smith and Cowan	1990		This study
Stone				Smith 1988
Woodruff	Smith and Cowan	1990		This study
ILLINOIS				
Cass				Mohlenbrock and Ladd 1977
Coles				Mohlenbrock and Ladd 1977
Douglas	G. Jones	1966	Illinois	
Douglas	G. Jones	1966	Florida	
Jackson				Mohlenbrock and Ladd 1977
Jersey	G. Fritz	1990		This study
Madison				Mohlenbrock and Ladd 1977
Massac	R. Mohlenbrock			
Morgan				Mohlenbrock and Ladd 1977
Piatt				Mohlenbrock and Ladd 1977
Randolph				Decker 1986
St. Clair?	W. Welsch	1862	Illinois	
St. Clair	H. Eggert	1875	Missouri	
St. Clair	H. Eggert	1876	Texas	
St. Clair	H. Eggert	1892	Missouri	
St. Clair	H. Eggert	1893	Harvard	
St. Clair	H. Eggert	1893	Missouri	
St. Clair	J. Kellogg	1901	Missouri	
Union				Mohlenbrock and Ladd 1977
KENTUCKY				
Powell	W. Booth	1990		This study
LOUISIANA				
Bossier	L. Baker	1990		This study
St. Helena	C. Allen	1971	LSU	
Tensas	G. Fritz	1990		This study
W. Feliciana	A. Martin	1972	LSU	

Table 4.1—*Continued next page*

TABLE 4.1—*Continued*

County	Collector	Year	Herbarium[a]	Reference
MISSISSIPPI				
Claiborne	K. Rogers	1978	Tennessee	
Forrest	K. Rogers	1971	Tennessee	
Rankin	S. Jones	1970	Georgia	
MISSOURI				
Barry	J. Steyermark	1955	Missouri	Steyermark 1963
Christian		1990		This study
Douglas	J. Steyermark	1957	Georgia	Steyermark 1963
Greene	Smith and Cowan	1990		This study
Howell				Steyermark 1963
McDonald				Steyermark 1963
Newton				Steyermark 1963
Ozark	Smith and Cowan	1990		This study
Polk	Smith and Cowan	1990		This study
St. Louis	G. Engelmann	1846	Missouri	Steyermark 1963
St. Louis	G. Engelmann	1846	Missouri	
St. Louis	Muehlenbach	1961	Missouri	
St. Louis	Muehlenbach	1964	Missouri	
St. Louis	Muehlenbach	1972	Missouri	
Taney		1990		This study
Texas	J. Steyermark	1956	Harvard	Steyermark 1963
Wright	Smith and Cowan	1990		This study
OKLAHOMA				
Adair	M. Hoffmann	1990		This study
Cherokee	B. Meyer	1990		This study
Mayes	D. Dickson	1990		This study
TEXAS				
Bell				Mahler 1988
Brazos	D. S. Correll		Texas	
Brazos				Decker 1986
Burleson				Decker 1986
Calhoun	Hartman and Smith		Texas	
Comal	Lindheimer		Texas	
DeWitt	D. S. Correll	1962	SWLS	
DeWitt	D. S. Correll		Texas	
DeWitt	Tharp		Texas	
Fayette				Decker 1986
Goliad				Decker 1986
Gonzales				Decker 1986
Grimes				Decker 1986
Lee				Decker 1986
Madison				Decker 1986
Refugio				Decker 1986
Robertson				Decker 1986
San Jacinto				Decker 1986
San Patricio				Jones 1975
Sutton	Reed		Texas	
Travis	A. T. Erwin	1938		Erwin 1938
Travis	Barkley		Texas	
Travis	Tharp		Texas	
Travis	Strandtmann		Texas	
Washington				Decker 1963

[a]Requests for information regarding free-living *C. pepo* specimens were sent to the following herbaria: University of Alabama, Auburn University, University of Arkansas, University of Florida, Florida State University, University of Georgia, Emory University, University of Illinois, Southern Illinois University, University of Kentucky, Indiana University, DePauw University, Louisiana State University, Southwest Louisiana State University, Harvard University, University of Mississippi, University of Missouri, Missouri Botanical Gardens, University of North Carolina, Duke University, University of Cincinnati, The Charleston Museum, University of Tennessee, University of Texas, Virginia Tech University, University of West Virginia, National Museum of Natural History, Smithsonian Institution.

cultural settings. Are they narrowly adapted weeds that are restricted to the specific soybean field settings proposed by the Asches? Are they tightly tied to a companion weed role for domesticated *C. pepo*? Or are they more generally adapted weeds of the agroecosystem? Do these gourds occur only in human-disturbed habitats such as agricultural fields and dumps, or do they also exist in unmodified natural habitat settings, either close to or far removed from gardens and agricultural fields? Obviously, the further removed from agricultural contexts that such gourds can be shown to grow, the stronger the case for the range extension of a wild gourd. Conversely, the more tightly tethered they are to specific soybean field habitats of the agroecosystem, the stronger the case for an escaped ornamental.

A second test of the Asches' new narrow niche soybean field weed explanation targets any indications that *Cucurbita* gourds were present beyond Texas prior to the hypothesized post-World War II initial appearance of the soybean field agroecosystem habitat into which they suggest the ornamental gourds escaped. The presence of free-living *Cucurbita* gourds in eastern North America prior to the late 1940s would clearly undermine the "new soybean field habitat" hypothesis.

A related and intriguing question involves the possible existence of long-established relict populations of *Cucurbita* gourds outside of Texas. Does the ongoing range extension of a wild gourd in the East have its origins only in Texas populations, or are there other source areas where relict populations of a wild eastern *Cucurbita* gourd could have existed, hiding in plain sight, for a long period of time?

Although the Asches (1992) devote considerable attention to establishing the absolute pre-World War II absence of noncultivated *pepo* gourds in eastern North America, a case can still be made that indigenous wild gourds were present in some areas of the region much earlier, and went unrecorded because they were assumed to be escapes from cultivation.

Excavation at the Hontoon Island site on the St. Johns River in Florida, for example, has documented the long-term presence of a wild morphotype gourd in northeast Florida from A.D. 800 to A.D. 1750 (Decker and Newsom 1988).

Collecting efforts by several prominent botanists in the St. Louis area in the nineteenth century also provide strong evidence for the presence of a wild morphotype free-living gourd in eastern North America long before soybeans became a major crop in the region. In 1846, George Engelmann, an authority on cacti and "ranked among the foremost of botanical workers" (Spaulding 1909a:125), grew several plants in St. Louis from "Texas seeds provided by Nicholas Riehl," depositing one of them, labeled *Cucurbita ovifera* var. *pyriformis,* in the Missouri Botanical Garden. This specimen has recently been annotated as *C. pepo* ssp. *texana* by Thomas Andres. At the same time that Engelmann was growing imported Texas gourd seeds, he was also collecting local free-living *Cucurbita* plants. Two 1846 specimens labeled "*Cucurbita*" were deposited by Engelmann in the Missouri Botanical Garden, each carrying the notation "Naturalized, St. Louis, along fences and fields." Just across the river in the American Bottom region of the Mississippi alluvial valley (St. Clair County, Illinois), W. Welsch collected a specimen in the period 1862–1871 which he labeled *C. pepo.* Five additional nineteenth-century records of free-living *Cucurbita* in St. Clair County are provided by Henry Eggert, who arrived in the St. Louis area in 1875. Eggert collected specimens of free-living *Cucurbita* in 1875, 1876, 1892, and 1893 (two). Another plant, labeled *C. ovifera,* and subsequently annotated as *C. pepo* ssp. *texana* by Thomas Andres, was collected by J. Kellogg near Fish Lake in the American Bottom in 1901 (Table 4.1).

Eggert labeled three of the five plants he collected in 1875, 1876, and 1893 as *Cucurbita ovifera* var. *pyriformis,* the same term applied by Engelmann to the plant he grew in 1846 from "Texas seeds." Eggert's 1876 specimen, currently in the University of Texas Herbarium, was subsequently annotated as *Cucurbita texana* by L. H. Bailey. Eggert's 1893 specimen of *C. ovifera* var. *pyriformis* is of particular interest. Containing the notation "Prairies and waste places," the herbarium sheet includes two pressed fruits, one in an early stage of development, the other almost mature. The almost mature fruit (Figure 4.8) is small (7.3 cm in length), thin-walled, green striped, and pyriform in shape, with a small peduncle scar (diame-

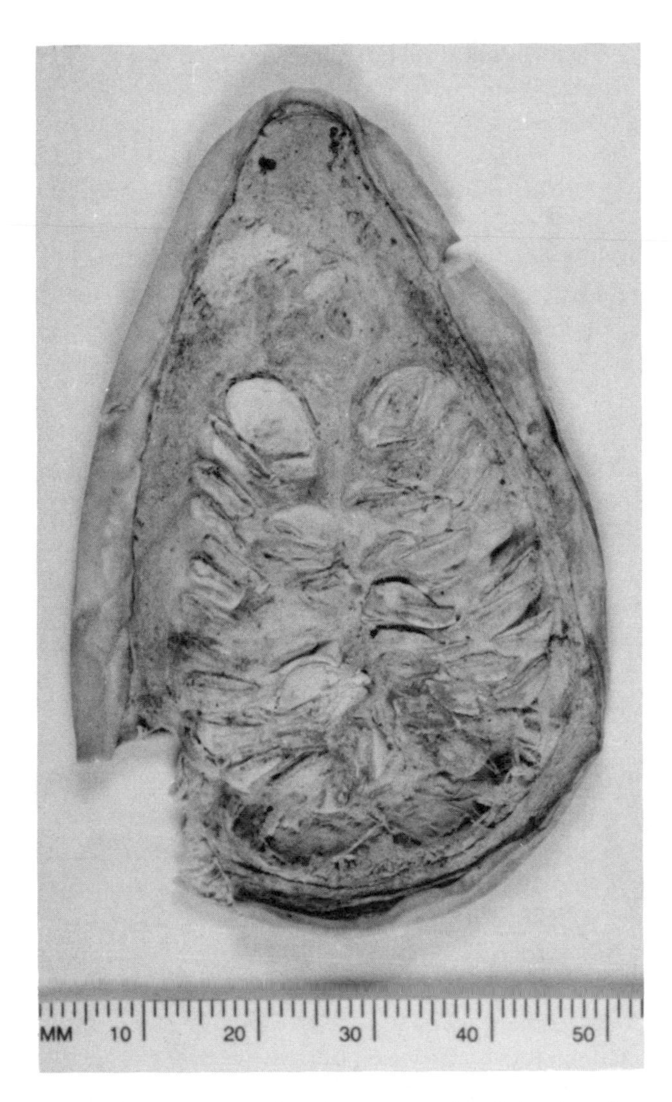

Figure 4.8 Cross section of an almost mature free-living *Cucurbita* gourd collected by H. Eggert in August of 1893.

ter 6 mm). One observable seed measures 8.9 mm by 6.2 mm. In terms of fruit and seed size, this specimen falls comfortably within the wild morphotype category. Eggert labeled the other two free-living *Cucurbita* specimens he collected as *Cucurbita ovalis* (1893) and *Cucumis* (1892), subsequently annotated as *C. pepo* ssp. *texana?* by Thomas Andres.

Like Engelmann, Eggert was a highly respected botanist. He was a vice president of the International Association of Botanists, and a charter member of the Engelmann Botanical Club. He "collected assiduously all around St. Louis for a considerable distance, and his collection probably represented the flora of this district better and more completely than any other ever made" (Spaulding 1909b:254). At his death Eggert's herbarium was estimated to contain about 60,000 specimens. In 1891 he published a 16-page "catalogue of plants growing in a radius of about 40 miles around St. Louis" (Eggert 1891). Included as the only *Cucurbita* entry among the 1,100 different species and varieties listed by Eggert in his catalogue of the flora of the St. Louis area was *Cucurbita ovifera* var. *pyriformis*.

The specimens collected by Henry Eggert, along with those obtained by Engelmann, Welsch, and Kellogg, strongly support the presence of a well-established, widely distributed free-living gourd in the St. Louis area during the period 1846–1901. While it is possible that some of the plants grown from "Texas seed" by Engelmann in 1846 escaped and proliferated outside of cultivation in the St. Louis area, to subsequently be collected by Eggert and others, it would seem much more likely that the plant was indigenous to local floodplain habitats. The inclusion of *C. ovifera* var. *pyriformis* in Eggert's 1891 catalogue indicates that the most active botanist of the time considered it to be an integral part of the native flora. These early records from the St. Louis area thus likely reflect the presence of an indigenous wild gourd close to the northern extent of the known present-day range of free-living gourds within eleven years of when *C. texana* was first collected in Texas.

When combined with the Hontoon Island material and other pre- and post-4,000 B.P. archaeobotanical specimens (e.g., Gilmore 1931, plate 24a), these Missouri plants provide strong evidence that wild morphotype *Cucurbita* gourds were not absent from the East prior to the expansion of soybean field habitats after the Second World War.

While herbarium sheets and archaeobotanical assemblages point toward both the existence of long-established populations of wild gourds and the ongoing range expansion of these plants rather than the dispersal of an escaped ornamental turned specialized soybean field weed, a far more telling test of these two alternative explanations takes place as modern populations of free-living gourds are studied first hand.

The Niche and Habitat of Free-Living Cucurbita *Gourds in Eastern North America*

Although *C. texana* has been of interest to botanists for more than a century and a half, and has been the subject of considerable genetic analysis since 1985, few detailed descriptions of its niche or habitat are available. Most of the extant habitat information regarding this plant and other free-living *Cucurbita* gourds in the eastern United States consists of brief field observations, often in the form of locational descriptions written on herbarium sheets. The vast majority of these brief notations provide support for, and are clarified by, perhaps the single most important characteristic of these wild eastern gourds: their distribution is almost exclusively limited to river and stream valley areas that are within the reach of spring floodwaters, because floodwaters move the buoyant gourds produced by plants the previous fall downstream to new locations. Perhaps the role of floodwaters as a primary means of seed dispersal for these gourds, and the attendant implications for their niche and habitat range, were so obvious as to not require comment, for we can find no mention of this fact in the literature.

All aspects or phases of this floating gourd dispersal process were observed at different locations along a number of stream and river courses in the Missouri and Arkansas Ozarks during November 1990. *Cucurbita* gourds (hard-walled fruits) were observed still attached to vines growing on the ground and suspended above ground, and where they had fallen after separation from vines, resting well within the reach of spring floodwaters. In at least one location, last year's gourd crop was also observed in transit, floating within an abandoned navigational lock chamber at Batesville, Arkansas. Dry, brown, and brittle gourds were also observed in a wide variety of locations where they had been carried by floodwaters. They were often caught against bushes or other floodwater "filters," occasionally trapped inside hollow stumps, sometimes simply lying on the ground where they were left by receding floodwaters, and frequently found in piles of wood and other debris, both small and large, deposited by powerful spring floods.

This last depositional context, which often places the gourds in juxtaposition with a wide variety of objects discarded by humans and subsequently picked up and spatially consolidated by floodwaters, is probably the source of the occasional observations that this plant can be found growing in dumps and waste places, and the conclusion that it therefore represents escapes from cultivation. The river or stream valley location and taphonomic context of such deposits serves to identify them as an inherent component of floodplain environments, rather than having directly resulted from human disposal activities. A final typical location for such flood deposited gourds to flourish is along roadside ditches and bridge approaches adjacent to river and stream valleys.

Bringing the process of seed dispersal full circle, vines of plants in a number of locations were found to originate in the fragmented remains of a dry brown gourd deposited by floodwaters of the previous spring.

With the exception of a few specimens that exhibited clear indications of introgression with domesticated plants, and contained only a few large seeds, the more than 400 gourds collected during the present study each contained a large number of small seeds (usually from 100 to 200) that were almost always extremely bitter tasting. This production of a large number of bitter tasting seeds per fruit is contrary to the Asches' characterization of this plant as lacking these attributes of a successful pioneering species (Asch and Asch 1991:53–54), and is consistent with a successful adaptation to natural floodplain environments.

Herbarium Sheet and Published Habitat Descriptions
Given this brief discussion of the primary mechanism of seed dispersal of these gourds, and their resultant restriction to river and stream bottom areas reached by floodwaters, consider the following herbarium sheet and published habitat descriptions, which constitute almost all of the previously available habitat information regarding this plant: "lowland" (St. Clair County, Illinois, 1875); "waste places in East Carondolet" (St. Clair County, Illinois, 1892); "near Fish Lake, Illinois, escaped to the woods" (St. Clair County, Illinois, 1901); "right of way of the Manufac-

turers Railway, in the jungle-like bushes at the Mississippi River Bank" (St. Louis County, Missouri); "White River in broad valley bottom—swaley depression between river and valley bottom" (Barry County, Missouri); "wooded floor of valley. . .along Roubidoux Creek" (Texas County, Missouri); "edge of rubbish heap" (Douglas County, Illinois); "along dry wash of. . .Beaver Creek. . .valley in shade" (Douglas County, Missouri); "15–20 plants in a one-half-mile stretch of roadside ditch near Zion Creek" (Greene County, Alabama); "growing along a fence row in low bottomland" (Marion County, Arkansas); "common along the White River bottoms" (Prairie County, Arkansas); "on gravel bar near Woolum" (Searcy County, Arkansas); "river bottoms near Fulton" (Hempstead County, Arkansas); "common along river bottoms (Arkansas River) near Conway" (Faulkner County); "Also seen along the Red River bottoms;" and "growing on piles of debris and driftwood along bank of Guadalupe River" (DeWitt County, Texas);

> It was located. . .on the flood plain of the Colorado River, where it grows luxuriantly in the thickets. (Erwin 1938:253)

> Occurs in rich woods of valleys and moist low ground along streams, gravel bars, and low woodland, sometimes along railroads. (Steyermark 1963:1426)

> Cultivated species found usually in woods in valleys and gravel bars near streams, rarely along railroads. (Steyermark 1963:1428)

> In debris and piles of driftwood, often climbing into trees, along several rivers. (Correll and Johnston 1979:1510)

> The fruits are common in flood drift along the Illinois River, and it is a very abundant weed of some flood-plain fields of the Illinois Valley. (Asch and Asch 1985:157)

Field research carried out over the past several years in the western Ozarks by Michael P. Hoffman and colleagues, and in the eastern Ozarks in the fall of 1990 by Bruce Smith and C. Wesley Cowan, provides considerable additional information regarding the niche and habitat of this plant, particularly in smaller streams and in river settings little influenced by human development or agroeconomies.

The Western Ozarks

In Benton County, Arkansas, directly adjacent to the southwest Missouri counties surveyed by Julian Steyermark in the 1950s, gourd populations have been studied at a number of locations along Osage, Little Sugar, and Spavinaw creeks. All three are spring-fed clear-flowing streams that are part of drainage systems that flow westward out of the Ozarks, eventually joining tributaries of the Arkansas River. Gourd plants were found growing in sand and gravel bars and banks adjacent to streams, often in open sunny locations with abundant subsurface water. Downstream along the Illinois River in Cherokee County, Oklahoma, *Cucurbita* gourds are reported as locally abundant, and have been collected from stream-side gravel bar settings.

Similarly, the plant recently collected in Powell County, Kentucky was "found growing on a sunny open sand bar" along the upper reaches of the Red River, with "no garden plots or farms for approximately 18 miles upstream" (C. Ison, personal communication, 1990).

Bryant Creek and the Gasconade River

Along Bryant Creek, near the headwaters of the North Fork of the White River drainage in Ozark County, Missouri, a single plant was found growing in a sand and gravel bar setting. A 4-m-long vine with three laterals of .6, 1.5, and 3 m was growing in full sun against a backdrop of willow and sycamore, about 4 m from the water's edge. Extending both along the ground and up into a ragweed plant, the vine had produced eight green-striped pyriform fruits.

Several counties north of Bryant Creek, along a 3.5 km portion of the upper reaches of the north flowing Gasconade River in Wright County, Missouri, gourd plants were found in three slightly different habitat setting variations on the open, full sun stream-edge gravel and sand bar contexts described above. In the Wilber Allen Wildlife area, a single plant having diverging vine lengths of 3, 6, and 6 m was rooted along an inside bend of the river edge in silt-loam soil. Growing in heavy underbrush, two of the vines extended on and through vegetation toward the river (and sunlight), while the third grew 3 m up into a

Figure 4.9 Ovoid fruits of a free-living *pepo* gourd, Wilber Allen Wildlife Area, Gasconade River, Wright County, Missouri.

Figure 4.10 Ovoid fruits of a free-living *pepo* gourd growing up into a sycamore tree, Gasconade River, Wright County, Missouri.

honey locust tree. The plant had produced 29 ovoid ivory-colored fruits (Figure 4.9).

One hundred meters upstream, two additional plants were discovered growing on opposite sides of a small low elevation gravel and sand island that supported a stand of 6-m-high sycamore trees. On the active channel side of the island a plant rooted 3 m from the river's edge and extending 5 m up into a sycamore had produced 18 ovoid ivory fruits, all of which were still attached to the vine (Figure 4.10).

Extending along a course gravel surface 4 m from the still-water cut-off channel on the opposite side of the island, a second 4.5-m-long vine with laterals of .9, .6, .4, .3, and .2 m was located in light shade, and held 10 green striped pyriform fruits (Figure 4.11).

Two additional plants were discovered 3.5 km upstream along the Gasconade River, also growing in a back channel chute situation. The chute was dry, however, more than 50 m from the river, and clogged with flood-deposited debris piles (which yielded 10 gourds deposited during the previous spring floods). Growing out of a hollow stump that had trapped the parent gourd, a 4-m-long vine with a single 1 m lateral had produced 11 small spherical ivory-colored fruits. Growing in full sun about 30 m away, a 4-m-long vine extending up onto a log had produced three large fruit, providing the first of only two obvious indications we observed of introgression with a domesticated *C. pepo* plant (Figure 4.12). The nearest likely area of garden cultivation was the town of Manes, approximately 2,500 m away.

Figure 4.11 Green-striped pyriform fruits of a free-living *pepo* gourd growing in gravel along a cut-off channel of the Gasconade River, Wright County, Missouri.

The Buffalo River

There are few areas in eastern North America that compare with the Buffalo River in terms of providing an opportunity to assess the ability of the free-living *pepo* gourd of eastern North America to exist in natural floodplain settings far removed from agricultural

Figure 4.12 *Cucurbita* gourd showing clear evidence of introgression with cultivar *C. pepo*, Gasconade River, Missouri.

habitats. The Buffalo River Valley, and the Ozarks in general, also represent an excellent potential heartland and refuge area for wild *Cucurbita* gourds outside of Texas, where subsequent to Native American occupation of the area, a wild indigenous gourd could have grown in relative obscurity until Steyermark collected it in the 1950s.

The Buffalo River witnessed a long history of occupation by Native American populations (Wolfman 1979), and archaeobotanical assemblages from bluff shelters (Fritz 1986) and river valley contexts (Fritz 1990) document their collection of wild plants and cultivation of domesticated crops, including *C. pepo*.

Extending 70 miles (148 river miles) across Newton, Searcy, and Marion counties (Figure 4.13), the Buffalo was designated a National River in 1972, and for the past 20 years the entire main valley corridor (96,000 acres) has been kept free in large measure of human habitation, much less any agricultural fields. The Park Service has allowed continued ownership of land for pasture, and some valley habitation at the community of Boxley on the Buffalo above Ponca (Figure 4.13). While there are small family farms along the tributaries of the Buffalo River today, the river's watershed of 1,388 square miles (about the size of Rhode Island) is heavily forested, has the lowest per capita income in the state, and remains one of the least populated regions in Arkansas, with Newton County averaging 9 persons per square mile (Pitcaithley 1987:89).

Even before the Buffalo River was restored to a natural state two decades ago, the 150-year span of Euroamerican occupation in the area did not involve sizable population or substantial agricultural pursuits, and the floodplain landscape of the valley has changed little since 1900:

> The Buffalo River is not one of Arkansas's major waterways; it has only occasionally been used for transportation purposes and has never been considered suitable for industrial use. It is regionally and nationally unique because it has never been dammed, and its banks have not been extensively developed. (Pitcaithley 1987:1)

As Pitcaithley points out in his detailed short history of the river (1987), the valley was neither a route to anywhere else, nor was it easily accessible or overly attractive to potential settlers. While the narrow valleys of the region possessed rich alluvial soil suitable for farming, cultivation was hampered by the frequency of flash floods after periods of heavy rainfall (Pitcaithley 1987:4, 62). The first pioneering homesteaders didn't reach the mouth of the Buffalo until about 1830. By 1840 small farms were scattered the length of the valley, but by 1860, even though the area had experienced the population boom typical of frontier areas, it was still sparsely settled, with a population density of 6.6 individuals per square mile and farms averaging 28 improved acres (Pitcaithley 1987:26–27).

Following widespread devastation during the Civil War years at the hands of loosely organized bands of brigands, the area prospered during the late 1800s as population grew and the average improved acreage per farm increased to 34 acres by 1890: "Buffalo River farmers grew wheat, corn, cotton, potatoes, oats,

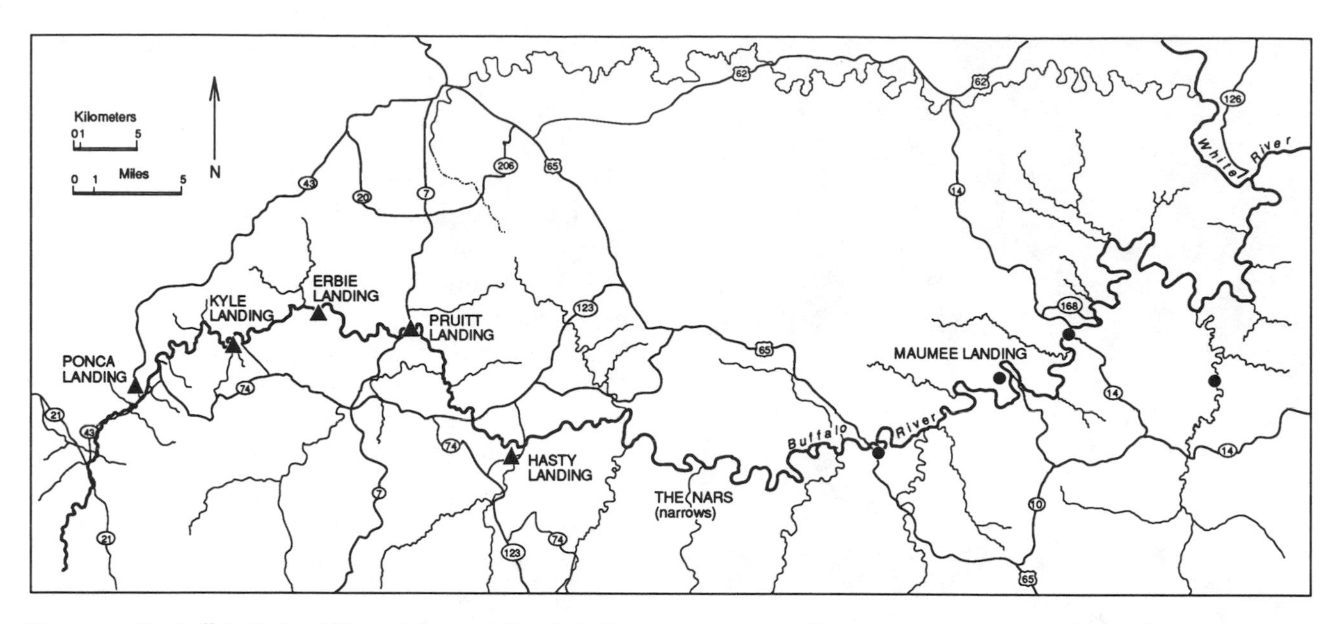

Figure 4.13 The Buffalo National River, Arkansas. Triangles indicate areas where free-living *pepo* gourds were not located during November 1990 survey. Circles indicate locations of *pepo* gourd populations described in the text.

and even some tobacco and rice" (Pitcaithley 1987:55). Cotton was a main crop in the 1880s, but in the following decade the area returned to diversified farming and "by the turn of the century the mountain farmer along the Buffalo raised wheat for bread and corn for feed" (Pitcaithley 1987:56).

The Buffalo River region witnessed a peak of prosperity through the turn of the century up to about 1920, due to a half century (1870–1920) of lumbering for downstream markets (primarily walnut and cedar), and the World War I era boom in lead and zinc mining, which collapsed by 1920. With the loss of the mining and timbering industries, population declined steadily over the next half century, with a total decrease from 1920 to 1970 of 56 percent (Pitcaithley 1987, table 3).

Through this boom and bust cycle based on the extraction of natural resources from the region, farming remained, right up until 1970, small scale and low intensity, consisting essentially of scattered single family farms.

In summary, the well-documented history of Euroamerican occupation of the Buffalo River Valley provides a picture of sparse population, limited agricultural development, and with the exception of the substantial impact of logging in the uplands, little modification of the forested landscape. The area did not witness a post-World War II boom in soybean cultivation, and throughout its history would have provided few opportunities for survival of a narrow agricultural niche *pepo* weed. Because of its relative isolation and limited Euroamerican occupation, the Buffalo River Valley does, on the other hand, provide a good opportunity to establish the presence (perhaps long-term) of populations of an indigenous wild gourd far removed, both temporally and spatially, from modern agricultural settings.

The Buffalo River Valley also provides an opportunity to consider the general habitat requirements of the free-living *Cucurbita* gourd. Pruitt marks a transition point for the river as it widens out from a narrow rocky valley floor with swiftly flowing spring floods into a broader floodplain having reduced floodstage water velocity and a more meandering main channel with associated sand and gravel bank, bar, and island formations.

As one might expect, given the forgoing habitat descriptions for free-living gourds, extensive survey in the narrow rocky floodplain above Pruitt yielded no evidence of the plant, while populations were common on the sand and gravel bars and islands in the areas surveyed below the Nars (Figure 4.13).

Along the inside bank of a large horseshoe bend in the river at South Maumee Landing, a rain-shortened search yielded a single 1.5-m-long vine growing about 40 m from the river's edge in sandy loam soil. Extending up through herbaceous weeds and a small willow, the plant had produced 21 ovoid ivory-colored fruits.

Twenty-five kilometers upstream, just east of the Highway 65 bridge, a large plant was found in a habitat setting of numerous braided gravel swales cut during flood stage. These were interspersed with gravel and sand hummocks that paralleled the river and supported small stands of sycamore, birch, and willow. Flood debris was frequently tangled in the branches of these small trees. Growing on the edge of one of these interfluve hummocks in full sun, the plant had five major vines and numerous laterals (total vine length >30 m) extending in rich confusion along the ground and up onto low plants (ragweed, *Amaranthus, Solidago, Datura*) and into the low limbs of adjacent trees, covering an area of perhaps 150 sq m. The vine had produced 25 elongate, sausage-shaped ivory-colored fruit.

Forty-one river kilometers downstream from this location (16 km from Maumee South), several hundred meters east of the Highway 14 bridge, a similar habitat setting was encountered on the north side of the river valley. Separated from the steep limestone bluffs to the north by a cane break, and the seasonally low river channel to the south by a 40-m-strand of gravel, a set of three 5–7-m-wide and 50 to 70-m-long flood interfluve sand ridge features paralleled the river and extended 1.5 to 2.0 m above intervening flood chute dry gravel swales. Representing a sequence of former shore lines, these ridges and their intervening "fosses" gouged out by flood action, are typical of gravel bars in the Ozarks (Deitz 1952). Each of these sandy flood interfluve ridges had small flood debris piles, a light scatter of young sycamore, willow, maple, and birch, and a sometimes dense understory of *Xanthium*, smartweed, and other species. While no free-living *Cucurbita* gourd plants were found on the two northernmost ridges, a total of 11 plants producing 105 fruits were growing along a 40-m-stretch of the ridge closest to the river (Figure 4.14):

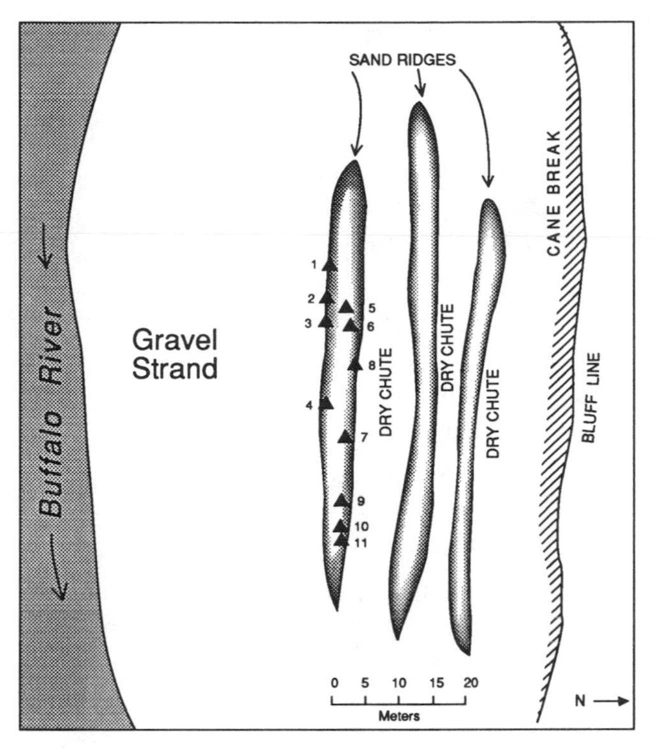

Figure 4.14 Schematic map of the distribution of wild morphotype *pepo* gourd plants along an interfluve sand ridge just south of the Highway 14 bridge, Buffalo River, Newton County, Arkansas.

1. A plant with two main vines, 6 m and 4 m in length, growing up onto *Polygonum* and *Xanthium*, producing five small ivory-colored egg-shaped fruits;

2. A plant having four main vines and numerous laterals, rooted in a willow thicket but growing downslope toward the river in full sun, producing 16 large ivory-colored egg-shaped fruits;

3. Rooted at the base of a young maple and growing 3 m up into it, a vine producing nine spherical ivory-colored fruits;

4. A plant producing 11 small to medium ivory-colored egg-shaped fruits, extending along the ground and up onto several ragweed plants;

5. A vine growing in shade up onto *Xanthium* and producing six small egg-shaped ivory-colored fruits;
6. Rooted in sandy soil, a 4 m vine growing up into *Xanthium* and producing two green-striped pyriform fruits;
7. A heavily shaded plant growing through bushes and onto a willow, producing 27 large egg-shaped ivory-colored fruits;
8. Growing in a small open area in partial sun up onto *Xanthium,* a vine producing 13 small egg-shaped ivory-colored fruits;
9. Rooted at the base of and extending 2 m up into the broken branches of a maple, a vine producing seven fruits;
10. A 2-m-long vine growing up into a dead maple, producing four pyriform ivory-colored fruits;
11. A 3-m-long vine growing onto bushes in partial shade and producing five egg-shaped ivory-colored fruits.

With a total of 11 plants producing 105 fruits growing within 40 m of each other, this location provided the best example described to date of a concentrated "population" in a natural habitat setting that would have provided ample opportunity for multiple interplant pollen transfer by bee vectors. The more commonly observed pattern of a greater spatial separation and isolation of plants along stream valley corridors would perhaps indicate a typically lower frequency of pollen transfer between plants.

This naturally occurring population, generating over 100 fruits and 10,000 to 20,000 seeds in a 200 sq m area, contradicts the Asches' characterization of this plant as neither occurring in abundance in natural disturbance settings nor producing the large number of seeds per unit area typical of weedy pioneer plants. To the contrary, this concentrated population and its high level of fruit and seed production indicate a plant well adapted to survive in natural floodplain environments.

Although we did not survey the final 50 river km of the Buffalo River where it passes through the Lower Buffalo Wilderness Area before joining the White River, a local resident recounted seeing gourds along Big Creek (Figure 4.13).

The White River

Moving downstream from the isolated and pristine river valley landscape of the Buffalo, with its steep limestone bluffs and streamside gravel and sand bar gourd habitats, the larger floodplain of the White River provided the opportunity to observe the habitats of *Cucurbita* gourds in closer proximity to agroeconomies and sandy alluvial terrace settings. Populations of free-living gourds were encountered at each of the three locations that were investigated along the valley of the White River.

Downstream about 150 to 200 m from the Highway 9 bridge near Sylamore (Izard County), Arkansas, six plants were located growing in partial shade to full sun along the edge of the sandy soil first terrace of the White River. Although 2 m above the level of the gravel river bed, the first terrace was within reach of spring floodwaters, as evidenced by flood debris suspended from trees and in drift piles.

A single plant bearing only two small spherical ivory-colored fruits was found growing on the ground 3 m in from the terrace edge in partial shade. A second isolated plant with a 4 m vine bearing eight small egg-shaped ivory-colored fruits was located about 70 m away growing up into *Solidago* along the edge of the terrace.

About midway between these two plants, a concentration of four plants was located in heavy undergrowth of *Datura, Chenopodium berlandieri,* and *Xanthium.* One of the plants showed evidence of introgression with domesticated *C. pepo,* bearing three large yellow to bright orange, green striped, melon-shaped fruits, while an intertwined vine of another plant produced nine ivory-colored pyriform fruits.

Ten meters away a second pair of plants with vines totaling 20 and 27 m in length had radiated over an 8 by 10 m area, growing through and onto *Ipoemea, Xanthium, Datura,* a 2.5-m-high *Chenopodium berlandieri* plant, and extending 3 m up into an adjacent mulberry tree and several small maples. These plants produced 24 flat spherical ivory-colored fruits and 53 medium to large egg-shaped ivory-colored fruits, respectively. Providing another clear example of abundant fruit and seed production in a naturally disturbed habitat setting, these two plants produced 77 fruit and approximately 8,000–15,000 seeds within an 80 sq m area.

Sixty kilometers downstream at Batesville, Arkansas, 200 m south of the spillway dam and lock, a similar level of fruit production was shown by a *Cucurbita* gourd plant found growing in first terrace fine silt soil 100 m from the river's edge. Rooted in an area where a number of intact and broken gourds from the previous spring's floodwaters were partially imbedded in the ground, the four main vines of the still green and flowering plant extended 4, 5, 5, and 7 m along the ground and up into ragweed. A total of 37 immature and mature pyriform ivory-colored fruits were counted.

One hundred meters upstream a second plant bearing 10 flat spherical ivory-colored fruits was found growing up into a bush on the edge of the first terrace.

Another 80 km downstream, where the Highway 64 bridge crosses the White River west of Augusta, Arkansas, six green and still flowering *Cucurbita* gourd plants were found growing at the flood line along the grassy embankment slope of the bridge approach. Two of the plants had produced mature fruit (six green striped spherical and two ivory-colored pyriform gourds, respectively), while the other four all exhibited green immature fruit. One of the plants was rooted in a fragmented brown gourd.

While these six plants growing along a grassy embankment periodically mowed by the highway department provide an example of one type of human modified environmental setting into which eastern *Cucurbita* gourds typically encroach, it is also well documented as an agricultural weed in floodplain fields along the Arkansas and Red River, and along the lower Mississippi River in Louisiana.

Cucurbita Gourds as Agricultural Weeds

The extent to which *Cucurbita* gourds have become a problem in low-lying bottomland fields in Arkansas and Louisiana is detailed by the publications describing efforts to control it (Oliver et al. 1983; Boyette et al. 1984). A recent study of herbicide effectiveness focused on "dense uniform infestations of Texas gourd" in soybean fields in three Arkansas locations, with density counts ranging from 32 plants per square meter at Fulton, Arkansas through 43 per sq m at Conway, Arkansas, to 129 plants per sq m at Garland, Arkansas (Oliver et al. 1983). In late November, 1990, near St. Joseph in Tenasas Parish, Louisiana, Gayle Fritz observed tens of thousands of free-living *pepo* gourds littering the ground in an "infested" bottomland cotton and soybean field located between the levee and Mississippi River (G. Fritz, personal communication, 1991). On a much smaller scale, Louis Baker collected several dozen *pepo* gourds of a variety of shapes in November of 1990 from fallow floodplain fields along Willow Chute, an old channel of the Red River just north of Shreveport (Bossier Parish) in northwest Louisiana (Frank Schaumbach, personal communication, 1990).

The Niche and Habitat of Free-Living *Cucurbita* Gourds

In summary, based on limited published references, information extracted from herbarium sheets, and field studies of the past two years, an initial outline of the most obvious aspects of the niche and habitat of free-living *Cucurbita* gourds in eastern North America can be offered.

Contrary to the Asches' characterization of this plant as lacking an adaptation to extreme disturbed habitats and not growing in places where few other plants thrive, free-living *Cucurbita* gourds of eastern North America exhibit a highly successful specific adaptation to stream and river valley floodplain habitats. Riding often violent floodwaters through annually reworked landscapes, these gourds are the first to occupy and grow in stream-side gravel and sand bar, bank, and terrace settings where few other plant species can match their colonizing abilities. The buoyant gourds and their cargo of seeds are often trapped by weeds, bushes, and other floodwater "filters," or are deposited, along with other floodborne materials, in debris piles.

Whether the context of deposition is stream-side sand and gravel bars and hummocks, as documented along the smaller streams and rivers of the Ozarks (for example, the Buffalo, White, Illinois, and Gasconade rivers and Bryant, Osage, Little Sugar, and

Spavinaw creeks), or the higher elevation sandy terraces and levee ridges of larger river valleys (the Arkansas, Red, Alabama, and Mississippi rivers), these gourd populations effectively colonize open edge areas within the constantly reworked floodplain landscape. These sunny, unoccupied, or edge area gravel and sand bars and hummocks, along with river and terrace edge banks, constitute the extremely disturbed habitat setting within which *Cucurbita* gourds are strong and successful competitors.

With rapidly growing vines that can extend for more than 30 m along the ground and climb 3 to 4 m up into trees and other vegetation, these gourds also exhibit impressive means of reaching sunlight and displaying their blossoms to pollen vector bees.

Although sometimes occurring in concentrated populations, as documented above, individual plants can also often be dispersed widely along floodplain corridors. With their capacity for self-pollination, however (with both staminate and pistillate flowers produced on the same plant), these free-living *Cucurbita* gourds are not disadvantaged by this potential for wide floodwater dispersal of individual plants. The Asches (1992:49, 53) cite Bailey (1929:69) in arguing for the self-sterility of individual gourd plants, concluding that as a result the plant is poorly adapted to exist in such dispersed patterns in natural floodplain habitats. Such self-sterility, however, goes unmentioned in more recent treatments of *Cucurbita*. While self-sterility may exist in some highly inbred cultivars, self-compatibility has been the usual observation by Decker-Walters in her extensive experience with several species of *Cucurbita* (personal communication, 1991), and self-pollination of a var. *texana* plant was successful in the recent research of Kirkpatrick, Decker-Walters, and Wilson (Kirkpatrick et al. 1985:290). In addition, in light of the recent documentation of bees serving as pollen vectors over distances of at least 1,300 m (Kirkpatrick and Wilson 1988), it is a mistake to characterize individual plants as necessarily reproductively isolated from other members of dispersed floodplain populations of free-living *Cucurbita* gourds.

These obvious adaptational aspects of free-living eastern *Cucurbita* gourds: buoyant hard-walled fruit functioning for efficient seed dispersal by floodwaters, aggressive climbing and growth characteristics of vines, prolific fruit and seed production, and self-compatibility, all suggest long-term evolution within, and adaptation to, river floodplain environments.

These gourd population's adaptations to naturally disturbed floodplain landscapes also prove advantageous for encroaching on and colonizing human maintained floodplain habitats such as bridge approaches, drainage ditches, and agricultural fields. Their well-documented, remarkable ability to chronically infest bottomland fields within the reach of floodwaters certainly qualifies them as floodplain agricultural weeds of the first rank. But as indicated by the foregoing habitat descriptions, it would be a mistake to characterize these free-living gourds rather narrowly as agricultural weeds, and certainly in error to view them as somehow tightly associated with the cultivation of soybeans. On the contrary, it is important to begin any characterization of these eastern *Cucurbita* gourds with a recognition of their successful evolutionary adaptational responses to naturally disturbed floodplain environments, and to then view their success in anthropogenic contexts as the simple expansion of the plant into areas where human activities have encroached within the reach of spring floodwaters and in effect expanded, sometimes dramatically, the habitat of the plant.

Conclusions

Is the gourd found along stream and river valleys of the East a wild plant indigenous to the region and a wild ancestor of *C. pepo,* or is it a foreign introduction and an escape from cultivation? In our view, the preponderance of the evidence to date provides strong support for the presence in the region of an indigenous wild *Cucurbita* gourd with a broad geographical range. In addition, genetic and archaeological research since 1980, along with recent field studies in the Ozarks, all point to the independent domestication of *C. pepo* in eastern North America.

By acknowledging the long developmental separation between the Mexican and eastern North American branches of the species, proponents of a single Mesoamerican origin for *C. pepo,* like the Asches, are forced to transport a Mesoamerican domesticated *pepo* gourd into eastern North America by 7,000 B.P.

The main arguments offered in support of this early gourd being a Mesoamerican domesticate as opposed to an indigenous wild plant (even though for 4,000 years it looked wild and showed none of the morphological changes that could be expected under cultivation) rest on the inability of present-day gourds to exist in the wild, beyond the confines of eastern North American agroecosystems.

Setting aside the existence of *C. texana* in east Texas river floodplains over the past 150 years, the Asches argue that the populations of free-living *Cucurbita* gourds growing today in the East are recent, substantially constrained escapes from cultivation that can survive only in narrow soybean field weed habitats, which in turn have emerged only since the Second World War. The absence of such row crop weed habitats in prehistory, and the associated absence in modern *pepo* gourds of a set of pioneer plant adaptations essential for survival in the wild, when combined, support the position that the early gourds in the East could not have existed as indigenous wild plants, outside of cultivation.

Based on recent field research on populations of free-living *Cucurbita* gourds in the Ozarks, however, and their documented long-term presence in the St. Louis area over a century ago, these supporting arguments for an early introduction from Mexico, and a single Mesoamerican origin of domesticated *C. pepo*, can be rejected. In more than two dozen separate locations in the Ozarks, free-living *Cucurbita* gourds were observed growing in natural floodplain habitats far removed from soybean fields or agricultural weed habitats. *Pepo* gourds were both common and locally abundant along the Buffalo River Valley, which has been devoid of agriculture for two decades.

Given that these Ozark gourds have existed in natural floodplain habitats for at least 40 years, and perhaps much longer, it is not surprising that they do in fact exhibit the attributes of pioneering plants that they were said to lack. They are in fact well adapted to colonizing specific habitats in constantly changing naturally disturbed river valley floodplain environments, rather than narrowly confined to row crop field settings.

Obviously these gourds can and do infest floodplain fields, and the post-World War II expansion of floodplain row crop farming has no doubt facilitated the ongoing range expansion of this wild gourd. It is in fact possible that in some areas (e.g., the Lower Illinois Valley?) where navigational dams have destroyed its natural habitat, these eastern gourds may exist today only where human activities have created expanded floodplain habitats.

Varying frequency and intensity of introgression with cultivated *C. pepo* crops can also be expected in these free-living gourd populations, depending upon the relative presence, from year to year, of cultivar varieties within bee vector range of different floodplain locations. But clearly this gourd does not depend upon human habitats and an agroeconomy to survive, nor does it have any tightly tethered reliance upon a companion weed role to cultivar *C. pepo*. To the extremely limited extent to which these wild gourds qualify as a companion weed to *C. pepo*, they are an unobligated, untethered, occasional companion, from a distance.

Although it will be difficult to demonstrate an unbroken presence of this wild gourd in the Ozarks from prehistory up to the present day, it is clear that they are clearly well adapted to colonize specific naturally disturbed settings in mid-latitude river floodplains, and can successfully compete in the wild without human assistance. If this wild gourd can exist today as an integral component in eastern floodplain ecosystems, there is no reason it could not have been present 7,000 years ago.

Recent isozymic research (Decker-Walters et al. 1992) provides additional and compelling evidence that the free-living *Cucurbita* gourd we collected in the Ozarks (now classified *C. pepo* ssp. *ovifera* var. *ozarkana*) is both wild and long indigenous to the region, and represents the ancestral progenitor of the eastern North American domesticate lineage (*C. pepo* ssp. *ovifera* var. *ovifera*, including the crooknecks and scallops). While about half of the 20 populations of var. *ozarkana* analyzed (Table 4.1) exhibited signs of limited and in many cases recent introgression from cultivars, they also shared a unique, largely coherent isozyme profile, which usually included the characteristic allele *Idh-2m,* and set them apart from var. *texana*. This distinctive profile is strong evidence that Ozark populations developed their own genetic identity in relative isolation over a long period of time. Var. *ozarkana* also possesses the isozyme patterns expected of the wild progenitor of the eastern domesticate lineage: isozyme alleles in the domesticates are generally a subset of those in the progenitor, and the two taxa are genetically similar, with a genetic identity value of .89 (Decker-Walters et al. 1992).

In conclusion, it is interesting to note the extent to which the niche of wild eastern *Cucurbita* gourds within river valleys of the East overlaps with that of *Chenopodium berlandieri* and *Iva annua,* two indigenous plants brought under domestication in eastern North America (see Chapters 7 and 8 for niche and habitat descriptions for *C. berlandieri* and *Iva annua*). In addition to invading bottomland fields and qualifying as floodplain agricultural weeds, all three plants also occupy distinctive colonizing niches within the constantly reworked landscape of river valley floodplains. The shared habitat of these three species, their similar profiles as successful colonizers of naturally disturbed floodplain landscapes (and thus pre-adapted to domestication—see Chapter 2), along with the shared temporal framework of initial appearance of domesticated forms of each species, all combine to form a strong and straightforward contextual argument for the combined independent domestication of a triad of indigenous seed plants in eastern North America.

Acknowledgments

We would like to thank all of those individuals who provided us with information and gourds that they had collected from various locations across eastern North America, including Louis Baker, Wallace Booth, Don Dickson, Dan Dourson, Gayle Fritz, Cecil Ison, Barbara Meyer, Frank Schaumbach, Craig Sheldon, and the many Benton County, Arkansas residents who in connection with taking the University of Arkansas course "Indians of Arkansas and the South," made collections along local creeks. We would also like to acknowledge those individuals who responded to our letter of inquiry sent out to herbariums, with particular thanks to Dennis Kearns of the University of Texas Herbarium, Michael Canoso of the Harvard University Herbaria, and George Yatskievych of the Missouri Botanical Garden. We also gratefully acknowledge the funding support provided by the Research Opportunity Fund of the National Museum of Natural History.

Literature Cited

Asch, D. L., and N. B. Asch

1985 Prehistoric Plant Cultivation in West-Central Illinois. In *Prehistoric Food Production in North America,* edited by R. I. Ford. Museum of Anthropology, Anthropological Papers No. 75. University of Michigan, Ann Arbor, Michigan.

1992 Archaeobotany. In *Geoarchaeology of the Ambrose Flick Site*, edited by Russell Stafford. Center for American Archaeology, Research Series 10. Kampsville, Illinois.

Bailey, L. H.

1929 The Domesticated *Cucurbitas. Gentes Herb.* 2:62–115.

Boyette, G., E. Templeton, and L. R. Oliver

1984 Texas Gourd *(Cucurbita texana)* Control. *Weed Science* 32:649–655.

Chomko, S. A., and G. Crawford

1978 Plant Husbandry in Prehistoric Eastern North America: New Evidence for its Development. *American Antiquity* 43:405–408.

Correll, D. S., and M. C. Johnston

1979 *Manual of the Vascular Plants of Texas.* University of Texas, Dallas.

Cowan, C. W.

1990 Prehistoric Cucurbits from the Cumberland Plateau of Eastern Kentucky. Paper presented at the 47th annual meeting of the Southeastern Archaeological Conference, November 7–10, Mobile, Alabama.

Cowan, C. W., H. E. Jackson, K. Moore, A. Nickelhoff, and T. Smart

1981 The Cloudsplitter Rockshelter, Menifee County, Kentucky: A Preliminary Report. *Southeastern Archaeological Conference Bulletin* 24:60–75.

Decker, D.

1985 Numerical Analysis of Allozyme Variation in *Cucurbita pepo. Economic Botany* 39:300–309.

1986 A Biosystematic Study of *Cucurbita pepo.* Ph.D. dissertation, Texas A&M University, College Station.

1988 Origin(s), Evolution, and Systematics of *Cucurbita pepo* (Cucurbitaceae). *Economic Botany* 42:4–15.

Decker, D., and L. Newsom
1988 Numerical Analysis of Archaeological *Cucurbita* Seeds from Hontoon Island, Florida. *Journal of Ethnobiology* 8:35–44.

Decker, D., and H. G. Wilson
1986 Numerical Analysis of Seed Morphology in *Cucurbita pepo*. *Systematic Botany* 11:595–607.

1987 Allozyme Variation in the *Cucurbita pepo* Complex: *C. pepo* var. *ovifera* vs. *C. texaxa*. *Systematic Botany* 12:263–273.

Decker-Walters, D.
1990 Evidence for Multiple Domestications of *Cucurbita pepo*. In *Biology and Utilization of the Cucurbitaceae*, edited by D. Bates and C. Jeffrey, pp. 96–101. Cornell University Press, Ithaca, New York.

Decker-Walters, D., T. Walters, W. Cowan, and B. Smith
1992 Isozymic Characterization of Wild Populations of *Cucurbita pepo*. Paper presented at the 15th annual conference of the Society of Ethnobiology, Smithsonian Institution, Washington, D.C.

Deitz, R.
1952 The Evolution of a Gravel Bar. *Annals of the Missouri Botanical Garden* 34:249–254.

Doran, G., D. Dickel, and L. Newsom
1990 A 7,290-Year-Old Bottle Gourd from the Windover Site, Florida. *American Antiquity* 55:354–359.

Eggert, H.
1891 *Catalogue of the Phaenogamus and Vascular Cryptogamous Plants in the Vicinity of St. Louis, Missouri*. Published by author, St. Louis.

Erwin, A. T.
1938 An Interesting Texas Cucurbit. *Iowa State College Journal of Science* 12:253–255.

Flannery, K. (editor)
1986 *Guila Naquitz*. Academic Press, Orlando, Florida.

Ford, Richard I.
1981 Gardening and Farming Before A.D. 1000: Patterns of Prehistoric Cultivation North of Mexico. *Journal of Ethnobiology* 1:6–27.

1985 Patterns of Prehistoric Food Production in North America. In *Prehistoric Food Production in North America*, edited by R. I. Ford. Museum of Anthropology, Anthropological Papers No. 75. University of Michigan, Ann Arbor.

Fritz, G.
1986 *Prehistoric Ozark Agriculture—the University of Arkansas Rockshelter Collections*. Ph.D. dissertation, Department of Anthropology, University of North Carolina at Chapel Hill.

1990 Archaeobotanical Remains from the Dirst Site, Buffalo National River, Arkansas. In *Archaeological Investigations at* 3MR80-Area D in the Rush Development Area, Buffalo National River, Arkansas, edited by G. Sabo, pp. 153–177. U.S. Department of the Interior Southwest Cultural Resources Center Professional Papers 38. Santa Fe, New Mexico.

Gilmore, M.
1931 Vegetal Remains of the Ozark Bluff-Dweller Culture. *Papers of the Michigan Academy of Science, Arts, and Letters* 14:83–102.

Heiser, C. B.
1985a Some Botanical Considerations of the Early Domesticated Plants North of Mexico. In *Prehistoric Food Production in North America*, edited by R. I. Ford, pp. 57–72. Museum of Anthropology, Anthropological Papers No. 75. University of Michigan, Ann Arbor.

1985b *Of Plants and People*. University of Oklahoma Press, Norman.

1986 Domestication of Cucurbitaceae: *Cucurbita* and *Lagenaria*. Paper presented at the World Archaeological Congress, September 1–7, Southampton, England.

1989 Domestication of Cucurbitaceae: *Cucurbita* and *Lagenaria*. In *Foraging and Farming*, edited by D. Harris and G. Hillman, pp. 472–480. Unwin Hyman, London.

1990 New Perspectives on the Origin and Evolution of New World Domesticated Plants: Summary. *Economic Botany* (supplement) 44(3):111–116.

Jones, F.
1975 *Flora of the Texas Coastal Bend*. Mission Press, Corpus Christi, Texas.

Jones, F., and G. Fuller
1955 *Vascular Plants of Illinois*. Illinois State Museum Scientific Papers 6. Springfield.

Kay, Marvin
1986 Phillips Spring: A Synopsis of Sedalia Phase Settlement and Subsistence. In *Foraging, Collecting, and Harvesting: Archaic Period Subsistence and Settlement in the Eastern Woodlands,* edited by S. W. Neusius, pp. 275–288. Center for Archaeological Investigations, Occasional Paper 6. Southern Illinois University, Carbondale.

Kay, M., F. King, and C. Robinson
1980 Cucurbits from Phillips Spring: New Evidence and Interpretations. *American Antiquity* 45:806–822.

Kirkpatrick, K., D. Decker, H. G. Wilson
1985 Allozyme Differentiation in the *Cucurbita pepo* Complex: *C. pepo* var. *medullosa* vs. *C. texana. Economic Botany* 3:289–299.

Kirkpatrick, K., and H. G. Wilson
1988 Interspecific Gene Flow in *Cucurbita: C. texana* vs. *C. pepo. American Journal of Botany* 75:519–525.

Mahler, W.
1988 *Shinner's Manual of the North Central Texas Flora.* Botanical Research Institute of Texas, Fort Worth.

Mohlenbrock, R., and D. Ladd
1977 *Distribution of Illinois Vascular Plants.* Southern Illinois University Press, Carbondale.

Nee, Michael
1990 The Domestication of *Cucurbita* (Cucurbitaceae). *Economic Botany* 44 supplement (3):56–68.

Oliver, L., S. Harrison, and M. McClelland
1983 Germination of Texas Gourd *(Cucurbita texana)* and its Control in Soybeans *(Glycine max). Weed Science* 31:700–706.

Pitcaithley, D.
1987 *Let the River Be: A History of the Ozark's Buffalo River.* Southwest Cultural Resources Center, National Park Service, Santa Fe, New Mexico.

Prentice, G.
1986 Origins of Plant Domestication in the Eastern United States: Promoting the Individual in Archaeological Theory. *Southeastern Archaeology* 5:103–119.

Smith, E. B.
1978 *An Atlas and Annotated Checklist of the Vascular Plants of Arkansas.* University of Arkansas Press, Fayetteville.

1988 *An Atlas and Annotated Checklist of the Vascular Plants of Arkansas,* 2nd edition. University of Arkansas Press, Fayetteville.

Spaulding, P.
1909a A Biographical History of Botany at St. Louis, Missouri III. *Popular Science Monthly* 74 (February).

1909b A Biographical History of Botany at St. Louis, Missouri IV. *Popular Science Monthly* 74 (March).

Steyermark, J.
1963 *Flora of Missouri.* Iowa State University Press, Ames, Iowa.

Whitaker, T., and H. Cutler
1986 Cucurbits from Preceramic Levels at Guila Naquitz. In *Guila Naquitz,* edited by K. Flannery, pp. 275–280. Academic Press, Orlando, Florida.

Wilson, H. G.
1989 Discordant Patterns of Allozyme and Morphological Variation in Mexican *Cucurbita. Systematic Botany* 14:612–623.

1990 Gene Flow in Squash Species. *Bioscience* 40:449–455.

Wolfman, D.
1979 *Archaeological Assessment of the Buffalo National River.* Arkansas Archaeological Survey Research Report 18. Fayetteville.

Yarnell, R.
1983 Prehistory of Plant Foods and Husbandry in North America. Paper presented at the annual meeting of the Society for American Archaeology, Pittsburgh.

Yarnell, R., and M. J. Black
1985 Temporal Trends Indicated by a Survey of Archaic and Woodland Plant Food Remains from Southeastern North America. *Southeastern Archaeology* 4:93–106.

III. Premaize Farming Economies in Eastern North America

CHAPTER 5

❑ ❑ ❑

The Role of *Chenopodium* as a Domesticate in Premaize Garden Systems of the Eastern United States

Introduction

The long and complex developmental trajectory of prehistoric plant husbandry in the mid-latitude Eastern Woodlands of the United States can be usefully subdivided by a sequence of four archaeologically visible transitions spaced roughly at intervals of 400 years:

1. By about 2,000 (± 200) B.P. there was an increase in the importance of four indigenous annual starchy seed-bearing plants: maygrass *(Phalaris caroliniana)*, little barley *(Hordeum pusillum)*, knotweed *(Polygonum erectum)*, and goosefoot *(Chenopodium)* (Asch and Asch 1983; Johannessen 1984; Crites 1984).
2. At about 1,600 B.P. (A.D. 350) another annual starchy seed-bearing crop (maize) was first introduced into the Eastern Woodlands, becoming a minor constituent in garden or field plots and remaining below or just above the level of archaeological visibility (Yarnell 1983). [AMS dating now places this introduction at ca. A.D. 200.]
3. At about 1,200 B.P. (A.D. 750) maize abruptly moved above the level of archaeological visibility in many areas and began to be present in ubiquitous abundance (Smith 1986).
4. Stable carbon isotope studies indicate that maize apparently did not become a major dietary component of prehistoric populations in the Eastern Woodlands until about 800 B.P. (A.D. 1150) (Boutton et al. 1984).

While providing convenient chronological points of reference, these four archaeologically visible transitions offer only limited insight into the gradual evolutionary processes leading from small garden plot cultivation to maize-dominated field agriculture. In

This chapter originally appeared in *Southeastern Archaeology* (1985) 4:51–72.

order eventually to gain an adequate understanding of the relevant variables, both natural and cultural, that were involved in this developmental process it is necessary to first address a number of interrelated and quite basic questions (having quite elusive answers) regarding the changing relationships between human populations and their plant food sources.

Analysis of well-dated archaeobotanical collections from different areas of the Midwest and Southeast can provide reliable lists of the different species of plants that were utilized by human populations through time, and strontium analysis can indicate the relative dietary contribution of plants versus animals (Price et al. 1984).

Establishing the relative dietary importance of various plant species is more difficult, however, as it involves various assumptions concerning the representativeness of archaeobotanical samples. Stable carbon isotope analysis can indicate the relative contribution of maize to the diet of prehistoric populations of the interior Eastern Woodlands (Price et al. 1984), but species-specific measures of dietary intake are otherwise lacking.

Similarly, it is often difficult to confidently place plant species along the continuum of human-plant relationships from wild through weedy and cultivated to domesticated status. Archaeological indications of the amount of time, energy, and land invested in the cultivation of crops by a prehistoric population are also hard to come by, making it difficult to accurately place a prehistoric population along the continuum of plant husbandry from gathering of wild species through small garden plot cultivation to different levels of small-scale horticulture and field agriculture.

Because of the difficulties inherent in attempting to answer even these basic questions concerning the relative importance and level of human manipulation of plant species, as well as the degree of investment of energy and land in cultivation, the nature and timing of the 2,600 to 750 B.P. developmental trajectory of plant husbandry in the Eastern Woodlands is still only poorly known.

One of the most important phases of this developmental process centers on the first of the transitions mentioned above. By ca. 2,000 B.P. populations in a number of different areas of the Midwest and Southeast appear to have substantially increased their dependence upon the starchy seeds of several indigenous annuals (erect knotweed, maygrass, little barley, and goosefoot) (Table 5.1).

The increased archaeological representation of these indigenous starchy-seeded annuals quite likely reflects a greater emphasis upon plant husbandry centering on these quasi-cultigens and cultigens, rather than simply a greater exploitation of wild stands. This apparent increased commitment of time, energy, and land to tending and cultivating plants, while difficult to measure, is nonetheless intriguing, since it would represent both a foreshadowing of and preadaptation to the introduction of maize at 600–550 B.P. (A.D. 350–400).

Rather than representing a processual prime mover from Mesoamerica on which the entire subsequent cultural developmental sequence of the Eastern Woodlands could be conveniently hung, maize may well have initially been incorporated as just another starchy-seeded annual within already well established systems of plant husbandry. It is possible that many of the cultural and technological readjustments associated with an increased role for field crops had already occurred prior to the introduction of maize.

Determining the composition and scale of the ca. 2,000–1,600 B.P. plant husbandry systems of the East into which maize was adopted—accurately placing them and their constituent plant species along the dual continuums of wild to domesticated and gathering to field agriculture—will be accomplished slowly as researchers identify and piece together disparate fragments of relevant information.

This article addresses one such small aspect of the larger issue of premaize plant husbandry systems by documenting the presence of a domesticated variety of *Chenopodium* in the Eastern Woodlands by 2,000 B.P. (Smith 1984).

The Continuum of Human-Plant Relationships

The continuum of human-plant relationships can be usefully considered to consist of a sequence of intergrading categories of increasing dependence by plant

Table 5.1. The Relative Abundance of Starchy-Seeded and Oily-Seeded Annuals in Premaize Archaeobotanical Seed Assemblages from Five Geographical Areas of the Eastern Woodlands

Category	Eastern Tennessee[a]		Central Kentucky[b]		Central Tennessee[c]		American Bottom[d]		West-central Illinois[e]	
	No.	%	No.	%	No.	%	No.	%	No.	%
Starchy-seeded annuals										
Chenopod	1,431	90	18,444	86	2,553	49	201	26	1,183	8.6
Maygrass			1,150	5.3	1,868	36	222	29	5,154	37.5
Knotweed			5		475	9	4	0.5	4,374	31.8
Little Barley									2,047	14.9
TOTAL	1,431	90	19,599	91.3	4,896	94	427	55.5	12,758	92.8
Oily-seeded annuals										
Sunflower	14	0.9	218	1.0	12	0.2			46	0.3
Sumpweed	26	1.6	512	2.4	3	0.1	1	0.1	39	0.3
Squash	*		11	0.5	*					
Bottle Gourd	*		*		*				6	0.1
TOTAL	40	2.5	741	3.9	15	0.3	1	0.1	91	0.6
TOTAL IDENTIFIED SEEDS	1,627		21,471		5,194		761		13,741	

* = Rind fragments present.

[a]Seed count information from Long Branch phase sites (2,300–1,800 B.P.) (Chapman and Shea 1981, table 4).

[b]Seed count information for the Salts Cave site (Vestibule JIV, Levels 4-11, 2,600–2,200 B.P.) (Yarnell 1974, table 16.5; Gardner 1984).

[c]Seed count information from the McFarland and Owl Hollow sites, late McFarland and early Owl Hollow phases (1,900–1,700 B.P.) (Crites 1978b, table 2.2; Kline et. al. 1982, table 6).

[d]Seed count information for Cement Hollow and Hill Lake phases (2,100–1,700 B.P.) (Johannessen 1981, table 29; 1983, table 21).

[e]Seed count information for the Smiling Dan site (2,100–1,700 B.P.) (Asch and Asch 1983, table 17.7).

varieties on human populations (Figure 5.1a), reflecting increasing human intervention in, or manipulation of, a plant's life cycle (Ford 1979:316; see Rindos 1984 for a different approach to this continuum).

Wild Status Plants

One end of this continuum is marked by wild-status plants. Wild plants grow outside the human-made habitat and are unable to compete successfully with weedy plants in settings that are constantly disturbed by humans (de Wet and Harlan 1975:319). While wild plants can constitute an important human food source, even intensive harvesting of natural stands does not have any appreciable genetic impact on the exploited species since it is the seeds that escape the harvester that comprise the next generation (Harlan et al. 1973:313–314). Wild species are frequently aggressive colonizers of unvegetated open areas, and it is the creation of such colonization opportunities by human populations that provides the context for the evolution, from wild colonizers, of weedy varieties of plants adapted to continuous habitat disturbance.

Weedy Plants

In contrast to wild plants, weeds are well adapted to human-maintained habitats and cannot compete successfully with wild plants in the absence of continuing disturbance (Harlan et al. 1973:319). This inability to compete in undisturbed situations constitutes the boundary condition for the wild to nonwild transition. Areas initially opened up to colonization as a result of human activities will be invaded by both wild and weedy plant varieties. In the absence of continued disturbance in subsequent years weedy plants will disappear from plant communities as succession proceeds (Harlan and de Wet 1965). If the area con-

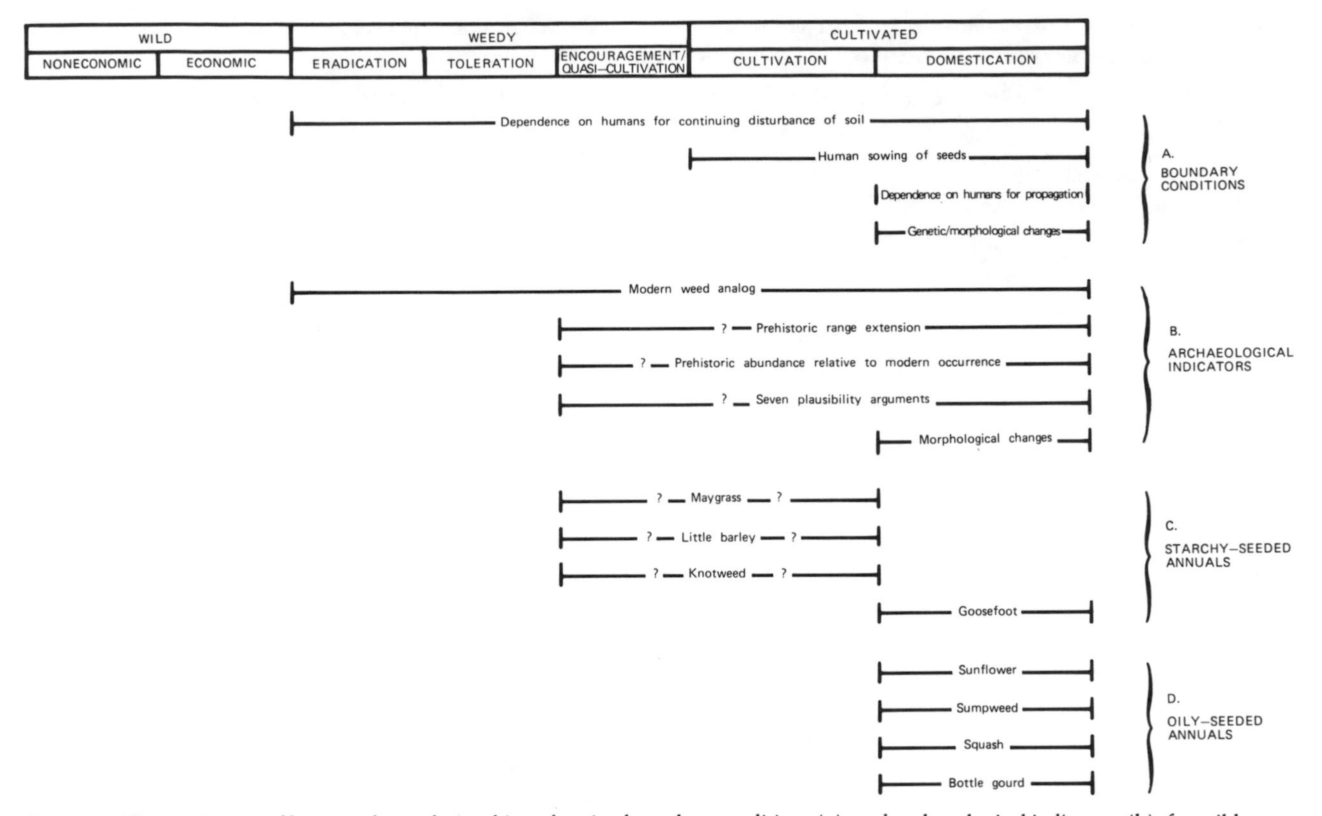

Figure 5.1 The continuum of human-plant relationships, showing boundary conditions (a), and archaeological indicators (b), for wild, weedy, and cultivated varieties of plants, as well as the placement along the continuum of premaize starchy-seeded (c) and oily-seeded (d) quasi-cultigens and cultigens of the Eastern Woodlands of North America.

tinues to be disturbed, however, succession will not proceed and weedy plants, with no further human intervention, will predominate to the relative exclusion of wild forms. This continual human intervention, which retards successional progression and favors weedy plants, encompasses a wide range of activities that result in the disturbance of the upper layer of soil so critical to the germination of weed seeds (Bye 1981:112) (turning the soil in garden plots or fields, construction of facilities such as houses and pits, and formation of refuse piles and pathways). The human activities that repeatedly disturb soil are most often not carried out deliberately to create a weedy habitat, and therefore usually represent an inadvertent and incidental intervention in the life cycle of weedy and wild plants.

The attitude of human populations toward the weedy plants that colonize disturbed ground situations can range from dislike and active eradication, through toleration and benign neglect, to various degrees of encouragement and utilization. Weeds with economic potential are obviously the most likely candidates for active encouragement within disturbed habitats, with such possible encouragement including the expansion of disturbed soil areas and the selective removal of competing plants.

Cultivated Plants

Along this symbiotic continuum the boundary between encouraged or "tended" weeds (Ford 1979:316) and cultivated plants can be marked by the intentional sowing of harvested seeds (Ford 1979:316; Asch and Asch 1982). While intentional propagation represents a clearly defined and logical boundary condition and beginning point for "cultivation," it is not a simple matter to distinguish between a "quasi-cultigen" (Yarnell 1983) (encouraged/tended weeds) and a cultigen in an archaeological context, where the intentional sowing of seed cannot be documented directly. Because of the current lack of consensus concerning the quasi-cultivated vs. cultivated status of a number of indigenous eastern North American plants species, the rather awkward term "quasi-cultigen/cultigen" will appear in reference to plant varieties that have been placed on both sides of the weed-cultigen boundary line (Figure 5.1c).

Domesticated Plants

The transition from cultivation to domestication occurs when, as a result of selective pressures within the human-made habitat, plants undergo genetic alteration resulting in observable (phenotypic) morphological changes. These genetic/morphological changes, reflecting further adaptation to the human-made environment and the selective pressures of harvesting and sowing, also result in plants which, unlike weeds, cannot survive in disturbed settings without continued human attention and intervention (including artificial propagation) over and above simple soil disturbance (de Wet and Harlan 1975:100).

Having identified boundary conditions for "cultivation" and "domestication" along this symbiotic continuum it is worth pointing out that these two terms are often used interchangeably, and with good reason. Certainly annual seed plants cannot be domesticated without also being cultivated, and if cultivation is defined in terms of the intentional planting of harvested seeds, then domestication can be viewed as the result of cultivation.

Placing the Plants of Premaize Garden Systems along the Wild to Domesticated Continuum

With the exception of plant assemblages sometimes recovered from dry cave and rock shelter deposits, archaeobotanical collections are predominantly composed of carbonized and often fragmentary reproductive plant parts such as nuts, fruits, and seeds. The limited view of prehistoric plant husbandry systems provided by such collections does not allow one to distinguish easily or confidently between plants harvested from wild stands and weeds adapted to disturbed soil situations. There are, however, a number of different comparisons to modern plant populations that have been employed in arguments for the nonwild (weed or cultivated) status of plants represented in archaeobotanical collections (Figure 5.1b).

Modern Weed Analogs

If the modern counterpart to a plant represented in archaeobotanical collections characteristically occurs in disturbed soil situations as opposed to undisturbed successional sequences, this would suggest that the plant may also have been a weed and/or cultigen prehistorically. Based on its present-day abundance in disturbed soil situations, for example, *Chenopodium missouriense* was also quite likely a weed within human-made habitats by 2,000 B.P. (Asch and Asch 1980:67), as was maygrass (Cowan 1978:272).

Prehistoric Range Extension

If a plant is represented in archaeological sites situated outside of the known present-day range of its closest modern analog, such a prehistoric range extension suggests human intervention (including sowing) over and above simple soil disturbance, and cultivated status for the prehistoric plant. Prehistoric range extension has been cited as evidence for the circa 2,000 B.P. cultivation of both marshelder *(Iva annua)* (Yarnell 1972; Asch and Asch 1978:320–321) and maygrass *(Phalaris caroliniana)* (Cowan 1978:282: Asch and Asch 1980:157; Yarnell 1983).

Archaeological Abundance Relative to Modern Occurrence

If a plant is relatively abundant in archaeobotanical assemblages and its closest modern counterpart is relatively rare in wild stands, such "evidence for a level of utilization that could not be sustained by gathering from natural stands alone" (Asch and Asch 1982:2) may suggest quasi-cultivated or cultivated status. Both erect knotweed *(Polygonum erectum)* and a chenopod *(Chenopodium bushianum)* have been proposed as cultigens on the basis of this archaeological criteria (Asch and Asch 1980:157–158, 1982:15).

"Plausibility Arguments"

In addition to these three (briefly discussed) primary criteria of prehistoric quasi-cultivation or cultivation, Asch and Asch (1982:2) have identified seven plausibility arguments that provide somewhat weaker evidence in support of the prehistoric cultivation of a plant: (a) economic importance; (b) lack of barriers to artificial propagation; (c) the cultivation of similar species elsewhere; (d) prehistoric association with known cultigens (see Gilmore 1931a:85 for an early example of this argument); (e) ethnohistoric evidence of cultivation; (f) a documented temporal increase in abundance prehistorically; and (g) an increase in population levels and sociopolitical complexity, implying an agricultural base.

Morphological Change

The final and most unequivocal archaeological indicator of cultivation of a plant is morphological change reflecting domestication. This is the single instance in which a boundary condition along the wild to domesticated continuum can be observed directly in the archaeobotanical record (Figure 5.1a, b). The status of both marshelder *(Iva annua)* and sunflower *(Helianthus annuus)* as prehistoric cultigens-domesticates in the Eastern Woodlands rests on such a documented morphological change—an increase in achene size (Yarnell 1972, 1978; Asch and Asch 1978).

Premaize Plant Husbandry Systems

Based on the set of archaeological indicators of varying strength and utility briefly outlined above, researchers have reconstructed the premaize plant husbandry systems of ca. 2,000 B.P. by placing both probable indigenous plant constituents and introduced tropical domesticates along the wild-to-domesticated continuum.

Three such tropical cultigens—tobacco (*Nicotiana* sp.), bottle gourd *(Lagenaria siceraria),* and squash *(Cucurbita pepo)*—had been introduced into the Eastern Woodlands prior to 2,000 B.P., and were domesticated crops within premaize gardens (Kay et al. 1980; Conard et al. 1984; Asch and Asch 1983; Johannessen 1984) [see Chapters 2, 3, and 4 regarding the identification of *C. pepo* as an indigenous domesticate rather than an introduced crop].

In addition to being a source of containers and floats as well as flesh, bottle gourd and squash provided oily seeds, and have been placed, along with sumpweed and sunflower, in an "oily-seeded domesticates" category (Asch and Asch 1983:685). The seeds of these four species are not very abundant in archaeobotanical assemblages dating to ca. 2,000 B.P., even though they represent domesticated crops (Yarnell 1974; Crites 1978b, 1984; Chapman and Shea 1981; Asch and Asch 1982, 1983; Table 5.1).

The starchy seeds of a number of plant species, in contrast, are recovered in greater numbers, reflecting the dramatic subsistence pattern change first reflected in the archaeological record at ca. 2,000 B.P. when some Eastern Woodland populations substantially increased their dependence upon the starchy seeds of indigenous annual plant species (Table 5.1). While this "starchy seed rise" has been observed in a number of areas of the Southeast and Midwest, it is most clearly documented in the detailed archaeobotanical sequences developed for west-central Illinois (Asch and Asch 1982, 1983), the American Bottom east of St. Louis (Johannessen 1984), central Kentucky (Yarnell 1974; Gardner 1984), and the Little Tennessee, Duck, and Elk river valleys of eastern and central Tennessee (Crites 1978a, 1978b, 1984; Kline et al. 1982; Chapman and Shea 1981; Figure 5.2).

Erect knotweed *(Polygonum erectum),* maygrass *(Phalaris caroliniana),* and goosefoot *(Chenopodium)* are the primary "participants" in this increased utilization of annual starchy seed plants, with little barley *(Hordeum pusillum)* recently added to this list for the west-central Illinois area (Asch and Asch 1983). Maygrass and little barley were crops that matured in late spring, while knotweed and goosefoot would have been harvested in the fall.

Researchers have placed each of these four starchy-seed annuals in the quasi-cultivated (Yarnell 1983) or cultivated (Asch and Asch 1982) category along the wild-to-domesticated continuum, based on their meeting one or more of the three primary archaeological criteria and set of seven plausibility arguments outlined above.

Maygrass has been assigned cultigen status primarily on the basis of its prehistoric range extension (Cowan 1978:285; Asch and Asch 1980:157; Yarnell

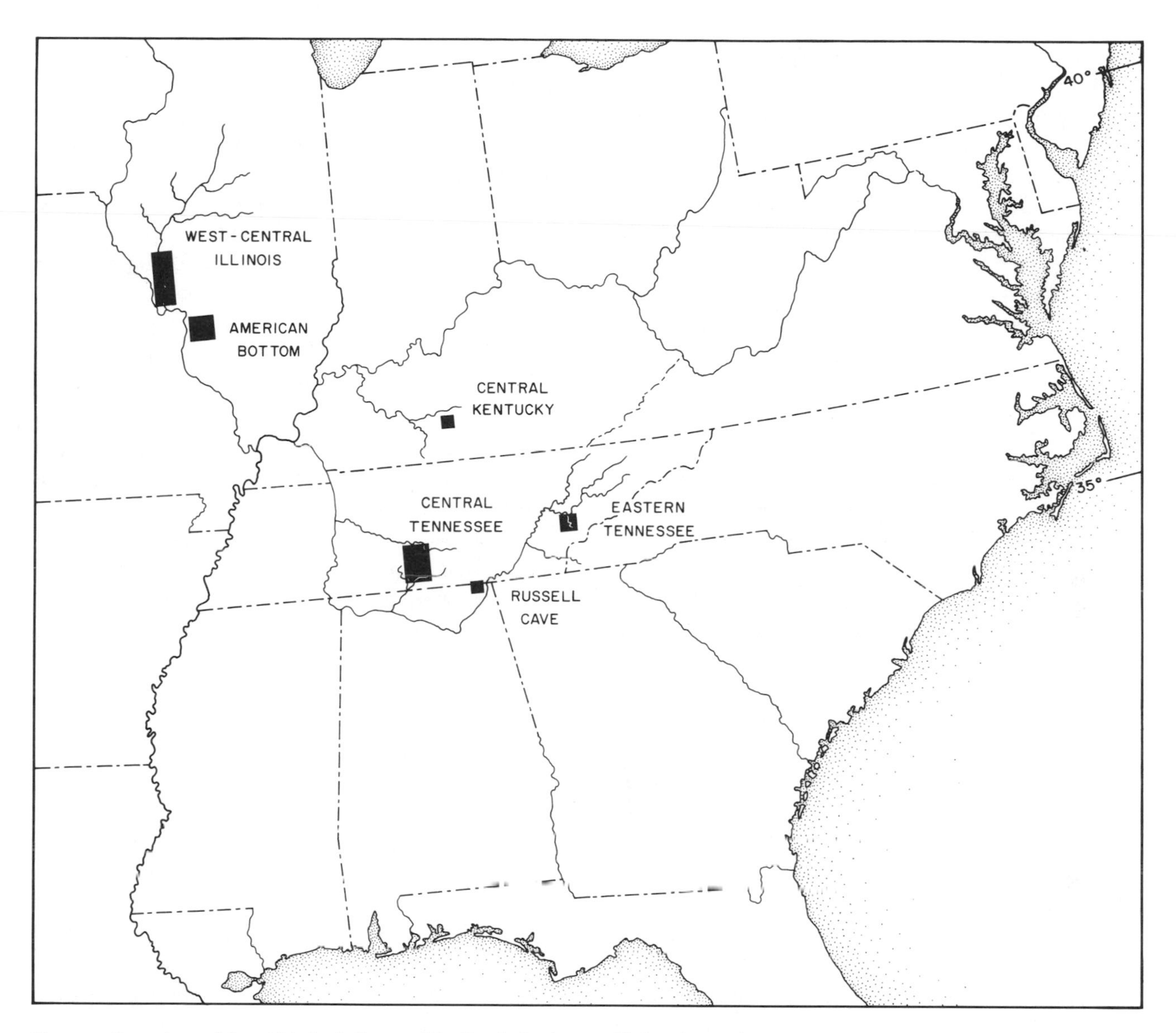

Figure 5.2 Six regions of the mid-latitude Eastern Woodlands that have yielded evidence of premaize plant husbandry systems (see Tables 5.1–5.3).

1983:12; Crites and Terry 1984:115), although Asch and Asch (1982:13) also mention its archaeological abundance relative to modern occurrence in west-central Illinois.

Similarly, the relative archaeological abundance of little barley compared both to prehistoric cultigens and to its present-day occurrence forms the basis for its identification as a premaize cultigen in west-central Illinois (Asch and Asch 1983:687).

Erect knotweed and *Chenopodium bushianum* have also been assigned cultivated status primarily on the basis of their sparse present-day occurrence combined with abundant and ubiquitous representation in premaize archaeobotanical assemblages (Asch and Asch 1980:158, 1982:10, 15, 1983:686). In providing a detailed summary of the supporting evidence for considering *Chenopodium bushianum* as having been cultivated in prehistoric garden plots, Asch and Asch

(1977) concluded that certain morphological characteristics first observed in the 1930s (Gilmore 1931b; Jones 1936) in archaeobotanical samples reflected habitat-related variability present in wild populations. More recently, however, the Asches have accepted an alternative causal hypothesis—that domestication is "the most probable explanation" for the morphological characteristics in question (Asch and Asch 1982). The value of these morphological characteristics as indicators of domestication, and the strength of the case for the presence of a domesticated variety of *Chenopodium* in premaize plant husbandry systems of the east, will be discussed in detail in subsequent sections of this article.

Taken together, these starchy seed annuals would appear to have been the focus of a marked expansion of premaize plant husbandry systems by ca. 2,000 B.P. Even though they have not been shown to exhibit unequivocal morphological changes indicating domestication and are thus placed "lower" along the wild-domesticated continuum, their ubiquitous and relatively abundant recovery suggests that they may have provided a dietary contribution equal to or greater than the oily-seeded domesticates (Table 5.1).

The apparent absence of obviously domesticated varieties of maygrass, little barley, knotweed, and goosefoot in ca. 2,000 B.P. archaeobotanical assemblages stands in contrast with their proposed quasi-cultivated or cultivated status and economic importance, suggesting the possibility that morphological changes associated with domestication may in fact be present but pass largely unnoticed in the partial remains found archaeologically. It is this possibility—of unnoticed morphological indicators of domestication—which will be addressed below, with specific reference to *Chenopodium*.

Morphological Indicators of Domestication in Chenopodium

The specific morphological characteristics identified by Wilson (1980, 1981a, 1981b; Wilson and Heiser 1979) as comprising the adaptive syndrome of domesticated *Chenopodium* will be briefly outlined within the more general framework of morphological change associated with the process of domestication that has been established by Harlan and his coworkers (Harlan et al. 1973; de Wet and Harlan 1975). In addition, the potential value of each of these morphological changes as archaeological indicators of domestication will be considered in terms of their probability of being both present and observable in archaeobotanical assemblages of *Chenopodium*.

The interrelated morphological changes, which together comprise an adaptive response to the artificial environment created by human intervention in the life cycle of *Chenopodium*, can be described under five general headings: (1) infructescence compaction, (2) loss of natural shatter mechanisms, (3) uniform maturation of fruit, (4) increased perisperm food reserves for germination and seedling growth, and (5) loss or reduction of germination dormancy. Of these five automatic consequences of adding the crucial step of planting to the annual cycle of encouraging, harvesting, and storing of a seed crop, the first two involve modification of the location and degree of retention of ripened fruits on individual plants, while the final three have to do with changes in the fruits themselves.[1]

Infructescence Compaction

In wild varieties of *Chenopodium* the flower clusters (inflorescences), which develop into fruiting heads (infructescences), are relatively small and distributed diffusely along the main stem and branches of individual plants (see Wilson 1980a, fig. 1c). Domesticated forms, on the other hand, have fewer and larger infructescences that are more compacted (denser) and terminalized (located at the ends of branches) (Harlan et al. 1973:316, 318; Wilson and Heizer 1979:199; Wilson 1981a:234).

This trend toward the concentration of fruit at the ends of branches, which could be consciously selected for during harvest, could also automatically develop without such deliberate selective pressure. The simple fact that "the plant that contributes the most seed to a harvest is likely to contribute more offspring to the next generation" (Harlan et al. 1973:318) would tend, over the years, to favor those plants that have their fruits visible and easily accessible during harvest.

Except for rare occasions when infructescence fragments are preserved in dry cave deposits (e.g.,

Wilson 1981a; Cowan et al. 1981), this morphological indicator of domestication would not be observable in archaeobotanical assemblages of the eastern United States.

Loss of Natural Shatter Mechanisms

The loss of natural mechanisms of seed dispersal, which is the most diagnostic of the morphological characteristics that separate domesticated and weedy forms of a seed-bearing species (Harlan et al. 1973:314–315), is also strongly and automatically selected for during harvest. Plants that retain their seeds through the harvest season are more likely to contribute to the next generation than are those that continue to disperse seeds through natural shatter mechanisms.

As with infructescence compaction, the loss of natural shatter mechanisms is an unlikely archaeobotanical indicator of domestication in the Eastern Woodlands since it can be observed only on those rare occasions when infructescence fragments are recovered intact (Wilson 1981a:234).

As Harlan and others note (1973:315): "The establishment of nonshattering traits is, genetically, one of the easiest and simplest steps in the entire process" of domestication, with a delay in harvesting a month or so after seed dispersal producing "a relatively nonshattering population" in a single generation.

Uniform Maturation of Fruit

While the sequential flowering and resultant staggered maturation of fruit that is typical of wild varieties of *Chenopodium* has a selective value in the natural environment, harvesting by human populations strongly selects against maturation of fruits over a long period of time (Harlan et al. 1973:316). Within the artificial environment of garden or field plots, plants that have simultaneous flowering and uniform maturation of fruit synchronized to an annual harvest by human populations will make a greater contribution to the gene pool of the next generation.

The uniform maturation of the fruit of domesticated *Chenopodium* within prehistoric garden or field plots should be reflected both by a relatively narrow range of variation in fruit size and a lower percentage of immature fruit in comparison to the harvest from wild stands, since simultaneous flowering produces more uniformly sized fruits that reach maturation within a narrower range of time (Wilson 1981a:234).

For a variety of reasons this morphological change has limited potential value at present as an archaeobotanical indicator of domesticated status. The range of variation in fruit size and frequency of immature fruits through the harvest season is not well documented for wild, weedy, and domesticated varieties of *Chenopodium* at either the individual plant or population level. In addition, the limited number of size range studies (fruit diameter) that have been carried out do not indicate dramatically lower standard deviation values for domesticated chenopods *(Chenopodium berlandieri* ssp. *nuttalliae)* (Asch and Asch 1977:36; Hugh Wilson, personal communication, 1982).

Establishing the range of variation in the size of fruits produced by modern wild varieties of *Chenopodium* in the eastern United States to use as a comparative baseline for archaeobotanical assemblages is complicated even further by the fact that *Chenopodium bushianum,* a common wild variety of the Northeast and Midwest, exhibits simultaneous late-season flowering and relatively uniform maturation of fruit, leading Wilson (1981a:233, 238) to suggest its participation in a prehistoric weed-crop complex, with resultant genetic interaction.

Finally, unless entire plants or infructescences were recovered from an archaeological context for comparative analysis, it would be difficult to demonstrate that a relatively narrow range of size variation and few immature fruits in a archaeobotanical assemblage of *Chenopodium* constituted a representative sample of a domesticated population, rather than a nonrepresentative sample of a wild or weedy population resulting either from a nonrandom harvesting method or differential preservation.

Increased Perisperm Food Reserves for Seed Germination and Seedling Growth

While the prepared seedbeds of prehistoric garden or field plots would have both encouraged germination and served to reduce competition from nondomesticated species, they would also have witnessed intense

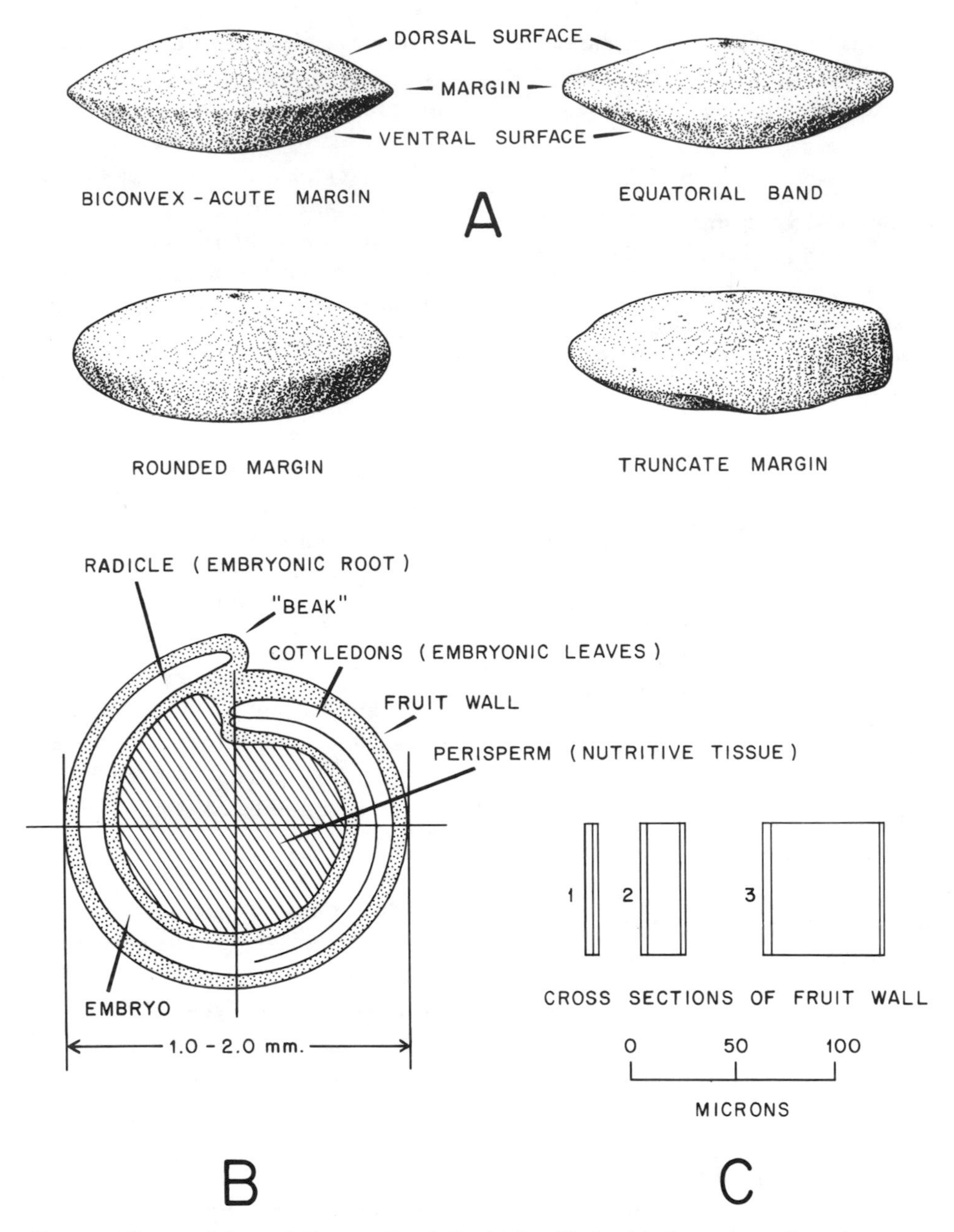

Figure 5.3 The morphology of *Chenopodium* fruits: (a) four idealized fruit margin configurations—biconvex-acute margin, rounded margin, equatorial band, and truncate margin (it should be noted that while the truncate margin fruit illustrated has a partially truncate margin similar to that of the "chia" domesticated variety, fruits of at least one domesticated variety—"huauzontle"—have fully truncate margins); (b) cross section of a *Chenopodium* fruit, showing internal structure; (c) variation in the fruit coat structure of (1) *Chenopodium berlandieri* ssp. *nuttalliae* cv. "huauzontle" (no outer seed coat present between pericarp and inner seed coat); (2) Russell Cave *Chenopodium* fruits and *Chenopodium berlandieri* ssp. *nuttalliae* cv. "chia" fruits (thin, 6–21 micron outer seed coat between pericarp and inner seed coat); and (3) wild varieties of present-day *Chenopodium* (thick, 40–55 micron outer seed coat).

competition between seedlings of the same species: "The first seeds to sprout and the most vigorous seedlings are more likely to contribute to the next generation than the slow or weak seedlings" (Harlan et al. 1973:318). The relative vitality and selective "fitness" of *Chenopodium* seedlings within prehistoric garden or field plots would have been determined by the amount of food reserves stored within seeds and available during germination and initial growth. The food reserves of seeds are stored both in perisperm and in embryonic leaves (cotyledons) (Figure 5.3b), and selection within artificial garden plot environments would have favored those seeds with increased perisperm and/or cotyledon volume.

Such selective pressure for increased food reserves is usually reflected by an increase in the external dimensions of seeds or fruits, paralleled by an increase in food reserves encapsulated within. The case for the prehistoric domestication of both sunflower and marshelder in the eastern United States, for example, rests on a documented increase in achene size (Yarnell 1972, 1978). Such an increase in external dimensions and stored food reserves can also, of course, be the result of deliberate human selection.

A similar argument linking increased external size with prehistoric domesticated status was proposed for *Chenopodium* a half-century ago by Gilmore (1931b) and Jones (1936). Their analysis of archaeobotanical assemblages from Newt Kash Rock Shelter, Kentucky, and the Hill site in Nebraska identified *Chenopodium* fruits having a larger diameter than those of any modern representative of the genus. David and Nancy Asch re-examined the large fruits in question, however, and established that they were pokeweed *(Phytolacca americana)* rather than *Chenopodium,* and concluded that "*Chenopodium* seeds from all other Eastern North American archaeological contexts apparently fall within the span of sizes found among modern members of the genus" (1977:15).

Comparison of modern wild and domesticated varieties of *Chenopodium* does not in fact indicate a significantly larger size for the domesticates in terms of maximum external diameter. Indeed, of the three varieties of *Chenopodium berlandieri* ssp. *nuttalliae* cultivated in Mexico, both "'huauzontle,' a broccoli-like vegetable crop with small (1.2–1.7 millimeters), horn colored fruits," and "'quelite,' a spinach-like vegetable crop with small brown fruits" are comparable to wild forms in terms of fruit diameter, while the third—"'chia' a grain crop with large fruits (1.5–2.3 millimeters)" (Wilson 1981a:237)—has a range of variation in fruit diameter that overlaps extensively with larger-fruited wild forms (Asch and Asch 1977:36). The fruits of domesticated varieties of *Chenopodium* are larger than those of wild varieties, however, when the basis of comparison is internal volume rather than maximum external diameter measurements. Domesticated varieties of *Chenopodium* have a characteristic rectanguloid cross section that is quite different from the biconvex, rounded, or equatorial banded shapes of wild fruits (Figure 5.3a; also see Wilson 1981a, fig. 2). This change in cross section results in domesticated fruits having a greater internal volume than wild fruits of the same diameter.

Although detailed comparative internal volume studies of wild and domesticated forms of *Chenopodium* have yet to be carried out, the increase associated with domesticated status is not great, and certainly not comparable to that documented for marshelder and sunflower (Yarnell 1972, 1978).

As to why *Chenopodium* fruits did not increase further in size, Harlan, de Wet, and Price suggest that increased food reserves (and larger seeds) are automatically selected for, but only up to a point:

> The plant that produces the greatest number of seeds also has an advantage and this factor may not be compatible with the largest seeds. Eventually a balance is reached in which selection is continuous for a large number of seeds yielding competitive seedlings. (Harlan et al. 1973:318)

In addition, insect predation may exert a selective pressure for smaller, more numerous seeds.

The embryo of *Chenopodium* fruits encircles the perisperm (Figure 5.3b), and the shift to a rectanguloid cross section and flattening or "truncation" of the seed coat margin in domesticated varieties is also associated with a change in cross section of the adjacent embryonic leaves (cotyledons) from the thick, round cross section typical of wild forms to the more mature elliptical shape present in domesticates (Wilson 1981a, fig. 2).

The set of associated morphological changes briefly described above (rectanguloid truncate margin cross section, elliptical cross-section cotyledons, increased internal fruit volume and perisperm food reserves), when taken together, certainly represent excellent archaeological indicators of domesticated status *Chenopodium* fruits.

It is not possible, however, to convincingly link this set of morphological characteristics in an exclusive cause-and-effect relationship with selective pressure for seed germination and seedling growth. There are a number of other potential causal variables that could be involved: (a) deliberate human selection for greater yield; and (b) selective pressure for a reduction in germination dormancy, as reflected by a reduction in seed coat thickness.

Loss or Reduction in Thickness of Outer Epiderm

In wild or weedy populations of *Chenopodium* mechanisms of germination dormancy function to prevent the premature germination of mature seeds that have been naturally dispersed and are present in the soil. Dormancy mechanisms in fact often block germination in at least a portion of the seeds dispersed by a plant beyond the next growing season, ensuring the presence of seeds in the soil the subsequent spring even if the intervening year's plants fail to mature and bear fruit.

While there are a variety of chemical or physiological inhibitors that can serve to delay germination in seeds, the most obvious block in the case of *Chenopodium* is the simple physical presence of a thick, black, hard outer seed coat (outer epiderm or testa). In addition to mechanically restricting the enlargement and germination of the embryo, such hard outer seed coats are largely impermeable to water and gasses and serve to block the passage of external elements that are essential prerequisites for embryo emergence and development. Outer seed coats also protect embryos from drying out and insect damage.

Although the physical dormancy produced by such thick outer seed coats is of obvious selective advantage in wild and weedy plants, it is nonadaptive in domesticated crops unless it is of short duration, breaking down between harvest and planting time: "Automatic selection pressures are very strong for seeds that come up when planted; therefore dormancy is much reduced in cultivated races" (Harlan et al. 1973:319). Human populations in effect assume the role of germination and insect block in such situations, being responsible for maintaining the proper temperature, moisture and insect controls for seeds in storage until a favorable season for germination. The prehistoric storage of containers of seeds in subground pits may reflect this concern with temperature, moisture, and insect control.

This much reduced germination dormancy in domesticates is often reflected either in a thinner outer epiderm, or even its complete loss, and such is the case with *Chenopodium* (Wilson and Heiser 1979:199). Among the Mexican domesticates, the "chia" variety of *Chenopodium berlandieri* ssp. *nuttalliae* retains a thin outer seed coat, while the "huauzontle" variety lacks an outer layer entirely (Wilson 1981a:236, 237).

This distinctive morphological change associated with reduced germination dormancy, which represents an excellent archaeological indicator of domestication, is invariably associated, perhaps causally, with the set of morphological changes briefly described above as reflecting selective pressure for seedling vitality (truncate margin, elliptical cross-section cotyledons, increased perisperm food reserves). As a result, it would seem reasonable to combine the morphological indicators of increased perisperm and loss or reduction of outer epiderm outlined above, and to expand this set of morphological traits to include the loss or reduction in thickness of external epiderm. This expanded set of associated morphological changes could then be considered as reflecting a general adaptive response to increased selective pressure among seedlings within prepared seed beds for both a loss or reduction in germination dormancy and perhaps an increase in perisperm food reserves for seed germination and seedling growth.

It is worth noting at this point that when these morphological changes associated with planting appear, they do not constitute a barrier to genetic exchange between the newly created "domesticate," and wild and weedy representatives of the same species. Nor does it necessarily signal the disappearance of such wild and weedy varieties from fields, garden plots, or archaeobotanical assemblages. Domesticated seed crops in general are almost invariably accompanied by weedy varieties which when crossed with them produce viable hybrids (Harlan et al. 1973:311–313, 319–321). *Chenopodium* is no exception in this regard, with both the South American and Mexican domesticates having related sympatric weed forms with which they freely exchange genetic materials, producing highly fertile hybrids (Wilson 1981a, 1981b; Wilson and Heiser 1979). Because of this "complete interfertility and sympatry" of the domesticates and their weed forms (Wilson and Heiser 1979:203), morphological characteristics associated with domestication can be expected to also occur in companion weed populations that mimic the associated cultivars (Harlan et al. 1973:319–321).

If *Chenopodium* was present as a domesticate in premaize plant husbandry systems of the East, it was in all likelihood accompanied by wild varieties of *Chenopodium* that would colonize anew each year, as well as by a companion weed form that would also exhibit some of the morphological characteristics of domestication. Since any year's harvest from such a garden plot might well include fruits from wild, weedy, and domesticated forms of *Chenopodium,* it would not be surprising if resultant archaeobotanical assemblages, representative or otherwise, often exhibited a wide range of variation in the occurrence and relative strength of the morphological characteristics associated with domestication.

Archaeological Indicators of Domestication in Chenopodium

Of the five morphological indicators of domestication briefly discussed above, two (infructescence compaction and loss of shatter mechanisms) can be of value in those rare situations (e.g., dry cave and rock shelter contexts) where infructescences are preserved intact.

In his analysis of several such rock shelter *Chenopodium* collections from the Ozarks, Wilson (1981a) identified the Mexican cultivar *Chenopodium berlandieri* ssp. *nuttalliae* cv. "huauzontle" on the basis of infructescence compaction and loss of shatter mechanisms, as well as the presence of fruits entirely lacking an outer epiderm. The white perisperm could be observed through the thin translucent inner seed coat (inner epiderm), giving the fruits a distinctive horn or white color, in contrast to the black color of wild fruits having a thick black outer epiderm. More recently, Fritz (1984) has noted the presence of pale-colored "huauzontle" fruits in additional Ozark rock shelter archaeobotanical collections. The Ozark rock shelter collections studied by Wilson and Fritz quite likely date after A.D. 1000, leading Ford (1981) to conclude that *Chenopodium berlandieri* ssp. *nuttalliae* (cv. "huauzontle") was introduced only as far east as the Ozarks from Mexico very late in the prehistoric sequence (well after A.D. 1000). David and Nancy Asch have recently suggested, however, that a variety of chenopod entirely lacking an outer epiderm was being cultivated in west-central Illinois by ca. 1,350 B.P. (A.D. 600) (1984:137–138).

In contrast to the morphological indicators that are rarely observable, a set of morphological changes associated with selective pressure for reduced germination dormancy and increased seed germination and seedling growth represent clear archaeological indicators of domestication that can be observed on individual fruits within archaeobotanical assemblages.

There is, however, an important caveat to assigning domesticated status to individual *Chenopodium* fruits in archaeobotanical assemblages on the basis of a truncate margin and a thin outer epiderm (the complete absence of an outer epiderm constitutes unequivocal evidence for domesticated status). Modern wild *Chenopodium bushianum* plants in the Eastern Woodlands have been observed to produce a small percentage of fruits that resemble those of a domesticated variety (the thin-testa "chia" variety of the Mexican domesticate *Chenopodium berlandieri* ssp. *nuttalliae*) (Asch and Asch 1977). These thin-testa truncate-margin reddish fruits may represent a small investment by wild stands in reduced dormancy fruits that will germinate quickly in the next growing season.

The Asches once made a collection from small, late-germinating plants on a Willow Island, Illinois mud flat in which such reddish, thin-testa fruits apparently predominated (Asch and Asch 1977:20). In subsequent field collecting, however, they have not located any populations producing more than a small percentage of reddish thin-testa fruits (Asch and Asch

1982:10, 1984:693), indicating that "The conditions under which the thin seed coat morph predominated in the wild are unusual" (Asch and Asch 1984:694).

Building a Case for Domesticated Chenopodium *in Premaize Plant Husbandry Systems*

Determining whether or not a domesticated variety of chenopod having a thin outer epiderm is represented in archaeobotanical collections is complicated by the possible occurrence of such low frequency thin-testa fruits of wild plants, and necessitates careful consideration of a number of aspects of sample size and composition, rather than the simple documentation of the presence of thin-testa fruits. The poor preservation of diagnostic morphological characteristics, the possible co-occurrence of fruits from wild, weedy, and domesticated varieties of *Chenopodium,* and the small mixed-deposit samples of *Chenopodium* fruits that are characteristic of archaeobotanical assemblages in the Eastern Woodlands further complicate any effort to demonstrate the presence of domesticated varieties.

Only a relatively weak case for domestication could be made, for example, on the basis of a small archaeobotanical sample of *Chenopodium* in which thin-testa fruits constituted only a small percentage. As the sample size of *Chenopodium* in an archaeobotanical assemblage grows, and the relative percentage of thin-testa fruits increases, a stronger case for domestication can be established. A large archaeobotanical collection composed exclusively of thin-testa rectanguloid cross-section *Chenopodium* fruits would constitute a strong case for the presence of a domesticated variety.

Two premaize archaeobotanical assemblages containing *Chenopodium* fruits have recently been proposed as constituting evidence for the presence of a domesticated variety of *Chenopodium.* Asch and Asch (1983:692–693) observed that while the archaeobotanical assemblage from the Smiling Dan site in west-central Illinois (1,900–1,700 B.P.) contained typical biconvex thick-testa wild chenopod fruits, such wild fruits were "somewhat less common" than fruits having a truncate margin and a thin outer seed coat (the chenopod sample size being 1,183). Because of the predominance of thin seed-coat fruits in the assemblage, it was concluded that a domesticated variety of chenopod was represented, rather than phenotypic variability in wild populations (Asch and Asch 1983:694).

In addition, similar thin-testa truncate-margin fruits have been recovered in association with rounded margin thick-testa wild fruits in Early Woodland contexts (ca. 2,500 B.P.; Gardner 1984) at Salts Cave, Kentucky, leading Yarnell (1983) to conclude that "*Chenopodium bushianum* was in some stage of the domestication process."

Even though the thin-testa chenopod fruits in both the Smiling Dan and Salts Cave archaeobotanical assemblages quite likely indicate the presence of a thin-testa domesticated variety of *Chenopodium,* it is possible to build a stronger and more detailed case for the presence of a domesticated chenopod in premaize plant husbandry systems of the Eastern Woodlands.

Constructing a stronger and more detailed case for domestication would involve, on the one hand, the establishment of a comparative data base by accurately describing and quantifying in modern wild and domesticated chenopod varieties those morphological characteristics that indicate domesticated status (e.g., a truncate margin and a "thin" outer epiderm).

At the same time, archaeobotanical collections that satisfy the following set of criteria should be sought out and analyzed: (a) recovered from a primary storage context as opposed to a secondary depositional situation; (b) recovered from a well-controlled temporal context; (c) consisting of a large number of fruits; (d) including well-preserved fruits having morphological characteristics relevant to taxonomic assignment and domesticated status intact; and (e) constituting a domesticated stand harvest, with very limited representation, if any, of either weedy or wild chenopod fruits.

Such chenopod assemblages would be expected to consist almost entirely of thin-testa truncate margin fruits that would on the one hand compare quite closely to the Mexican domesticate *Chenopodium berlandieri* spp. *nuttalliae* cv. "chia," while being distinctively different in morphology from modern wild varieties of chenopods of the eastern United States.

The number of documented archaeobotanical assemblages of *Chenopodium* fruits that appear to meet the set of optimal conditions outlined above is extremely limited (Asch and Asch 1977:35), with chenopod collections from Ash Cave, Ohio (Andrews 1877), and Russell Cave, Alabama (Miller 1960), holding the most promise for building a strong and detailed case for the presence of a domesticated variety of *Chenopodium* in premaize plant husbandry systems of the Eastern Woodlands.

The Russell Cave Chenopodium *Assemblage*

Situated in extreme northeastern Alabama (Figure 5.2), Russell Cave was designated a National Monument in 1961, and was the location of a large-scale archaeological research project carried out by the National Park Service (NPS) the following year. Published in 1974, John Griffin's report on the NPS field season of 1962 remains the only comprehensive account of the ten-millennium-long sequence of intermittent human occupation documented at this important site.

The NPS project was not, however, the only archaeological research carried out at Russell Cave. With National Geographic Society funding, Carl F. Miller, then with the River Basin Survey, Smithsonian Institution, conducted three field seasons of work (totaling 10 months) at the site (1956–1958), opening up a 30 ft (9.14 m) square block unit situated against the northwestern wall of the cave (Figure 5.4). The only published accounts of Miller's excavations, however, are a series of brief references to his field seasons that appeared in the *Annual Reports of the Bureau of American Ethnology* (Stirling 1957, 1958; Roberts 1959, 1960), as well as several popular accounts (Miller 1956, 1958) and a number of brief, narrowly focused articles (Miller 1957a, 1957b, 1960, 1962, 1965). Miller's work at Russell Cave remains undocumented in yet another respect. The large volume of cultural material he excavated and deposited with the Department of Anthropology, National Museum of Natural History, over twenty years ago remains uncatalogued.[2]

While Griffin's site report fortunately balances in most respects the absence of available information concerning Miller's excavation at Russell Cave, there are a number of aspects of Miller's research findings that were not replicated by the NPS's work at the site. Most significant in this regard was Miller's mention, in several of his brief publications, of the recovery of what he describes as a small charred basket containing carbonized chenopod seeds.

During Miller's first field season a 30-ft-long (9.14 m) trench was excavated against the northwestern wall of the cave (Miller 1956:545, 550; Griffin 1974:5–6) and

> at about seven feet we came upon a basket. The basket was saucer shaped, about 10 inches in diameter, and made of coiled strands of grass fiber sewed together. It was filled with small seeds, probably some wild grain the cave men gathered and ate. Both seeds and basket were charred. . . . Since it was late in the evening when we found the basket, I decided to wait until morning before trying to dig it out. I did pick up a small loose section of it and some of the seeds. . .but when we entered the cave the next morning, we were dismayed to find it gone—basket, seeds, and all. During the night someone had vandalized the cave, not only destroying the basket but also gouging several large holes in the side and floor of our pit. (Miller 1956:555)

An idealized vertical profile included in the same *National Geographic* article includes the following caption at the 7 ft "2600 B.C.–3000 B.C." level: "Stone pestle and charred remains of basket containing seeds" (Miller 1956:544). The only cultural context provided for the "basket" was Miller's consistent statements that no ceramics were recovered below a depth of 5 ft (1.52 m) (Miller 1956:544, 554; 1958). In a two-page article published in 1960, however, Miller places the "basket" and seeds within an Early Woodland (early ceramic) context (Griffin 1974:105):

> During the first season's work in Russell Cave, the charred remains of a small hemispherically-shaped basket were found filled with equally charred *Chenopodium* seeds. The seeds were later identified by experts in the United States Department of Agriculture as belonging to this plant family. Their presence on the Early Woodland horizon, about 5,000 years ago, indicate that these people knew the potential of these wild uncultivated seeds as a staple food source, harvested them by means of seed beaters and baskets and converted them into food. (Miller 1960:31–32)

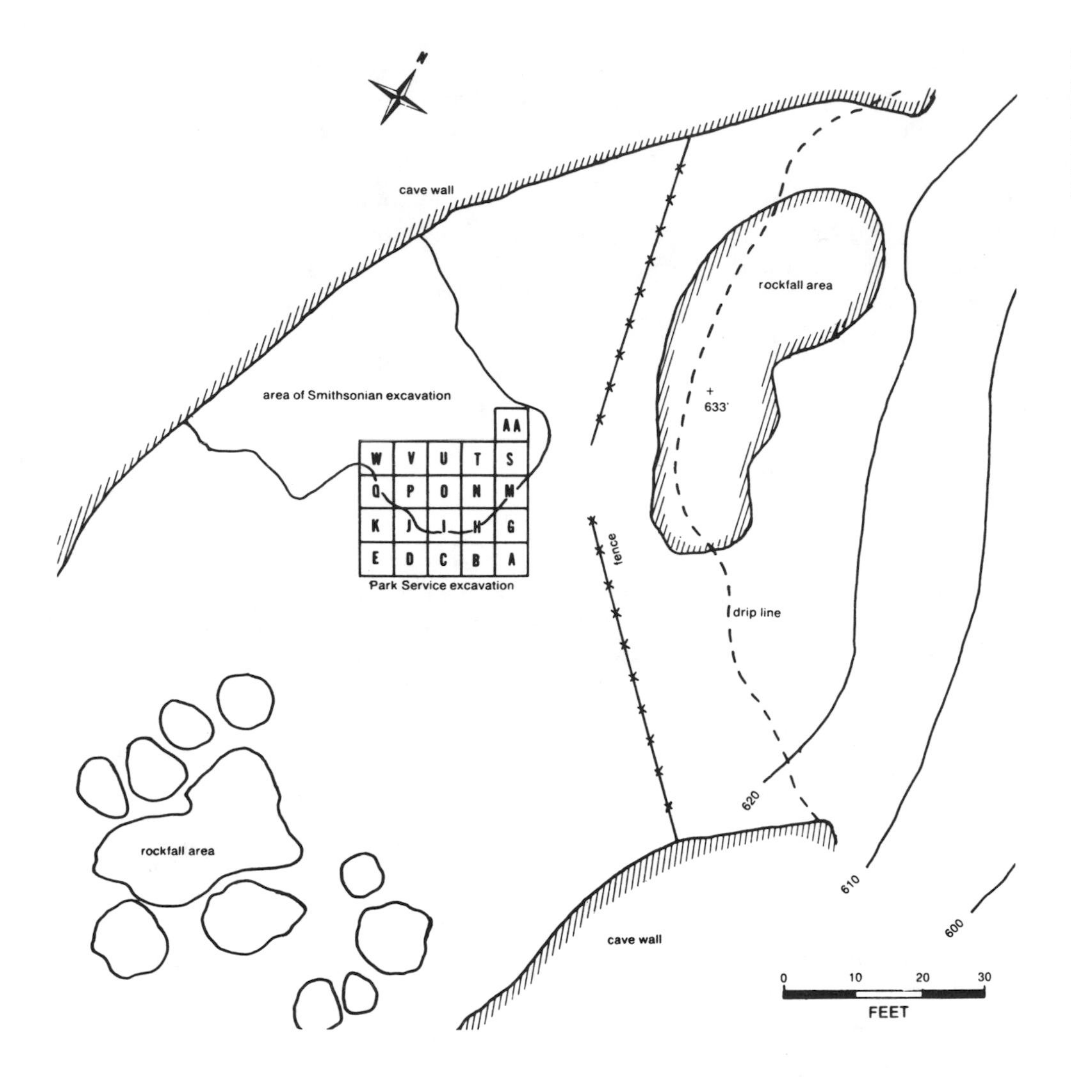

Figure 5.4 The location of the areas excavated by Miller (1956–1958) and John Griffin (1962) within Russell Cave (reproduced from Griffin 1974, fig. 5).

The above quotations constitute the entire published record of the recovery of the "basket" and fruits at Russell Cave.

In summary, during excavation of a trench against the northwestern wall of Russell Cave in 1956, Carl Miller uncovered at a depth of 7 ft (2.38 m) (a full 2 ft below ceramic-bearing strata) what he believed to be a charred hemispherical basket of coiled grass filled with carbonized fruits. Only a "small loose section of it [the basket] and some of the seeds" survived an overnight visit to Russell Cave by pothunters, however, with the seeds later identified as *Chenopodium* in Miller's 1960 article.

Rediscovery: The Basket and Its Temporal and Cultural Context

In August 1982, a search of the 38 drawers of unaccessioned, uncatalogued Russell Cave collections at the National Museum of Natural History turned up the charred "basket" section and fruits excavated and initially reported by Miller. A cigar box (Tampa Nugget Sublimes) bearing the longhand inscription "Basket F.S. [Field Specimen] 23" was found to contain a brown paper bag similarly labeled "F.S. 23" and filled with what at first appeared to be a quite diverse mixture of charred plant remains. An initial hand sort of this ethnobotanical assemblage into general categories, however, indicated that over 30 percent (by weight) of the collection consisted of carbonized *Chenopodium* fruits (Table 5.2). Oak, wax myrtle, and other leaves that lined the "basket" were also present (two fruits were found to have been fused to leaf fragments during carbonization), along with a total of 240 stem fragments of big bluestem grass *(Andropogon gerardi).* These remaining fragments of the "coiled strands of grass fiber sewed together" averaged 130 mm in length (range: 50–342 mm) and 23.7 mm in width (range: 10–41 mm). Ninety-four smaller grass stem fragments (1–4 mm in diameter) were also present. It is quite possible that the "basket" described by Miller was in fact what remained of the grass lining of a storage pit. The use of big bluestem grass for this purpose has been documented (e.g., Funkhouser and Webb 1929:47–49).

The fragmentary remains of the "basket" provided a much-needed temporal context for the fruits it contained. After being weighed, measured, and photographed, the larger grass stem fragments (4.3 g) were submitted to the Smithsonian Institution Radiocarbon Laboratory, yielding a radiocarbon age of 1,975 ± 55 B.P. (S.I. 5502).[3]

This single date compares favorably to the two radiocarbon dates of 1,995 ± 180 B.P. and 2,100 ± 200 B.P. obtained by Griffin (1974:13) for Layer D at Russell Cave. The base of Layer D was situated at a depth of about 4 ft (1.39 m) below the cave floor (Griffin 1974:7), but a full 3 ft (.91 m) above Miller's reported stratigraphic point of recovery of the "basket." This apparent vertical separation can be explained by the likely lined storage pit context of the *Chenopodium* assemblage. Griffin describes numerous pits dug down from Layer D into the Late Archaic layers below. Griffin (1974:6) also discusses the discrepancies in Miller's vertical stratigraphic records.

Yielding a material cultural assemblage dominated by Long Branch Fabric Marked ceramics (48 percent of the 1,168 sherds), along with Mulberry Creek Plain (40 percent) and Wright Check Stamped (9 percent) (Griffin 1974:35), Layer D contained few features other than circular pits excavated into underlying Archaic layers. Griffin considered the "poorly delineated" (1974:109) Layer D to represent a fall-winter hunting camp that was seasonally occupied by small groups, perhaps coming up into Doran Cove along Widow's Creek from more permanent village settlements situated within the valley of the Tennessee River, which flows 8 miles (12.8 km) to the east and southeast of Russell Cave (Griffin 1974:113). It is such seemingly fully sedentary Woodland villages of the mid-latitude Southeast that have consistently yielded evidence of large-scale storage of plant foods in subground "silos" (Smith 1986).

Table 5.2. General Material Categories, Field Specimen 23, Russell Cave, Alabama

Material category	Weight (g)
Clay fragments	1.6
Carbonized wood fragments	1.7
Carbonized leaf fragments (including oak and wax myrtle)	.7
Carbonized hickory nut fragments	1.2
Carbonized large big bluestem stem fragments	4.3
Carbonized small big bluestem stem fragments	1.6
Carbonized *Chenopodium* fruits	19.1
Small debris (<.2 mm) (sand, clay, carbonized plant remains)	13.3
Large debris (>.2 mm) (sand, clay, carbonized plant remains)[a]	17.5
Total weight, archaeobotanical assemblage	61.0

[a]Two Smithsonian volunteers (Donna Karolick and Allan Gotschall) spent over 100 hours in initial hand sorting of this archaeobotanical assemblage. The large debris category represents what was remaining after this hand sorting was ended due to diminishing returns. While this large debris category contains split cane, leaf, and seed fragments, its potential information content is both quite limited and redundant relative to the sorted material categories.

Initial Processing and General Condition of the Fruits

After careful hand separation from other ethnobotanical materials and transferral to shallow specimen boxes for viewing at a magnification of 40x, all of the fruits were found to have been carbonized.

Judging from the "fusion" of several of the fruits to leaves lining the basket or pit during carbonization, as well as the carbonized nature of the associated grass, this carbonization process occurred in situ, after the fruits had been placed in a pit for storage and covered in some manner.

That a substantial amount of heat was involved in the carbonization process is indicated by the "popped" condition of the vast majority of the fruits. This heating caused swelling of the perisperm and eventually resulted in the rupturing of the seed coat, most commonly around the margin of the fruits. While there was considerable variation in the degree to which this margin splitting extended around the circumference of the fruits, it was not uncommon for the seed coat to have split completely, with the dorsal and ventral surfaces separated somewhat as the perisperm swelled upward. Marginal rupturing, while often obliterating fruit-margin and cross-sectional information, did provide ample opportunity to observe external seed coats in cross section, and both dorsal and ventral surfaces often retained pericarp morphological information.

Not all of the fruits were ruptured during carbonization, however, and a total of 581 unruptured or minimally damaged fruits were located and sorted from the ruptured seeds for further analysis. The explanation as to why slightly more than 1 percent of the *Chenopodium* fruits did not rupture involves a number of factors. Based on their characteristic concavo-convex cross section, 36 of the unruptured fruits were immature and lacked sufficiently developed perisperm and moisture content to rupture the seed coat. In other unruptured fruits the integrity of the seed coat appears to have been broken prior to carbonization, allowing the perisperm and embryo to dry out and thus avoid rupturing. Most commonly, such breaks in the seed coat are small cracks, but in 11 of the fruits, small (0.4 mm) circular holes suggest insect activity. Other fruits, however, had no obvious openings in the external seed coat, and their unruptured condition remains unexplained, perhaps reflecting microenvironmental variation within the basket of fruits during carbonization.

In spite of their age, carbonized state, history of recovery, handling, and storage prior to rediscovery, the fruits were surprisingly well preserved. Distortion in the unruptured fruits (swelling to a rounded cross section) was limited, and the vast majority of both the ruptured and unruptured specimens retained their outer pericarp layer. Calyx and stem structures were also occasionally observed to be still adhering to the fruits.

Unruptured Fruits

To facilitate future restudy of the 581 unruptured fruits, a number was assigned to each, and a storage system was designed to allow their individual observation with minimal disturbance. Each of the 581 numbered fruits was then observed at magnifications of 40x to 80x under a bright raking light, and information concerning a number of variables was recorded.

DIAMETER. The diameter of each fruit was measured along two perpendicular axes, the first extending to the tip of the quite prominent "beak," and the second axis placed at a 90 degree angle to the first (Figure 5.3b). The "beak diameter" was fairly consistently 0.1 mm greater than the measurement taken along the other axis, and, with only a few exceptions, represented the maximum dimension of each fruit.

MARGIN CONFIGURATION. The margin configuration of each fruit was recorded with reference to the four idealized categories shown in Figure 5.3a (biconvex-acute margin, rounded margin, equatorial band, truncate margin). The fruit margin adjacent to both the enclosed embryonic root (radicle) and leaves (cotyledons) (Figure 5.3b) is flattened or truncated in the "huauzontle" cultivar variety of *Chenopodium berlandieri* ssp. *nuttalliae*. The margin area adjacent to the radicle in the "chia" cultivar variety of *Chenopodium berlandieri* ssp. *nuttalliae,* however, is often less flattened than the margin area adjacent to the cotyledons, and a clear transition from distinctly truncate to more equatorial banded is often present at a point directly across from the fruit's "beak" (Figures 5.3a, 5.6b).

PERICARP PATTERNING. The dorsal and ventral surfaces of each fruit were examined for evidence of pericarp morphology. Ventrally the presence or absence of a distinctive radiating elongated-alveolate pattern was recorded (Asch and Asch 1977:16). On the dorsal surface the presence of a reticulate (netlike) patterned pericarp was recorded as: (a) prominent, (b) present, (c) faint, or (d) absent.

MATURITY, DISTORTION. In addition, the maturity and level of distortion of each fruit, along with the presence of calyx and stem structures, was noted.

Table 5.3. Estimation of the Number of Ruptured *Chenopodium* Fruits in the Russell Cave Archaeobotanical Assemblage

Screen no.	Mesh size (cm)	Size class[a]	100 Fruit weight (g)	Sample weight (g)	Estimated number of fruits[b]
35	0.500	A	.00765	0.09	1,276
30	0.595	B	.01129	0.23	2,137
25	0.707	C	.01793	1.18	6,681
20	0.841	D	.02817	3.47	12,418
18	1.000	E	.04593	7.56	16,560
16	1.190	F	.06125	6.17	10,173
14	1.410	G	.07211	0.22	405
12	1.680				

[a]Each size class consists of the fruits that were caught in the corresponding mesh size, having passed through the next largest screen in the sequence.

[b]The total estimated number of ruptured fruits is 49,650.

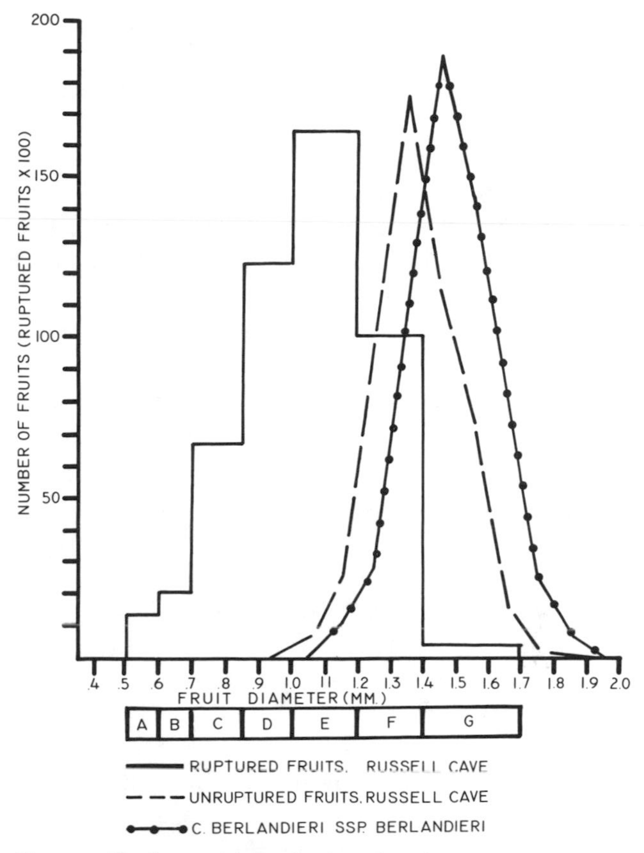

Figure 5.5 The frequency distribution of size (maximum fruit diameter) for three samples of *Chenopodium* fruits: unruptured and ruptured carbonized *Chenopodium* fruits from Russell Cave, together with the modern wild species *Chenopodium berlandieri* ssp. *berlandieri* (information on the modern wild species obtained from Hugh Wilson, personal communication, 1982). The capital letters indicate size classes for the ruptured fruits (see Table 5.3). The unruptured Russell Cave fruits had a mean diameter of 1.33 mm, and the population of 525 fruits measured had a standard deviation of 0.13.

Ruptured Fruits

In an effort both to estimate the quantity of ruptured fruits that were present and to extract some information concerning the diameter of individual specimens, a standard geological screen series (W. S. Tyler) was employed to sort the ruptured fruits by size. This mechanical sorting was undertaken only after a procedure was developed that allowed the fruits to be sequentially passed through a graded series of sieves with no resultant fragmentation or abrasion.

Once the ruptured fruits had been sorted by size, 100 fruits from each size class were weighed to the nearest hundredth of a milligram on a Mettler analytical balance, and these 100 fruit weight values were then divided into the total weight values for each size class to arrive at fruit count estimates (Table 5.3).

Scanning Electron Microscopy

A number of unruptured and ruptured *Chenopodium* fruits from Russell Cave, along with both carbonized and uncarbonized fruits from a range of modern domesticated and wild varieties of *Chenopodium* were viewed and photographed with the aid of a scanning electron microscope. While scanning electron microscopy was employed primarily to measure external seed coat thickness in both archaeobotanical and modern specimens, it also proved to be very valuable in observing other morphological characteristics, including pericarp patterns and the cross-section margin configuration of fruits.

Fruit Size

A total of 525 of the 545 unruptured mature fruits yielded diameter measurements (Figure 5.5). There are two potential sources of bias in the size range histogram shown in Figure 5.5. The first involves the likelihood that the *Chenopodium* fruits were reduced in size as a result of carbonization, and the second relates to whether the 525 fruits that were measured constitute a representative sample of the total archaeobotanical collection under study.

Although controlled carbonization of modern specimens of both *Iva annua* and *Helianthus annuus* have been conducted (Yarnell 1972, 1978:296; Asch and Asch 1978:326), similar shrinkage figures for *Chenopodium* have not been reported in the literature. Carbonization of a sample of 50 modern wild fruits of *Chenopodium* ranging in size from 1.0 to 1.4 cm produced an average reduction in maximum diameter of 5 percent. If applied as a correction factor this shrinkage rate would shift the size distribution of the unruptured *Chenopodium* fruits in Figure 5.5 slightly to the right, into close agreement with that of the wild variety *Chenopodium berlandieri* ssp. *berlandieri* (Hugh Wilson, personal communication, 1982).

Although the sample of 525 unruptured mature measurable fruits represents slightly more than 1 percent of the estimated 49,650 ruptured fruits present in the collection (Table 5.3), it would appear to constitute a representative sample (at least in terms of size) of the total archaeobotanical collection, judging from the similarity in Russell Cave size range curves illustrated in Figure 5.5. Although the histogram of ruptured fruit diameter in Figure 5.5 forms a "shadow" to the left of the unruptured fruits, this is to be expected. The rupturing of external seed coats almost invariably resulted in the fragmentation and loss of a portion of the fruit margin, reducing the maximum diameter of the fruits. Visual inspection of the ruptured fruits confirmed that they were fragmentary or marginless.

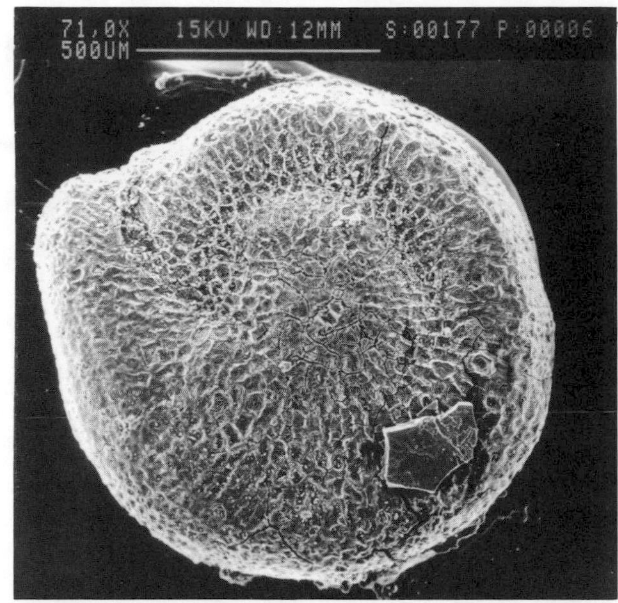

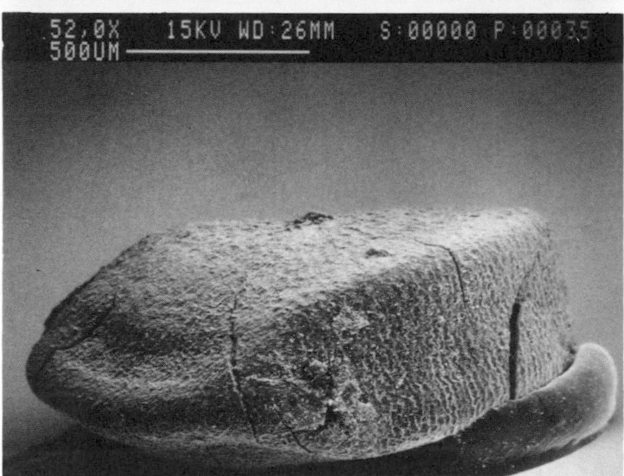

Figure 5.6 Scanning electron micrographs of Russell Cave *Chenopodium* fruits 177 (top), and 33 (bottom), showing the distinctive reticulate-alveolate dorsal pericarp pattern and the truncate to equatorial band transition of the margin.

Pericarp Morphology

A total of 523 of the 545 unruptured mature fruits were coded for pericarp patterning. All of the coded fruits exhibited a distinctive radiating elongated-alveolate pattern on their ventral surface.

With the exception of five fruits (nos. 119, 349, 410, 452, 528) that appeared to have an essentially smooth dorsal pericarp, all of the fruits coded for pericarp patterning displayed a reticulate pericarp on their dorsal surface (Figure 5.6). On the basis of this reticulate pattern, which was prominent on 403 fruits, present on 89, and faint on an additional 26, the Russell Cave *Chenopodium* assemblage can be assigned to section *Chenopodium*, subsection *Cellulata* of the genus *Chenopodium* (Wilson 1980).

Margin Configuration

A total of 387 fruits were sufficiently undistorted to be coded for cross section and margin configuration, with 353 (92 percent) having a truncate margin (Figures 5.3a, 5.6). Twenty-five fruits exhibited an equatorial band, while eight had a rounded margin, and one had a biconvex cross section and acute margin.

Outer Epiderm Thickness Measurements

In order to establish a modern analog data base for comparison with the Russell Cave *Chenopodium* assemblage, a total of 38 modern wild and domesticated chenopod fruits were sectioned, and outer epiderm thickness values were determined with the aid of a scanning electron microscope. Twenty-two of the 38 modern fruits were carbonized prior to taking outer epiderm measurements to establish if carbonization might influence outer epiderm thickness values. The process of carbonization neither reduced or increased the thickness of external seed coats.

Outer epiderm thickness values for modern fruits of the three wild species *Chenopodium bushianum*, *Chenopodium berlandieri*, and *Chenopodium missouriense* ranged from 40 to 57 microns (Figures 5.7a, 5.8).

In contrast, 20 fruits of the Mexican domesticate *Chenopodium berlandieri* ssp. *nuttalliae* cv. "chia" were found to have much thinner outer seed coats, ranging in thickness from 9 to 21 microns (Figures 5.7b, 5.8).

SEM observation of 20 fruits from the Russell Cave collection yielded outer epiderm thickness values ranging from 7 to 16 microns (Figures 5.7c, 5.8).

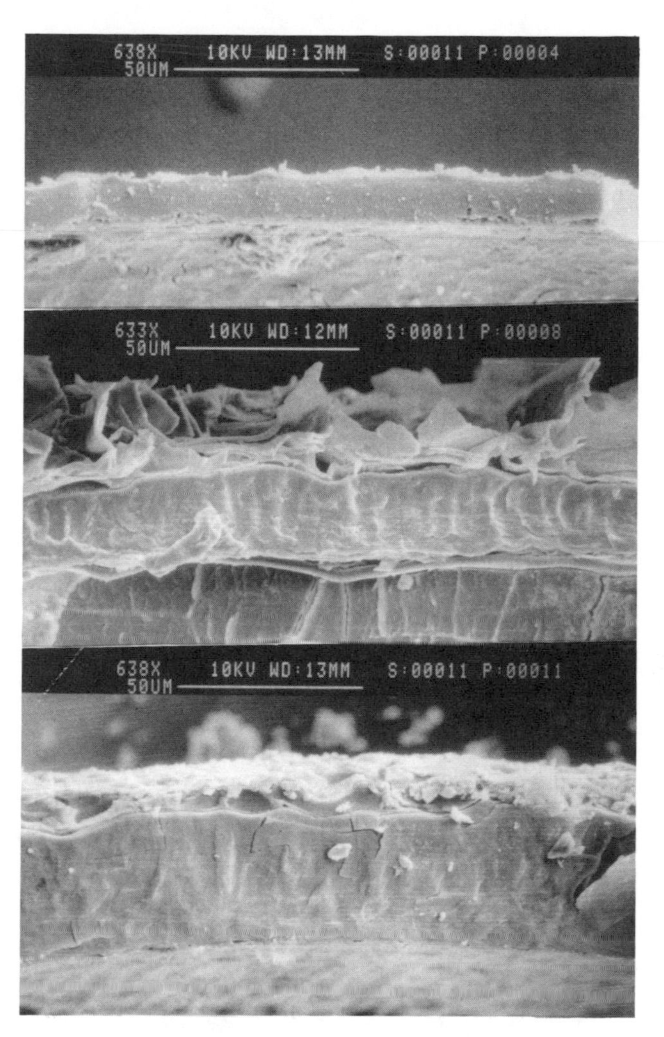

Figure 5.7 Scanning electron micrographs showing outer seed coats in cross section: (top) Russell Cave fruit 2–4 (14 microns thick); (middle) *Chenopodium berlandieri* ssp. *nuttalliae* cv. "chia" fruit 11–9 (20 microns thick); (bottom) *Chenopodium berlandieri* fruit 6–1 (41 microns thick).

The Strength of the Case for Domestication

The 50,000 *Chenopodium* fruits recovered from Russell Cave comprise an almost unique archaeobotanical assemblage that meets the set of conditions outlined above as constituting a strong case for the presence of a domesticated variety of chenopod in premaize plant husbandry systems of the Eastern Woodlands. Representing a dated storage context rather than secondary deposition, the large sample included well preserved fruits having intact morphological characteristics relevant to assessing taxonomic assignment and domesticated status.

Based on an analysis of morphological attributes, this *Chenopodium* assemblage probably represents a portion of the fall harvest of a domesticated stand by the inhabitants of Russell Cave, with only very limited representation of weedy or wild varieties. Ninety-two percent of the analyzed fruits exhibited the distinctive truncate margin characteristic of domesticated varieties of *Chenopodium*. In addition, the range of variation in outer epiderm thickness measurements for Russell Cave chenopodium fruits was comparable to that of the modern Mexican domesticate *Chenopodium berlandieri* ssp. *nuttalliae* cv. "chia" and considerably below that of three present-day indigenous wild varieties of the Eastern

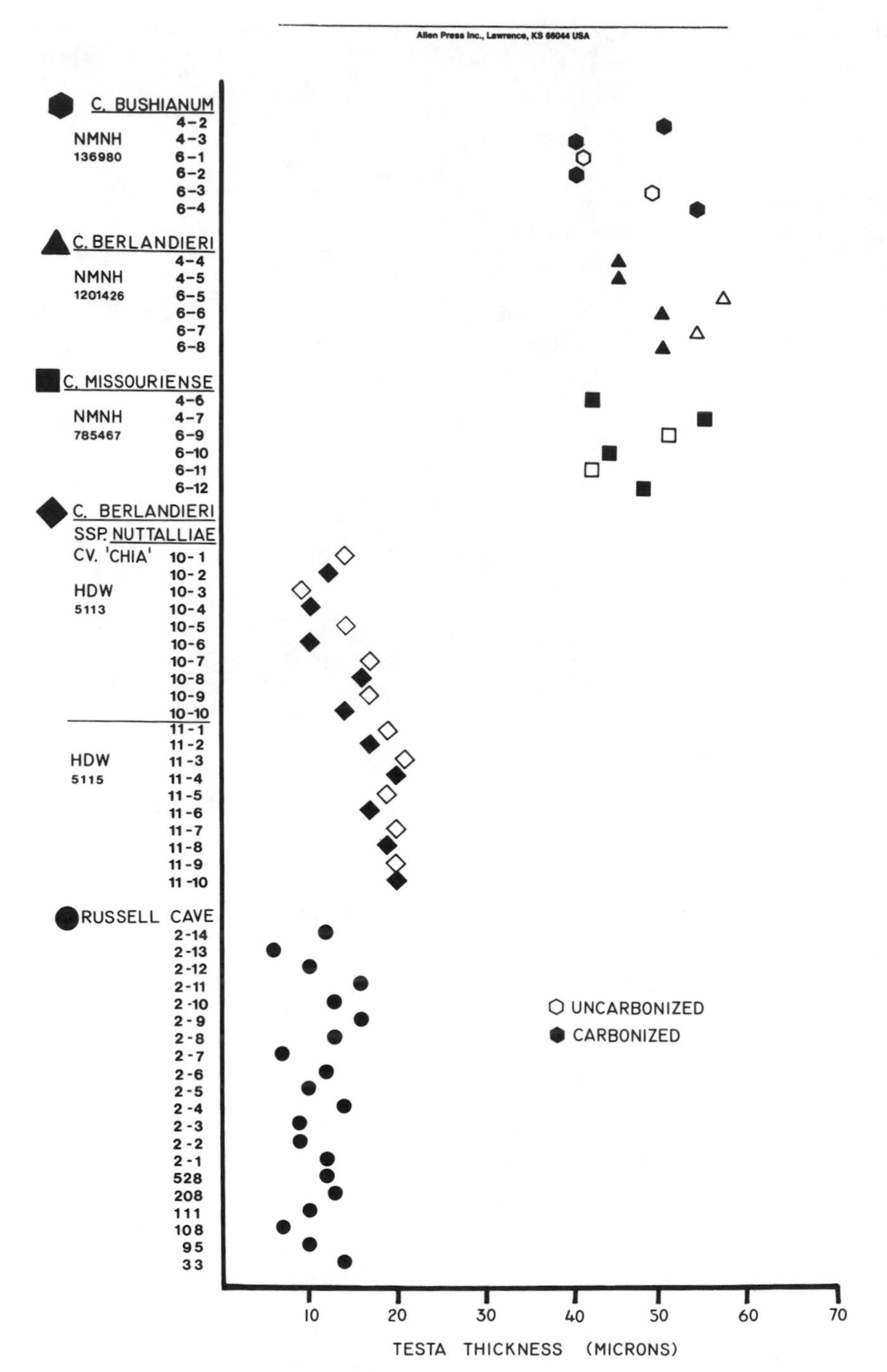

Figure 5.8 The relative thickness of the outer seed coat of modern domesticated and wild varieties of *Chenopodium,* compared to the Russell Cave *Chenopodium* assemblage. Numbering sequence along the vertical axis provides SEM mount information. Fruits from modern wild varieties were obtained from NMNH Department of Botany collections (catalog numbers indicated). Fruits of the Mexican domesticate "chia" were provided by H. Wilson (HDW 5113 and 5115 collected from cultivated fields near Opopeo and Zirahuen, Michoacan, Mexico, 1983). Some of the modern specimens were carbonized in order to establish that the process of carbonization does not result in any significant reduction in the thickness of the outer seed coat.

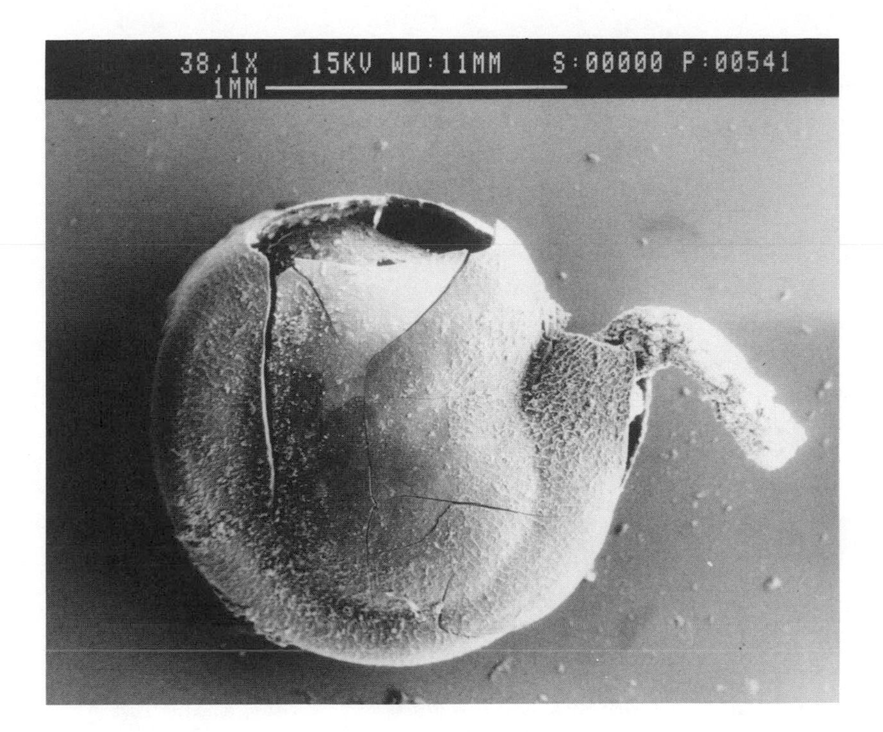

Figure 5.9 Scanning electron micrograph of Russell Cave *Chenopodium* fruit 541 showing emergent radicle (embryonic root) suggesting germination during storage.

Woodlands. The consistent, almost universal, occurrence of these interrelated morphological characteristics (thin outer epiderm and truncate margin) rules out phenotypic variability in wild varieties of *Chenopodium* as a cause. It instead reflects a general adaptive response to selective pressure for both seedling vitality and a reduction in germination dormancy.

Less substantial evidence of a loss of germination dormancy is provided by the three Russell Cave *Chenopodium* fruits, which appear to have been in an initial stage of germination at the time of carbonization (Figure 5.9). It is of course possible that the extension of the embryonic root (radicle) from these three fruits was the fortuitous result of carbonization.

This clear evidence of selective pressure for increased seedling vitality and a reduction in germination dormancy constitutes a compelling argument that *Chenopodium* was a domesticated crop as well as a wild invader and weed within the garden or field plots of the ca. 2,000 B.P. inhabitants of Russell Cave.

Discussion

In addition to establishing the presence of a domesticated variety of *Chenopodium* in northeast Alabama by 2,000 B.P., the present analysis provides additional support for the premaize domesticated chenopod arguments put forth for other areas of the Eastern Woodlands (e.g., west-central Illinois and central Kentucky).

It also establishes guidelines for future analysis and should encourage the close scrutiny of prehistoric chenopod assemblages for morphological indicators of domestication. The recognition and documentation of such morphological indicators of domestication in *Chenopodium* assemblages from different regions will eventually result in a far better understanding of the timing and geographical location of this domestication process.

Based on the apparent presence of a domesticated thin-testa variety of *Chenopodium* in central Kentucky (Salts Cave) by ca. 2,500 B.P., and both west-central Illinois (Smiling Dan) and northeastern Alabama (Russell Cave) by 2,000 B.P., it is reasonable to postulate that by 2,500–2,000 B.P. this crop had been added to premaize plant husbandry systems over a broad area of the mid-latitude interior drainage system of the Eastern Woodlands (west of the Appalachian Wall, from 35 to 40 degrees north latitude) (Figure 5.2).

It still remains to be established, however, whether this thin-testa domesticated variety of *Chenopodium,* which had been added to crop systems in the East by 2,500–2,000 B.P., was introduced from Mexico or was, alternatively, the result of an independent process of domestication (Asch and Asch 1983:695; Fritz 1984).

It may well prove quite difficult to establish which of these alternatives is correct on the basis of archaeobotanical evidence, and the present-day taxonomy of the genus *Chenopodium* provides little help in resolving this question of Mexican versus an indigenous origin. The presence of a reticulate dorsal pericarp allows the confident inclusion of the Russell Cave *Chenopodium* collection within section *Chenopodium* subsection *Cellulata* of the genus *Chenopodium,* but finer taxonomic assignment is more difficult.

Russell Cave is located in a blank space on the map of the geographical distribution of the wild representatives of subsection *Cellulata* in eastern North America. Forming population systems with considerable genetic affinity rather than being biologically distinct (Wilson and Heiser 1979; Wilson 1980:260), *Chenopodium bushianum* and *Chenopodium berlandieri* are sympatric over an extensive area of the Midwest from Arkansas north through Wisconsin and Michigan. The range of *C. bushianum* extends to the Northeast (north of Russell Cave), while that of *C. berlandieri* extends east along the Gulf Coast (south of Russell Cave) and then northward along the Atlantic Coastal Plain (east of Russell Cave) (Hugh Wilson, personal communication 1982). In addition to being recovered from a location between the known modern geographical ranges of *C. bushianum* and *C. berlandieri,* the Russell Cave *Chenopodium* fruits are morphologically distinct from the fruits of these two wild representatives of subsection *Cellulata* in a number of important respects (margin configuration, reticulate-alveolate pericarp and seed coat, and outer seed-coat thickness). On the other hand, the Russell Cave *Chenopodium* collection compares favorably with *C. berlandieri* in terms of fruit diameter (Figure 5.5), and not with the larger fruited *C. bushianum.*

Since *Chenopodium berlandieri* has also been identified as the taxon from which the Mexican domesticated forms *(C. berlandieri nuttalliae)* most likely initially evolved (Wilson and Heiser 1979), it is possible to assign the Russell Cave archaeobotanical assemblage of *Chenopodium* fruits to *Chenopodium berlandieri* without favoring either developmental explanation (indigenous domestication or introduction from Mexico). While future archaeobotanical research documenting the process of domestication of *Chenopodium* in both Mexico and the Eastern Woodlands may eventually result in the Russell Cave *Chenopodium* being designated as a cultivar variety of *Chenopodium berlandieri* ssp. *nuttalliae,* such a taxonomic assignment is not possible at this time.

It should similarly prove difficult to establish what developmental relationship, if any, existed between the "chia"-like thin-testa variety of Chenopodium berlandieri of premaize plant husbandry systems, and the "huauzontle"-like variety of the same species which was present in the Eastern Woodlands perhaps as early as 1,300 B.P., a full seven centuries later (Asch and Asch 1984:137–138). The "huauzontle"-like variety quite likely represents an introduction from Mexico (Wilson 1981a; Fritz 1984). Alternatively, it might be the product of an independent process of domestication in the East, either parallel with, or developing out of, the "chia" variety.

Establishing the early presence of a domesticated variety of *Chenopodium berlandieri* in the mid-latitude Eastern Woodlands not only implies a greater degree of human interest and manipulation, but also provides support for the proposition that the "starchy seed rise" of ca. 2,100–1,700 B.P. corresponded with either the initial addition or substantial expansion of a variety of regionally distinct field cropping strategies that were complementary to already existing systems of small garden-plot cultivation.

Although potential indications of this shift to field cropping systems are not abundant in the archaeological record, changes in settlement patterns and

Table 5.4. Apparent Geographical Variation in the Relative Importance of Starchy-seeded Annuals in Premaize Plant Husbandry Systems

Annual	Eastern Tennessee (%)	Central Kentucky (%)	Central Tennessee (%)	American Bottom (%)	West-Central Illinois (%)
Oily-seeded annuals	2.7	3.7	0.3	0.2	0.8
Starchy-seeded annuals					
Chenopod	97.3	90.6	52.0	47.0	9.2
Maygrass	0.0[a]	5.7	38.0	51.9	40.1
Knotweed	0.0[a]	0.02	9.7	0.9	34.0
Little Barley	0.0	0.0	0.0	0.0	15.9
Total	97.3	96.3	99.7	99.8	99.2

NOTE: Information taken from Table 5.1

[a]Maygrass and knotweed seeds are present in subsequent Connestee phase (1,800–1,400 B.P.) components (Chapman and Shea 1981).

plant food storage facilities (Smith 1986) as well as archaeobotanical indicators of land clearing activities (Kline et al. 1982:63; Johannessen 1984:201) are compatible with such an expansion of plant husbandry systems.

At least five and perhaps six different quasi-cultigens and/or cultigens could potentially have been involved in various premaize field crop systems. Little barley and maygrass would have provided a spring yield in May and June (Cowan 1978:267; Asch and Asch 1983:691), their harvest overlapping with the planting period of the three fall maturing crops—goosefoot, knotweed, and marshelder. David and Nancy Asch (1978:319) have proposed that marshelder "is more appropriate as a field crop than as a small crop in a garden." Sunflowers, too, may have been grown in a field context, although they could also have been advantageously grown in small garden plots. If planting of these three and perhaps four fall crops was scheduled for mid to late June, a multiple cropping system that mixed spring-maturing and fall-maturing species in the same fields would have been possible.

Although oily-seeded domesticates consistently comprised less than 4 percent of the quasi-cultigen/cultigen seed count in premaize archaeobotanical assemblages (Table 5.4), they were a consistent and important component in mid-latitude plant husbandry systems. Because of the possibility that oily seeds are less likely to be preserved (particularly in open air sites), and the larger size of the former, this low percentage does not provide a reasonable projection of the relative dietary contribution of starchy- versus oily-seeded plants. Richard Yarnell estimates, for example, that taken together, the two dominant oily-seed annuals (sunflower and sumpweed) were equal in importance to the two dominant starchy-seeded annuals (chenopod and maygrass) in the diet of the 2,600–2,200 B.P. inhabitants of Salts Cave (personal communication, 1984).

Table 5.4 also shows that there was clear geographical variation in the relative importance of the starchy-seed annuals. This variation suggests that regional field cropping strategies differed considerably in terms of diversity (the number and relative importance of quasi-cultivated, cultivated, and domesticated crops).

In eastern Tennessee *Chenopodium* is the only starchy-seed annual present in 2,000–1,800 B.P. archaeobotanical assemblages. Based on a preliminary survey of these assemblages by the author, thin-testa domesticated varieties of chenopod were not grown, suggesting that premaize field cropping systems in this region centered on a single fall-maturing starchy-seed quasi-cultigen/cultigen *(Chenopodium),* along with two oily-seeded domesticates (sunflower and marshelder).

In contrast, thin-testa *Chenopodium* fruits are present in the premaize Salts Cave archaeobotanical assemblages of central Kentucky, suggesting that, if present, field cropping systems in this region incorporated domesticated, weedy, and perhaps quasi-cultigen/cultigen varieties of this starchy-seed crop. As in eastern Tennessee, the fall harvest in the central Kentucky region centered on a starchy-seed annual *(Chenopodium),* and two oily-seed domesticates (sunflower and marshelder).

Central Kentucky cropping systems were more diverse than those of eastern Tennessee, however. Maygrass, a spring harvested starchy-seed quasi-cultigen/cultigen was grown, and apparently was approximately equal in importance to marshelder, behind the co-dominants sunflower and *Chenopodium.* Rather than being from the quasi-cultigen/cultigen *Polygonum erectum,* the knotweed seeds in the Salts Cave archaeobotanical assemblage probably represent weed contamination in the chenopod harvest (Richard Yarnell, personal communication, 1984).

In the American Bottom and central Tennessee regions, however, maygrass appears to have been approximately equal in importance to *Chenopodium,* suggesting both a greater diversity and a greater balance in the field cropping systems between spring and fall maturing starchy-seed quasi-cultigen/cultigens (thin-testa chenopod fruits have yet to be documented in these two regions). This may represent a multicropping strategy focusing on two field crops—maygrass and goosefoot.

Knotweed was present as a secondary fall harvest starchy-seeded quasi-cultigen/cultigen. The low representation of marshelder and sunflower in these assemblages is likely a result of differential preservation rather than limited utilization.

Judging from Table 5.4 the west-central Illinois region supported the most diverse and generalized premaize plant husbandry system. Although several varieties of chenopod, including a thin-testa domesticated variety were present in fall harvests, knotweed was the dominant fall starchy-seed crop. As in the American Bottom and central Tennessee there is an approximate balance in the relative importance of spring versus fall harvested starchy-seeded field crops, again suggesting a multicropping strategy. Little barley joins maygrass in the spring harvest, with knotweed-chenopodium and maygrass-little barley apparently forming, respectively, fall and spring primary-secondary starchy-seeded crop pairs. Oily-seeded annuals are present, and their importance is probably underrepresented.

As additional archaeobotanical assemblages from different areas of the Eastern Woodlands are analyzed, it will be possible to determine whether the patterns just outlined actually reflect different field-cropping strategies, or are simply a function of differential preservation, recovery, or seasonality of occupation of the sites being studied.

It will also be possible to assess the relative strength of the more general proposition that field-cropping strategies focusing on starchy-seeded annuals were added to small garden plot cultivation systems by ca. 2,600–2,000 B.P., well in advance of the introduction of maize into the Eastern Woodlands.

As Asch and Asch have pointed out (1982), there is a role to be played both by healthy skepticism and informed but speculative approaches in attempting to gain an understanding of prehistoric plant husbandry. Much of the foregoing discussion concerning the existence, initial emergence, and apparent regional (and temporal?) variation in postulated field cropping systems certainly falls under the rubric of informed speculation. Such speculative propositions will, it is hoped, stimulate additional consideration of the possibility of premaize field-cropping systems and the indigenous plant species which might have been involved.

Notes

1. *Chenopodium* produces small indehiscent "seed-like" fruits, each consisting of a seed enclosed by a very thin adherent pericarp. Since this thin tissuelike pericarp is the only thing distinguishing fruits from seeds, the terms "fruit" and "seed" are used interchangeably.

2. In anticipation of producing a final report of his work at Russell Cave, Miller has as yet not provided the National Museum with copies of field notes or catalogs. As a result, the collection remains undocumented and uncatalogued.

3. The Libby half-life of 5,568 years was employed, yielding an uncorrected age of 1,955 ± 55 B.P. A small quantity of the sample carbon dioxide was sent to Geochron for $^{12}C/^{13}C$ analysis. This analysis indicated that the sample was from a C_3 pathway plant, necessitating only a small correction, and changing the age to 1,975 ± 55 B.P. (Robert Stuckenrath, personal communication, 1982). To obtain a second radiocarbon age determination directly from the *Chenopodium* fruits, a .734-g-sample of ruptured

specimens was submitted to Beta Analytic, Inc. for particle accelerator (AMS) analysis. The sample yielded a $^{12}C/^{13}C$ corrected date of 2,340 ± 120 B.P. (Beta 11882).

Acknowledgments

Successive drafts of this paper benefited immeasurably from the generous comments, suggestions, and corrections patiently offered by a number of individuals, including David and Nancy Asch, C. Wesley Cowan, Gayle Fritz, George Milner, Vincas Steponaitis, Richard Yarnell, Hugh Wilson, and Wilma Wetterstrom. I would like to thank Hugh Wilson, in particular, for providing collections of fruit and essential information concerning the Mexican domesticated varieties of *Chenopodium*. The research described herein also benefited from the contributions of time and expertise by a number of other individuals. Donna Karolick and Allen Gotchall, Smithsonian volunteers, contributed long hours in the demanding and meticulous hand sorting of *Chenopodium* fruits. Scott Wing, U.S. Geological Survey, and C. Wesley Cowan provided identification of leaf and grass stem fragments. Dr. Robert Stuckenrath, Smithsonian Institution Radiation Biology Laboratory provided the radiocarbon determination so essential to interpretation of the archaeobotanical assemblage. George Robert Lewis drew Figures 5.2 and 5.3. Catherine Valentour, Anthropological Conservation Laboratory, piloted the NMNH Cambridge Stereoscan S100 scanning electron microscope during the present research project. Darla Hawkins typed the manuscript.

Literature Cited

Andrews, E. B.

1877 Report on Exploration of Ash Cave in Benton Township, Hocking County, Ohio. *Peabody Museum of American Archaeology and Ethnology. Tenth Annual Report of the Trustees* 2:48–50.

Asch, D. L., and N. B. Asch

1977 Chenopod as Cultigen: A Re-evaluation of Some Prehistoric Collections from Eastern North America. *Midcontinental Journal of Archaeology* 2:3–45.

1978 The Economic Potential of *Iva annua* and its Prehistoric Importance in the Lower Illinois Valley. In *The Nature and Status of Ethnobotany*, edited by R. I. Ford, pp. 300–341. Museum of Anthropology, Anthropological Paper No. 67. University of Michigan, Ann Arbor.

1980 Early Agriculture in West-central Illinois: Context, Development and Consequences. Paper presented at the School of American Research, advanced seminar on the Origins of Plant Husbandry in North America, March 2–8, 1980, Santa Fe. Center for American Archaeology, Archeobotanical Laboratory Report 35.

1982 Chronology for the Development of Prehistoric Horticulture in West-central Illinois. Paper presented at the 47th annual meeting of the Society for American Archaeology, Minneapolis. Center for American Archeology, Archeobotanical Laboratory Report 46.

1983 Archeobotany. In *Excavation at the Smiling Dan Site*, edited by B. D. Stafford and M. B. Sant, pp. 635–725. Center for American Archeology, Report of Investigations 137, Contract Archaeology Program.

1984 Archeobotany. In *Archeological Testing Along FAS 603 in Pike County, Illinois. Part II, the Hill Creek Site*, edited by M. D. Conner, pp. 107–149. Center for American Archeology, Report of Investigations 151, Contract Archaeology Program.

Boutton, T. W., P. D. Klein, M. J. Lynott, J. E. Price, and L. L. Tieszen

1984 Stable Carbon Isotope Ratios as Indicators of Prehistoric Human Diet. In *Stable Isotopes in Nutrition*, edited by J. R. Turnlund and P. E. Johnson, pp. 191–204. American Chemical Society Symposium Series 258.

Bye, R. A., Jr.

1981 Quelites—Ethnoecology of Edible Greens—Past, Present, and Future. *Ethnobiology* 1:109–123.

Chapman, J., and A. B. Shea

1981 The Archaeobotanical Record: Early Archaic Period to Contact in the Lower Little Tennessee River Valley. *Tennessee Anthropologist* 6:61–84.

Conard, N., D. L. Asch, N. B. Asch, D. Elmore, H. E. Grove, M. Rubin, J. A. Brown, M. O. Wiant, K. B. Farnsworth, and T. G. Cook

1984 Accelerator Radiocarbon Dating of Evidence for Prehistoric Horticulture in Illinois. *Nature* 308:443–446.

Cowan, C. W.

1978 The Prehistoric Use and Distribution of Maygrass in Eastern North America: Cultural and Phytogeographical Implications. In *The Nature and Status of Ethnobotany,* edited by R. I. Ford, pp. 263–288. Museum of Anthropology, Anthropological Paper No. 67. University of Michigan, Ann Arbor.

Cowan, C. W., H. E. Jackson, K. Moore, A. Nickelhoff, and T. L. Smart

1981 The Cloudsplitter Rockshelter, Menifee County, Kentucky: A Preliminary Report. *Southeastern Archaeological Conference Bulletin* 24:60–75.

Crites, G. D.

1978a Paleoethnobotany of the Normandy Reservoir in the Upper Duck Valley, Tennessee. M.A. thesis, Department of Anthropology, University of Tennessee, Knoxville.

1978b Plant Food Utilization Patterns during the Middle Woodland Owl Hollow Phase in Tennessee: A Preliminary Report. *Tennessee Anthropologist* 3:79–92.

1984 Middle Woodland Paleo-Ethnobotany of the Eastern Highland Rim. Paper presented at the 5th annual Mid-South Archaeological Conference, Pinson Mounds, Tennessee.

Crites, G. D., and R. D. Terry

1984 Nutritive Value of Maygrass, *Phalaris caroliniana. Economic Botany* 38:114–120.

Ford, Richard I.

1979 Paleoethnobotany in American Archaeology. In *Advances in Archaeological Method and Theory,* vol. 2, edited by M. Schiffer, pp. 285–336. Academic Press, Orlando, Florida.

1981 Gardening and Farming before A.D. 1000: Patterns of Prehistoric Cultivation North of Mexico. *Ethnobiology* 1:6–27.

Fritz, G. J.

1984 Identification of Cultigen Amaranth and Chenopod from Rockshelter Sites in Northwest Arkansas. *American Antiquity* 49:558–572.

Funkhouser, W. D., and W. S. Webb

1929 *The So-called "Ash Caves" in Lee County Kentucky.* University of Kentucky Reports in Archaeology and Anthropology 2. Lexington.

Gardner, P.

1984 Plant Food Subsistence at Salts Cave, Kentucky: New Evidence. Paper on file, Research Laboratories of Anthropology, University of North Carolina, Chapel Hill (see *American Antiquity* 52:357–366).

Gilmore, M. R.

1931a Vegetal Remains of the Ozark Bluff-Dweller Culture. *Papers of the Michigan Academy of Science, Arts and Letters* 14:83–102.

1931b Ethnobotany Laboratory Report 11(2). Unpublished report on file at the Museum of Anthropology, University of Michigan, Ann Arbor.

Griffin, J. W.

1974 *Investigations in Russell Cave.* Publications in Archaeology 13. National Park Service, U.S. Department of the Interior, Washington, D.C.

Harlan, J. R., and J. M. J. de Wet

1965 Some Thoughts about Weeds. *Economic Botany* 19:16–24.

Harlan, J. R., J. M. J. de Wet, and E. G. Price

1973 Comparative Evolution of Cereals. *Evolution* 27:311–325.

Johannessen, S.

1981 Plant Remains from the Truck 7 Site. In *Archaeological Investigations of the Middle Woodland Occupation at the Truck* 7 and Go Kart South Sites, by A. C. Fortier, pp. 116–130. University of Illinois at Urbana-Champaign, Department of Anthropology FAI 270 Archaeological Mitigation Project Report 30.

1983 Plant Remains from the Cement Hollow Phase. In *The Mund Site,* by A. C. Fortier, F. A. Finney, and R. B. LaCampagne, pp. 94–103. American Bottom Archaeology FAI-270 Site Reports 5.

1984 Paleoethnobotany. In *American Bottom Archaeology,* edited by C. J. Bareis and J. W. Porter. University of Illinois Press, Urbana.

Jones, V. H.

1936 The Vegetal Remains of Newt Kash Hollow Shelter. In *Rock Shelters in Menifee County, Kentucky,* edited by W. S. Webb and W. D. Funkhouser, pp.

147–167. University of Kentucky Reports in Archaeology and Anthropology 3(4). Lexington.

Kay, M., F. B. King, and C. K. Robinson
1980 Cucurbits from Phillips Spring: New Evidence and Interpretations. *American Antiquity* 45:806–823.

Kline, G. W., G. D. Crites, and C. H. Faulkner
1982 *The McFarland Project: Early Middle Woodland Settlement and Subsistence in the Upper Duck River Valley in Tennessee.* Tennessee Anthropological Association, Miscellaneous Paper 8.

Miller, C.
1956 Life 8,000 Years Ago Uncovered in an Alabama Cave. *National Geographic Magazine* 110:542–558.

1957a Field Impressions of the Archeology of Russell Cave, Northern Alabama. *Eastern States Archaeological Federation Bulletin* 16:10–11.

1957b Radiocarbon Date from an Early Archaic Deposit in Russell Cave, Alabama. *American Antiquity* 23:84.

1958 Russell Cave: New Light on Stone Age Life. *National Geographic Magazine* 113:426–437.

1960 The Use of Chenopodium Seeds as a Source of Food by the Early Peoples in Russell Cave, Alabama. *Southern Indian Studies* 12:31–32.

1962 Napier-like Vessel from Russell Cave, Alabama. *Southern Indian Studies* 14:13–15.

1965 Paleo-Indian and Early Archaic Projectile Point Forms from Russell Cave, Northern Alabama. *Anthropological Journal of Canada* 3:2–5.

Price, T. D., M. J. Schoeninger, and G. J. Armelagos
1984 Bone Chemistry and Past Behavior: An Overview. *Journal of Human Evolution* 13(7).

Rindos, D.
1984 *The Origins of Agriculture.* Academic Press, Orlando, Florida.

Roberts, F. H. H. Jr.
1959 *Seventy-Fifth Annual Report of the Bureau of American Ethnology,* 1957–1958. Washington, D.C.

1960 *Seventy-Sixth Annual Report of the Bureau of American Ethnology,* 1958–1959. Washington, D.C.

Smith, Bruce D.
1984 Chenopodium as a Prehistoric Domesticate in Eastern North America: Evidence from Russell Cave, Alabama. *Science* 226:165–167.

1986 The Archaeology of the Southeastern United States from Dalton to de Soto (10,500 B.P.–500 B.P.). In *Advances in World Archaeology,* vol. 5, edited by F. Wendorf and A. E. Close., pp. 1–92. Academic Press, Orlando, Florida.

Stirling, M. W.
1957 *Seventy-Third Annual Report of the Bureau of American Ethnology,* 1955–1956. Washington, D.C.

1958 *Seventy-Fourth Annual Report of the Bureau of American Ethnology,* 1956–1957. Washington, D.C.

de Wet, J. J. M, and J. R. Harlan
1975 Weeds and Domesticates: Evolution in the Man-Made Habitat. *Economic Botany* 29:99–107.

Wilson, H.
1980 Artificial Hybridization among Species of *Chenopodium* sect. *Chenopodium. Systematic Botany* 5:253–263.

1981a Domesticated *Chenopodium* of the Ozark Bluff Dwellers. *Economic Botany* 35:233–239.

1981b Genetic Variation among South American Populations of Tetraploid *Chenopodium* sect. *Chenopodium* subsect. *Cellulata. Systematic Botany* 6:380–398.

Wilson, H. D., and C. B. Heiser, Jr.
1979 The Origin and Evolutionary Relationships of "Huauzontle" (*Chenopodium nuttalliae* Stafford), Domesticated Chenopod of Mexico. *American Journal of Botany* 66:198–206.

Yarnell, R.
1972 *Iva annua* var. *macrocarpa:* Extinct American Cultigen? *American Anthropologist* 74:335–341.

1974 Plant Food and Cultivation of the Salts Cavers. In *Archaeology of the Mammoth Cave Area,* edited by Patty Jo Watson, pp. 113–122. Academic Press, New York.

1978 Domestication of Sunflower and Sumpweed in Eastern North America. In *The Nature and Status of Ethnobotany,* edited by R. I. Ford, pp. 289–299. Museum of Anthropology, Anthropological Papers No. 67. University of Michigan, Ann Arbor.

1983 Prehistoric Plant Foods and Husbandry in Eastern North America. Paper presented at the 48th annual meeting of the Society for American Archaeology, Pittsburg.

CHAPTER 6

❑ ❑ ❑

Chenopodium berlandieri ssp. *jonesianum*: Evidence for a Hopewellian Domesticate from Ash Cave, Ohio

Introduction

In July of 1876 Ebenezer Baldwin Andrews, an eminent geologist and member of the faculty of Marietta College, wrote to Asa Gray of Harvard University, asking for his assistance in the identification of a sample of small black seeds that he had recently recovered during excavations in Ash Cave, Hocking County, Ohio. After being identified as *Chenopodium album* L.[1] by Gray, the seeds were cataloged along with the other Ash Cave materials, which had been sent by Andrews to Harvard's Peabody Museum in October of 1876. Andrews's four-page handwritten account of his excavations at Ash Cave, including marginal illustrations, accompanied the specimens sent to Peabody Museum. This account appeared, without illustrations and with minor editorial changes, in the Museum's annual report of the following year (Andrews 1877).

The 1877 report on Ash Cave has recently been acknowledged as the earliest significant report of prehistoric plant remains in the Eastern Woodlands (Yarnell 1983), and the Ash Cave *Chenopodium* assemblage has been mentioned numerous times as one of the few well documented examples of deliberate prehistoric storage of this indigenous starchy-seeded annual (for example, Asch and Asch 1977).

It is somewhat surprising, therefore, that the Ash Cave chenopod seeds were not the subject of a more detailed analysis at any time during the 107 years that have elapsed since they were excavated by Andrews, identified by Gray, and added to the collections of the Peabody Museum.

This article provides such an analysis. The archaeological and temporal contexts of the Ash Cave chenopod assemblage are established, the morphological characteristics of the fruits/seeds[2] are described, and the question of taxonomic assignment is addressed.

This chapter originally appeared in *Southeastern Archaeology* (1985) 4:107–133.

Such a discussion of morphology, taxonomic assignment, and cultural-temporal context is warranted because the Ash Cave assemblage constitutes one of the largest and best preserved archaeobotanical *Chenopodium* collections yet recovered in the Eastern Woodlands. In addition, the Ash Cave *Chenopodium* fruits consistently exhibit a set of interrelated morphological characteristics that represent unequivocal evidence of domestication (Smith 1984, 1985). Because of their excellent state of preservation, the Ash Cave *Chenopodium* fruits offer an unusual opportunity to document the morphology of a prehistoric domesticated variety of this species, as well as providing information concerning its temporal and spatial distribution.

The Andrews Excavation

Located in a state park 5 km (3 miles) east and south of the town of South Bloomingville, on the west side of a narrow tributary valley of the east fork of Queer Creek, which flows into Salt Creek (NE¼ of NW¼ of Section 26, Township 11, Range 18, Hocking County; Figure 6.1), Ash Cave (33Ho1) is better characterized as a rock shelter than cave:

> This cave is simply a large recess under a high sandrock (of the Waverly Sandstone Series) bordering a small stream. The top of the ledge is perhaps 100 feet high above the stream. (Andrews 1877:48)

The eastward-facing shelter overhang was described as protecting an area approximately 30 ft (9.1 m) wide and extending 100 ft (30.5 m) along the base of the ledge (Figure 6.2), with this protected area containing a very dry "ash belt" deposit (Andrews 1877:48).

Beginning just within the outer edge of the ash deposit, Andrews excavated a narrow trench that extended approximately 25 ft (7.6 m) to the back wall of the shelter (Figure 6.3). The ash deposit was found to vary in thickness from 2.0 to 2.5 ft (61–67 cm), and to overlay a clean (and presumably culturally sterile) sand layer. At first no stratigraphy was noticed by Andrews in the ash layer, with only occasional animal bone fragments and chert flakes being recovered from the trench. But as the back wall of the shelter was approached, a well-defined cultural deposit was encountered.

Of unspecified thickness and extending out at least 3 ft from the back wall of the shelter, the cultural deposit was located from 4 to 6 inches (10–15 cm) below the surface of the ash deposit (Figure 6.4), and consisted of "A confused mass of sticks for arrows, stalks of coarse grasses, food bones in great variety, bits of pottery, flints, nuts, corn-cobs, etc., etc." (Andrews 1877:48). This brief description by Andrews corresponds quite closely to the Peabody Museum specimens attributed to this deposit: two corn kernels, four corn cobs (catalog no. 11.025), three undiagnostic grit-tempered ceramic body sherds (11.027), two gourd fragments (11.028), nut fragments (11.029), chert flakes (11.030), seven *unio* shells (11.031), five arrow-shaft(?) fragments (11.032), sphagnum moss (11.033), and animal bones (11.035–11.036).

Underneath this "layer of refuse" (Andrews 1877:48) and separated from it by a 1.5–2.0 ft (46–61 cm) thick ash deposit was a human burial (11.023–11.024) placed in a sitting position and situated about 3 ft (91 cm) from the back wall of the shelter (Figure 6.4). Andrews did not mention any recognition of a pit outline, and concluded that the overlying refuse deposit was distinct from, and more recent than, the burial.

The burial had been placed in a cavity in the sand against a small loose rock (11.026?) and protected by a largely decayed bark covering (11.034?). No artifacts or items of clothing were recovered in association with the burial.

Adjacent to the skeleton, however, a mat of ferns, leaves, grass, and a piece of coarse cloth (11.037, 11.038) had been placed directly on the sand layer and was found to cover an unlined pocket in the sand containing a quantity of small black seeds (11.039) (Andrews 1877:49). Andrews does not specify either the size or shape of the "pocket" containing the seeds. Judging from the quantity of seeds recovered, however, the pocket or pit likely held a volume of approximately 26.5 liters (24 quarts). In his handwritten account of the excavation (Peabody Museum accession file 76–7) Andrews mentions "a large quantity of small black seeds." In the published version this has been changed to read "about three pecks of small

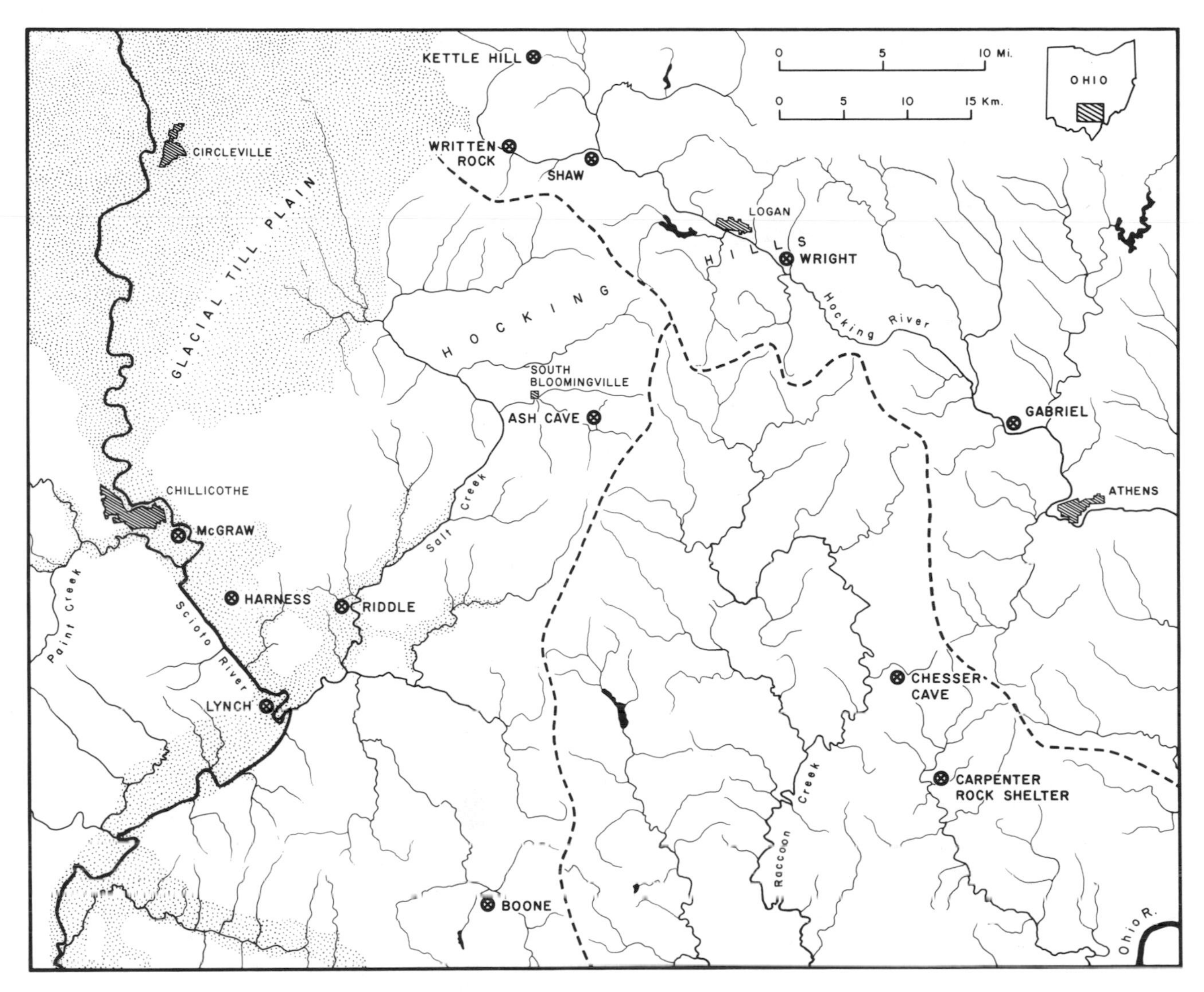

Figure 6.1 The Chillicothe, Ohio, area, showing the transition from the glacial till plain to the unglaciated Hocking Hills region, as well as the location of Ash Cave and other Woodland period sites discussed in the text. Note the demarcation of different drainage catchments.

black seeds" (24 quarts, 0.75 bushel) (Andrews 1877:49). This change was not indicated on the original handwritten report, and it is not known whether it was made by Andrews or by someone at the Peabody Museum. The source for the change may have been the letter written to Gray in which Andrews places the volume of recovered seeds at "a half bushel or more." Judging from the sample of approximately 25,000 seeds (0.064 liter) sent by Andrews to the Peabody Museum (see discussion below) the three pecks of seeds placed in the pocket or pit likely numbered about 9,600,000. The *Chenopodium* assemblage that is the subject of the present study would thus appear to represent about a 0.3 percent sample of over nine million seeds recovered from an undisturbed, below-ground context.

Although this large quantity of harvested seeds was located "quite near" the burial and may have been deposited along with it, a clear cultural and temporal association between the two cannot be established. Neither the burial nor the pit itself yielded any diagnostic materials that would have allowed the accurate cultural or temporal assignment of the *Chenopodium* assemblage.

Figure 6.2 A sketch of Ash Cave made by E. B. Andrews in 1876, showing the location of his excavation trench. From Andrews's original four-page handwritten account of his excavation (Peabody Museum accession file 76–7). (Photograph by Hillel Burger, courtesy of Peabody Museum, Harvard University.)

Subsequent Excavations by Wilson, Moorehead, and Goslin

Further information concerning the cultural deposits and stratigraphy of Ash Cave are provided by the accounts of excavations carried out after the pioneering work of Andrews.

In July of 1895 Thomas Wilson, Smithsonian Institution, and Warren K. Moorehead, Ohio State University, began the exploration of caves in Ross, Highland, and Hocking Counties, Ohio. Two days were devoted to excavating in Ash Cave:

> Picnic parties, relic collectors and others have pretty well dug over the entire deposit. . . . The strata are

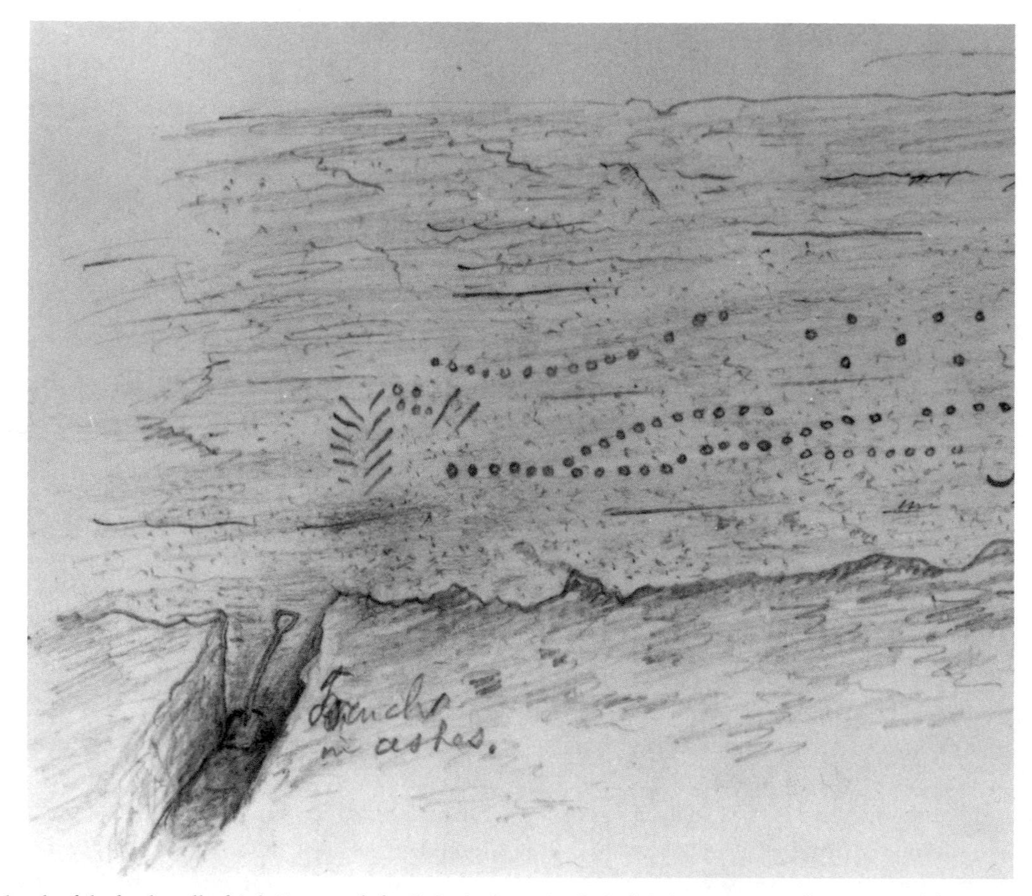

Figure 6.3 A sketch of the back wall of Ash Cave made by E. B. Andrews in 1876, showing a pattern of grooves and dots, as well as the intersection of the "trench in ashes" with the back wall of the cave (Peabody Museum accession file 76–7). (Photograph by Hillel Burger, courtesy of Peabody Museum, Harvard University.)

Figure 6.4 A sketch by E. B. Andrews showing the stratigraphic relationship between the "layer of refuse" and the underlying burial and chenopod assemblage adjacent to the back wall of Ash Cave (Peabody Museum accession file 76–7). (Photograph by Hillel Burger, courtesy of Peabody Museum, Harvard University.)

> greatly disturbed. We understand from people living near the cave that three human skeletons have been found at 4 or 5 feet depth, and that several whole pottery vessels, a sack of seeds, and two or three pairs of sandals were exhumed by some gentleman from New York, but we could not ascertain his name.
>
> We understood that the sack of seeds buried with one of the individuals was of very coarse texture, and in size about 24 x 18 inches.
>
> We think from our examination that the cave was used as a habitation site for a considerable length of time, and that three or four burials were made in the sand at the back of the cave against the rear wall. The ashes covering the sand at the place of burial, are about three feet in depth. We found a number of human bones at the rear of the cave, and from their state of preservation the skeletons uncovered by previous explorers must have been perfect.
>
> We have excavated extensively in Ash Cave and found the relics of man very numerous throughout the ashes and the cave dirt on the floor. These relics consisted of small fragments of pottery, flakes and scales of flint, bones hardened or burned by fire, broken stones, arrowheads, blocks of partly worked flint, bone awls charcoal, burnt stone, etc. About 400 specimens were found in two days' digging.
>
> The pottery is thin, made of local clays and superior in texture to that usually found in Ohio. The decorations are simple incised lines cut with a sharp flint or pointed stick. (Moorehead 1895:311)

Although Moorehead does not give a very detailed account of the 1895 excavation in Ash Cave, he does provide some stratigraphic information indicating that the ash layer, which was "about three feet in depth at the rear of the cave," and already "greatly disturbed" in 1895, overlay a deposit of "cave dirt on the floor." Cultural material was recovered both from the ash layer and the underlying "cave dirt" deposit.

Moorehead also mentions the recovery of human skeletal elements from the rear of the cave, and provides local accounts of prior excavation of three burials, several whole ceramic vessels, sandals, and a "sack of seeds." While the "sack of seeds" suggests that these descriptions most likely refer to the work of Andrews in 1876, the mention of multiple burials, intact ceramic vessels, and sandals may indicate a mixed local history account of Andrews's excavation and subsequent digging in the deposits. There may have been additional burials removed from the rear of Ash Cave between 1876 and 1895.

During the weekend of April 6–8, 1928, Robert Goslin spent a total of 16 hours excavating in Ash Cave. His subsequent letter to H. C. Shetrone, Director of the Ohio State Museum (accession number 899, Ohio Historical Society), in which he describes his work, when combined with his sketch maps (Figures 6.5 and 6.6), materials recovered, and an ethnobotanical report by Melvin Gilmore (1932), provide substantial stratigraphic information and support the description of Ash Cave cultural deposits given by Andrews 50 years earlier (Goslin 1952:24–26).

Goslin's initial excavation trench was oriented roughly east-west and situated toward the southern end of the rock shelter (Figure 6.5a). Excavation was begun at the eastern edge of the ash layer, and "Nothing showed up until about three o'clock when I was carrying my trench towards the back wall and about two feet below the surface I came across a skeleton of an infant" (Goslin 1928; Figures 6.5, 6.6). Goslin completed the excavation of the 5 by 5 ft unit containing the burial and an adjacent 5 by 10 ft unit to the

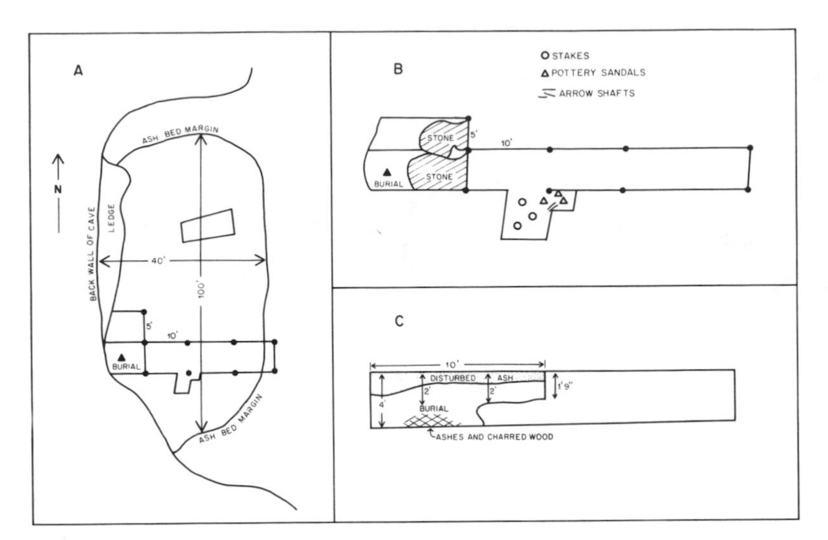

Figure 6.5 Sketch maps (not to scale) of Ash Cave excavations, April 6–8, 1928. Redrawn from Robert Goslin's originals (Ohio Historical Society accession number 899): (a) overall diagram of Ash Cave; (b) "view from above" of excavation units; (c) "cross section looking north."

north "without finding other things of interest" (Goslin 1928).

Goslin then tested various parts of the shelter "without results." The front part of the cave yielded "some fiber chords [sic] and much grass which had been disturbed before" (Goslin 1928) before uncovering several sandals, gourd rinds, arrow shafts, a pocket of broken pottery, and three wooden stakes, all at a depth of only 2–3 inches (5–8 cm) along the southern edge of the excavation trench (Figure 6.5).[3]

Goslin's recovery of artifactual materials at a depth of only 2–3 inches below ground surface corresponds closely to the "layer of refuse" described by Andrews as being located 4–6 inches (10–15 cm) below the surface of the ash deposit. In addition, the burial recovered by Goslin is similar to the one excavated by Andrews in terms of location (within 3 ft [91 cm] of the back wall of the shelter), depth (2 ft [61 cm] below the surface of the ash deposit), and position (flexed, with the head and torso elevated). Both burials were accompanied by masses of vegetal material.

Goslin's findings concerning Ash Cave are compatible with those of Andrews in documenting the presence of a thin, shallow, refuse layer overlying a series of relatively shallow burials placed close to the back wall of the shelter. But Goslin's work did not provide any additional information concerning the cultural or temporal placement of the refuse layer or the burials, nor did it establish if the burials were contemporary with the overlying cultural deposit.

The Temporal Context of the Ash Cave Deposits

Two radiocarbon dates provide information concerning the temporal context of both the shallow refuse layer in Ash Cave and the *Chenopodium* assemblage and burial excavated by Andrews. In addition, these two dates confirm Andrews's belief that the thin habitation deposit was more recent than the underlying chenopod deposit (and perhaps the burial).

As a result of renewed interest in Ash Cave, Raymond Baby submitted a radiocarbon sample comprised of big bluestem grass *(Andropogon gerardi)* to the University of Michigan Memorial Phoenix Project Laboratory. The following information was recorded on the information sheet that accompanied the sample:

> Date of collection and name of collector: April 8, 1928, Robert M. Goslin
> Occurrence and stratigraphic position in precise terms: About two or three inches of ashes and dust overlay the refuse layer, among which I found a pocket of about 100 pieces of broken pottery, grass, a nice assortment of

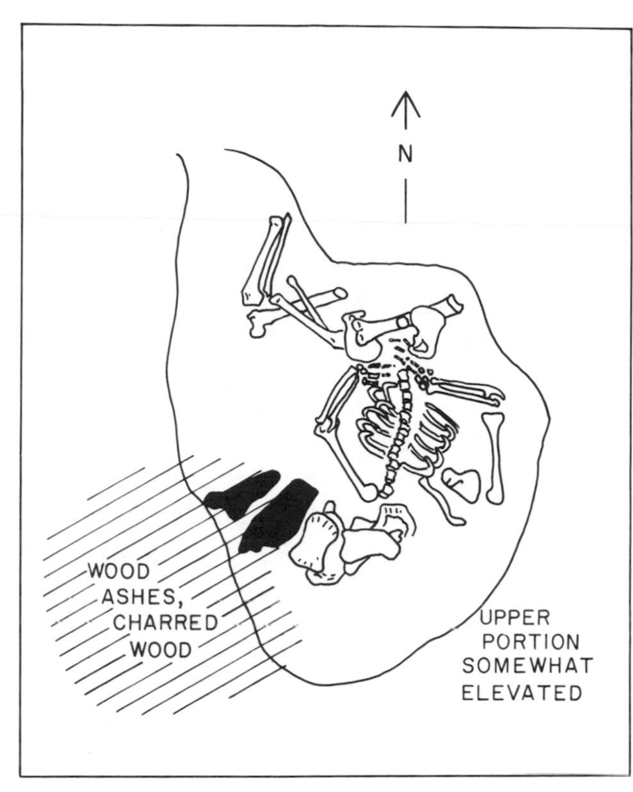

Figure 6.6 A sketch of the burial excavated from Ash Cave by Robert Goslin in April of 1928. Redrawn from Goslin"s original (Ohio Historical Society accession number 899). No scale given.

> gourd rinds, one piece of pumpkin, 2 sticks wrapped with a grass fibre; one having a feather wrapped on it and in place. Among this refuse I found the remnants of another sandal. (Ohio Historical Society Accession Number 899)

Gilmore's ethnobotanical report (1932) contains an itemized description of the plant remains recovered by Goslin from Ash Cave in 1928. Based on this report (and an earlier brief inventory by Gilmore [Ohio Historical Society accession number 899]) the sample of big bluestem grass submitted by Baby for radiocarbon dating was undoubtedly selected from catalog number 899/1 (the only *Andropogon gerardi* identified by Gilmore) "a bundle of broken pieces of culms and leaves of big bluestem grass" (Gilmore 1932). Taken together, the radiocarbon sample information sheet and Gilmore's report establish that the sample was recovered from the shallow refuse layer of the cave approximately 20 ft (6.1 m) from the back wall (Figure 6.5).

The sample in question (M-465) indicated an age of 1,170 ± 200 B.P.: A.D. 780 (Crane and Griffin 1962:188), dating the Ash Cave habitation refuse layer to the Late Woodland period.

In order to establish the temporal context of the Ash Cave *Chenopodium* assemblage in the absence of any associated diagnostic artifacts, the age of the fruits/seeds themselves was determined. A sample consisting of 100 fruits/seeds (weighing 0.025 g) was submitted to Beta Analytic, Inc. for a particle accelerator (AMS) radiocarbon date. The sample indicated an age of 1,720 ± 100 B.P.: A.D. 230 (Beta 11346).

Although having a temporal overlap of 50 years at the two sigma range, these two dates place the shallow habitation refuse deposits in Ash Cave a full five centuries later in time than the chenopod assemblage, which, at A.D. 230, falls comfortably within the Middle Woodland time period. While the burial excavated by Andrews adjacent to the chenopod assemblage may also date to ca. A.D. 230, flexed burials unaccompanied by associated cultural material are typical of the Late Woodland period interments in caves in the Hocking Hills, and the burial might alternatively be associated with the overlying ca. A.D. 780 refuse layer.

The Cultural Context of the Ash Cave Chenopod Assemblage

The absence of diagnostic artifacts in the relatively sparse material culture assemblage associated with the Ash Cave chenopod deposit does not permit its easy assignment to a specific prehistoric cultural entity. Given its Middle Woodland temporal placement, however, a consideration of the surrounding regional environmental and cultural landscape at A.D. 230 allows its general cultural context to be established.

Ash Cave is situated close to environmental boundaries of several kinds, at several different scales. The abrupt and dramatic transition from the low flat glacial till plain of northern and central Ohio to the dissected westernmost extension of the Appalachian plateau in southern Ohio is located only 20 km northwest of Ash Cave (Figure 6.1; Murphy

1975:25). Forty kilometers to the west of Ash Cave, at Chillicothe, the Scioto River, flowing south to the Ohio River, crosses this boundary between two major physiographic provinces, forming a broad and fertile valley averaging 3.2 km in width and flanked by steep wooded slopes.

South of Chillicothe two main tributaries join the Scioto River: Paint Creek from the west and Salt Creek from the east. The Salt Creek drainage catchment area extends to the south, east, and northeast of the Scioto River, into the Hocking Hills, to within several kilometers of the Hocking River (Figure 6.1). Ash Cave, situated on a seasonal tributary of the east fork of Queer Creek, is less than 8 km from the Salt Creek-Hocking River drainage divide, and only 3 km from the Raccoon Creek drainage catchment (Figure 6.1).

In cutting down through Mississippian and Pennsylvania sandstones and shales, these streams have produced a highly dissected, dendritic patterned landscape of deep, steep-walled valleys and coves having numerous rocky overhangs (Murphy 1975:25–27; Black 1979:19).

This dramatic environmental contrast between the glacial till plain and broad valley of the Scioto on the one hand, and the rugged landscape of the Hocking Hills on the other, is paralleled by a similar contrast in the cultural landscape of the area during the Middle Woodland period (Black 1979:19).

The Paint Creek and Scioto River valleys in the vicinity of Chillicothe contain a number of classic Hopewell geometric earthworks and mounds (for example, Harness, Seip, Baum) (Brose and Greber 1979). In association with these large geometric earthworks, a number of authors have postulated communities consisting of numerous small habitation sites or farmsteads (Prufer 1965, 1975b:311–316; Shane 1971:144–145; Black 1979:23). The McGraw site, located just to the southeast of Chillicothe (Figure 6.1), is the only such farmstead in this area to have been excavated (Prufer 1965). Numerous other small Hopewell sites have been identified, however, on the basis of surface collections from restricted midden or refuse areas (Shane 1971). The Ilif Riddle 1 site (Figure 6.1) is typical of such small surface-collected Hopewell sites. Located in the lower Salt Creek Valley, the Riddle site yielded over 200 McGraw series grit-tempered sherds, 75–100 prismatic Flint Ridge blades, and several mica fragments from a 10 by 10 m midden area (Mark Seeman, personal communication, 1985). Within the Lynch site complex at the confluence of Salt Creek and the Scioto River, Lynch 4 also appears to be a small farmstead similar to the McGraw site (Prufer 1975B:311).

The large Middle Woodland earthworks and numerous associated small habitation sites of the Paint Creek, Scioto River, and lower Salt Creek valleys stand in stark contrast to the barely visible Hopewell presence in the Hocking Hills to the east. Moving up into the narrow, steep-sided valleys and coves of the Salt Creek, Raccoon Creek, and Hocking River drainages, one finds only scattered, infrequent evidence of the apparently short-term occupation of rock shelters and open-air sites during the Middle Woodland period (Black 1979:21, 24–25).

In contrast, the numerous small ridge- and bluff-top Adena mounds (Murphy 1975:41; Black 1979:19, 24,n25), as well as the open-air multicomponent sites within the Hocking drainage that have produced surface collections, attest to a strong Early Woodland (Adena) and Late Woodland (A.D. 500–1000) presence. The Middle Woodland period is represented to a limited extent in surface collections from the Gabriel and Wright sites (Figure 6.1), along with other multicomponent sites, primarily by Flint Ridge blade fragments (Murphy 1975:221, 253, 276, 341–332). Boone Rockshelter (Figure 6.1; Mills 1912, fig. 6; Prufer 1968:14), situated along with Ash Cave in the Salt Creek drainage, is the only site in the Hocking Hills region to have yielded diagnostic Hopewell ceramics (Mark Seeman, personal communication, 1985).

The numerous other caves and rock shelters that have been excavated within 40 km of Ash Cave (including Kettle Hill Cave, Written Rock, Shaw Rock Shelter, Chesser Cave, and Carpenter Rock Shelter [Figure 6.1]) also indicate a very limited occupation of the Hocking Hills by Hopewell populations during the Middle Woodland period.

Stratigraphically distinct Woodland components have rarely been observed in the rock shelters of the Hocking Hills, due to either the prehistoric and/or historic disturbance of cave deposits, or the failure on the part of early excavators to look for distinct cultural strata (Murphy 1975). Based on recovered artifact assemblages, however, Kettle Hill Cave (Murphy 1975:310–311), Written Rock Shelter (Murphy 1975:318), Carpenter (Murphy 1975:323), and Chesser Cave (Prufer 1975a:49) were occupied primarily, if not exclusively, during the Late Woodland period. To judge from the ceramics recovered from the two test squares excavated at the Shaw Rock Shelter, it too had a primarily Late Woodland occupation, but a "distinct Middle Woodland component is also present, evidenced by the well made Flint Ridge bladelets" (Murphy 1975:318).

Ash Cave, with a thin Late Woodland habitation layer underlain by a very limited Middle Woodland component that lacked diagnostic artifacts, can be seen to fit quite comfortably into the general pattern of occupation of the Hocking Hills during the Woodland period, as briefly outlined above.

A number of alternative hypotheses have been proposed to explain the limited nature of the evidence for a Middle Woodland presence in the Hocking Valley (Black 1979:24): (a) the continuation of Adena in the area into the Middle Woodland period, constituting a co-existent cultural entity that temporally overlapped with Hopewell populations in the Scioto River Valley to the west; (b) the existence of non-Hopewellian Woodland populations (Prufer's Scioto tradition) whose material culture is difficult to distinguish from that of the preceding Early Woodland and subsequent Late Woodland periods (Murphy 1975:228; Shane and Murphy 1975:332–333); and (c) a reduced occupation and utilization of the Hocking Hills during the Middle Woodland period.

Even though, as Black points out (1979:25), it is not possible to demonstrate conclusively which of these hypotheses is correct, the reduced Middle Woodland presence appears to be the most reasonable explanation of the available archaeological evidence. With Hopewell settlements concentrating along primary rivers and streams, the Hocking Hills would have witnessed only short-term, perhaps seasonal, occupation of rock shelters and open-air sites.

In summary, within the general context of its surrounding environmental and cultural landscape, I would propose that Ash Cave was utilized infrequently during the Middle Woodland period, and that the chenopod assemblage represents either the deliberate storage of this cultigen, or a deposit associated with the adjacent burial. In addition, it does not seem unreasonable to look downstream, to the lower Salt Creek or Scioto River Valley, for the Hopewell populations that were utilizing Ash Cave at around A.D. 230.

The Ash Cave Chenopodium *Assemblage*

General Description

After a century of storage in a capped glass bottle within the Peabody Museum collections, the Ash Cave *Chenopodium* assemblage was found to be in an excellent state of preservation.

With a volume of 0.064 l and a total weight of 8.765 g, the contents of the storage jar consisted almost entirely of *Chenopodium* fruits, along with a small amount of sand and gravel, as well as a number of insect fragments. The *Chenopodium* fruits were neither sorted from the sand and gravel nor individually counted to obtain a total fruit count value. A sample of 1,000 fruits sorted for diameter measurements, however, weighed 0.308 g, yielding a total fruit estimate of 28,458 when divided into the total weight value for the assemblage. Because of the sand and gravel component of the assemblage, a lower total fruit estimate of 25,000 is proposed.

Although chenopod fruits exhibit considerable variation in external form (Figure 6.7a), they share a common internal organization, with centrally located perisperm (nutritive tissue) encircled by the embryonic plant. The embryonic plant is situated directly adjacent to the internal margin of the fruit wall, and the radicle (embryonic root) often forms a "beak" by extending out beyond the arc of circumference of the "fruit wall" (Figure 6.7b).

The "fruit wall," which encapsulates the perisperm and plant embryo, consists of three layers: (a) a thin outer pericarp layer which covers, and is often adherent to, (b) the testa or outer epiderm, which in turn covers (c) a thin inner epiderm layer.

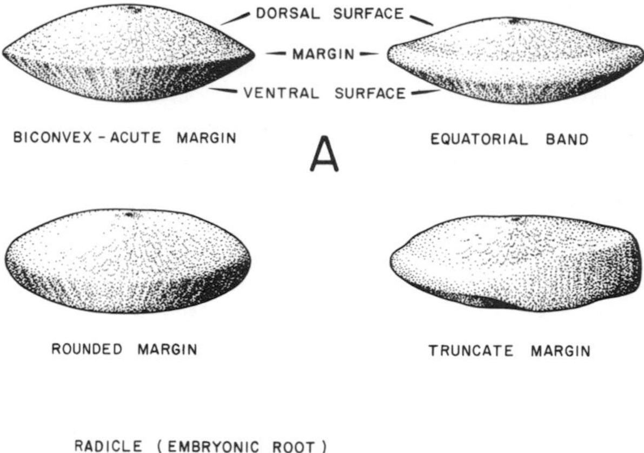

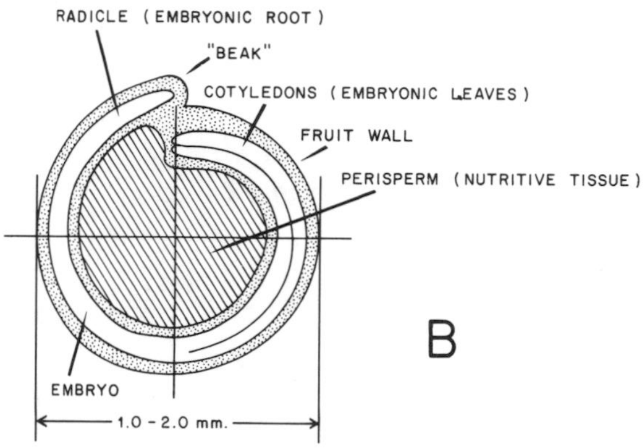

Figure 6.7 The morphology of *Chenopodium* fruits: (a) four idealized fruit margin configurations—biconvex-acute, rounded, equatorial band, and truncate; (b) cross section of a *Chenopodium* fruit, showing internal structure.

With the exception of the perisperm (which had broken down into a dry powdery mass), all of the above mentioned fruit structures were preserved intact in the Ash Cave specimens.

The fruits were uncarbonized, desiccated, and the majority were unfragmented. The outer pericarp layer was at least partially intact in most cases, and very little distortion of the fruits appears to have taken place. Circular holes in the testa, suggesting insect activity, were rare. The most frequently observed distortion was a cracking and partial separation of the radicle "beak" away from the fruit.

Maximum Fruit Diameter

A size range curve was established for the Ash Cave *Chenopodium* assemblage based on maximum diameter measurements taken on a sample of 1,000 fruits (Figure 6.8, Table 6.1, line 13). With a mean maximum fruit diameter value of 1.87 mm, the Ash Cave *Chenopodium* assemblage contains considerably larger fruits than the previously analyzed Russell Cave archaeobotanical assemblage (Smith 1984, 1985 [Chapter 5]).

Located approximately 400 km (250 miles) to the southwest of Ash Cave, Russell Cave yielded approximately 50,000 carbonized *Chenopodium* fruits dating several hundred years earlier than Ash Cave (1,975 ± 55 B.P., 2,340 ± 120 B.P.), and having a mean maximum fruit diameter value of 1.32 mm (Figure 6.8, Table 6.1, line 12).

This variation in fruit diameter between the Ash Cave and Russell Cave *Chenopodium* assemblages appears to conform to geographical trends of variation in fruit size among closely related modern *Chenopodium* populations in the eastern United States. Based on the presence of a reticulate pericarp pattern (discussed in more detail below), both the Russell Cave and Ash Cave *Chenopodium* assemblages can be assigned to subsection *Cellulata* of section *Che-*

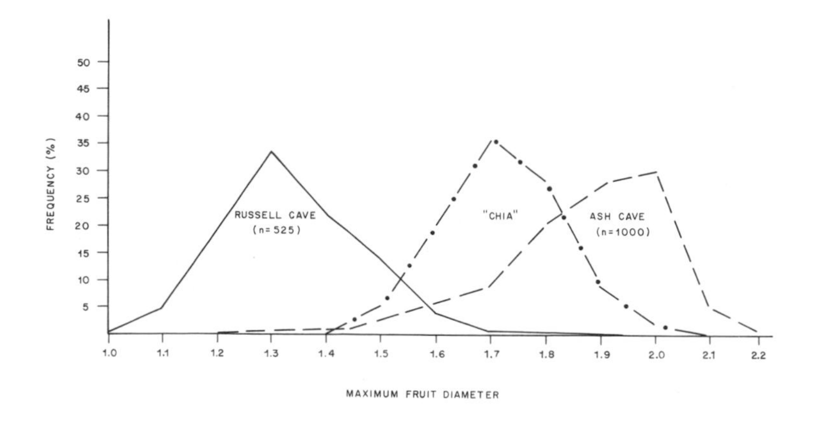

Figure 6.8 The frequency distribution of size (maximum fruit diameter) for three *Chenopodium* populations: Russell Cave, Alabama; modern *C. berlandieri* ssp. *nuttalliae* cv. "chia" (HDW 5116); and Ash Cave, Ohio (Table 6.1, lines 11–13). (Illustration by Ellen Paige.)

nopodium of the genus *Chenopodium* (Wilson 1980). Some modern eastern North American *Chenopodium* populations belonging to subsection *Cellulata* appear to exhibit a general south-to-north cline of variation toward increasing fruit diameter (Table 6.1, lines 1–8).

The pattern of geographical variation in fruit size shown in Figure 6.8 and Table 6.1 is more than a simple latitudinal cline, however, and conforms to the present day geographical range of two species belonging to subsection *Cellulata. Chenopodium berlandieri* var. *zschackei,* which produces relatively small fruits, comparable in size to the Russell Cave assemblage, has a geographical range that encompasses all of the continental United States west of the Mississippi River, with an eastward extension in the upper Midwest that includes Wisconsin, Michigan, Illinois, and part of Indiana. In addition, the range of *C. berlandieri* var. *zschackei* extends east of the Mississippi along the Gulf and Atlantic coastal plain. The plant occurs infrequently in portions of Mississippi, Alabama, Georgia, Florida, and the Carolinas. *Chenopodium bushianum,* a larger-fruited species, has a geographical range that includes much of the northeast and midwest United States. In addition, populations of *C. bushianum* recently collected in Hardin County, Tennessee (BDS 101), Cullman and Tuscaloosa Counties, Alabama (BDS 105, BDA 108), and Cherokee County, South Carolina (Table 6.1, line 5) indicate a southeastern extension of geographical range of this species. *C. bushianum* and *C. berlandieri* are sympatric over an extensive area of the Midwest from Arkansas north through Wisconsin and Michigan. Within this area of overlapping geographical distribution there are apparent intergrading populations. This, when combined with the complete genomic compatibility of the two taxa, indicates that "They might be better treated as a single species" (Wilson 1980:253).

For comparison, fruit diameter information for three modern Mexican populations of *C. berlandieri* (belonging to subsection *Cellulata*) are presented in Table 6.1 (lines 9–11) and Figures 6.8 and 6.9. Two populations of the cultivar *C. berlandieri* ssp. *nuttalliae* cv. "chia" having very similar size range curves (Figure 6.8, Table 6.1, lines 10, 11), are comparable in size to the more northerly modern nondomesticated populations of *C. bushianum.* Line 9 of Table 6.1 provides maximum fruit diameter information for a companion agricultural weed population of *C. berlandieri* ssp. *berlandieri* found growing within 50 m of a cultivated "chia" population (HDW 5113, Table 6.1, line 10 [Hugh Wilson, personal communication, 1983]). The size range curve for this agricultural weed is closely comparable to several of the maximum fruit diameter curves of eastern North American populations of *C. bushianum* (Figure 6.9; Table 6.1, lines 2–5).

Based on the foregoing discussion and the information contained in Figures 6.8–6.9 and Table 6.1, three points can be made concerning the fruit diameter of modern and archaeobotanical *Chenopodium* populations belonging to subsection *Cellulata.* First, modern populations of *C. bushianum* and *C. berlandieri* in the eastern United States exhibit geographical variation in fruit size. Second, the Russell Cave and Ash Cave archaeobotanical chenopod assemblages conform to this general latitudinal cline of variation. Third, both the present-day Mexican cultivar "chia" and a companion chenopod weed population produce fruit comparable in diameter to those of nondomesticated populations of *C. bushianum* in the eastern United States. This similarity of size range curves indicates that in contrast to marsh elder *(Iva annua)* and sunflower *(Helianthus annuus)* (Yarnell 1972, 1978), fruit/seed size does not provide a reliable indication of domesticated status in archaeobotanical assemblages of *Chenopodium* (Smith 1985).

Pericarp Morphology

It is quite fortunate that the Ash Cave chenopod assemblage was both well preserved and well curated, resulting in the frequent preservation of the fruits' thin outer pericarp layer, since this pericarp layer carries morphological characteristics of taxonomic value.

Wilson (1980) has demonstrated, through artificial hybridization, the taxonomic validity of subsection *Cellulata* within section *Chenopodium* of the genus *Chenopodium.* In addition to including the New World domesticated chenopod taxa of Mexico *(C. berlandieri* ssp. *nuttalliae)* and South America *(C. quinoa)* (along with their related sympatric weed

Table 6.1. Maximum Fruit Diameters of Modern Undomesticated, Modern Domesticated, and Prehistoric Domesticated Populations of *Chenopodium* Belonging to Subsection *Cellulata*

Collection	Taxonomic designation	Habitat	Geographical location	Approximate north latitude
1. BDS 57	*C. berlandieri/ bushianum*	disturbed ground	Mississippi County, Arkansas	
2. FMNH 965303[a]	*C. bushianum*	bluff edge, Lamine River	Cooper County, Missouri	38.5°
3. BDS 39	*C. bushianum*	disturbed ground	Fulton County, Pennsylvania	39.5°
4. NUAP 267[a]	*C. bushianum*	beneath bridge, Illinois River	Scott County, Illinois	39.5°
5. BDS 82	*C. bushianum*	beneath bridge, Broad River	Cherokee County, South Carolina	35°
6. NUAP 646–649[a]	*C. bushianum*	high bank, Willow Island	Greene County, Illinois	39.5°
7. BDS 20	*C. bushianum*	disturbed ground	Wayne County, Michigan	42.5°
8. FMNH 880919[a]	*C. bushianum*	riverbank, among trees	Cass County, North Dakota	47°
9. HDW 5114	*C. berlandieri* ssp. *berlandieri*	weed, within 50 m of HDW 5113	Opopeo, Michoacán, Mexico	
10. HDW 5113	*C. berlandieri* ssp. *nuttaliae* cv. "chia"	domesticated variety, cultivated	Opopeo, Michoacán, Mexico	
11. HDW 5116	*C. berlandieri* ssp. *nuttaliae* cv. "chia"	domesticated variety, cultivated	Opopeo, Michoacán, Mexico	
12. NMNH 528403	*C. berlandieri* ssp. *jonesianum*	domesticated variety, rehistoric	Russell Cave, Jackson County, Alabama	35°
13. Peabody 11039	*C. berlandieri* ssp. *jonesianum*	domesticated variety, prehistoric	Ash Cave, Hocking County, Ohio	39.5°

[a]Source: Asch and Asch 1977, table 2.

[b]Column headings are in millimeters; tabulated frequencies are percentages.

forms), subsection *Cellulata* also contains four modern eastern North American nondomesticated taxa (*C. berlandieri, C. bushianum, C. boscianum* [Gulf Coast], and *C. macrocalycium* [North Atlantic Coast]).

The single most important and most consistent morphological characteristic that serves to define subsection *Cellulata* is a reticulate (netlike), alveolate (pitted) dorsal pericarp pattern (Wilson 1980:254, figs. 1–4). Figure 6.10a–f shows this distinctive pericarp pattern on fruits from modern Mexican and eastern North American chenopod populations belonging to subsection *Cellulata*.

Fruits of both the Russell Cave and Ash Cave chenopod assemblages were also found to exhibit this distinctive honeycomb dorsal pericarp pattern (Figure 6.11a–f), allowing their confident assignment to subsection *Cellulata*. The pericarp patterning of the Russell Cave and Ash Cave fruits is also quite similar to that of modern taxa assigned to subsection *Cellulata* in terms of the size, shape, and orientation of the individual "honeycomb" elements.

Organized in a general radiating pattern, the individual elliptical honeycomb pericarp elements of modern members of subsection *Cellulata* range in size from 40 to 100 microns, with their long axis oriented radially (Figure 6.10a–e). The individual honeycomb elements of the dorsal pericarp of Ash Cave and Russell Cave have a similar range of variation in size, and are also organized in a radial pattern (Figure 6.11a–e).

When viewed in cross section the pericarp layer of modern members of subsection *Cellulata* exhibit a complex structure and vary in thickness from a minimum of 2–3 microns at the center of each "honeycomb" element to as much as 20–25 microns at the "ridges" between honeycomb elements (Figure 6.10f). In contrast, the pericarp layer of Ash Cave and

Table 6.1.—*Extended*

			Maximum fruit diameter												
		Standard	Frequency distribution												
Sample size	Mean (mm)	deviation (mm)	1.0	1.1	1.2	1.3	1.4	1.5	1.6	1.7	1.8	1.9	2.0	2.1	2.2
100	1.37	0.08			5	34	44	17							
25	1.46	0.07				4	48	36	12						
100	1.55	0.08					6	48	35	9	2				
100	1.55	0.05					2	46	48	4					
100	1.57	0.09					7	41	34	15	3				
100	1.61	0.13					5	26	46	14	3	3	2. 1		
100	1.74	0.10					4	16	35	35	7	2	1		
25	2.08	0.08										8	20	56	15
50	1.50	0.10			2	4	20	46	22	6					
50	1.71	0.10						6	20	36	32	6			
50	1.71	0.09						6	20	36	28	8	2		
525	1.32	0.13	1	5	20	34	22	14	3	0.3	0.3				
1,000	1.87	0.15			0.5	0.8	2.4	5.1	9.2	21.8	24.7	29.6	5.1	0.8	

Russell Cave fruits is less complex, with much less prominent "ridges," and ranges in thickness from 1 to 7 microns (Figure 6.11f). This difference in the height of pericarp ridges of modern and archaeobotanical fruits is quite likely at least partially the product of deterioration of the latter over the past 1,700 years.

In summary, the dorsal pericarp layer of chenopod fruits from Ash Cave (and Russell Cave) is quite simi-

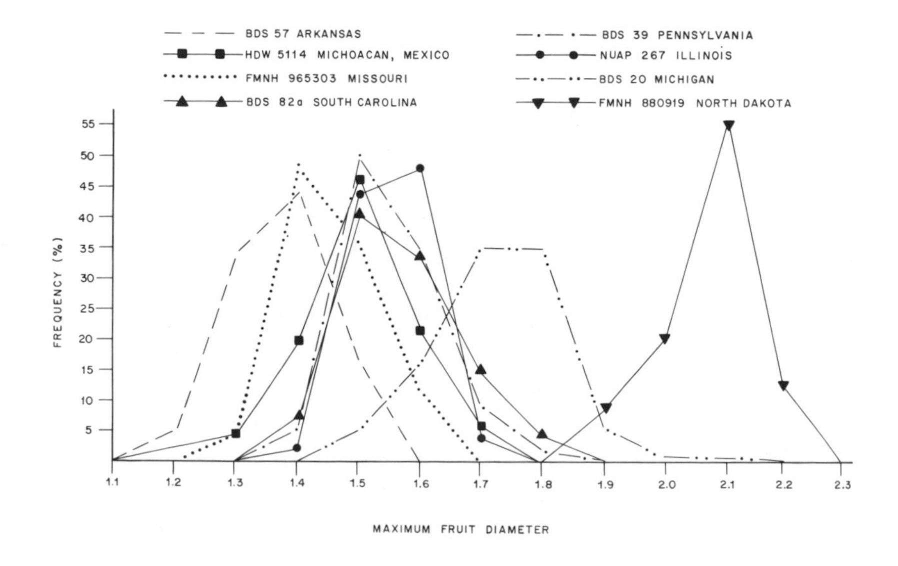

Figure 6.9 The frequency distribution of size (maximum fruit diameter in mm) for eight modern *Chenopodium* populations belonging to subsection *Cellulata* (Table 6.1).

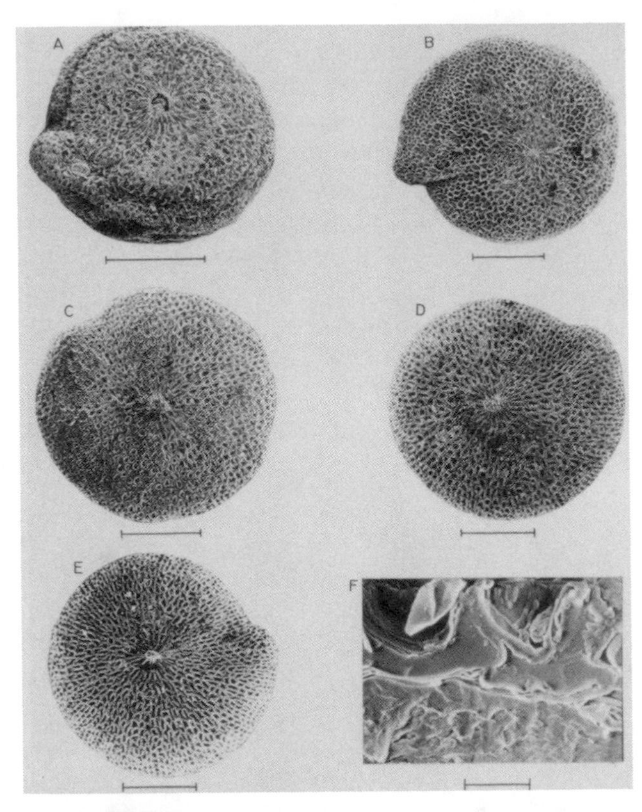

Figure 6.10 Photomicrographs showing the reticulate-alveolate dorsal pericarp patterning of modern taxa belonging to subsection *Cellulata:* (a) *C. berlandieri* ssp. *nuttalliae* cv. "huauzontle," HDW 5119, Tlaxcala Mexico; (b) *C. berlandieri* ssp. *nuttalliae* cv. "chia," HDW 5116, Michoacan, Mexico; (c) *C. berlandieri* ssp. *berlandieri,* HDW 5114, Michoacan, Mexico; (d) *C. bushianum,* BDS 39, Fulton County, Pennsylvania; (e) *C. bushianum,* BDS 20, Wayne County, Michigan; (f) *C. berlandieri* ssp. *berlandieri,* HDW 5114, Michoacan. Mexico—cross section showing pericarp and outer epiderm layers (scale bars a–e, 500 microns; f, 20 microns; HDW = Hugh Wilson catalog number; BDS = Bruce D. Smith catalog number).

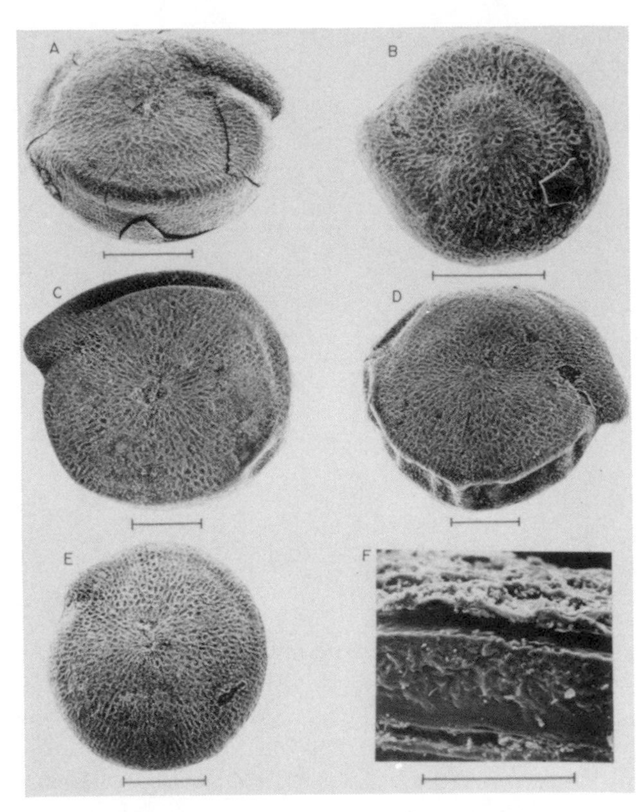

Figure 6.11 Photomicrographs showing the reticulate-alveolate dorsal pericarp patterning of prehistoric taxa belonging to subsection *Cellulata:* (a) Russell Cave, Alabama, mount 5, specimen 1; (b) Russell Cave, Alabama, mount 5, specimen 3; (c) Ash Cave, Ohio, mount 21, specimen 5; (d) Ash Cave, Ohio, mount 21, specimen 9; (e) Ash Cave, Ohio, mount 29, specimen 7; (f) Ash Cave, Ohio, mount 45, specimen 5 (round-margin fruit)—cross section showing pericarp, outer epiderm, and inner epiderm layers (scale bar a–e, 500 microns; f, 50 microns).

lar to that exhibited by modern taxa belonging to subsection *Cellulata* in terms of having a reticulate or "honeycomb" pattern, with individual "honeycomb" elements ranging in size from 40–100 microns, and having their long axis organized in a radial pattern.

Testa or Outer Epiderm

A number of different characteristics of testa morphology are of potential value in distinguishing between the fruits of domesticated and nondomesticated taxa belonging to subsection *Cellulata,* including testa thickness, testa (fruit) margin configuration, and testa surface morphology.

Chenopod fruits from modern populations and from the Ash Cave assemblage were mounted and viewed with the scanning electron microscopes of the National Museum of Natural History in order to investigate these micromorphological characteristics.

TESTA THICKNESS. In nondomesticated populations of *Chenopodium* the thick outer epiderm or testa serves to block the passage of external elements that are essential prerequisites for embryo emergence and development, thus preventing the premature germination of mature seeds that have been naturally dispersed and are present in the soil.

Nine separate populations of *Chenopodium* from the Eastern Woodlands of North America were sampled to establish the range of variation in testa thickness that could be expected for fruits of nondomesticated taxa (Figure 6.12i–q). Seven of the nine populations (Figure 6.12i–o) were quite similar in terms of testa thickness values, with means ranging from 43 to 52 microns. A population of *C. bushianum/berlandieri* from Arkansas (Figure 6.12p) and a population of *C. gigantospermum* from Kentucky had larger mean testa thickness values of 60 and 69 microns, respectively.

In contrast to the clear geographical cline in maximum fruit diameter documented for nondomesticated populations of *Chenopodium* belonging to subsection *Cellulata* (Table 6.1), testa thickness appears to remain consistent across a broad geographical area of the eastern United States.

While the physical dormancy produced by such thick outer seed coats is of obvious selective advantage in wild and weedy populations, it is nonadaptive in domesticated crops, where selection pressures are very strong for seeds that germinate soon after planting. The strong selection pressure for reduced dormancy in domesticates often results in a thinner outer epiderm, or even its complete loss. Such is certainly the case for the Mexican domesticated varieties of *Chenopodium* (Wilson and Heiser 1979:199). The "chia" variety of *C. berlandieri* ssp. *nuttalliae* has a thin outer epiderm, while the "huauzontle" variety lacks an outer seed coat entirely (Wilson 1981:236, 237).

A sample of 20 fruits of the modern thin-testa domesticate "chia" yielded testa thickness values ranging from 9 to 21 microns, with a mean of 16 microns (Figure 6.12c). This represents a 62 percent reduction in testa thickness in comparison with the modern nondomesticated population of *Chenopodium* having the smallest mean testa thickness value (Figure 6.12i).

Previous analysis of a sample of 20 *Chenopodium* fruits from Russell Cave, Alabama had shown them to be comparable to the domesticate "chia," with a mean testa thickness value of 11 microns (Figure 6.12a) (Smith 1984, 1985).

Twenty fragmented fruits from the Ash Cave *Chenopodium* assemblage were found to range in testa thickness from 11 to 21 microns, with a mean of 14.9 microns (Figure 6.12b).

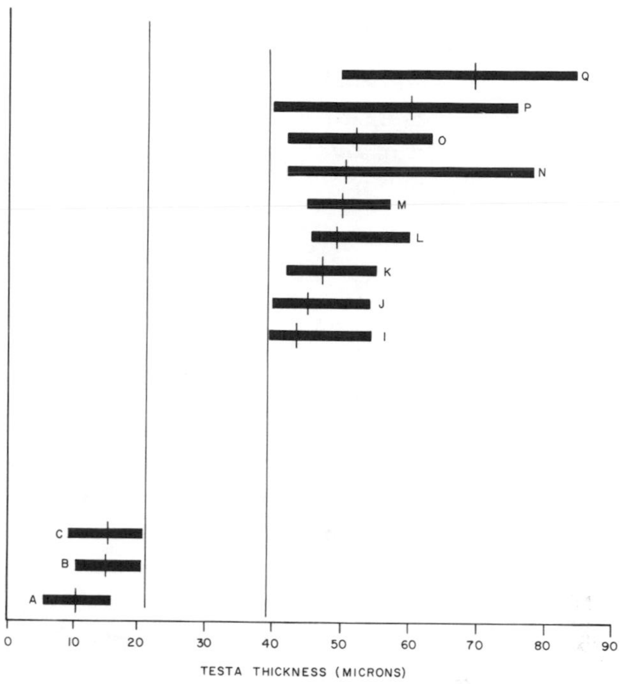

Figure 6.12 Testa thickness in microns for 12 modern and prehistoric *Chenopodium* populations: (a) Russell Cave, NMNH 528403, Jackson County, Alabama; (b) Ash Cave, Peabody 11039, Hocking County, Ohio; (c) *C. berlandieri* ssp. *nuttalliae* cv. "chia," HDW 5113, 5115, Michoacan, Mexico; (i) *C. bushianum*, BDS 20, Wayne County, Michigan; (j) *C. bushianum*, Jackson County, Missouri; (k) *C. missouriense*, NMNH 785467, Jackson County, Missouri; (l) *C. bushianum*, BDS 82a, Cherokee County, South Carolina; (m) *C. berlandieri*, NMNH 1201426, Jackson County, Missouri; (n) *C. bushianum*, BDS 39, Fulton County, Pennsylvania; (o) *C. bushianum*, CAA 646, Greene County, Illinois; (p) *C. berlandieri/bushianum*, BDS 57, Mississippi County, Arkansas; (q) *C. gigantospermum*, BDS 48, Menifee County, Kentucky. (HDW = Hugh Wilson catalog number; BDS = Bruce Smith catalog number; CAA = Center for American Archaeology, Kampsville, Illinois catalog number; NMNH = National Museum of Natural History catalog number.)

The Ash Cave *Chenopodium* assemblage thus clearly compares favorably in terms of testa thickness with both the modern Mexican "chia" population and the prehistoric domesticate recovered from Russell Cave, while contrasting dramatically with the modern thick-testa nondomesticated populations studied (Figure 6.12).

TESTA COLOR. The Ash Cave fruits are without exception matte black in color. They are translucent, however, and dark red when illuminated from behind. In comparison, the fruits of "chia" are predominately red in color, but black-colored fruits often occur (Wilson 1981:237).

TESTA/FRUIT MARGIN CONFIGURATION. A reduction in testa thickness, indicative of a reduction in germination dormancy, is also associated, perhaps causally, with a change in the margin configuration of *Chenopodium* fruits.

The fruits of domesticated varieties of *Chenopodium* have a truncate margin, resulting in a characteristic rectanguloid cross section, which is quite different from the biconvex, rounded, or equatorial banded margins of nondomesticated fruits (Figures 6.7a, 6.13). This margin truncation extends around the full circumference of the fruits of the "testa-less" modern Mexican domesticate "huauzontle." In contrast, the margin of thin-testa fruits of the "chia" variety is truncated along that portion of the fruit adjacent to the embryonic leaves, while often having a more rounded or equatorally banded margin adjacent to the radicle or embryonic root (Figures 6.7a, 6.11).

With the exception of a small number of fruits having a rounded margin (discussed later), the Ash Cave *Chenopodium* fruits all exhibited a distinctive truncate margin and rectanguloid cross section (Figure 6.13d) very similar to that of the modern Mexican domesticate "chia."

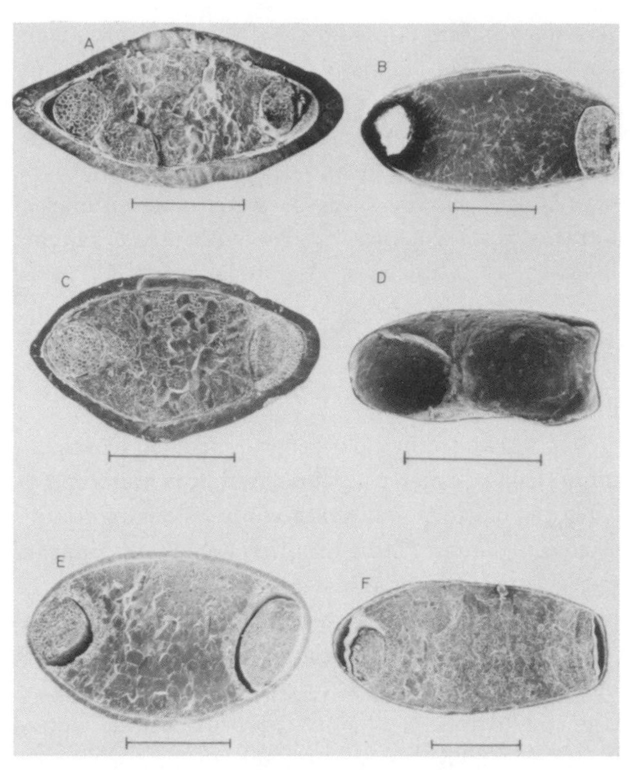

Figure 6.13 Cross-section photomicrographs of six *Chenopodium* fruits, showing variation in morphology and testa thickness: (a) *C. gigantospermum,* BDS 48, Menifee County, Kentucky; (b) *C. berlandieri* ssp. *nuttalliae* cv. "chia," HDW 5116, Michoacan, Mexico; (c) *C. berlandieri/bushianum,* BDS 57, Mississippi County, Arkansas; (d) *Chenopodium* from Ash Cave, mount 48, specimen 1, Peabody Museum 11039, Hocking County, Ohio; (e) *C. berlandieri* ssp. *berlandieri,* HDW 5114, Michoacan, Mexico; (f) *C. bushianum* (red-morph fruit), BDS 82a, Cherokee County, South Carolina (scale bar 500 microns).

THE "RED MORPH" PROBLEM. The two interrelated morphological characteristics briefly discussed above (truncate margin, thin testa) would appear to represent clear archaeological indicators of domestication that could be observed on individual fruits within archaeobotanical assemblages. It is not possible, however, to assign domesticated status to individual *Chenopodium* fruits in archaeobotanical assemblages on the basis of thin testa and truncate margin characteristics alone (the complete absence of an outer epiderm would constitute unequivocal evidence for domesticated status).

Within modern, nondomesticated eastern North American populations of *Chenopodium* belonging to subsection *Cellulata,* a small percentage of fruits produced by individual plants closely resemble the thin-testa truncate-margin fruits of the Mexican cultivar *C. berlandieri* ssp. *nuttalliae* cv. "chia."

Although the possible underlying factors controlling the low frequency production of such reddish colored "red morph" fruits have never been studied, they may represent a small investment in reduced dormancy fruits that will germinate quickly in the next growing season (Williams and Harper 1965).

Such thin-testa reddish fruits were found to predominate in the harvest obtained by David and Nancy Asch from a population of small, late-germinating *C. bushianum* plants growing on a mud flat of Willow Island, in the lower Illinois River (Asch and Asch 1977:20). Subsequent field collecting has not identified any populations producing more than a small percentage of reddish thin-testa fruits (Asch and Asch 1982:10, 1985:377, 693), indicating that "The conditions under which the thin seed coat morph predominated in the wild are unusual" (Asch and Asch 1985:377).

Three random samples of 500 fruits, drawn from three separate modern populations of *Chenopodium* belonging to subsection *Cellulata* (Table 6.1; lines 1, 3, 5) were found to contain, respectively, 9, 12, and 16 red morph fruits. A single 1.2-m-high plant harvested from a fourth population (Table 6.1, line 7) in October 1984 yielded a total of 3,236 fruits (weighing 4.623 g), including 113 immature fruits and 47 red morph thin-testa fruits.

This admittedly quite limited study of red morph fruit frequency in modern populations of *Chenopodium* belonging to subsection *Cellulata* suggests that, with the exception of those rare situations where conditions encouraging higher production of red morph fruits are present, they will represent only 1–3 percent of a population's fruit production. A similar frequency of production (1.26–2.64 percent) of thin-testa "brown morph" fruits was observed for five populations of *C. album* in the United Kingdom (Williams and Harper 1965:142).

A total of 26 red morph fruits from three modern populations of *C. bushianum,* including the lowground Willow Island population studied by Nancy and David Asch, were sectioned, and testa thickness measurements were taken with the help of a scanning electron microscope (Figure 6.14d–f). These red morph fruits were found to be comparable in terms of testa thickness with those recovered from Ash Cave and Russell Cave, as well as those of the modern Mexican domesticate "chia" (Figure 6.14).

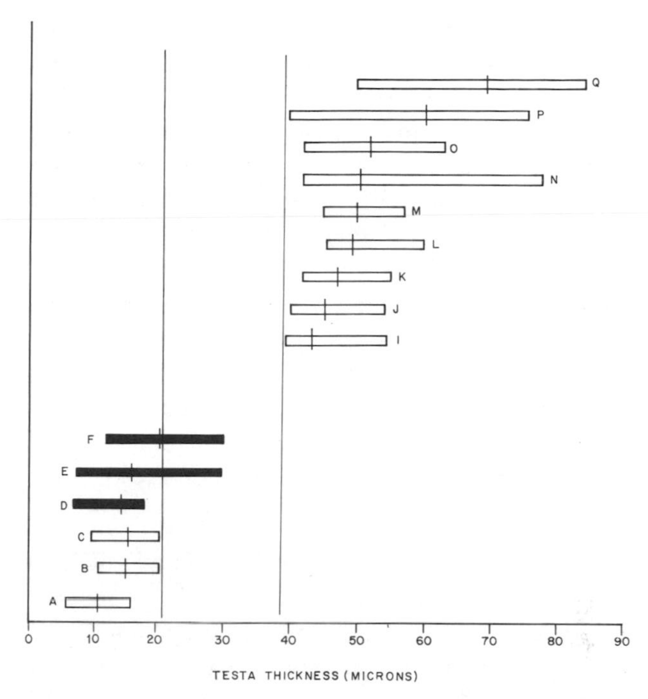

Figure 6.14 Testa thickness in microns for red-morph fruits from three *Chenopodium* populations compared with 12 modern and prehistoric *Chenopodium* populations: (a) Russell Cave, NMNH 528403, Jackson County, Alabama; (b) Ash Cave, Peabody 11039, Hocking County, Ohio; (c) *C. berlandieri* ssp. *nuttalliae* cv. "chia," HDW 5113, 5115, Michoacan, Mexico; (d) *C. bushianum* (red-morph fruits), BDS 20, Wayne County, Michigan; (e) *C. bushianum* (red-morph fruits), CAA 653, low ground Willow Island, Greene County, Illinois; (f) *C. bushianum* (red-morph fruit), BDS 82a, Cherokee County, South Carolina; (i) *C. bushianum,* BDS 20, Wayne County, Michigan; (j) *C. bushianum,* Jackson County, Missouri; (k) *C. missouriense,* NMNH 785467, Jackson County, Missouri; (l) *C. bushianum,* BDS 82a, Cherokee County, South Carolina; (m) *C. berlandieri,* NMNH 1201426, Jackson County, Missouri; (n) *C. bushianum,* BDS 39, Fulton County, Pennsylvania; (o) *C. bushianum,* CAA 646, Greene County, Illinois; (p) *C. berlandieri/bushianum,* BDS 57, Mississippi County, Arkansas; (q) *C. gigantospermum,* BDS 48, Menifee County, Kentucky. (For sample statistics, see caption of Figure 6.20; for key to abbreviations, see caption to Figure 6.12).

Red morph fruits were also found usually to exhibit a rounded but sometimes truncate margin configuration similar to that of the domesticate "chia" (Figure 6.13f).

Establishing the presence of a thin-testa domesticated variety of *Chenopodium* in archaeobotanical assemblages is obviously complicated by the ability of nondomesticated chenopod populations to produce small frequencies of red morph fruits that are sometimes almost indistinguishable from the fruits produced by domesticated stands of *Chenopodium.* Hence it is hard to argue for domestication on the basis of a small sample of *Chenopodium* in which only a few thin-testa fruits are present.

While the potential impact of this red morph problem declines in importance as both the size of an archaeobotanical *Chenopodium* assemblage and the frequency of thin-testa fruits increases (Smith 1985), the identification of morphological characteristics that could be used to differentiate between wild-weedy red-morph and domesticated fruits would be of obvious value in establishing the presence or absence of a domesticated variety of *Chenopodium* in the small-sample situations that are common in the Eastern Woodlands.

Based on an admittedly small data set, it appears that the relative smoothness of the surface of the outer epiderm or testa might well represent such a critical morphological characteristic that would allow nondomesticated red morph and domesticated fruits to be differentiated (Asch and Asch 1985:377).

TESTA SURFACE MORPHOLOGY. When the pericarp is removed from the fruits of modern nondomesticated taxa belonging to subsection *Cellulata*, the distinctive reticulate-alveolate pattern in the pericarp layer can be seen to extend down into the testa in the form of numerous depressions or pits (Figure 6.15). This pattern of pits separated by an irregular netlike pattern of ridges conforms closely to the overlying pericarp pattern (Figure 6.15d).

Although often less distinct and more variable, the same pattern of irregular netlike ridges and depressions can be observed on thin-testa "red morph" fruits, particularly toward the fruit margin (Figure 6.16a–b) (Asch and Asch 1977:20).

In contrast, the testa surface of "chia" fruits is essentially smooth, as is the testa surface of the Ash Cave fruits. While Ash Cave and "chia" fruits lack depressions and netlike ridges on the testa surface, they often do have small dimples or bumps evident on the testa surface, measuring from 20 to 40 microns in diameter (Figure 6.16c–f).

INNER EPIDERM. A comparison of the inner epiderm layer of Ash Cave fruits with that of fruits of modern domesticated and nondomesticated taxa belonging to subsection *Cellulata* did not identify any clear pattern of morphological variation. As a result, the inner epiderm does not have any diagnostic value in determining the domesticated status of *Chenopodium* fruits recovered from archaeobotanical assemblages, but it is worth a brief description here for several reasons. First, the inner epiderm layer is subject to considerable deterioration and resultant variation in appearance in archaeobotanical specimens. Second, the inner epiderm, if well preserved, might be mistaken for a dorsal pericarp layer exhibiting a reticulate-alveolate pattern.

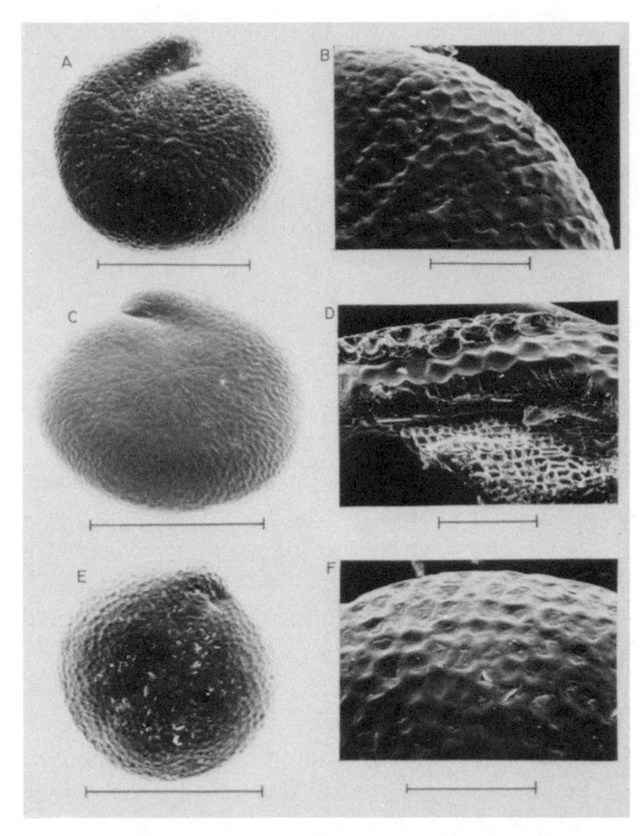

Figure 6.15 Photomicrographs of the testa surface of modern nondomesticated *Chenopodium* fruits belonging to subsection *Cellulata*, showing the distinctive alveolate morphology: (a) *C. berlandieri* ssp. *berlandieri*, HDW 5114, Michoacan, Mexico; (b) *C. berlandieri* ssp. *berlandieri*, HDW 5114, Michoacan, Mexico, detail of a; (c) *C. bushianum*, BDS 39, Fulton County, Pennsylvania; (d) *C. berlandieri* ssp. *berlandieri*, HDW 5114, Michoacan, Mexico, cross section showing pericarp, testa, and inner epiderm layers; (e) *C. berlandieri/bushianum*, BDS 57, Mississippi County, Arkansas; (f) *C. berlandieri/bushianum*, BDS 57, Mississippi County, Arkansas, detail of e (scale bars: a, e, 1 mm; b, d, f, 200 microns; c, 500 microns).

For those Ash Cave fruits in which it could be observed, the thin (about 2–4 microns) inner epiderm layer showed considerable variation in preservation. While it was not at all unusual to observe inner epiderm segments of Ash Cave fruits that showed little deterioration in comparison to modern fruits (com-

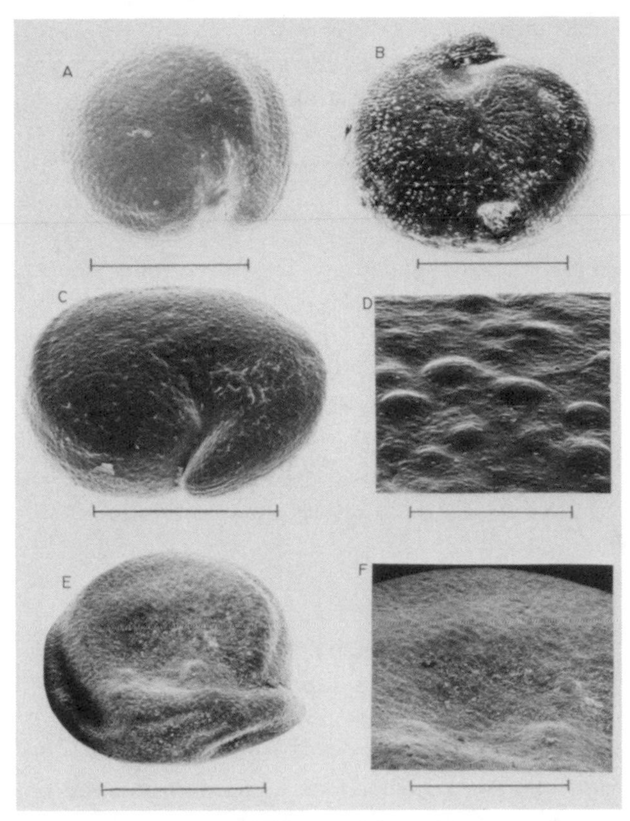

Figure 6.16 Photomicrographs of the testa surface of modern domesticated, red-morph, and Ash Cave fruits: (a) *C. bushianum,* CAA 653, Greene County, Illinois, low ground Willow Island red-morph fruit; (b) *C. bushianum,* BDS 20, Wayne County, Michigan, red-morph fruit; (c) *C. berlandieri* ssp. *nuttalliae* cv. "chia," HDW 5116, Michoacan, Mexico; (d) *C. berlandieri* ssp. *nuttalliae* cv. "chia," HDW 5116, Michoacan, Mexico, detail of c; (e) Ash Cave, mount 47, specimen 5, Peabody Museum 11039, Hocking County, Ohio; (f) Ash Cave, mount 47, specimen 5, Peabody Museum 11039, Hocking County, Ohio (scale bars: a–c, e, 1 mm; d, 500 microns; f, 100 microns).

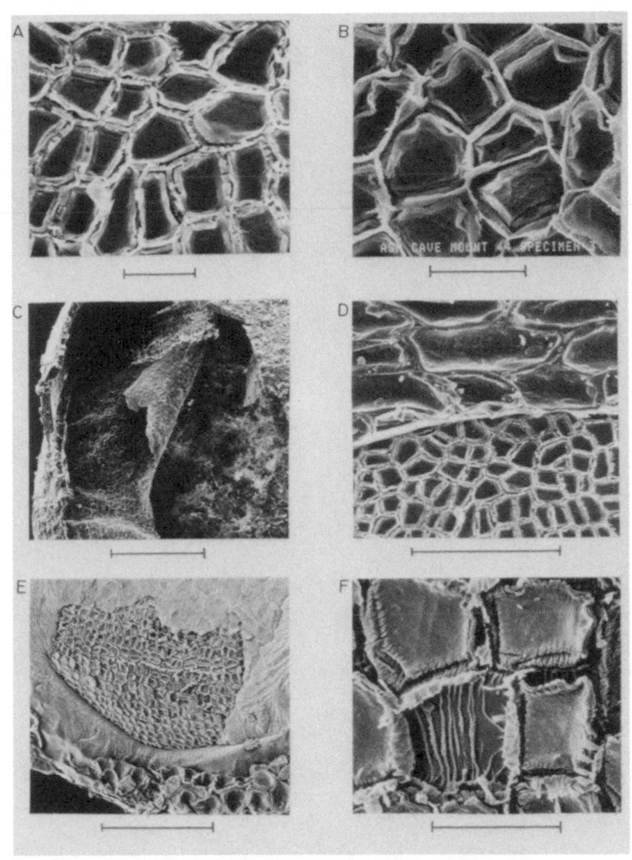

Figure 6.17 Photomicrographs of modern and archaeological *Chenopodium* fruits, showing inner epiderm morphology: (a) *C. bushianum,* CAA 653 Greene County, Illinois; (b) Ash Cave, mount 44, specimen 3, Peabody Museum 11039, Hocking County, Ohio; (c) Ash Cave, mount 29, specimen 2, Peabody Museum 11039, Hocking County, Ohio, folded inner epiderm structure showing loss of "cup" structures; (d) *C. bushianum,* CAA 653, Greene County, Illinois, comparison of pericarp patterning with underlying inner epiderm layer, note scale difference; (e) *C. berlandieri* ssp. *berlandieri,* HDW 5114, Michoacan, Mexico, cut section showing pericarp layer, outer and inner epiderm; (f) *C. berlandieri* ssp. *berlandieri,* HDW 5114, Michoacan, Mexico, detail of e, showing missing "cup" structure (scale bars: a, b, f, 20 microns; c, e, 200 microns; d, 100 microns).

pare Figure 6.17a with 6.17b), it was far more common to find that the inner epiderm layer had deteriorated considerably. This deterioration consistently involved the separation and scaling of upper "cup-like" polygonal structures from an underlying smooth layer (Figure 6.17c–d). The loss of such a structure from a modern specimen is shown in Figure 6.17f. As a result, the inner epiderm may well be present, but only as a smooth and largely unpatterned layer in archaeobotanical specimens.

In contrast, if intact, these upper polygonal honeycomb structures provide the inner epiderm layer with a resemblance to the reticulate-alveolate dorsal pericarp pattern of taxa belonging to subsection *Cellulata.* As shown in Figures 6.15d and 6.17d–e, however, any possible confusion of pericarp and inner epiderm layers can be avoided easily due to the difference in scale. While the honeycomb pericarp elements of taxa belonging to subsection *Cellulata* range in size from 40 to 100 microns, inner epiderm polygonal structures are rarely larger than 20 microns.

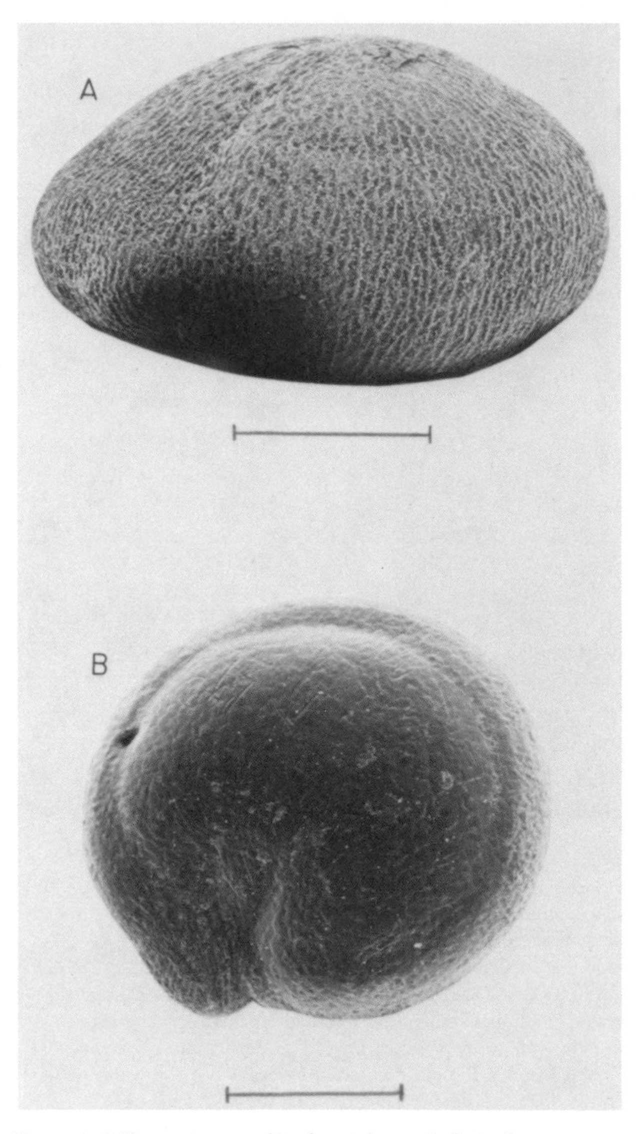

Figure 6.18 Photomicrographs of round-margin fruits from the Ash Cave *Chenopodium* assemblage: (a) mount 29, specimen 7, side view showing oval cross section, rounded margin, and prominent pericarp; (b) mount 45, specimen 1, dorsal oblique view showing testa surface morphology (pericarp scraped off) (scale bars a–b, 500 microns).

SMALL, ROUND-MARGIN FRUITS. During microscopic examination of the Ash Cave *Chenopodium* assemblage two quite different kinds of fruits were observed. In addition to the truncate-margin thin-testa fruits, which comprised the vast majority of the assemblage, a total of 144 (out of about 25,000) fruits were visually distinct from the "chia-like" fruits in a number of respects.

The most obvious distinguishing characteristics of these minority fruits were their oval cross section and rounded margin configuration (Figure 6.18a), distinct pericarp layer (almost always intact), and a far less prominent "beak" (compare the truncate-margin fruits in Figure 6.11c–d with the rounded-margin fruit in Figure 6.11e).

Finally, the rounded-margin fruits were significantly smaller than the truncate-margin fruits in the Ash Cave assemblage, with very little overlap in the size range curves of the two fruit types (Figure 6.19a).

These small rounded-margin fruits, morphologically so distinct from the larger truncate-margin thin-testa specimens that predominate in the Ash Cave assemblage, may reflect either variation within the thin-testa Ash Cave taxa, or the accidental inclusion in the harvest of fruits from wild/weedy taxa belonging to subsection *Cellulata*.

The intriguing possibility that the rounded-margin fruits might represent a sympatric weed companion is suggested by several morphological characteristics. If derived from a wild population, the rounded-margin fruits would be expected to be comparable to modern eastern North American taxa belonging to subsection *Cellulata* in terms of testa surface morphology (distinct reticulate-alveolate pattern; Figure 6.15), and testa thickness (40–70 microns; Figure 6.12i–p). In contrast, if rounded-margin fruits were from a companion weed to a domesticated variety of chenopod, with genetic interaction occurring, they could be expected to be morphologically intermediate in some respects between nondomesticated and domesticated taxa, to "carry some characteristics of a given domesticated form" (Wilson 1981:238)

The presence of such a weed would not be unexpected, since most seeds crops have companion weed races—hybrids between wild and cultivated taxa. Such hybrids

> are fertile, often compete successfully with their domestic parent for disturbed habitats, and are capable of natural seed dispersal. Such hybrids are frequently encountered as weeds in and around cultivated fields where wild and domestic races are sympatric. (de Wet and Harlan 1975:106)

Indeed, both the South American *(C. quinoa)* and Mexican *(C. berlandieri* ssp. *nuttalliae)* domesticates

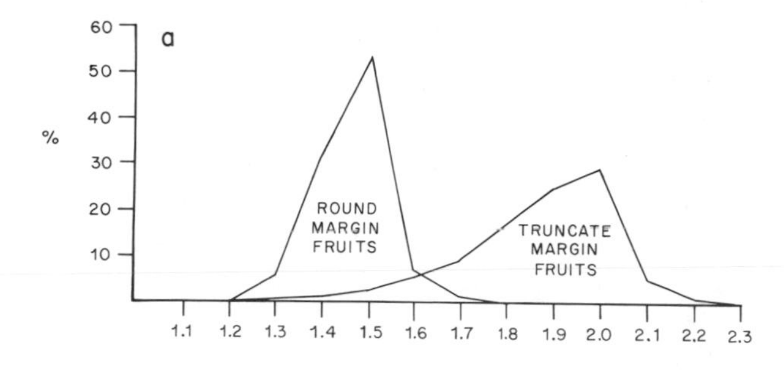

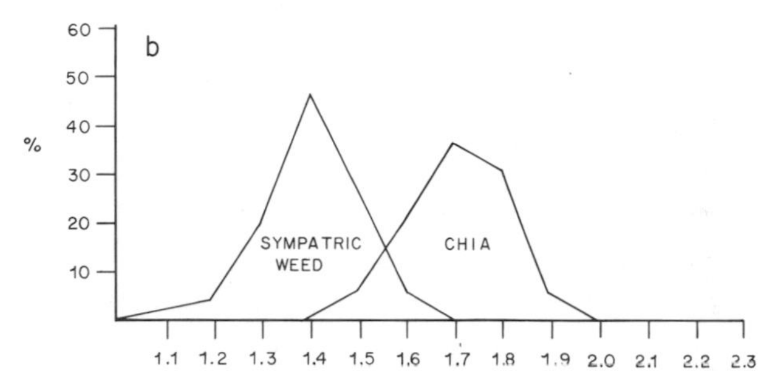

Figure 6.19 A paired comparison of maximum fruit diameter values for four *Chenopodium* populations belonging to subsection *Cellulata:* (a) frequency distributions of truncate- and round-margin fruits in the Ash Cave assemblage (truncate-margin fruit statistics from Table 6.1 [line 13], rounded-margin fruit statistics: sample size 144, mean 1.46 mm, range 1.3–1.7 mm, standard deviation .07); (b) frequency distributions of two *Chenopodium* populations from Michoacan, Mexico—HDW 5113 is a field crop of the cultivar *C. berlandieri* ssp. *nuttalliae* cv. "chia" (Table 6.1 [line 10]), while HDW 5114 is an agricultural weed *(C. berlandieri* ssp. *berlandieri)* harvested within 50 m of HDW 5113 (Table 6.1 [line 9]).

are "accompanied throughout their ranges by conspecific weed populations with which they hybridize" (Wilson 1981:238).

In a recent article, Hugh Wilson predicted the presence of a similar companion weed form in the Eastern Woodlands:

> If prehistoric cultivation of a domesticated chenopod was extensive in eastern North America, a companion weed could have been generated, either as a direct derivation of the domesticate, or through genetic interaction with a local, wild species. (Wilson 1981:238)

Although the fruit morphology of sympatric, conspecific companion weed forms of *Chenopodium* has not been studied in detail, they could be expected to be morphological intermediates and to exhibit some characteristics of the domesticated adaptive syndrome (Wilson 1981; Smith 1985). Wilson has noticed a partial loss of germination dormancy, for example, in some accessions of *C. berlandieri* from Mexico (Wilson and Heiser 1979:203).

In considering the likelihood that the Ash Cave rounded-margin fruits may represent a companion weed form, a modern analog is provided by a collection of *C. berlandieri* ssp. *berlandieri* fruits made by Hugh Wilson (HDW 5114) in Mexico in 1983. The plant collected by Wilson (HDW 5114) was growing as an agricultural weed within 50 m of a cultivated field of "chia" (HDW 5113).

Figure 6.19b illustrates a fruit-size relationship between the cultivar population of "chia" and the smaller-fruited weed population of *C. berlandieri* ssp. *berlandieri* that is quite similar to that shown in Figure 6.19a for the Ash Cave truncate- and rounded-margin fruits. Both the modern Mexican weed and the Ash Cave round-margin fruits lack the prominent "beak" of the domesticated chenopods.

While the fruits of the modern Mexican weed clearly exhibit the distinctive reticulate-alveolate testa surface characteristic of nondomesticated taxa belonging to subsection *Cellulata* (Figure 6.15a–b), the rounded-margin Ash Cave fruits can be seen to have an essentially smooth testa surface when the pericarp layer is removed (Figure 6.18b).

A final interesting comparison is shown in Figure 6.20, with both the modern Mexican weed (Figure 6.20h) and the rounded-margin Ash Cave fruits (Figure 6.20g) having testa thickness values intermediate between domesticated and nondomesticated taxa of eastern North America belonging to subsection *Cellulata*.

The intermediate testa thickness values, smooth testa surface morphology, and oval cross section of the Ash Cave rounded-margin fruits strongly suggest that they may represent a sympatric companion weed that carries some morphological characteristics of the domesticate with which it forms a weed-cultigen complex. Alternatively, the limited number of rounded-margin fruits observed in the Ash Cave assemblage may simply reflect morphological variation within the thin-testa Ash Cave chenopod taxon.

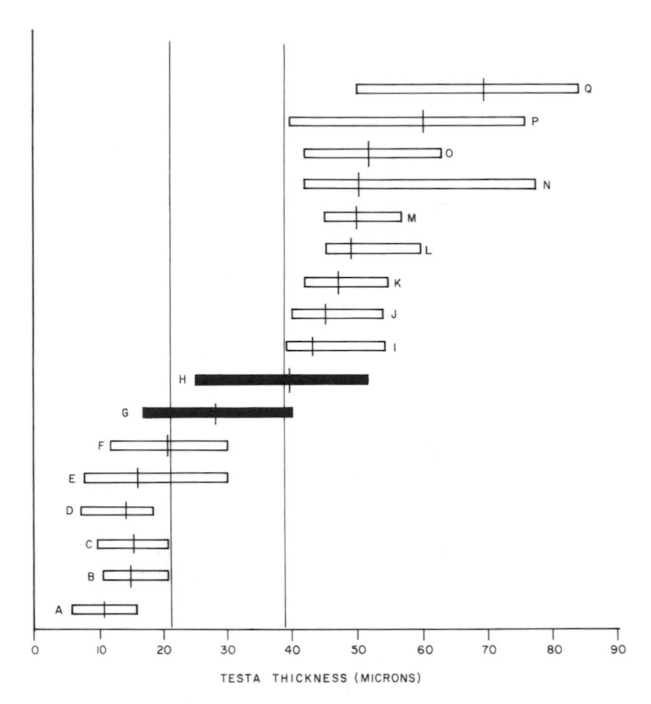

Figure 6.20 Testa thickness in microns for a modern sympatric agricultural weed population of *C. berlandieri* ssp. *berlandieri,* and the small rounded-margin fruits from Ash Cave, compared with 12 modern and prehistoric *Chenopodium* populations: (a) Russell Cave, NMNH 528403, Jackson County, Alabama (sample size 20, range 6–16, mean 11.25); (b) Ash Cave, Peabody 11039, Hocking County, Ohio (sample size 20, range 11–21, mean 14.9); (c) *C. berlandieri* ssp. *nuttalliae* cv. "chia," HDW 5113, 5115, Michoacan, Mexico (sample size 20, range 9–21, mean 16.25); (d) *C. bushianum* (red-morph fruits), BDS 20, Wayne County, Michigan (sample size 12, range 7–18, mean 14); (e) *C. bushianum* (red-morph fruits), CAA 653, low ground Willow Island, Greene County, Illinois (sample size 9, range 7–29, mean 16.7; (f) *C. bushianum* (red-morph fruit), BDS 82a, Cherokee County, South Carolina (sample size 8, range 12–30, mean 22); (g) Ash Cave (round-margin fruits), Peabody 11039, Hocking County, Ohio (sample size 15, range 17–40, mean 28.3); (h) *C. berlandieri* ssp. *berlandieri,* HDW 5114, Michoacan, Mexico, growing as a weed within 50 m of HDW 5113 (sample size 12, range 25–52, mean 39.6); (i) *C. bushianum,* BDS 20, Wayne County, Michigan (sample size 8, range 39–54, mean 43; (j) *C. bushianum,* Jackson County, Missouri (sample size 6, range 40–54, mean 45.6); (k) *C. missouriense,* NMNH 785467, Jackson County, Missouri (sample size 6, range 42–55, mean 47); (l) *C. bushianum,* BDS 82a, Cherokee County, South Carolina (sample size 5, range 43–60, mean 49.2); (m) *C. berlandieri,* NMNH 1201426, Jackson County, Missouri (sample size 6, range 45–57, mean 50.16); (n) *C. bushianum,* BDS 39, Fulton County, Pennsylvania (sample size 9, range 42–78, mean 51); (o) *C. bushianum,* CAA 646, Greene County, Illinois (sample size 9, range 42–63, mean 52); (p) *C. berlandieri/bushianum,* BDS 57, Mississippi County, Arkansas (sample size 9, range 41–76, mean 60.1); (q) *C. gigantospermum,* BDS 48, Menifee County, Kentucky (sample size 8, range 50–84, mean 69).

The Case for Domestication: Summary of a Comparative Morphological Analysis

The *Chenopodium* assemblages recovered from Russell Cave, Alabama and Ash Cave, Ohio fill a critical role in documenting the presence of a "chia-like" domesticated variety of chenopod in the Eastern Woodlands by the Middle Woodland time period.

Although only half as large as the Russell Cave assemblage, the Ash Cave chenopod assemblage is the more significant of the two. The uncarbonized, well-preserved nature of the Ash Cave fruits permitted a detailed morphological study to be undertaken, resulting in a comprehensive description of the micromorphology of the fruits.

This detailed understanding of the morphology of the prehistoric specimens in turn has allowed a six-sided comparative study to be undertaken that involved the fruits of (a) the Middle Woodland truncate-margin thin-testa chenopod taxa represented at Ash Cave and Russell Cave; (b) the Middle Woodland rounded-margin chenopod taxa represented in the Ash Cave assemblage; (c) the modern Mexican grainlike cultivar *C. berlandieri* ssp. *nuttalliae* cv. "chia"; (d) a modern Mexican nondomesticated agricultural weed of *C. berlandieri* ssp. *berlandieri*; (e) the modern nondomesticated taxa of eastern North America belonging to subsection *Cellulata (C. bushianum, C. berlandieri);* and (f) the low-frequency "red morph" produced by the nondomesticated taxa of eastern North America belonging to subsection *Cellulata.*

The results of this six-sided comparative analysis of fruit morphology can be summarized as follows:

1. Based on the presence of a distinctive reticulate-alveolate dorsal pericarp pattern, both the truncate-margin thin-testa taxa and the round-margin taxa of the prehistoric eastern United States can be assigned to subsection *Cellulata,* along with the Mexican domesticate *C. berlandieri* ssp. *nuttalliae* and several modern nondomesticated taxa of the eastern United States.
2. In terms of maximum fruit diameter, the prehistoric truncate-margin thin-testa taxa represented at Ash Cave and Russell Cave conform to apparent geographical clines of variation in

modern nondomesticated taxa of the eastern United States belonging to subsection *Cellulata*.

3. The overlapping size ranges of the modern Mexican cultivar "chia" and modern nondomesticated taxa of the eastern United States belonging to subsection *Cellulata* demonstrate that fruit diameter cannot be used as a morphological indicator of domestication.

4. Domesticated status in *Chenopodium* can be established on the basis of four apparently interrelated morphological characteristics involving the testa. Linked to underlying selective pressures related to the adaptive syndrome of domestication (Smith 1985), these morphological characteristics are: (a) a substantial reduction in testa thickness relative to that of modern eastern North American nondomesticated taxa (21 microns or less); (b) a truncate margin configuration; (c) an essentially smooth testa surface, sometimes exhibiting small bumps 20 to 40 microns in diameter, but not exhibiting a reticulate-alveolate surface pattern; and (d) a more prominent beak.

5. The third of these morphological indicators of domesticated status (smooth testa surface) can serve to distinguish in small sample archaeobotanical situations the fruits of domesticated taxa from the low frequency red morph fruits produced by nondomesticated eastern North American taxa.

6. Based on the consistent occurrence of the four morphological indicators of domestication listed above, the Ash Cave truncate-margin thin-testa fruits, like the Russell Cave fruits, can be assigned domesticated status.

7. Because they exhibit intermediate testa thickness values, smooth-testa surface morphology, and an oval cross section, the Ash Cave rounded-margin fruits quite likely represent a morphologically intermediate weed companion to this truncate-margin thin-testa taxon.

In providing a detailed morphological description of the Ash Cave fruits and documenting their domesticated status, the comparative analysis briefly summarized above should provide a basis for future identification and analysis of the truncate-margin thin-testa domesticated chenopod taxon in archaeobotanical assemblages of the Eastern Woodlands. To facilitate easier reference to, and discussion of, this prehistoric domesticate, it is reasonable at this point to consider an appropriate taxonomic assignment.

Taxonomic Considerations: Chenopodium berlandieri *ssp.* jonesianum

Two primary issues should be addressed in considering the assignment of a taxonomic name to the truncate-margin thin-testa "chia-like" *Chenopodium* taxon identified at Russell Cave and Ash Cave.

First and most importantly is the question of origins: was this domesticate introduced from Mexico, or was it the result of a developmental process indigenous to the Eastern Woodlands? Second, the current taxonomic framework for modern eastern North American chenopods belonging to subsection *Cellulata* needs to be considered, particularly in light of the apparent conspecific nature of *C. bushianum* and *C. berlandieri* in the Eastern Woodlands.

The presence of a reticulate-alveolate pericarp certainly allows the assignment of the Ash Cave and Russell Cave fruits to section *Chenopodium* subsection *Cellulata* of the genus *Chenopodium*.

On the basis of the analysis of the Russell Cave assemblage, taxonomic assignment had been extended down to the species level, with the truncate-margin thin-testa taxon assigned to *C. berlandieri* (Smith 1984, 1985). At the time, the assignment to *C. berlandieri* seemed preferable to *C. bushianum* for a number of reasons. Russell Cave is situated south of the known geographical range of *C. bushianum,* and in terms of fruit size (maximum diameter), the Russell Cave assemblage compared favorably with *C. berlandieri* and not with the larger fruited *C. bushianum*.

Even though Ash Cave is situated well within the geographical range of *C. bushianum* and the Ash Cave assemblage is comparable to *C. bushianum* in terms of fruit size, it is still appropriate to apply the species designation *C. berlandieri* to the Ash Cave assemblage. In light of the apparent conspecific nature of *C. berlandieri* and *C. bushianum* it is likely that as the systematics of eastern North American chenopods belonging to subsection *Cellulata* is revised, *C. bushianum* will be relegated to subspecific taxonomic status within *C. berlandieri.*

C. berlandieri is by far the broader taxon, subsuming an extensive geographical variety of wild, weedy, and domesticated forms of *Chenopodium,* both present-day and prehistoric, while the *C. bushianum* label is restricted in application to the large-fruited chenopod populations of the Northeast, Midwest, and mid-latitude Southeast that belong to subsection *Cellulata.*

In anticipation of a likely reassessment of the taxonomic position of *C. bushianum* within subsection *Cellulata* in the eastern United States, it is prudent to assign the truncate-margin fruits from Ash Cave and Russell Cave to the more firmly established taxon *C. berlandieri* and to avoid any extension of the taxon *C. bushianum.* David and Nancy Asch, working in west-central Illinois, have already begun to refer to all prehistoric material assignable to subsection *Cellulata* as *C. berlandieri,* and have dropped the use of the term *C. bushianum* (Asch and Asch 1985).

In turning to the problem of origins, it is tempting, on the one hand, to simply assign the prehistoric truncate-margin thin-testa fruits identified at Russell Cave and Ash Cave to *C. berlandieri* ssp. *nuttalliae* cv. "chia," since the prehistoric specimens of the Eastern Woodlands are morphologically almost indistinguishable from the fruits of the modern Mexican cultivar (they can be distinguished morphologically only on the basis of the relative thickness and structural complexity of the pericarp layer). It would seem, on the other hand, to be rather precipitous to confirm through taxonomic assignment, and in the absence of any supporting evidence, a historical developmental process that spans 2,500 km and 2,000 years (the cultivar "chia" has yet to be demonstrated to have a time depth in Mexico comparable to that of the truncate-margin thin-testa taxon of the prehistoric Eastern Woodlands).

At the same time, it is certainly possible that rather than resulting from a prehistoric range extension of "chia" into the eastern United States, the truncate-margin thin-testa fruits may be the product of an independent process of domestication. Certainly the recent documentation of the independent nature of the development of weed-crop *Chenopodium* complexes in South America and Mexico (Wilson and Heiser 1979) lends some support to the proposition that an independent process of domestication may have taken place in the Eastern Woodlands.

As a result it appears most appropriate at this point to establish a separate and distinct taxon at the subspecies level that would incorporate all of the prehistoric chenopod material from the eastern United States that conforms morphologically to the thin-testa truncate-margin fruits recovered from Ash Cave and Russell Cave. This taxon, it should be emphasized, does not include those chenopod fruits documented in the Ozarks (Wilson 1981), and recently reported in Illinois (N. Asch and D. Asch 1985:148) that completely lack a testa, and which appear to be morphologically quite similar to *C. berlandieri* ssp. *nuttalliae* cv. "huauzontle."

The establishment of such a distinct taxon, while facilitating easier reference to the "chia-like" chenopod fruits recovered from prehistoric archaeological sites in the eastern United States, does not lend support to either the diffusion or independent-development hypotheses.

The subspecific name *Chenopodium berlandieri* ssp. *jonesianum* is therefore proposed as a "neutral" taxonomic label. The subspecies name *jonesianum* was selected in recognition of the important archaeobotanical contributions made by Volney H. Jones. Jones was one of the first to suggest that *Chenopodium* might have been a prehistoric domesticate in the eastern United States (Jones 1936; see Smith and Funk 1985 for a formal botanical description of *C. berlandieri* ssp. *jonesianum*).

Discussion: Hopewellian Plant Husbandry Systems

The existence, nature, and relative importance of Hopewellian plant husbandry systems have been interrelated topics of consistent discussion over the past quarter century (Caldwell 1958; Griffin 1960;

Struever 1962; Yarnell 1963, 1983; Struever and Vickery 1973; Ford 1979, 1981; Asch and Asch 1982). Until recently, and in the absence of much relevant archaeobotanical information, this discussion has focused almost exclusively on the relative importance of the introduced tropical cultigen maize. Stable carbon isotope studies, however, have failed to indicate maize consumption by either Havana (Illinois) or Hopewell (Ohio) populations (Bender et al. 1981). In addition, the timing of the introduction of maize into the Eastern Woodlands is currently under review, and it appears likely that maize was not even present in the East prior to A.D. 200 (Yarnell 1983; Smith 1985). Continued direct dating of purported Middle Woodland maize by the AMS method should resolve this issue within the next several years.

As the perceived relative importance of maize as a Hopewellian cultigen has faded, detailed archaeobotanical sequences developed for a number of different regions of the mid-latitude Eastern Woodlands have illuminated the central role of indigenous annual starchy- and oily-seed-bearing plants in Hopewellian (Middle Woodland) premaize plant husbandry systems of west-central Illinois (Asch and Asch 1982, 1985), the American Bottom east of St. Louis (Johannessen 1984), central Kentucky (Yarnell 1974; Gardner 1984), and the Little Tennessee, Duck, and Elk river valleys of eastern and central Tennessee (Crites 1978a, 1978b, 1984; Kline et al. 1982; Chapman and Shea 1981).

Of the six indigenous plant species identified in one or more of these regions, two starchy-seeded quasi-cultigen/cultigens (little barley *[Hordeum pusillum]* and maygrass *[Phalaris caroliniana]*) would have provided a spring harvest, while the remaining two starchy-seeded annuals (knotweed *[Polygonum erectum]*, a quasi-cultigen/cultigen, and goosefoot *[C. berlandieri* ssp. *jonesianum]*, a domesticate) along with two oily-seeded domesticates (marshelder *[Iva annua]* and sunflower *[Helianthus annuus]*) were fall-maturing crops (Smith 1985).

Based on the variable abundance of these six indigenous seed crops in archaeobotanical assemblages, there was apparently considerable interregional variation in their relative importance and in the composition of Middle Woodland premaize plant husbandry systems.

The oily-seeded domesticates, if present at all, occur only in small frequencies, and are likely to be underrepresented in each of the regions studied. A nondomesticated variety of *Chenopodium* and maygrass are the only starchy-seeded annuals represented in eastern Tennessee, and maygrass joins *Chenopodium* to form a spring-fall starchy-seeded crop complex in central Tennessee, central Kentucky, and the American Bottom. While knotweed occurs as a secondary fall-maturing starchy-seeded crop in the American Bottom and central Tennessee, it appears to have been more important than *Chenopodium* in west-central Illinois. Little barley joins maygrass as a spring-harvest starchy-seeded crop in west-central Illinois [see Chapter 5 for a fuller discussion of these patterns of regional variation].

Although relatively few Ohio Hopewell archaeobotanical assemblages containing indigenous seed plants have been recovered and analyzed (Ford 1979, table 29.1), the Ash Cave *Chenopodium* assemblage indicates that Ohio Hopewell plant husbandry systems included part of the maygrass-chenopod spring-fall starchy-seeded crop complex that has been identified in eastern and central Tennessee, central Kentucky, the American Bottom, and west-central Illinois. In addition, it is likely that the oily-seeded domesticates (sumpweed and sunflower) were also grown, along with knotweed and perhaps little barley. A more detailed understanding of the relative importance of various cultigens in Ohio Hopewell plant husbandry systems, however, must await the publication of analyses of larger and more representative archaeobotanical assemblages such as the forthcoming Harness 28 (33RO186) report.

While future research holds the promise of both extending and establishing the accuracy of these apparent patterns of regional variation in premaize plant husbandry systems, it will at the same time prove quite difficult to determine the scale of human investment in, and the relative dietary importance of, plant cultigens in Hopewellian societies.

I have recently put forward the proposal that many, if not all, of these indigenous plant species were grown by 2,000 B.P. as part of field-cropping strategies that were complementary to already existing small garden-plot cultivation practices (Smith 1985 [Chapter 5]), and that maize may well have initially been incorporated as just another starchy-seeded annual within already well-established indigenous field cropping systems.

It seems quite likely that the by now three-decade old debate concerning the relative importance of maize in Hopewellian diets will be replaced by an equally long-term discussion focusing on the existence of Hopewellian field-cropping systems and the dietary importance of indigenous cultigens.

While there is certainly a wide range of archaeological data that can be brought to bear on these interrelated questions, one of the most obvious and potentially most valuable lines of research involves scrutinizing the micromorphology of the fruits and seeds of the various indigenous quasi-cultigens/cultigens currently denied domesticated status. It may be that little barley, maygrass, and knotweed, as well as other species, do in fact exhibit morphological characteristics consistent with domestication (see, for example, the recent discussion of knotweed by Nancy and David Asch [1985:139–147]). Morphological changes reflecting the adaptive syndrome of domestication provide clear evidence of substantial intervention in the life cycle of cultivated plants, and by implication, their greater economic importance.

Another line of inquiry that could potentially strengthen the case for Hopewellian field-cropping systems and the economic importance of indigenous cultigens would involve the controlled planting, cultivation, and harvesting of modern weed forms of these plant species to establish potential yield values, combined with nutritional studies such as that recently done for maygrass (Crites and Terry 1984).

If these cultigens are found to have impressive potential yield values in comparison to maize (e.g., Asch and Asch 1978, table 6) and to be relatively easy to process, they would be more easily perceived as potential field crops of economic importance rather than as minor dietary items in small-garden husbandry systems.

Notes

1. Authorities for all species mentioned in the text are as follows: *Andropogon gerardi* Vitman; *C. album* L.; *C. berlandieri* Moq.; *C. berlandieri boscianum* (Moq.) Wahl; *C. berlandieri* spp. *jonesianum* Bruce Smith; *C. berlandieri* spp. *nuttalliae* (Stafford) Wilson and Heiser; *C. berlandieri* spp. *zschackei* Zobel; *C. bushianum* Aellen; *C. gigantospermum* Aellen; *C. macrocalycium* Aellen; *C. missouriense* Aellen; *C. quinoa* Wild; *Helianthus annuus* L.; *Hordeum pusillum* Nutt.; *Iva annua* L. ssp. *macrocarpa* (Blake) Jackson; *Phalaris caroliniana* Walt.; *Polygonum erectum* L.

2. *Chenopodium* produces small indehiscent "seedlike" fruits, each consisting of a seed enclosed by a very thin, usually adherent, pericarp. Since this thin pericarp is the only thing distinguishing fruits from seeds, the terms "fruit" and "seed" are used interchangeably.

3. The materials recovered by Goslin from Ash Cave in 1928 have been assigned Ohio Historical Society catalog numbers 899/1–25. Catalog numbers 899/26–42 have been assigned to Ash Cave materials recovered in 1956, apparently through surface collection, by Jack Shaffer, Robert Goslin, and Raymond Baby.

Acknowledgments

I would like to thank a number of individuals who provided me with invaluable assistance during research on the Ash Cave *Chenopodium* assemblage. Stephen Williams, Peabody Museum, approved and facilitated the loan of the Ash Cave *Chenopodium* assemblage and provided access to documentary records of Andrews's 1876 excavations. James L. Murphy, Ohio State University, generously investigated the Ash Cave records of the Ohio Historical Society in response to my initial inquiry. Bradley K. Baker, Collections Manager (Archaeology), Ohio Historical Society, provided the essential information contained both in the OHS Ash Cave accession file and the Ash Cave-Hocking County file. Mark Seeman graciously provided invaluable information and insights concerning the prehistory of the region. David and Nancy Asch and Hugh Wilson generously provided modern *Chenopodium* fruits for analysis. David Asch, Gary Crites, Vicki Funk, James L. Murphy, Mark Seeman, and Richard Yarnell read the manuscript and offered numerous corrections of fact and interpretation. Walter Brown of the NMNH SEM lab trained me to operate the Cambridge Stereoscan 100, and Susann Braden and Heidi Wolf were kind enough to charge the SEM filament before each session. The scientific illustrators of the Department of Anthropology (George Robert Lewis, Ellen Paige, Britt Griswold) drew many of the illustrations, and Victor E. Krantz photographed the illustrations. The manuscript was typed by Darla Hawkins.

Collections

While it might be tempting to provide a different, more "modern," mode of storage for the Ash Cave *Chenopodium* fruits than the small airtight glass bottle in which they have been kept, it is difficult to argue with success. The fruits were found to be exceptionally well preserved after being stored for over a century in the small bottle, with little apparent damage or deterioration.

The assemblage is currently curated, along with the other materials recovered from Ash Cave, in the Ohio archaeology storage area of Peabody Museum, Harvard University (catalog number 11.039). During the research project described here, a total of 150 of the 25,000 Ash Cave chenopod fruits were removed from the collection. One hundred of these were submitted as a radiocarbon sample, and 50 were mounted for SEM viewing. These mounted fruits (SEM mounts 21, 29, 30, 31, 44, 45, 47, 48) are at present in the possession of the author, and will eventually be returned to the Peabody Museum for curation with the rest of the Ash Cave assemblage.

In addition, a small vial of chenopod fruits "collected by E. B. Andrews from Ash Cave" are in the Ash Cave collections of the Ohio Historical Society (Catalog number 899/43). These were apparently obtained by Raymond S. Baby from the Peabody Museum sometime after 1956.

Both the Ash Cave and Russell Cave *Chenopodium* assemblages, it should be mentioned, provide clear evidence of the research potential of museum collections, and underscore the importance of careful long-term curation of archaeobotanical materials. Largely unnoticed and essentially unstudied for 107 and 28 years, respectively, the Ash Cave and Russell Cave *Chenopodium* assemblages have contributed to the recently emerging new perspective on Middle Woodland plant husbandry systems and the "eastern agricultural complex."

Literature Cited

Andrews, E. B.
1877 Report on Exploration of Ash Cave in Benton Township, Hocking County, Ohio. *Peabody Museum of American Archaeology and Ethnology, Tenth Annual Report of the Trustees* 2:48–50.

Asch, D. L., and N. B. Asch
1977 Chenopod as Cultigen: A Re-evaluation of some Prehistoric Collections from Eastern North America. *Midcontinental Journal of Archaeology* 2:3–45.

1978 The Economic Potential of *Iva annua* and its Prehistoric Importance in the Lower Illinois Valley. In *The Nature and Status of Ethnobotany,* edited by R. I. Ford, pp. 301–341. Museum of Anthropology, Anthropological Paper No. 67. University of Michigan, Ann Arbor.

1982 A Chronology for the Development of Prehistoric Horticulture in West-Central Illinois. Paper presented at the 47th annual meeting of the Society for American Archaeology, Minneapolis. Center for American Archaeology, Archaeobotanical Laboratory Report 46.

1985 Archeobotany. In *Excavations at the Smiling Dan Site,* edited by B. D. Stafford and M. B. Sant, pp. 327–401. Center for American Archaeology, Kampsville Archaeological Center, Research Series 2.

Asch, N. B., and D. L. Asch
1985 Archeobotany. In *The Hill Creek Homestead,* edited by Michael D. Conner, pp. 115–170. Center For American Archaeology, Kampsville Archeological Center, Research Series 1.

Bender, M. M., D. A. Baerreis, and R. L. Steventon
1981 Further Light on Carbon Isotopes and Hopewell Agriculture. *American Antiquity* 47:346–353.

Black, D.
1979 Adena and Hopewell Relations in the Lower Hocking Valley. In *Hopewell Archaeology,* edited by D. Brose and N. Greber, pp. 19–26. Kent State University Press, Kent, Ohio.

Brose, D., and N. Greber (editors)
1979 *Hopewell Archaeology.* Kent State University Press, Kent, Ohio.

Caldwell, J. R.
1958 *Trend and Tradition in the Prehistory of the Eastern United States.* Memoirs of the American Anthropological Association 88.

Chapman, J., and A. B. Shea
1981 The Archaeobotanical Record: Early Archaic Period to Contact in the Lower Little Tennessee River Valley. *Tennessee Anthropologist* 6:61–84.

Crane, H. R., and J. B. Griffin
1962 University of Michigan Radiocarbon Dates VII. *Radiocarbon* 4:183–203.

Crites, Gary D.
1978a Paleoethnobotany of the Normandy Reservoir in the Upper Duck Valley, Tennessee. M.A. thesis, Department of Anthropology, University of Tennessee, Knoxville.

1978b Plant Food Utilization Patterns during the Middle Woodland Owl Hollow Phase in Tennessee: A Preliminary Report. *Tennessee Anthropologist* 3:79–92.

1984 Middle Woodland Paleo-Ethnobotany of the Eastern Highland Rim. Paper presented at the 5th annual Mid-South Archaeological Conference, Pinson Mounds, Tennessee.

Crites, G. D., and R. D. Terry
1984 Nutritive Value of Maygrass, *Phalaris caroliniana. Economic Botany* 38:114–120.

Ford, Richard I.
1979 Gathering and Gardening: Trends and Consequences of Hopewell Subsistence Strategies. In *Hopewell Archaeology,* edited by D. S. Brose and N. Greber, pp. 234–238. Kent State University Press, Kent, Ohio.

1981 Gardening and Farming Before A.D. 1000: Patterns of Prehistoric Cultivation North of Mexico. *Ethnobiology* 1:6–27.

Gardner, P.
1984 Plant Food Subsistence at Salts Cave, Kentucky: New Evidence. Paper on file at the Research Laboratories of Anthropology, University of North Carolina, Chapel Hill (see *American Antiquity* 1987 52:358–366).

Gilmore, M.
1932 The Ash Cave Site, Ohio Museum Archaeological Explorations. Ethnobotanical Report 33, Laboratory of Ethnobotany, University of Michigan. Accession file 899, Ohio Historical Society, Columbus.

Goslin, R. M.
1928 Letter to H. C. Shetrone. Accession file 899, Ohio Historical Society.

1952 Cultivated and Wild Food from Aboriginal Sites in Ohio. *Ohio Archaeologist* 2:9–29.

Griffin, J. B.
1960 Climatic Change: A Contributory Cause of the Growth and Decline of Northern Hopewellian Culture. *Wisconsin Archaeologist* 41:21–33.

Johannessen, S.
1984 Paleoethnobotany. In *American Bottom Archaeology,* edited by C. J. Bareis and J. W. Porter, pp. 197–214. University of Illinois Press, Urbana.

Jones, V. H.
1936 The Vegetal Remains of Newt Kash Hollow Shelter. In *Rock Shelters in Menifee County, Kentucky,* by W. S. Webb and W. D. Funkhouser, pp. 147–167. University of Kentucky Reports in Archaeology and Anthropology 3(4). Lexington.

Kline, G. W., G. D. Crites, and C. H. Faulkner
1982 *The McFarland Project: Early Middle Woodland Settlement and Subsistence in the Upper Duck River Valley in Tennessee.* Tennessee Anthropological Association, Miscellaneous Paper 8.

Mills, W. C.
1912 Archaeological Remains of Jackson County. *Ohio State Archaeological and Historical Quarterly* 21:175–214.

Moorehead, Warren King
1895 Preliminary Exploration of Ohio Caves. *The Archaeologist* 3(9):304–312.

Murphy, James L.
1975 *An Archeological History of the Hocking Valley.* Ohio University Press, Athens, Ohio.

Prufer, O. H.
1965 *The McGraw Site: A Study in Hopewellian Dynamics.* Cleveland Museum of Natural History, Scientific Publications (n.s.) 4.

1968 *Ohio Hopewell Ceramics.* Museum of Anthropology, Anthropological Paper No. 33. University of Michigan, Ann Arbor.

1975a Chesser Cave, a Late Woodland Phase in Southeastern Ohio. In *Studies in Ohio Archaeology,* edited by O. H. Prufer and D. H. McKenzie, pp. 1–62. Kent State University Press, Kent, Ohio.

1975b The Scioto Valley Archaeological Survey. In *Studies in Ohio Archaeology,* edited by O. H. Prufer and D. H. McKenzie, pp. 267–328. Kent State University Press, Kent, Ohio.

Shane, O. C., III
1971 The Scioto Hopewell. In *Adena: The Seeking of an Identity,* edited by B. K. Swartz, Jr., pp. 142–157. Ball State University, Muncie, Indiana.

Shane, O. C., III, and J. L. Murphy
1975a Survey of the Hocking Valley, Ohio. In *Studies in Ohio Archaeology,* edited by O. H. Prufer and D. H. McKenzie, pp. 329–356. Kent State University Press, Kent, Ohio.

Smith, Bruce D.
1984 *Chenopodium* as a Prehistoric Domesticate in Eastern North America: Evidence from Russell Cave, Alabama. *Science* 226:165–167.

1985 The Role of *Chenopodium* as a Domesticate in Pre-Maize Garden Systems of the Eastern United States. *Southeastern Archaeology* 4(1):51–72.

Smith, Bruce D., and V. A. Funk
1985 A Newly Described Subfossil Cultivar of *Chenopodium* (Chenopodiaceae). *Phytologia* 57:445–448.

Struever, S.
1962 Implications of Vegetal Remains from an Illinois Hopewell Site. *American Antiquity* 27:584–587.

Struever, S., and K. D. Vickery
1973 The Beginnings of Cultivation in the Midwest-Riverine Area of the United States. *American Anthropologist* 75:1197–1220.

de Wet, J. J. M., and J. R. Harlan
1975 Weeds and Domesticates: Evolution in the Man-Made Habitat. *Economic Botany* 29:99–107.

Williams, J. T., and J. L. Harper
1965 Seed Polymorphism and Germination. *Weed Research* 5:141–150.

Wilson, H. D.
1980 Artificial Hybridization among Species of *Chenopodium* sect. *Chenopodium. Systematic Botany* 5:253–263.

1981 Domesticated Chenopodium of the Ozark Bluff Dwellers. *Economic Botany* 35:233–239.

Wilson, Hugh D., and C. B. Heiser, Jr.
1979 The Origin and Evolutionary Relationships of "huauzontle" *(Chenopodium nuttalliae* Stafford), Domesticated Chenopod of Mexico. *American Journal of Botany* 66:198–206.

Yarnell, R. A.
1963 Comments on Struever's Discussion of an Early "Eastern Agricultural Complex." *American Antiquity* 28:547–548.

1972 *Iva annua* var. *macrocarpa:* Extinct American Cultigen? *American Anthropologist* 74:335–341.

1974 Plant Food and Cultivation of the Salts Cavers. In *Archaeology of the Mammoth Cave Area,* edited by Patty Jo Watson, pp. 113–122. Academic Press, New York.

1978 Domestication of Sunflower and Sumpweed in Eastern North America. In *The Nature and Status of Ethnobotany,* edited by R. I. Ford, pp. 289–299. Museum of Anthropology, Anthropological Papers No. 67. University of Michigan, Ann Arbor.

1983 Prehistoric Plant Foods and Husbandry in Eastern North America. Paper presented at the 48th annual meeting of the Society for American Archaeology, Pittsburg.

CHAPTER 7

The Economic Potential of *Chenopodium berlandieri* in Prehistoric Eastern North America

Introduction and Research Design

The tropical cultigen triad of corn, beans, and squash often is considered the core of prehistoric plant husbandry in eastern North America, and these three introduced crops were certainly of substantial economic importance during the late prehistoric period. But maize apparently was not introduced into the East any earlier than about A.D. 200–300 (Ford 1987; Chapman and Crites 1987; Yarnell and Black 1985:102; Smith 1985a:51), and food production systems dominated by maize did not develop in the Eastern Woodlands of North America until after A.D. 1000 (Smith 1985a:51, 1986).

When it first arrived in the Eastern Woodlands, maize was adopted as just another starchy seed crop within already established "premaize" plant husbandry systems that centered on a group of six plant species: marshelder *(Iva annua)*, sunflower *(Helianthus annuus)*, goosefoot *(Chenopodium berlandieri)*,[1] erect knotweed *(Polygonum erectum)*, maygrass *(Phalaris caroliniana)*, and little barley *(Hordeum pusillum)* (Yarnell and Black 1985). The two premaize oily-seed crops, sunflower and marshelder, were long-standing domesticates in the East, as was at least one of the starchy-seed crops—a thin-testa cultivar chenopod (Smith 1987a; Smith and Cowan 1987). When maize was initially introduced into the East at ca. A.D. 200–300, these three long-standing domesticates had recently been joined by a no-testa or naked cultivar chenopod (Fritz 1986), as well as by, perhaps, a domesticated variety of knotweed (Fritz 1987). Little barley and maygrass, both spring harvested starchy-seed crops, were also important in premaize food production economies, even though no archaeological specimens of either species have as yet been demonstrated to exhibit morphological changes associated with domestication.

This chapter originally appeared in *Ethnobiology* (1987) 7:29–54.

Variation in the relative abundance of seeds of these premaize cultigens in archaeobotanical assemblages from different regions of the East suggest both geographical and temporal diversity in their economic importance [Chapter 5]. These assemblages also reflect a fairly uniform increase in the importance of such regionally diverse premaize plant husbandry systems across the mid-latitude interriverine area of the East at about 2,500–2,000 B.P. [Chapter 5], and these indigenous seed crops appear to have continued in importance up through the post A.D. 1000 transition to maize agriculture. But they seem to have all but disappeared from view by the time of European contact. With the possible exception of Harriot's 1586 account from the Carolinas of sunflower and *melden* (Sturtevant 1965), and Le Page du Pratz's eighteenth-century description of *choupichoul* being grown by the Natchez (Smith 1987b [Chapter 10]), the ethnohistorical record is silent in regard to possible remnant premaize cultigens of the Eastern Woodlands.

In the past decade the developmental trajectory of these premaize plant husbandry systems, from initial emergence to eventual decline, has attracted increasing research interest because of their likely central role in the evolution of more complex Woodland period (2,500–1,200 B.P.) prehistoric societies in the East. One important aspect of gaining a better understanding of these early cultigens involves attempting to determine their economic potential through both nutritional analyses of seeds and present-day harvesting experiments to establish their potential yield. The nutritional composition of seeds has been established for *Polygonum erectum* (Asch and Asch 1985:361), *Chenopodium berlandieri/bushianum* (Asch and Asch 1985:361), *Phalaris caroliniana* (Crites and Terry 1984), *Iva annua* (Asch and Asch 1978), and *Helianthus annuus* (Earle and Jones 1962; Watt and Merrill 1963; Jones and Earle 1966:15). In addition, Seeman and Wilson (1984) and Murray and Sheehan (1984) provide nutritional composition information for other species of *Polygonum* and *Chenopodium*. Efforts to obtain potential harvest yield information for these six premaize crops, on the other hand, have proven difficult due to the apparent present-day absence of large wild or weedy stands suitable for harvest studies. Because of this difficulty modern harvest yield research has yet to be carried out on either maygrass or little barley. Potential harvest yield estimates for sunflower are derived from modern commercial field crop statistics (e.g., Martin and Leonard 1967:932–933). In addition, potential yield information for *Polygonum erectum* is limited to a single 25 sq ft (2.29 sq m) stand harvested by Murray and Sheehan during their study of five species of *Polygonum* in Illinois and Indiana (1984:288, 290–291). Similarly, harvest yield projections for *Chenopodium berlandieri/bushianum* are based on two isolated plants, each occupying about 1 sq m (Asch and Asch 1978:313). Unable to locate any populations of *Chenopodium berlandieri* large enough to sample repeatedly, Seeman and Wilson (1984) provide detailed harvest yield information for *Chenopodium missouriense* stands harvested in Indiana. *Iva annua* is the best documented of these six early cultigens in terms of both harvest rate values (based on 20 timed collections from seven stands) and harvest yield values (based on eight total yield collections from 1 sq m plots; Asch and Asch 1978 [see Chapter 8]).

In light of their central role in fueling prehistoric Woodland period cultural change, it is surprising that information concerning the prehistoric economic potential of these premaize seed plants is so limited, and that so few present-day wild stands have been harvested. In order to begin to fill this gap in our understanding of these prehistoric crop plants, and to learn more about their habitat requirements, present-day geographical ranges, and harvest yield potential, I began, in the fall of 1984, an annual fall harvesting circuit through the eastern United States. For the first two years, field studies focused on *Iva annua* and *Chenopodium berlandieri,* and this article reports the results of the 1984 and 1985 harvests in regard to *Chenopodium berlandieri.*

Methods

One of the main goals of the 1984–1985 fall harvesting project was to locate and harvest stands of *Iva annua* and *C. berlandieri* over a broad area of the Midwest and Southeast. Accordingly, I did not spend an extended period of time in any one area, and I selectively concentrated my survey efforts in those high probability target habitats I thought likely to support stands of either of these two species. Because marshelder was known to be a component of early successional floodplain communities, I selected a route of travel through the East that crossed and recrossed a number of river valleys. At each river crossing I surveyed up and down the floodplain for several miles in search of stands of *Iva annua* and *C. berlandieri.*

These two species frequently colonize recently disturbed soil in both floodplain and upland locations, and are often found growing on construction sites, in vegetable gardens and agricultural fields, as well as in a wide range of other settings, including vacant lots, abandoned gas stations, playgrounds, parking lots, cotton gin waste dumps, dredging spoil piles, and under bridges. Such disturbed ground habitat settings were easy to recognize from a moving car, and I frequently stopped to investigate them, particularly when chenopod plants were observed in passing. Although most chenopod stands occurring in such disturbed soil situations were dominated by the ubiquitous *C. missouriense,* pockets of *C. berlandieri* were occasionally located within larger stands of *C. missouriense.* By combining the deliberate surveys of river valley segments with more opportunistic surveys of upland disturbed soil situations as they were observed, I was able to investigate a large number of chenopod stands while still traveling over 200 miles per day.

Each *C. berlandieri* stand located was assigned a field catalog number and this, along with its geographical location, habitat setting, and plant characteristics, were recorded in a field notebook. Stand and individual plant characteristics that were described included: (1) stand size in square meters; (2) individual plant height in centimeters; (3) plant habit and habitat—the shape and habitat setting of plants (for example, slender and unbranched, growing in dense cover, light shade); (4) the color and condition of individual plants.

Harvesting was accomplished by hand stripping individual infructescences (fruit clusters) into a bag attached to the belt loops of the collector. Large plastic garbage bags served admirably in this capacity when simply twisted through belt loops and tied off, leaving both hands free for me to grasp an infructescence in a closed fist and strip it from the plant into the waist bag (see Seeman and Wilson 1984:305 for other harvesting options for *Chenopodium*). Harvest times for individual plants were recorded in most cases. Each evening, after removing the infructescence material harvested that day from the plastic waist bags, I placed it on newspaper overnight to partially dry, then repackaged it in paper bags for further drying. When I found a stand containing *C. berlandieri* plants that still retained leaves, a voucher specimen was pressed for the National Herbarium (NMNH). Occasionally I also collected entire plants after stripping their infructescences, returning them to the lab to be dried and weighed.

Subsequent to thorough drying and weighing in the laboratory, harvested fruit was winnowed from perianth, leaf, and stem fragments.[2] Cleaned fruits comprised from 65 to 79 percent of unwinnowed material (Table 7.1).

Results

During harvesting trips through the eastern United States in the autumn of 1984 and 1985, *C. berlandieri* stands were harvested in Maryland, Pennsylvania, Ohio, Michigan, Missouri, Tennessee, Arkansas, Alabama, and South Carolina. I recorded individual plant yield information for a total of 86 *C. berlandieri* plants (Table 7.1), and carried out timed harvest experiments on 16 *C. berlandieri* stands (Table 7.2).

Because of the elusive nature of present-day stands of *C. berlandieri,* the absence of previous research, and the need for additional harvest experiments on modern descendant populations of this important prehistoric cultigen, I provide the following brief, if admittedly tedious, descriptions of the stands harvested in 1984 and 1985.

Wayne County, Michigan (Catalog numbers 20–30, 87–88)

Catalog numbers 20–28 were assigned to individual *C. berlandieri* plants harvested in October 1984 from a disturbed soil setting in a vacant lot in Grosse Pointe, Michigan. Large, often dense (> 50 plants per sq m), stands of both *C. berlandieri* and *C. missouriense,* frequently mixed, occurred in both full

Table 7.1. Harvest Yield Information for 86 *Chenopodium berlandieri* Plants

Catalog number	Plant height (cm)	Plant weight (g)	Unwinnowed fruit weight (g)	Winnowed fruit weight (g)	Unwinnowed/ Winnowed ratio	Harvest time (min)
20	122	8.6		4.62		
21	127	18.2		27.19		
22	84	7.7		5.31		
23	112	13.1		12.88		
24	79	4.3		3.66		
25	46	4.7		3.73		
26	56	5.1		4.01		
27	61			14.76		
28	183			72.4		
39	152			109.0		7:00
40	124			11.14		0:45
41	58			2.98		0:15
42	142			19.17		1:00
82A	183	160.0		49.0		2:00
87	200	126		91.8		8:00
89A	58	16.8		12.6		1:00
89B	86	36.4		38.4		2:00
90A	105	220.0	75.7	55.2	73	2:00
90B	83	95.3	41.9	30.3	72	1:30
90C-1		5.4	3.3	2.3	73	0:15
90C-2		17.3	11.0	8.4	77	0:15
90C-3		24.1	22.6	17.3	77	
90C-4	48	1.4	.9	.7	78	
90C-5		10.5	4.0	2.8	70	
91	150	497.5		428.1		16:00
92A		3.2		2.6		
92B		18.5	11.7	8.5		
92C		18.5		16.7		
92D		12.5		8.6		
92E		21.1		18.6		
92F		10.7	5.6	3.9		
92G		26.1		31.5		
92H		23.1		20.5		
92I		60.0	30.4	19.7	65	
92J		82.2	103.8	71.6	69	
94A	125–150		93.3	71.7	77	8:00
94B	125–150		148.2	101.6	68	8:00
94C	125–150		144.7	100.8	70	11:00
94D	175–200			127.3		10:00
94E	125–150			89.1		6:00
94F	100–125			74.4		9:00
94G	100–125			10.8		1:00
94H	175–200		71.5	56.7	79	4:40
94I	150–175			9.7		1.40
94J	75–100			5.1		1:00
94K	75–100			20.6		1:00
94L	75–100			14.9		2:12
94M	100–125			58.5		2:30
94N	75–100			10.4		0:49
94O	75–100		21.4	16.6	78	1:19
94P	100–125			41.4		3:15
94Q	125–150			61.5		3:55
94R	125–150			76.0		6:30
94S	100–125			73.0		4:00
94T	100–125			33.3		1:30

Table 7.1—*Continued next page*

Table 7.1—*Continued*

Catalog number	Plant weight (cm)	Plant weight (g)	Unwinnowed fruit weight (g)	Winnowed fruit weight (g)	Unwinnowed/ winnowed ratio	Harvest time (min.)
94U	50–75			7.0		0:54
94V	100–125			34.0		5:15
94W	125–150			68.0		7:00
94X	100–125			79.2		5:00
94Y	75–100			32.2		1:30
94Z	175–200			84.6		4:17
94AA	75–100			36.2		4:00
94BB	100–125			80.5		4:30
94CC	175–200		230.6	175.5	76	11:00
94DD	100–125			32.2		2:00
94EE	75–100			15.0		2:27
94FF	75–100			9.1		0:15
94GG	75–100			1.5		0:30
94HH	75–100			9.5		1.13
94II	100–125			78.7		6:00
94JJ	125–150			66.8		5:10
95A	75–100			35.0		4:05
95B	75–100			26.8		2:13
95C	75–100			31.0		3:40
95D	125–150			129.8		10:10
95E	75–100			23.4		1.27
95F	75–100			13.4		0.30
95G	75–100			22.3		2.45
95H	175–200			30.5		3.34
95I	75–100			5.0		0.45
95J	125–150		83.2	54.4	65	4.30
96	100			49.2		3:00
101A	200	32.2		9.1		
101B	100			0.3		
105	86	5.0	2.6	2.0	77	
108	60		4.5	3.4	76	

sun and light shade. A mixed stand of *C. berlandieri* and *C. missouriense* covering an area of approximately 250 sq m and located in partial shade was assigned field number 30 and selectively harvested in October 1984. With an approximate average density of 50 plants per sq m (four plots counted, each 1 sq m), the stand contained about 12,000 plants, ranging in height from 30 to 130 cm, with few lateral branches. As is almost always the case in upland mixed stands, *C. missouriense* was the more abundant of the two species, comprising over three-fourths of the stand. Rather than being scattered randomly, *C. berlandieri* plants formed a number of discrete pockets within the larger stand. The largely leafless, distinctively mustard colored *C. berlandieri* plants had dark, large fruited, and glomerate infructescences. I could easily distinguish them, both by touch and sight, from the still leafed bright purple *C. missouriense* plants, making species selective harvesting of the stand relatively easy. Moving through the stand and selecting the largest and most visible infructescences, I was able to hand strip 300 *C. berlandieri* infructescences into a waist bag in 22 minutes, yielding 575 g of infructescence material and 460 g of clean fruit after winnowing.

Returning to this vacant lot in October of the following year I harvested an isolated, still green, 2-m-high *C. berlandieri* plant growing in full sun, and having 28 lateral branches (Catalog number 87). In addition, I harvested a 2 sq m cluster of 58 *C.*

Table 7.2. Harvest Yield Information for Sixteen Stands of *Chenopodium berlandieri*

Catalog number	Stand area (sq.m)	Number of plants	Fruit weight (g)	Harvest yield (kg/ha)	Harvest time (min)	Harvest rate (kg/hr)
30	250	4000(?)	459.8	—	22	1.250
39–42	1	4	142.3	1423	9	.949
57	>1	10	108.9	1089	9	.726
58	>1	9	83.3	833	6	.832
59	>1	7	121.3	1213	8:30	.856
82A	1	3	49.0	490	2	1.470
87	1	1	91.8	918	8	.689
88	2	58	112.6	563	5	1.350
89	1	2	51.0	510	3	1.020
90	1	4	95.9	959	4	1.438
91	1.5	1	428.1	2854	16	1.605
92	3	10	202.2	679	12	1.010
94 Top Ten Plants						
94CC	1	1	175.0	1750	11	.955
94D	1	1	127.3	1273	10	.764
94B	1	1	101.6	1016	8	.762
94C	1	1	100.8	1008	11	.540
94E	1	1	89.1	891	5	1.060
94Z	1	1	84.6	846	4:17	1.190
94BB	1	1	80.5	805	4:30	1.073
94X	1	1	79.2	792	5	.950
94II	1	1	78.7	787	6	.786
94R	1	1	76.0	760	6:30	.702
94 Top Ten Plants (clustered)[a]	10	10	1064.0	1064	71	.899
94 All 36 plants (clustered)	32	36	1860.0	580	148	.754
95D	1	1	129.8	1298	10	.778
95 All Ten Plants (clustered)	10	10	371.6	371	34	.655
96	1	1	49.2	492	3	.984
97	3.5	27	96.7	276	14	.414

[a]The individual plants comprising stands 94 and 95 were fairly widely scattered across a heavily overgrown soybean field (Figure 5). In order to estimate what the harvest yield values for each of the stands would be if the plants were not so scattered, "clustered" statistics were obtained by centering each plant in an arbitrary 1 meter square and then assuming that the one meter squares adjoined each other.

berlandieri plants (Catalog number 88, Table 7.2) found growing within a 15 by 20 m stand of *C. missouriense*.

Fulton County, Pennsylvania (Catalog numbers 38–42)

Catalog numbers 38 (herbarium voucher) and 39–42 were assigned to five individual plants harvested from a 1 sq m cluster of *C. berlandieri* found within a surrounding 4 sq m stand of *C. missouriense* growing in a construction site dirt pile.

Mississippi County, Arkansas (Catalog numbers 57–59)

A 5 by 7 m stand of *C. missouriense* was observed growing in full sun on a roadside dirt pile in November of 1984. Catalog numbers 57, 58, and 59 were assigned to three separate clusters of *C. berlandieri* located within the larger dense stand of *C. missouriense*. Each less than 1 sq m in area, these clusters contained 10, 9, and 7 plants respectively. The *C. berlandieri* plants ranged in height from 130 to 150 cm, had few lateral branches within a meter of the

Figure 7.1 Overgrown creek bottom vegetable garden location of *Chenopodium* stands 91 and 92. Pike County, Ohio.

ground, and could be visually distinguished from the surrounding, still green, *C. missouriense* plants by their grayish brown leafless condition and generally smaller size.

Cherokee County, South Carolina (Catalog numbers 82A–82Q)
Three clusters of *C. berlandieri* were observed growing adjacent to a bridge over the Broad River in November 1984. Two of the clusters (82A–C, 82D–G) were approximately 1 sq m in area, while the third (82H–Q) was 1 by 3 m in size. All of the plants were dead and had lost the majority of their fruit.

Prince Georges County, Maryland (Catalog numbers 89A, 89B)
In the spring of 1985 *C. berlandieri* seeds were scattered in turned soil adjacent to garden plots at the Museum Support Center, Smithsonian Institution, in Silver Hill Maryland. Only two plants (89A, 89B) grew to maturity, however, likely due to a very dry growing season and a failure to expose the seeds to freezing temperatures over the winter. Upon harvesting them in October, both plants were dry, brown, with partial seed loss, and were choked by a dense growth of grass and weeds.

Washington County, Maryland (Catalog number 90A, 90B, 90C1–90C5)
In November 1985, I observed *C. missouriense* growing in fill dirt along a road-edge guardrail. Within this stand I harvested a 1 by 2 m cluster of *C. berlandieri* containing two large (90A, 90B) and five smaller (90C1–90C5) plants. These seven plants were all dead, brown, and dry, with partial seed loss, and could be easily distinguished from the surrounding, still green *C. missouriense* plants. The two larger plants, along with two of the smaller plants (90C1, 90C2) occupied an area of less than 1 sq m.

Pike County, Ohio (Catalog numbers 91, 92A–92J)
Catalog number 91 was assigned in November 1985 to a single *C. berlandieri* plant growing in an overgrown creek bottom vegetable garden (corn, pumpkins, sunflowers, summer squash) (Figure 7.1). Occupying a 1 by 1.5 m area, the large (height 1.5 m), bright yellow leafless plant had numerous lateral branches with black terminal infructescences (Figure 7.2). Little seed loss was apparent. After hand stripping, the plant was felled and subsequently weighed in the laboratory. Occupying a 1 by 3 m area, a stand of ten *C. berlandieri* plants (92A–92J) was found growing in dense undergrowth in the same garden

Figure 7.2 A large *Chenopodium berlandieri* plant, field catalog number 91, which yielded 428 grams of seeds (surrounding weeds removed prior to photograph).

(Figure 7.3). The plants all had leafless yellow stalks and black infructescences, and could be quickly distinguished from the surrounding dense undergrowth.

Mississippi County, Missouri (Catalog numbers 94A–94JJ, 95A–95J, 96, 97)

Forty-six *C. berlandieri* plants (94A–94JJ; 95A–J) were found growing in an overgrown soybean field in November 1985. Separated from the main channel and sand bars of the Mississippi River on the east by a 20–30-m-wide black willow vegetation zone, and from the levee on the west by a 100-m-wide low-lying area of wet clay soil (Figure 7.4), the soybean field was about 70 m wide and paralleled the river for approximately 160 m. The mature but unharvested field of soybeans, which appeared to not have been sprayed with herbicides and to have been generally neglected, was thickly overgrown with weeds, including Johnson grass, wild bean *(Strophostyles helvola)* and both *C. missouriense* and *C. berlandieri.* In contrast to previously observed upland situations where these two chenopod species were found growing in the same locales, *C. missouriense* was not a dominant constituent of the weed plant community (see Seeman and Wilson 1984:305). *C. berlandieri,* on the other hand, was quite abundant in the field.

C. Wesley Cowan and I mapped, described, and harvested a rather scattered linear stand of 46 plants over a period of 4 hours (Figure 7.5). After being mapped and assigned a letter code, the height of each

Figure 7.3 The dense undergrowth setting of *Chenopodium berlandieri* stand 92.

Figure 7.4 Looking east from the Mississippi River levee toward the soybean field location of *Chenopodium berlandieri* stands 94 and 95. The main channel and opposite shore of the Mississippi River are visible in the background, along with exposed sand bars. The low clay soil area separating the levee from the soybean field is visible in the foreground.

plant was recorded, along with information regarding stalk and infructescence color and condition, the relative abundance of lateral branches, and the presence/absence of wild bean vines. Catalog number 96 was assigned to a single *C. berlandieri* plant located 20 m from the main channel of the Mississippi River. Growing in full sun on the sand beach of the river, and partially entangled by *Strophostyles helvola*, this specimen was brown and dry, with numerous lateral branches (Figure 7.6). After locating Catalog number 96 along the sandy beach edge of the Mississippi River, C. Wesley Cowan and I walked further south along the river for a distance of about 400 m within the relatively narrow (20–50 m) black willow vegetation zone which paralleled the river's edge. In the light shade setting under a willow canopy, *C. berlan-*

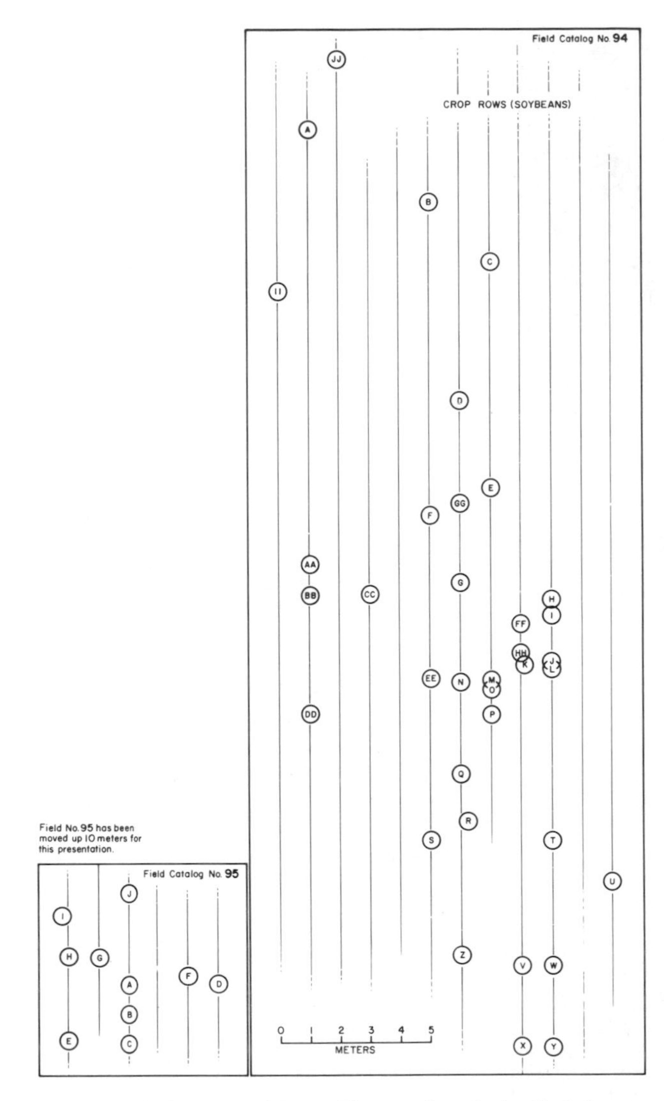

Figure 7.5 The location of the 46 *Chenopodium berlandieri* plants comprising stands 94 and 95.

dieri plants were quite common, often occurring in linear stands paralleling the river. In contrast to the habit of *C. berlandieri* plants observed growing in full sun, these partial shade understory plants had thin straight stalks and lateral branches with few leaves and small diffuse infructescences (Figure 7.7). In addition, these understory plants were still green and in full leaf. We harvested a stand of 27 plants occupying a 3.5 sq m area.

Figure 7.6 A *Chenopodium berlandieri* plant growing on the sandy bank of the Mississippi River.

Hardin County, Tennessee (**Catalog numbers 101A, 101B**)

Two *C. berlandieri* plants were found growing in a bank slump area along the Tennessee River and were harvested in November of 1985. Located in partial shade, both plants were similar in habit to the understory plants described above (97), having tall thin stalks and lateral branches with diffuse infructescences. Both plants were leafless and brown, with 101A exhibiting little apparent fruit loss, while 101B had lost most of its fruits.

Cullman County, Alabama (**Catalog number 105**)

In November 1985, I located a stand of four *C. berlandieri* plants in a flat floodplain area only 5 m from the edge of the Black Warrior River. All four plants were still green and were in excellent condition for herbarium specimens. Three of the four were pressed for the collections of the National Herbarium (NMNH), while the fourth was hand stripped for the present study.

Figure 7.7 The light shade black willow understory setting of *Chenopodium berlandieri* stand 97.

Tuscaloosa County, Alabama (Catalog number 108)
In November 1985, I located a single plant in thick undergrowth at the top of a steep sand bank down to the Black Warrior River. Growing in full sun, the plant was still green but leafless when harvested.

Discussion

The Habitat of *C. berlandieri* in the Eastern United States

The field research phase of this study confirmed what a number of previous researchers had noted—*C. berlandieri* is an elusive subject for harvest yield experiments since it is not generally abundant in the eastern United States, nor does it frequently occur in large stands (Seeman and Wilson 1984:303, 304; Asch and Asch 1977:25; Munson 1984). As a result of the field research reported here, the preferred habitat situation of this species can now be described with a greater degree of accuracy, which should facilitate the location of stands for future harvest yield studies.

The primary habitat of *C. berlandieri* in the eastern United States is river valley floodplains, particularly large meandering rivers such as the Mississippi River and its major tributaries. Within this river valley alluvium habitat zone *C. berlandieri* can be found growing in a number of different disturbed soil situations. Primary among these is as an abundant understory constituent of black willow river margin sand bank vegetation communities. These "naturally disturbed" river margin black willow zones are subject to annual scouring by floodwaters and deposition of alluvial, primary heavy fraction sandy soil. Within this shady understory setting, *C. berlandieri* plants are tall, slender, have small diffuse infructescences, and a quite low seed yield per square meter value (Table 7.2, Figure 7.8, Catalog number 97). Wahl (1954:44) briefly mentions this river margin understory habitat setting for *Chenopodium berlandieri*: "*C. bushianum* occurs most often as a weed of cultivated places but is found also in alluvium along streams and in waste places. The shade form is more delicate." Steyermark (1963:614) also mentions this habitat: *C. bushianum* "occurs in sandy fields and alluvial ground along rivers, waste places, wooded slopes, and dry open or shaded ground." In addition, a Field Museum of Natural History herbarium specimen identified as *C. bushianum* and examined by Asch and Asch (1977:36) carried the following habitat description: "riverbank among trees—Fargo, North Dakota."

While this understory setting can be considered the primary "natural" floodplain habitat of *C. berlandieri,* the plant also occurs in two other nearby "naturally disturbed" floodplain situations. Jackson's statement (1986:183) that "stands of chenopods also can be found along bank margins left bare by receding summer water level" corresponds with field observations made in November 1985 regarding *C. berlandieri.* As the water level of the Mississippi River receded during the summer and fall, the exposed

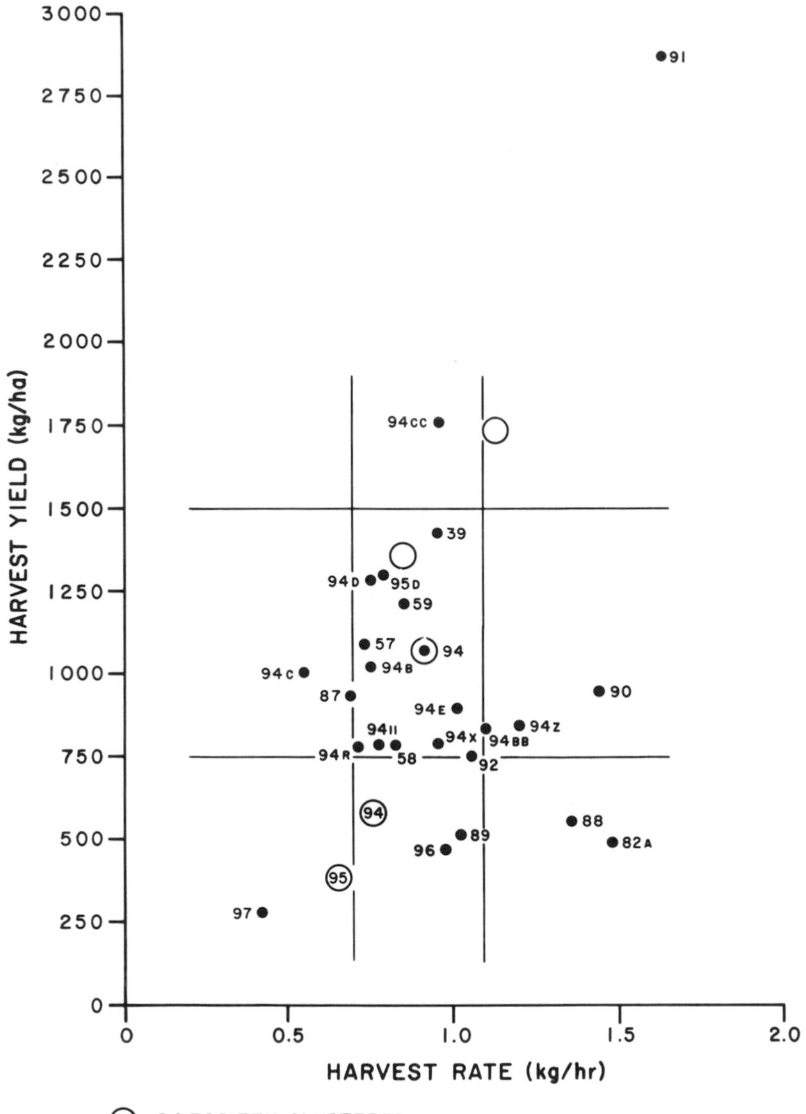

Figure 7.8 Harvest yield (kg/ha) and harvest rate (kg/hour) values for 16 *Chenopodium berlandieri* stands harvested in 1984 and 1985, compared with the single plant yields reported by Asch and Asch (1978). See Table 7.2 for explanation of "clustered" values. Range estimates for harvest yield values (750–1,500 kg per ha) and harvest rate (.7–1.1 kg per ha) are indicated.

river margin sand beach supports multibranched *C. berlandieri* plants with large terminal infructescences (Table 7.1, Figure 7.6, Catalog number 96). Unstable and actively eroding river and terrace banks and bluffs also support occasional, often solitary, *C. berlandieri* plants, particularly along the upper bank edge (Table 7.1, Catalog numbers 101, 105, 108; Asch and Asch 1977:20; Munson 1984:381, 383–384). It is important to note that while *C. berlandieri* can be recognized as occurring in at least three different "naturally disturbed" habitat settings within river valley floodplains, Steyermark's reference (quoted above) to the plant's association with sandy soil is quite perceptive in that *C. berlandieri* rarely occurs in other floodplain soils of heavier texture. Although Asch and Asch (1977) describe a late season stand of

Chenopodium berlandieri growing on an exposed mud flat in the lower Illinois River Valley, such stands are rare. During extensive surveys by Klein, Daley, and Wedum (1975) and by Munson (1984) of Illinois and Mississippi River Valley mud flats exposed during seasonal drying of shallow lakes, ponds, and sloughs, no chenopod species were noticed in mud flat vegetation communities.

In addition to occurring in these three overlapping naturally disturbed sandy soil floodplain situations, *C. berlandieri* can also be found in a variety of anthropogenically disturbed soil settings. It is often the dominant chenopod weed in river valley fields and gardens (Catalog numbers 91, 92, 94, 95). Wilson and Walters (n.d.) describe the habitat of *C. bushianum* as "disturbed ground, especially alluvial soil of agricultural areas," while Steyermark (1963:614) has it occurring ". . .in sandy fields and alluvial ground along rivers," and Wahl (1954:44) considers it as "A weed of cultivated places. . .," also found "in alluvium along streams and in waste places." While floodplain sandy soil fields and gardens can be considered the primary "anthropogenic" disturbed habitat setting for *C. berlandieri* in the eastern United States, it is also found in other disturbed soil and waste place settings in both river floodplains (under bridges, Catalog number 82) and upland areas (fields, gardens, construction sites, and along highways). *C. berlandieri* occurs only infrequently in upland settings, and when it does occur it is rarely alone, but almost invariably found growing in small scattered pockets, as a minor constituent within larger stands dominated by *C. missouriense* (Catalog numbers 20–28, 38–42, 57–59, 87–88, 90) (Asch and Asch 1977:25–26; Seeman and Wilson 1984:303–305).

In summary, *C. berlandieri* in the eastern United States is an early successional plant species inhabiting a variety of different natural and anthropogenically disturbed sandy soil situations within river valley floodplains. Within such river valley floodplain environments, it would not be at all unusual to find *C. berlandieri* inhabiting a number of adjacent habitats all situated within 50 m of each other (exposed low water river margin sand beaches, black willow understory, eroding terrace sand banks, and floodplain agricultural fields). It is also an infrequent minor constituent in upland disturbed soil area chenopod stands. It is thus within the meander belt of the Mississippi River and its tributaries, adjacent to active main channels, that stands of *C. berlandieri* suitable for harvest studies are most likely to be encountered. Within this area, poorly weeded and overgrown natural levee fields and gardens provide the best opportunity for locating extensive stands of multiple branched plants. That floodplain gardens and fields hold the best promise for future harvest yield studies of this plant is underscored by the results of the 1984–1985 fall harvest project. While only four of the sixteen stands harvested were situated in floodplain fields, those four stands (91, 92, 94, 95) yielded two-thirds of the total seed collected. Fortunately, such present-day floodplain fields and gardens also represent essentially the same setting within which prehistoric cultivation of *C. berlandieri* would have primarily occurred.

The Economic Potential of *C. berlandieri*

Yield values obtained for the 16 *C. berlandieri* stands harvested in 1984 and 1985 show considerable variation (Table 7.2, Figure 7.8), from a low of 276 kg per ha for field Catalog number 97—the 3.5 sq m black willow, partial shade stand, to a high of 2,854 kg per ha for stand 91, a single plant occupying 1.5 sq m and yielding 428 g of seed. This variability is to be expected because of considerable variation in both the amount of sun received by different stands and the degree of crowding and competition from surrounding plants. Variation in plant height to seed yield ratios and plant weight to seed yield ratios (Table 7.1) also reflect differing degrees of competition and sunlight. Substantial seed loss prior to harvesting was only occasionally a factor in harvest yield variation (82A, 101B). The long period seed retention of *C. berlandieri* is well documented (Wahl 1954; Seeman and Wilson 1984:303, 309). The application of fertilizers and herbicides played undocumented roles in influencing yield values.

In the only previous published consideration of the economic potential of *C. berlandieri,* Asch and Asch (1978:313) present yield values of 1,330 and 1,740 kg per ha for two tall multiple branched plants, each growing in an open full sun plot "in which the plants had maximum potential for vegetative growth." Because of the limited competition represented in the case of these two plants, Asch and Asch suggested that it was "probably unreasonable to extrapolate its production to a large stand where the chenopods would be competing with each other" (1978:313).

While the harvest yield values presented by Asch and Asch fall toward the upper end of the range of yield values recorded in the present study (Figure 7.8), I think that they do not represent unreasonably high approximations for prehistoric *Chenopodium* production in the Eastern Woodlands. Their average yield value (1,535 kg per ha) is about half that of the most productive (plant) stand in the present study (field Catalog number 91, yielding 2,854 kg per ha, and falls about midway along the documented range of variation—276–2,854 kg per ha). In addition, with the exception of stands 82a and 87, all of the plants harvested in the present study were competing with other plants, either as small pockets surrounded by larger stands of *C. missouriense,* or in dense undergrowth situations (Figures 7.1, 7.3, 7.9). As a result, with the exception of stand 91, the yield values obtained from the 16 stands harvested in 1984 and 1985, as presented in Figure 7.8, could be considered as defining the lower end of the range of likely prehistoric harvest yield levels for *C. berlandieri*—what might be expected from poorly maintained and overgrown fields. A relatively conservative range estimate of 750–1,500 kg per ha for prehistoric yield levels for *C. berlandieri* is proposed (Figure 7.8).

Harvest rate values obtained during this study also showed considerable variation, with stands 97 and 91 again providing minimum (.41 kg per ha) and maximum (1.6 kg per ha) values, respectively. Variability in harvest rate values was due primarily to differences in the size and degree of compactness, and terminalization of infructescences of harvested plants. The light shade plants of stand 97 had numerous small and diffuse infructescences that took far longer to locate and to strip than did the large compact fruit clusters of stands 91–92 and 94–95 (Figure 7.10). Surrounding chenopod plants and weeds, particularly wild bean *(Strophostyles helvola),* also substantially decreased harvest rates by making it more difficult to see and reach infructescences. A relatively conservative harvest rate range estimate of 0.7–1.1 kg per hour is proposed for *Chenopodium berlandieri* in the prehistoric Eastern Woodlands (Figure 7.8).

Figure 7.9 The dense weedy growth setting of *Chenopodium berlandieri* stand 94. An exposed sand bar of the main channel of the Mississippi River is visible in the background.

Figure 7.10 The typical infructescence form of *Chenopodium berlandieri* plants growing in full sun.

Harvest Yield Comparisons

In order to make any harvest yield comparisons between *C. berlandieri* and other crop plants of the prehistoric Eastern Woodlands of North America, it is first necessary to take into consideration the thick seed coat of the modern wild/weedy populations of *C. berlandieri* harvested in the present study. Ranging in thickness from 40–70 microns (Smith 1985b), the seed coat of wild/weedy fruits consists largely of non-nutritive fiber, and accounts for at least 30 percent of the seed weight (Seeman and Wilson 1984).

Of the two prehistoric domesticated varieties of *C. berlandieri* cultivated in the prehistoric Eastern Woodlands, one, *C. berlandieri* ssp. *jonesianum,* had a thin (> 20 microns) testa (Smith and Funk 1985), and the other lacked a testa entirely (Wilson 1981; Fritz 1986). Harvest yield rates at least 30 percent lower could therefore be expected for prehistoric cultivated stands producing the same number of fruits per hectare as modern wild/weedy stands. Reducing harvest yield values obtained during the present study by 30 percent (from 750–1,500 kg per ha to 525–1,050 kg per ha) results in a closer approximation of the nutritive yield of thin-testa and testaless prehistoric cultigen varieties (Figure 7.11).

Application of the same 30 percent reduction factor to two other modern thick testa chenopods *(C. ambrosioides* and *C. missouriense)* results in harvest yield estimates of 784 kg per ha and 331–538 kg per ha respectively (Figure 7.11). While *C. missouriense* falls at the lower end of the range of values obtained for *C. berlandieri, C. ambrosioides* falls within the *C. berlandieri* range.

The modified range estimate of 525–1,050 kg per ha for *C. berlandieri* is also quite comparable to harvest yield values reported for *C. quinoa,* the testaless domesticated chenopod of South America. Elmer (1942) reports average yields of 504–1,008 kg per ha, and White et al. (1955:535) present production figures of 493–896 kg per ha. Recent initial plantings of *C. quinoa* (varieties 407 and 407 black) in Colorado have resulted in yields of 331–805 kg per ha. Yields of 997 kg per ha were also obtained from five 1 sq m plots of *C. quinoa* variety 407 grown at Vadito, Colorado in 1985 (Johnson and McCamant 1986). Optimal yields in the 3,000–5,000 kg per ha range for *C. quinoa* have also been reported or projected (Elmer 1942:21; White et al. 1955:535; Johnson and McCamant 1986).

Because the harvest yield range estimate for *C. berlandieri* presented here is so comparable to the large acreage cultivated harvest yield estimates available for *C. quinoa,* it would not seem unreasonable to propose a harvest yield range estimate of 500–1,000 kg per ha for prehistoric chenopod cultivated in the Eastern Woodlands of North America. This range estimate of 500–1,000 kg per ha can in fact be considered rather conservative, if differential efficiency of harvest methods are considered. The cutting of either infructescences (cutting method) or whole plants (mass collection method) for subsequent drying and flailing for seed recovery have been shown to be more effective methods for harvesting *Chenopodium* than simple hand stripping (Seeman and Wilson 1984:308–309) in that they result in higher harvest yields. Mass collection is the harvest method employed today in South America by Quechua groups (Elmer 1942; Gade 1970), and may also have been the preferred prehistoric method in the East, judging from the stored sheaves of *C. berlandieri* recovered from dry caves and rock shelters (Seeman and Wilson 1984:309).

Nevertheless, the proposed harvest yield range estimate of 500–1,000 kg per ha for *C. berlandieri* is considerably higher than available harvest yield values for other premaize crop plants of the Eastern Woodlands. Asch and Asch (1978:310 [see Chapter 8]) obtained harvest yield values of 77–468 kg from eight 1 sq m plots of *Iva annua.* Murray and Sheehan (1984) report a harvest yield value of 1,115 kg per ha for *Polygonum erectum,* based on the collection of a 2.29 sq m stand. Assuming that the achene "shell" accounts for 50 percent of total knotweed achene weight, *P. erectum* would have a corrected harvest yield value of 557 kg per ha, at the lower end of the range proposed for *C. berlandieri* (Figure 7.11). Similarly, employing the 1959 United States and world average production figures for sunflower (645 kg per ha, 979 kg per ha; Martin and Leonard 1967:932) to define a harvest yield range estimate, and assuming that the achene "shell" accounts for 46 percent of the total achene weight (Asch and Asch 1978:314), the corrected harvest yield range estimate of 348–528 kg per ha would only slightly overlap with that of *C. berlandieri.*

Against the backdrop provided by the comparatively lower harvest yield levels of other prehistoric eastern cultigens, the hand-stripped yields of *C. berlandieri* stands underscore both its value as an uncultivated prehistoric plant food source and its preeminent potential as a cultigen in premaize gardens and fields. The high harvest yield levels of present-day uncultivated stands of *C. berlandieri* may help to explain why it was initially brought under domestication prehistorically, as well as its ubiquitous presence

Figure 7.11 Range estimate harvest yield values for ten seed plants. To facilitate comparison of yield estimates, paired bar graphs are presented for those species with thick "seed coats," indicating harvest yield range values with and without seed coats.

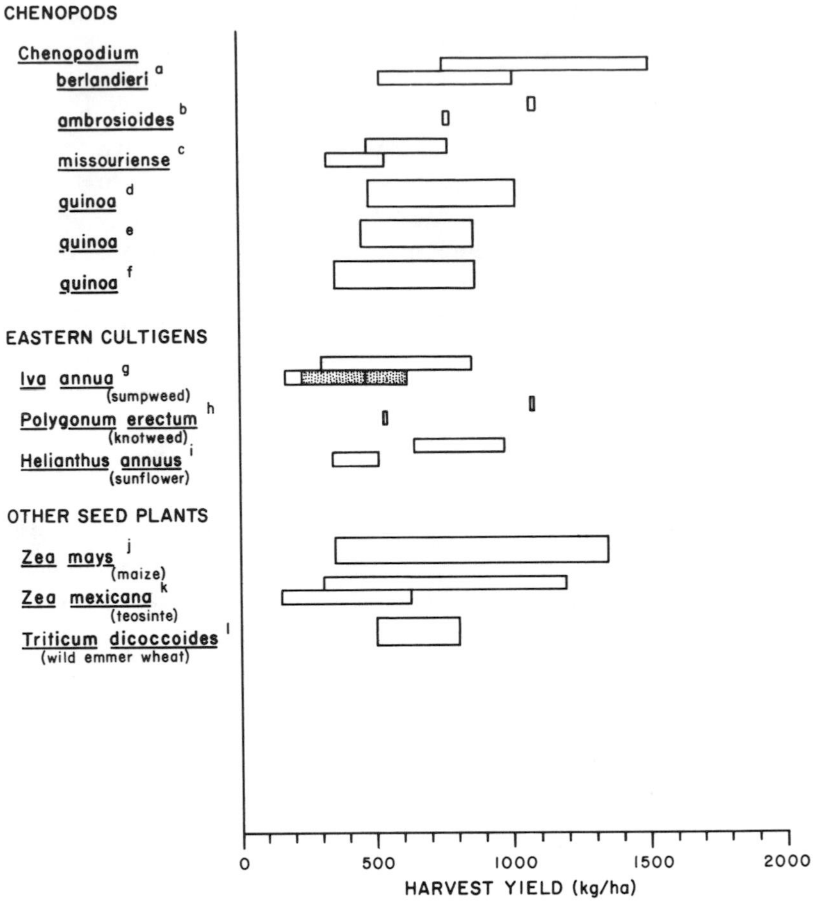

[a] Results of this study. Plants hand stripped, weighs represent clean seed. Range values 750–1,500 kg per ha and 525–1,050 kg per ha when corrected for 30 percent seed coat weight.

[b] Containing ascaridole, which kills internal parasites, *C. ambrosioides* is cultivated and machine harvested in Carroll County, Maryland for its oil. Yields of 1,000 lb of seed per acre (1,120 kg per ha) have been reported (Seeman and Wilson 1984:307).

[c] Hand-stripping, cutting, and mass collecting (whole plants cut, bundled, dried, and threshed, with fruits then winnowed) harvesting methods employed. Yields ranged from 473–769 kg per ha, 331–539 kg per ha when corrected for seed coat weight (Seeman and Wilson 1984).

[d] Quinoa plants cut, dried, flailed, and fruit winnowed (mass collecting method). Yields ranged from 504–1,008 kg per ha (Elmer 1942).

[e] Mass collecting method (White et al. 1955).

[f] Mass collecting method, machine threshed (Johnson and McCamant 1986).

[g] Best seven of eight 1 sq m plots. Hand-stripping method employed. Kernels considered 55 percent of total achene weight. Yields ranged from 300–850 kg per ha, 165–468 kg per ha when corrected for achene shell weight (Asch and Asch 1978).

[h] A single stand occupying 25 sq ft (2.29 sq m) harvested by hand-stripping. Assuming achene "shells" accounted for 50 percent of total achene weight, harvest yield values of 1,115 and 557 kg per ha (corrected for achene shell weight) were derived (Murray and Sheehan 1984).

[i] Martin and Leonard (1967:932) report that the 627 farms growing sunflower commercially in the United States in 1959 produced 623,000 bushels of seed from 25,732 acres, for an average yield of 24.2 bushels (576 lbs of seed) per acre (645 kg per ha). Assuming that achene shells comprise 46 percent of total achene weight, Asch and Asch (1978:314) derived a hulled clean seed yield value of 350 kg per ha. Martin and Leonard also reported typical yields of 784–1,120 kg per ha for the major producing states (Minnesota and North Dakota), and noted yields of 2,016 kg per ha from California. In addition they provide an average world production level of 979 kg per ha for the 17 million acres under cultivation in 1959. The average U.S. and world production levels (645 kg per ha and 979 kg per ha) are employed to define a range estimate, with associated range estimate yield values for cleaned seed being 348 kg per ha and 528 kg per ha.

[j] Harvest yield estimates reported by Will and Hyde (1917) for Indian groups of the Upper Missouri during the 1860s and 1870s. Yields of 8–30 bushels per acre (unshelled) allowed a range estimate of 358–1,344 kg per ha to be derived.

[k] Yields of 305–1,254 kg per ha reported by Robson et al. (1967:247). Assuming the fruit capsule (involucre) accounts for 50 percent of seed yield (Robson et al. 1967), a range estimate of 152–617 kg per ha was derived for clean seed.

[l] Hand-stripped, clean grain, range estimate for optimum yield, rainy year (Zohary 1969).

Table 7.3. The Nutritive Value of Seed of Cultigens of the Prehistoric Eastern Woodlands of North America (percent dry basis)

Species	Protein	Fat	Carb.	Fiber	Ash
Starchy-seeded					
Goosefoot[a]					
C. berlandieri	19.12	1.82	47.55	28.01	3.50
Maygrass[b]					
P. caroliniana	23.7	6.4	54.3	3.0	2.14
Knotweed[a]					
P. erectum	16.88	2.41	65.24	13.33	2.34
Oily-seeded					
Sumpweed[c]					
I. annua	32.25	44.47	10.96	1.46	5.80
Sunflower[d]					
H. annuus	24.00	47.30	16.10	3.80	4.00
Tropical crops					
Maize[c]					
Z. mays	8.9	3.9	70.20	2.0	1.2
Squash[c]					
C. pepo	29.0	46.7	13.10	1.9	4.9
Bean[e]					
P. vulgaris	22.0	1.6	60.8	4.3	3.6
Quinoa[f]					
C. quinoa	12.5	6.0	72.5	5.6	3.4

[a]Asch and Asch 1985:361.
[b]Crites and Terry 1984.
[c]Asch and Asch 1978.
[d]Watt and Merrill 1963.
[e]Wu Leung 1961.
[f]White et al. 1955.

and abundance in otherwise regionally variable premaize plant husbandry systems of the Eastern Woodlands (Smith 1985a:52). Although it has a far thicker seed coat than the prehistoric thin-testa domesticated form of chenopod that was cultivated prehistorically in the Eastern Woodlands, the present day wild-weedy form of *C. berlandieri* in the eastern United States—the subject of this study—represents an appropriate analog for the prehistoric domesticate in that it retains a number of characteristics of domestication (simultaneous inflorescence, extended seed retention) that strongly suggest that it represents the weedy descendant of the prehistoric domesticate (Wilson 1981). The seeds of *C. berlandieri* and other "starchy-seeded" crops have relatively low protein and fat content when compared to the "oily-seeded" crops—marshelder and sunflower (Table 7.3) and are high in carbohydrates. As a result, they have been considered to have been less important nutritionally than the "oily-seeded" annuals and to have been roughly comparable to maize in terms of food value. Analysis of the essential amino acid pattern of *C. quinoa,* however, has shown it to be exceptionally high in two essential amino acids (lysine and methionine), which make it extremely attractive as a source of protein and as a general source of human nutrition (White et al. 1955; Cusak 1984:23). Although it is yet to be documented, it is quite likely that *C. berlandieri* has a similar amino acid pattern to that of *C. quinoa.*

In addition to shedding light on its initial domestication and subsequent development as an important crop in premaize plant husbandry systems of the eastern United States, the documented economic potential and likely nutritional qualities of *C. berlandieri* also invite a comparison between *C. berlandieri* and maize, which came to dominate prehistoric eastern agricultural economies after A.D. 1000. Estimating prehistoric harvest yield values for maize is particularly difficult, due to the lack of documented analog situations involving nonhybrid maize cultivated and harvested without the benefit of fertilizers, draft animals, or machinery. Will and Hyde (1917:103, 108, 142) provide maize yield statistics for the 1860s and 1870s for Indian groups of the Upper Missouri area (South Dakota, North Dakota) that range from 8 to 30 bushels per acre (1867 Kansa, 19 bushels per acre; 1867 Yanktons, 30 bushels per acre; 1874 Sac and Fox, 20 bushels per acre; 1878 Kansa, 8 bushels per acre; 1878 Santee Sioux, 26 bushels per acre; 1878 average for Upper Missouri groups, 20 bushels per acre). While a bushel of shelled corn weighs 56 lbs (Martin and Leonard 1967:965) it is likely that the statistics extracted by Will and Hyde from historical records referred to bushels of corn on the ear. Since 70 lbs of corn on the ear yields 50 lbs of shelled corn (at a 15 percent moisture level, Martin and Leonard 1967), each "bushel" of the 1860s–1870s likely contained about 40 lbs of shelled corn. Employing this figure of 40 lbs of shelled corn per bushel, a range estimate of 358–1,344 kg per ha can be derived from Will and Hyde's 8–30 bushel per acre range (Figure 7.11). While acknowledged as being "no doubt generous as an estimate of prehistoric yields," average production figures of 10–45 bushels per acre (shelled) have been reported for the prehybrid maize grown prior to 1925 in the Black Warrior River Valley of west-central Alabama (Peebles 1978:402–403), providing a harvest yield range estimate of 627–2,822 kg per ha. The Will and Hyde maize yield data for the northern Plains in the 1860s–1870s probably represents a closer approximation than the Black Warrior River Valley values to the prehistoric Woodland levels of maize production in the Eastern Woodlands.

It will, of course, be difficult to establish with any degree of confidence the economic potential and actual yield of maize in the Eastern Woodlands during the A.D. 1000–1200 period of transition to maize agriculture. But yields in excess of 1,000 kg per ha (about 17 bushels per acre) would have been necessary before maize would represent an attractive alternative to *C. berlandieri* in terms of harvest yield alone. Potential yield obviously was not the only factor influencing crop selection, and maize would likely have required considerably less commitment of time and energy to harvest than *C. berlandieri* during the critical fall period of intensive hunting of deer and collecting of wild plant resources. While maize cobs could be picked, stripped, dried, and stored, *C. berlandieri* would have required cutting and drying of whole plants, followed by flailing and winnowing prior to storage. The continued cultivation of *C. berlandieri* after A.D. 1200 as a secondary field crop within maize-dominated field systems is not surprising, given its high yield and nutritional profile and the nutritional shortcomings of maize (Robson et al. 1976:246–247).

The obvious next step in pursuing the issue of harvest yield levels for *C. berlandieri* and the other prehistoric cultigens of the Eastern Woodlands is to grow them in relatively large cultivated stands in order to establish their economic potential in well-controlled field plot settings. When combined with ongoing morphological and quantitative analysis of archaeobotanical assemblages of these early prehistoric cultigens, such modern experimental studies should provide substantial illumination of the nature and importance of premaize food production systems in the Eastern Woodlands of North America.

Notes

1. Assignable to subsection *Cellulata* of the genus *Chenopodium* on the basis of its reticulate-alveolate pericarp, the species *Chenopodium berlandieri* subsumes an extensive geographical variety of wild, weedy, and domesticated forms of chenopod. Although still frequently referred to as a distinct species within subsection *Cellulata, Chenopodium bushianum,* a large-fruited chenopod of the Northeast, Midwest, and mid-latitude Southeast, has been shown to have considerable genetic affinity with *C. berlandieri,* rather than being biologically distinct (Wilson and Heiser 1979; Wilson 1980:260).

As the systematics of eastern North American chenopods belonging to subsection *Cellulata* is revised, it is highly likely that *C. bushianum* will be relegated to subspecific taxonomic status within *C. berlandieri.* In anticipation of this reassessment and reassignment of *C. bushianum,* the species designation *C. berlandieri* is used throughout this article, even when geographical location and fruit size would suggest the species label *C. bushianum*.

2. *Chenopodium* produces small indehiscent "seedlike" fruits, each consisting of a seed enclosed by a very thin adherent pericarp. Since this thin pericarp is the only thing distinguishing fruits from seeds, the terms "seed" and "fruit" are used interchangeably. Winnowing was accomplished in a two-step process. Collected infructescence material was first rubbed between the palms of the hands to dislodge fruits from attached perianths, and the material was then distributed along the top edge of an inclined cotton sheet. The lighter and more angular perianth, leaf, and stem fragments would adhere to the sheet while fruits would roll down the angled sheet to be collected at its base.

Acknowledgments

The harvest yield research described in this article was made possible by Smithsonian Institution Research Opportunity Fund grants 1233f505 and 1233f611, awarded by David Challinor, Assistant Secretary for Research. This timely support is gratefully acknowledged.

C. Wesley Cowan, Cincinnati Museum of Natural History, accompanied the author during the 1985 harvest season, and contributed substantially to the success of the project. Willard Van Asdall, Gayle Fritz, Richard I. Ford, and Richard A. Yarnell read early versions of this article, and their comments and suggestions for revision resulting in substantial improvements.

Literature Cited

Asch, D., and N. B. Asch

1977 Chenopod as Cultigen: A Re-evaluation of Some Prehistoric Collections from Eastern North America. *Midcontinental Journal of Archaeology* 2:3–45.

1978 The Economic Potential of *Iva annua* and its Prehistoric Importance in the Lower Illinois Valley. In *The Nature and Status of Ethnobotany,* edited by R. I. Ford, pp. 301–341. Museum of Anthropology, Anthropological Paper No. 67. University of Michigan, Ann Arbor.

1985 Archaeobotany. In *Smiling Dan,* edited by B. Stafford and M. B. Sant, pp. 327–399. Center for American Archaeology, Research Series 2. Kampsville, Illinois.

Chapman, J., and G. Crites

1987 Evidence for Early Maize *(Zea mays)* from the Icehouse Bottom Site, Tennessee. *American Antiquity* 52:352–354.

Crites, G., and R. D. Terry

1984 Nutritive Value of Maygrass. *Economic Botany* 38:114–120.

Cusak, D.

1984 Quinoa: Grain of the Incas. *The Ecologist* 14:21–31.

Earle, F. R., and Q. Jones

1962 Analysis of Seed Samples from 113 Plant Families. *Economic Botany* 16:221–250.

Elmer, L. A.

1942 Quinoa *(Chenopodium quinoa). East African Agricultural Journal* 8:21–33.

Ford, Richard I.

1987 Dating Early Maize in the Eastern United States. Paper presented at the annual meeting of the American Association for the Advancement of Science, February 17, Chicago.

Fritz, G.

1986 Prehistoric Ozark Agriculture. Ph.D. dissertation, Department of Anthropology, University of North Carolina, Chapel Hill.

1987 The Trajectory of Knotweed Domestication in Prehistoric Eastern North America. Paper presented at the 10th annual meeting of the Society for Ethnobiology, March 5–8, Gainesville, Florida.

Gade, D.

1970 The Ethnobotany of Canihua *(C. pallidicaule),* Rustic Crop of the Altiplano. *Economic Botany* 24:55–61.

Jackson, H. E.

1986 *Sedentism and Hunter-Gatherer Adaptations in the Lower Mississippi Valley.* Ph.D. dissertation, Department of Anthropology, University of Michigan, Ann Arbor. University Microfilms, International.

Johnson, D., and J. McCamant

1986 1985 Quinoa Research: Preliminary Report. Sierra Blanca Associates, Denver.

Jones, Q., and F. R. Earle
1966 Chemical Analysis of Seeds II: Oil and Protein Content of 759 Species. *Economic Botany* 20:127–155.

Klein, W., R. Daley, and J. Wedum
1975 *Environmental Inventory and Assessment of Navigation Pools* 24, 25, and 26, Upper Mississippi and Lower Illinois Rivers: A Vegetation Study. Missouri Botanical Garden, St. Louis.

Martin, J., and W. Leonard
1967 *Principles of Field Crop Production,* 2nd edition. Macmillian, London.

Munson, P.
1984 Weedy Plant Communities on Mud-Flats and Other Disturbed Habitats on the Central Illinois River Valley. In *Experiments and Observations on Aboriginal Wild Food Utilization in Eastern North America,* edited by P. Munson, pp. 379–386. Prehistoric Research Series 6(2). Indiana Historical Society, Bloomington.

Murray, P., and M. Sheehan
1984 Prehistoric *Polygonum* Use in the Midwestern United States. In *Experiments and Observations on Aboriginal Wild Food Utilization in Eastern North America,* edited by P. Munson, pp. 282–298. Prehistoric Research Series 6(2). Indiana Historical Society, Bloomington.

Peebles, C.
1978 Determinants of Settlement Size and Location in the Moundville Phase. In *Mississippian Settlement Patterns,* edited by B. Smith, pp. 369–416. Academic Press, Orlando, Florida.

Robson, J., R. I. Ford, K. Flannery, and J. E. Konlande
1976 The Nutritional Significance of Maize and Teosinte. *Ecology, Food and Nutrition* 4:243–249.

Seeman, M., and H. Wilson
1984 The Food Potential of *Chenopodium* for the Prehistoric Midwest. In *Experiments and Observations on Aboriginal Wild Food Utilization in Eastern North America,* edited by P. Munson, pp. 299–317. Prehistoric Research Series 6(2). Indiana Historical Society, Bloomington.

Smith, Bruce D.
1985a The Role of *Chenopodium* as a Domesticate in Pre-Maize Garden Systems of the Eastern United States. *Southeastern Archaeology* 4:51–72.

1985b *Chenopodium berlandieri* ssp. *jonesianum:* Evidence for a Hopewellian Domesticate from Ash Cave, Ohio. *Southeastern Archaeology* 4:107–133.

1986 The Archaeology of the Southeastern United States: From Dalton to de Soto, 10,500–500 B.P. In *Advances in World Archaeology,* vol. 5, edited by F. Wendorf and A. Close, pp. 1–92. Academic Press, Orlando, Florida.

1987a The Independent Domestication of Indigenous Seed-Bearing Plants in Eastern North America. In *Emergent Horticultural Economies of the Eastern Woodlands,* edited by William Keegan, pp. 3–47. Center for Archaeological Investigations, Occasional Papers No. 7. Southern Illinois University, Carbondale.

1987b In Search of *Choupichoul,* the Mystical Grain of the Natchez. Keynote address, 10th annual conference of the Society for Ethnobiology, March 5–8, Gainesville, Florida.

Smith, Bruce D., and C. W. Cowan
1987 Domesticated *Chenopodium* in Prehistoric Eastern North America: New Accelerator Dates from Eastern Kentucky. *American Antiquity* 52:355–357.

Smith, Bruce D., and V. Funk
1985 A Newly Described Subfossil Cultivar of *Chenopodium* (Chenopodiaceae). *Phytologia* 57:445–449.

Steyermark, J.
1963 *Flora of Missouri.* Iowa University Press, Ames, Iowa.

Sturtevant, W.
1965 Historic Carolina Algonkian Cultivation of *Chenopodium* or *Amaranthus. Southeastern Archaeological Conference Bulletin* 3:64–65.

Wahl, H.
1954 A Preliminary Study of the Genus *Chenopodium* in North America. *Bartonia* 27:1–46.

Watt, B., and A. Merrill
1963 *Composition of Foods.* U. S. Department of Agriculture Agricultural Handbook 8.

White, P., E. Alvistur, C. Dias, E. Vinas, H. White, and C. Collazos
1955 Nutrient Content and Protein Quality of Quinoa and Canihua: Edible Seed Products of the Andes Mountains. *Journal of Agriculture and Food Chemistry* 3:531–534.

Will, G., and G. Hyde
1917 *Corn among the Indians of the Upper Missouri River.* Torch Press, Cedar Rapids.

Wilson, H.
1980 Artificial Hybridization among Species of *Chenopodium. Systematic Botany* 5:253–263.

1981 Domesticated *Chenopodium* of the Ozark Bluff Dwellers. *Economic Botany* 35:233–239.

Wilson, H., and C. Heiser
1979 The Origin and Evolutionary Relationship of 'Huauzontle' (*Chenopodium nuttalliae* Stafford), domesticated Chenopod of Mexico. *American Journal of Botany* 66:198–206.

Wilson, H., and T. Walters
n.d. Vascular Flora of the Southeastern United States Chenopodiaceae. Ms. in possession of authors.

Wu Leung, W.
1961 *Food Composition Table for Use in Latin America.* A research project sponsored jointly by INCAP-ICNND, National Institutes of Health, Bethesda, Maryland.

Yarnell, R., and J. Black
1985 Temporal Trends Indicated by a Survey of Archaic and Woodland Plant Food Remains from Southeastern North America. *Southeastern Archaeology* 4:93–107.

Zohary, D.
1969 The Progenitors of Wheat and Barley in Relation to Domestication and Agricultural Dispersal in the Old World. In *The Domestication and Exploitation of Plants and Animals,* edited by P. Ucko and G. Dimbleby, pp. 47–66. Duckworth, London.

CHAPTER 8

The Economic Potential of *Iva annua* in Prehistoric Eastern North America

Introduction

During the investigation of modern *Chenopodium berlandieri* populations in the fall of 1984 and 1985, as described in the preceding chapter, *Iva annua* (marshelder, sumpweed) was also a focus of study (Figures 8.1, 8.2). Like *C. berlandieri, I. annua* was an important element in the premaize farming economies of eastern North America, and continued to be grown as a domesticated plant, at least in some areas, after the circa A.D. 1000 transition to maize-centered agriculture in the region. But as marshelder did not survive as a domesticated crop plant up into the historic period, modern wild populations offer the only present-day perspective on its potential value as a prehistoric crop plant.

In a landmark article, David and Nancy Asch (1978) combined a discussion on the archaeological record of marshelder in Illinois with a wide-ranging consideration of modern wild populations that included nutritional analysis of achenes,[1] and field research on the habitat settings, size, density, and harvest yield of *Iva annua* stands. The initial goal of the field research undertaken in 1984 and 1985 and described here was to first extend the research on modern wild stands of *Iva annua* begun by the Asches, with a particular emphasis on habitat characterization and the location and harvest, if possible, of populations large enough to allow sustained harvest yield experiments. Such harvest yield information, it was hoped, could then be combined with the results of similar research conducted on modern stands of *C. berlandieri* (Chapter 7), and allow a reasonably well informed consideration of the economic potential of the two seed-bearing plants as field crops in Hopewellian (A.D. 0–200) premaize farming economies.

Methods

As discussed in greater detail in the methods section of the preceding chapter, the 1984–1985 fall harvesting project primarily targeted habitat zones thought likely to support stands of *C. berlandieri* or *Iva annua,* with routes of travel selected accordingly, to

Figure 8.1 *Iva annua* achene clusters at the ends of branches, plant collected at Goodrich Landing, Louisiana (Catalog number 66).

intersect and often follow river valley floodplain landscapes. In addition to focusing on floodplain habitats, more opportunistic surveys of other habitat settings were also carried out. In the case of *Iva annua,* a wide range of locations within reach of floodwaters were investigated (marshelder achenes are dispersed by floodwaters), as well as upland agricultural fields, ponds, vacant lots, backyard gardens, fence rows, and drainage ditches.

Each *Iva annua* stand located was given a field catalog number, and essential information was recorded in a field notebook, including: (1) stand size in square meters, measured with a 100 m tape (Figure 8.3); (2) plant density, established by counting the number of plants growing in one or more 1 sq m plots; (3) plant height and habit (primarily the degree of lateral branching present); (4) the color and condition of individual plants, particularly the degree of

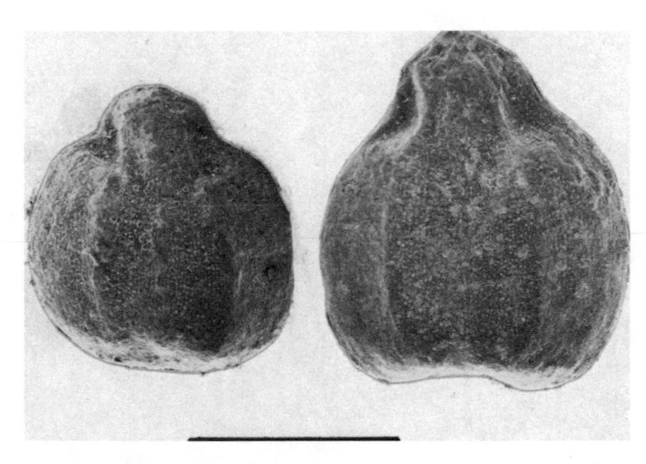

Figure 8.2 Scanning electron micrograph of *Iva annua* achenes from a plant harvested in East Carroll Parish, Louisiana (Catalog number 67) (Scale bar, 2mm).

leaf and achene retention; and (5) habitat setting–soil moisture conditions, presence of other plants, placement within floodplain landscapes.

Harvesting was carried out by using both hands to strip achenes from branches into plastic garbage bags attached to trouser belt loops (Figure 8.4). Harvested material, including achenes, stem and leaf fragments, and insects was partially dried overnight on newspaper and then transferred to paper bags the next morning to allow further drying. When good voucher specimens were encountered, they were pressed for the National Herbarium of the National Museum of Natural History (Figure 8.1). At one location (Catalog number 106), the plants from a harvest plot were collected after they had been hand stripped, and were dried, measured, and weighed. Subsequent to thorough drying and weighing, harvested achenes were winnowed from leaf, stem, and other plant parts.

Results

During research trips through the eastern United States in the autumn of 1984 and 1985, *Iva annua* stands were harvested in Kentucky, Tennessee, Missouri, Arkansas, Louisiana, Mississippi, and Alabama. Basic habitat, stand size, and plant density information was recorded for 19 populations (Table 8.1) and timed harvest experiments were carried out on 11 stands (Table 8.2). The following brief descriptions of the *Iva annua* stands studied in 1984–1985 provide additional basic information on habitat setting and plant characteristics.

Marshall County, Kentucky (Catalog number 50, 51)
Construction of the Interstate Highway 24 bridge over the Tennessee River produced a large "terrace" along the north side of the western approach to the bridge. This terrace extends 200 m along the bridge approach, and is 16–20 m wide. The soil was wet when visited in November 1984, and the abundant presence of flood deposited materials indicated that the terrace had been inundated by spring floodwater. There was occasional ground cover of low grass, and small trees reaching 6 m in height were scattered across the terrace. A stand of *Iva annua* averaging 8 m in width extended for 125 m along the terrace edge (Figure 8.5). Ten 1 sq m counting plots were spaced along the terrace edge stand, yielding an average of 71 plants per sq m. Using this average number of plants per square meter value, the total 125 by 8 m stand can be estimated to consist of about 71,000 plants. Ranging in height from less than a meter to 2 m, the plants in this large stand of marshelder had few lateral branches with most of their achenes located at the top of the main stalk. All of the plants were dead. They were leafless, dry, brittle, and brownish gray in color. They had lost most of their achenes, which could be seen in large numbers on the ground, and those remaining on plants were easily shaken free. The plants within a 7 by 4 m area were harvested by hand stripping into a waist fastened collection bag in 32 minutes, yielding 378 g of achenes (Catalog number 50).

Just to the north of the terrace described above, and 6 m lower in elevation, was a second, larger terrace, with a 10-m-wide stand of *Iva annua* extending for 100 m along a road parallel to a power line. As was the case with the upper terrace stand of marshelder, the plants were dead, dry, leafless, and brittle. Ranging in height from less than a meter to 2 m (most were in the 1 m range), the plants had few lateral branches but seemed to have somewhat better achene retention than stand 50. A 5 by 5 m area contained 1,283 plants, providing a total stand estimate

Figure 8.3 Measurement of *Iva annua* stand 93 in Mississippi County, Missouri.

of about 51,000 plants. The 5 sq m plot was harvested by hand-stripping into a collection bag in 29 minutes, yielding 570 g of achenes (Catalog number 51).

Obion County, Tennessee (Catalog numbers 53, 55, 56)

A single *Iva annua* plant (Catalog number 53) was found in November 1984 growing at the edge of a soybean field in Union City, Tennessee.

On the south side of the Obion River, just west of the Route 51 bridge, a backwater lake is bordered by a low, seasonally inundated zone extending back almost 100 m from the edge of the lake. When visited in November of 1984, this lake edge zone was covered by a stand of solid *Iva annua* (Catalog numbers 55, 56). Ranging in height from 30 to 100 cm and having few lateral branches, the plants comprising this stand were dry and brittle and had lost a majority of their achenes. Nearby, a second large stand of marshelder 15 m wide extended for 170 m along a drainage ditch. Most of the plants in this stand were unbranched and 30 to 70 cm in height. They too had lost many of their achenes. A total of 118 plants were counted in a 1 sq m area, providing a total stand estimate of 300,000 plants. The plants in a 5 by 10 m area (Catalog number 55) were harvested by hand-stripping into a waist bag in 70 minutes, yielding a total of 710 g of achenes. A second plot measuring 5 by 5 m (Catalog number 56) was harvested in a similar manner in 21 minutes, yielding 246 g of achenes.

Crittenden County, Arkansas (Catalog number 60)

On the south side of Interstate Highway 55 just west of the Mississippi River bridge, a small pond ap-

Figure 8.4 C. Wesley Cowan hand-stripping *Iva annua* achenes into a collection bag (stand 93, Mississippi County, Missouri).

Table 8.1. Stand Size and Plant Density Information for Nineteen *Iva annua* Populations (in rank order by yield)

Catalog number	Stand area (sq.m)	Density estimate (plants/sq.m)	Estimated total number of plants in stand
67	5,000	?	?
55, 56	2,550	118	301,000
60	2,000	53	106,000
93	1,000	100	100,000
50	1,000	71	71,000
51	1,000	51	51,000
103	1,000	45	45,000
106	600	67	40,200
99	560	75	42,000
107	150	30	4,500
67	30	51	1,530
70	4	11	43
75	4	6	25
64	?	43	?
53	Isolated plant in soybean field		
69	Two isolated plants, bank of Big Black River		
100	Isolated plant, bank of Tennessee River		
102	Isolated plant, bank of Tennessee River		
104	A few scattered plants along Town Creek		

proximately 1 ha in size was bordered on its north and east sides, in November of 1984, by a solid stand of *Iva annua* approximately 20 by 100 m in extent. Although the 1 to 2 m high plants in this stand were dead, dry, gray-brown in color, and leafless, achene retention was still good. A 1 sq m plot contained 53 plants, yielding a total stand estimate of 106,000 plants. The plants in a 5 by 5 m area were harvested by hand-stripping into a collection bag in 38 minutes, producing 858 g of achenes.

Chicot County, Arkansas (Catalog number 64)
A thin, 1-m-wide stand of *Iva annua* was found growing for a considerable distance along the margin of Lake Chicot, east of the junction of Highways 82 and 65. Growing in dense weeds to a height of 1 to 2 m, the plants varied considerably in appearance, from still green through purple to dry and brown. A 1 sq m plot contained 43 plants. The plants harvested from a 1 by 5 m area produced 172 g of achenes.

East Carroll Parish, Louisiana (Catalog number 67)
Just south of the boat launch at Goodrich Landing, 20 miles north of Tallulah, Louisiana, a small 30 sq m stand of *Iva annua* was located in November of 1984 on a natural levee of the main channel of the Mississippi River. The plants were 1 to 2 m high, still green, with leaves and achenes intact. A 1 sq m plot contained 51 plants, providing a total stand estimate of 1,530 plants. The plants growing in a 3 by 5 m area were harvested by hand-stripping into a collection bag in 27 minutes, yielding 568 g of achenes. In a low, seasonally inundated area several hundred meters west of stand 67, a 100 by 50 m stand of marshelder was found. The plants were dead and all achenes had been dropped. No collection was made.

Hinds County, Mississippi (Catalog number 69)
In November 1984, two small *Iva annua* plants were found growing along the Big Black River 3 miles west of Flowers, Mississippi.

Table 8.2. Harvest Yield Information for Nineteen *Iva annua* Populations (in rank order by yield)

Catalog number	Harvest area (sq.m)	Total Achene yield (g)	Chaff/Achene ratio	Achene yield (g/sq.m)	Achene yield (kg/ha)	Kernel yield (kg/ha)[a]	Harvest rate (kg achenes/hr)	Plant height (cm)
99	1	92	.69	92.1	921	672	1.10	150
106	1	74	.65	74.0	740	540	.88	107
70	4	190	.73	47.5	475	347	.66	100–200
67	15	568	1.00	37.9	379	277	1.26	100–200
64	5	172	.63	34.4	344	251	.68	100–200
60	25	858	.40	34.3	343	250	1.34	100–200
51	25	570	.23	22.8	228	166	1.16	<100–200
55	50	710	.34	14.2	142	104	.60	30–100
93	20	270	.35	13.5	135	99	.84	50–75
50	28	378	.19	13.5	135	99	.70	<100–200
56	25	246	.44	9.8	98	72	.70	30–70
A8[b]	1	85		85	850	621 (468)[c]		98
A3	1	84		84	840	613 (462)		149
A7	1	74		74	740	540 (407)		60
A6	1	61		61	610	445 (336)		42
A1	1	41		41	410	299 (226)		54
A2	1	40		40	400	292 (220)		65
A5	1	30		30	300	219 (165)		133
A4	1	14		14	140	102 (77)		114

[a]Kernels are estimated to represent 73% of the total weight of *Iva annua* achenes, based on the relative weight of 15 kernels (.0423 g) and their shells (.0158 g).

[b]The A1 through A8 entries are harvest yield values reported for 1-square-meter plots of *Iva annua* in Asch and Asch (1978:310).

[c]The values shown in parentheses are those presented by Asch and Asch, based on the assumption that kernels are 55% of the achene weight (1978:302, 312).

Rankin County, Mississippi (Catalog number 70)
In November of 1984, a small 1 by 4 m stand of *Iva annua* containing 43 plants was found along the slope up from the first to the second terrace of the Pearl River at Jackson, Mississippi. Ranging in size from 1 to 2 m, the plants were green, with leaves and achenes intact. Growing in the open, the marshelder plants of stand 70 had numerous lateral branches. All 43 plants were harvested by hand-stripping into a collection bag, producing 190 g of achenes.

Crenshaw County, Alabama (Catalog number 75)
A small stand of *Iva annua,* located in November 1984 in a low area along the Conecuh River, consisted of 25 plants growing in a 1 by 4 m area. Few achenes were left on the plants, and the stand was not harvested.

Mississippi County, Missouri (Catalog number 93)
Situated in a low, seasonally inundated area adjacent to the Mississippi River levee at the Dorena Ferry, this stand of *Iva annua,* in November of 1985, extended over an area 20 by 50 m in size (Figure 8.3). The plants comprising the stand were small, 50 to 75 cm in height, lacking lateral branches, and were dead, dry, and grayish brown in color. Considerable achene loss had occurred. A soil sample taken from a 100 sq cm area within the stand was found to contain 188 achenes (Figure 8.6). If projected to a 1 sq m area of ground, this loss would be almost 19,000 achenes, or about 29 g per sq m, given an achene count of 632 per gram for stand 93. A 1 sq m plot contained 100 plants, yielding a total stand estimate of 100,000 plants. The plants from a 2 by 10 m area harvested by hand-stripping into a collection bag in 19 minutes produced 270 g of achenes.

Crittenden County, Arkansas (Catalog number 99)
The location of stand 60, harvested in November of 1984, was revisited the following year. A dense stand of *Iva annua* still grew in the location, but was con-

Figure 8.5 *Iva annua* stand 50, in Marshall County, Kentucky.

siderably reduced in size, measuring only 14 by 40 m, in comparison to the 20 by 100 m stand present the previous year. Averaging 1.5 m in height, the plants were largely unbranched, dry, and leafless, but with excellent retention of achenes. A 1 sq m plot contained 75 plants, yielding a total stand estimate of 42,000 plants. The 1 sq m plot was harvested by hand-stripping into a collection bag in 5 minutes, yielding 92 g of achenes.

Hardin County, Tennessee (**Catalog numbers 100, 102**)
A single isolated marshelder plant (Catalog number 100) was located in November 1985, growing at the edge of the Tennessee River, just below Pickwick Dam. A single *Iva annua* plant (Catalog number 102) was located growing on the west bank of the Tennessee River just north of Savannah, Tennessee.

Acorn County, Mississippi (**Catalog number 103**)
A large, not very dense stand of *Iva annua* was located in November of 1985 in an overgrown field at the junction of Highways 72 and 45. Growing about 1 m above the floodplain of a small seasonal stream, the stand extended over an area 50 by 20 m in size. The plants ranged in size from 60 to 150 cm in height, and a 1 sq m plot contained 45 plants.

Colbert County, Alabama (**Catalog number 104**)
In November 1985, investigation of a small 4-m-wide creek (Town Creek) with a narrow floodplain located several *Iva annua* plants 3 to 5 m from the water"s edge among a dense growth of other annuals.

Jefferson County, Alabama (**Catalog number 106**)
In November 1985, a number of dense *Iva annua* stands were located in a low field next to a Uniroyal

Figure 8.6 *Iva annua* achenes on the ground, stand 93, Mississippi County, Missouri.

tire dealership in Bessemer, Alabama, just off Interstate Highway 59/20 (Exit 112). The plants of the various stands, which occurred in wet areas of the field and along shallow drainage ditches, ranged from green with leaves intact to gray with total achene loss. The largest of the stands covered an area of 20 by 30 m. A 1 sq m plot in this large stand contained 67 plants, providing a total stand estimate of 40,200 plants. The 1 sq m plot was harvested by hand-stripping into a collection bag in five minutes, yielding 74 g of achenes. The 67 plants harvested averaged 107 cm in height (range 69–161 cm), with a total dry weight, after harvesting, of 883 g.

Tuscaloosa County, Alabama (Catalog number 107)
In a backswamp area within the floodplain of the Black Warrior River, situated nearby the Foster Ferry Bridge (Routes 11 and 43), a 10 by 15 m stand of *Iva annua* was located in November of 1985. The plants were brown and leafless, but achene retention was good. A 1 sq m plot contained 30 plants. Only a single plant was harvested, due to canine activity.

Discussion

The Habitats of *Iva annua* in the Eastern Woodlands
Asch and Asch provide a good summary discussion of the habitats of *Iva annua* (1978:308–310) that is based both on previous accounts and their own field observations. My own observations in 1984 and 1985 are generally compatible with these earlier descriptions. Steyermark described the habitats of *Iva annua* in Missouri as "alluvial soils along streams, borders

Figure 8.7 *Iva annua*, growing in the drip line of a house, northeast Arkansas.

of ponds and sloughs, river bottom prairies and meadows, low fields in valleys, along roadsides and railroads" (1963:1536). During their research along the Illinois and Mississippi rivers between central Illinois and central Arkansas, Asch and Asch found marshelder growing in locations that are "(1) flooded in the spring and often wet throughout the growing season; (2) on recently disturbed soil; (3) open, growing where there is a cover of short grasses but less successful where it competes with other weeds and tall grasses" (1978:309).

Although the primary habitat range of *Iva annua* is defined by its mode of achene dispersal—seasonal floodwaters—it can also be found growing in a wide variety of upland settings. Asch and Asch noticed its ability to colonize dry upland habitats when record high floods dispersed some of their *Iva annua* seed stock from their laboratory to the adjacent lawn, where a stand grew for several years (1978:319). During the fall of 1984 and 1985 *Iva annua* was observed growing in a variety of upland settings above the reach of floodwaters, including a small stand next to a fast food restaurant in Tallulah, Louisiana, a few plants along the drip line of a house on Crowley's ridge in northeast Arkansas (Figure 8.7), an isolated plant in a soybean field in Union City, Tennessee (Catalog number 53), and several plants on a construction dirt pile in Jackson, Mississippi. While the dispersal of *Iva annua* achenes into such upland settings appears to be fairly common, and likely the result of human activities, colonization appears to be short lived, and to involve only small numbers of plants (Asch and Asch 1978:319). Small stands of marshelder consisting of only a few scattered plants were also found in other less than optimum habitat settings such as within the shady understory of narrow upland creek valleys (Catalog number 104).

Within the primary habitat range of *Iva annua,* as defined by the reach of seasonal floodwaters, Asch and Asch have attempted to further define the niche of *Iva annua* in terms of its inability to compete with plant species that are better adapted to either wet or dry soil conditions:

> *Iva* is mainly an edge species occurring between permanently wet and somewhat better drained soils, and its extension in either direction is limited by better adapted plant competitors. Thus, its distribution is linear, following microtopographic drainage contours. Stands are frequently only a few plants wide, and a 5-meter-wide stand would be exceptionally broad. (Asch and Asch 1978:309)

Although Asch and Asch do not provide any measurements of the ten *Iva annua* stands they studied, three of the four populations they harvested in Arkansas were large enough to allow two collectors to hand strip plants for from 45 to 170 minutes (1978:312). They conclude that in general, one could expect that: "The total area of a stand is small. We have never seen a continuous stand that occupied more than a fraction of a hectare" (1978:309). This is a rather open-ended definition of what qualifies as a "small" stand. Does, for example, a 50 by 50 m (¼ ha) stand, containing 125,000 plants, qualify as small?

The Asches' general characterization of *Iva annua* stands as "small" and linear is directly contradicted by the marshelder populations studied in 1984 and 1985. Of the 20 stands located and described, 12 were more than 5 m wide, and 8 were 1,000 sq m or more in size, comprised of an estimated 42,000 to more than 300,000 plants (Table 8.1). Only three of the stands located in 1984–1985 (Catalog numbers 64, 70, 75) conformed to the narrow linear stand configuration considered by the Asches to be generally characteristic for *Iva annua.* Each of these three cases fit the Asches' expectation nicely in that within the context of a relatively abrupt change in topography, the narrow linear bands of *Iva annua* were constrained between the wetter soils of lower elevations and the better drained soils of higher elevations, each with their respective strong plant competitors.

At the same time, the larger marshelder stands located in 1984 and 1985 do not necessarily contradict the Asches' general theory of *Iva annua* habitat and stand size being circumscribed by the relative wetness and elevation of soils and the presence of better adapted plant species. What these large stands do indicate, however, is that this optimum *Iva annua* habitat zone of seasonally inundated soils that remain wet well into the growing season is not restricted to narrow bands, but often extends over wide areas. In speculating about how such large stands of marshelder are initially established, and how long they might be expected to persist in particular locations within floodplain environments, it is worthwhile to begin with a consideration of several fairly obvious aspects of achene dispersal and deposition. Once achenes fall to the ground in the fall of the year, the floodwaters of the following spring provide the primary means of any subsequent dispersal. That area of a floodplain covered by water during any one year defines the total area within reach of floodborne marshelder achenes. The distance and direction that achenes will actually travel before being deposited will depend upon the force, duration, and direction of the moving waters. Many of the thousands of achenes carried by floodwaters over wide areas each year will be eventually deposited in high competition settings, where already established plant species or those better adapted to the specific edaphic conditions of the location will prohibit the establishment of new stands. While a large percentage of such achene "scouts" come to earth in unfriendly territory and fail to prosper, others will manage to find a favorable setting, and a new stand will be established. If the change in elevation or slope is steep where achenes come to rest, and the optimal *Iva* habitat zone (seasonally inundated soils that remain wet well into the growing season) is as a result laterally compressed, the *Iva* stands that are established will be narrow in width, as achenes deposited to either side of the zone fail to produce, and the stand will conform to the Asches' model of *Iva annua* occurrence in floodplain settings. There might also be a similarly narrow depositional process at work, with such compressed linear stands resulting from achenes being deposited, like a bathtub ring, along the sloping banks of watercourses or lakes by stronger currents or winds.

But the floodplains of major rivers in the eastern United States, particularly the Mississippi River and the lower reaches of some of its tributaries, are not landscapes of abrupt topographical change, of narrow and stable boundaries between wet and dry. Rather, they are landscapes of gradual change, both topographically and seasonally, containing vast swamp areas where the floodwaters of spring become the shallow and slowly receding slackwater lakes and ponds of summer. It is such slackwater areas that offer large expanses of optimum habitat and the two primary prerequisites for the development of large marshelder stands: (1) very low relief; and (2) poor drainage, resulting in wet soils long into the growing season, as floodwaters slowly evaporate and areas of standing water shrink in size. Rather than representing newly established colonies, the large marshelder stands observed in 1984–1985, containing from 42,000 to 300,000 plants, had undoubtedly existed for a number of years, with the annual cycle of short distance dispersal of achenes by rising waters allowing the growth of populations over wide areas as they expanded to the dry soil upper limit of their habitat zone.

While the slow and steady deposition of soil and the associated plant successional sequence eventually would close out such large stand habitat settings for *Iva annua,* there is no reason to consider them as having a short time span of existence. As long as the cycle of spring flooding and wet summer soils is maintained, I would suggest that large stands could exist for long periods of time. Unfortunately I have not, in the last five years, revisited any of the marshelder locations observed in 1984–1985, and as a result cannot provide any longitudinal perspective on the life span of large stands, other than the observation that stand 60/99 existed for at least two years.

The specific characteristics of topography and seasonal cycles of flooding and evaporation which combine to form large areas of optimum habitat for marshelder certainly occur in abundance in modern floodplains, both undisturbed and modified by human activities, and would, I think, have similarly been present, and occupied by sizable marshelder stands, in the river valleys of prehistory. It is possible, however, that marshelder stands, including ones as large as those observed in 1984–1985, have shorter life spans than suggested here—longitudinal studies are clearly called for. Based on the habitat setting and size of the marshelder populations I observed, however, I would challenge several of the main conclusions reached by the Asches regarding the distribution and relative abundance of marshelder in major floodplain environments and the species' negligible "total harvest potential" (1978:313). I do not agree with their position that "its distribution changes with each year's pattern of flooding" (1978:309) or with their conclusion that "the total area of a stand is small" (1978:309). While this small, short life-span characterization of *Iva* stands might be appropriate for the modern floodplain environment of the lower Illinois River, large and quite likely long life span stands occur today in abundance in the Lower Mississippi Valley. The existence of these stands, some of which were the subject of five person-hour harvests by the Asches, calls into question their conclusion that because of their limited number, small size, and shifting nature, wild *Iva annua* stands provided only ". . .a small (though useful) contribution to the annual food supply" (1978:315).

The Economic Potential of *Iva annua*

Achene yields obtained for the eleven *Iva annua* stands harvested in 1984 and 1985 show considerable variation (Table 8.2, Figure 8.8), from a low of 98 kg per ha recorded for field catalog number 56, a 25 sq m plot of small dead plants with substantial achene loss, to a high of 921 kg per ha for stand 99, a 1 sq m plot of 75 plants having excellent retention of achenes and yielding 92 g of achenes. The substantial impact of achene loss on harvest yield values was documented in stand 93, located in Mississippi County, Missouri, where a 100 sq cm area of ground contained 188 achenes or about 29 g of lost achenes per square meter. This rough estimate of achenes lost prior to harvesting (67 percent) would markedly increase the harvest yield estimate for stand 93 if it was added to the 13.5 g of achenes per sq m actually har-

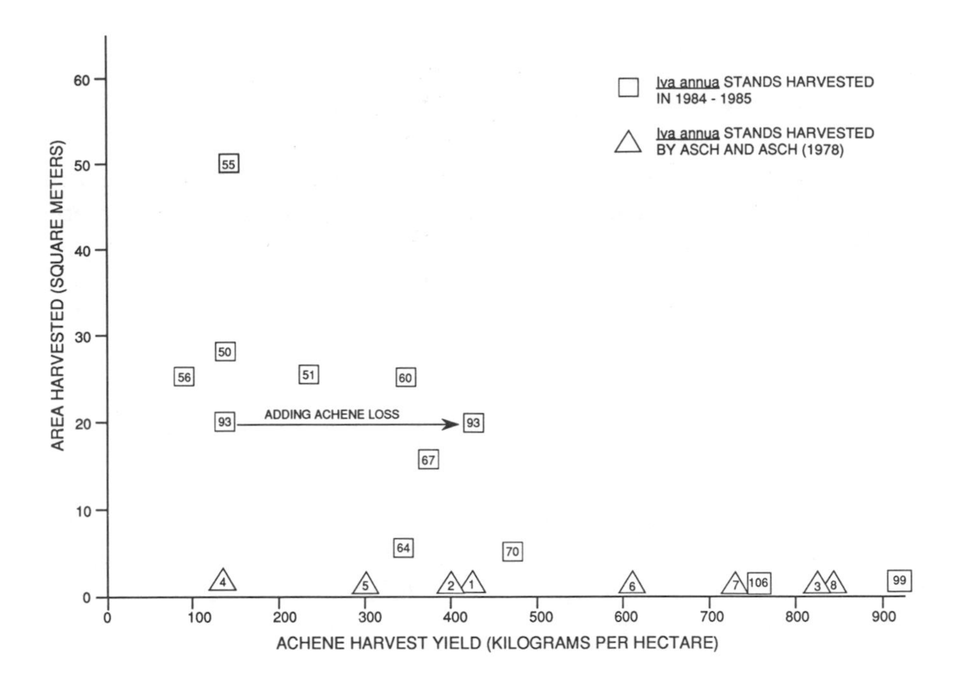

Figure 8.8 Harvest plot size and harvest yield values for 19 *Iva annua* populations.

vested (Figure 8.8). The other four lowest yielding marshelder stands harvested in 1984 and 1985 (50 and 51 in Marshall County, Kentucky; 55 and 56 in Obion County, Tennessee) all likely suffered similar levels of achene loss. These five low-yield stands were also located the furthest north of all the stands studied, and lost their achenes earlier in the fall harvest season. The Asches had a similar problem with early achene loss in northern populations (1978:310–311), and they estimated that their two lowest yielding plots (A4 and A5, Figure 8.8) experienced respective seed loss of 20 percent and 75 percent. The only previously published consideration of the economic potential of *Iva annua* was that conducted by the Asches, with eight 1 m plots yielding from 14 to 85 g of achenes (Table 8.2, Figure 8.8). The Asches results are comparable to the harvest yield values obtained in the present study, both in terms of northern latitude low-yield high-achene loss populations, and the range in harvest yield values for the remaining higher yield, lower loss stands.

If the high achene loss stands considered above (50, 51, 55, 56, A4, A5) are dropped from consideration, the remaining six stands in each of the two separate studies form quite similar sets in terms of both range of variation in achene harvest yield (Asches, 400–850 kg per ha; Smith, 343–921 kg per ha) and overall average yield (Asches, 641 kg per ha; Smith 533 kg per ha). The degree of comparability of the results of the two studies becomes even more pronounced when the larger size of four of the harvest plots in the current study are taken into consideration. Asch and Asch discuss the difficulty of sustaining high harvest yield values from large plots within wild, often variable, populations (1978:311) and this is well illustrated in Figure 8.8, where the larger harvest plots of 1984 and 1985 produced lower yields. These larger plot, lower yield values, which fall at about 350 kg per ha, could be considered, I think, as defining the lower end of the range of likely prehistoric harvest yield levels for domesticated *Iva annua*—what might be expected from poorly maintained and overgrown fields. At the same time, the three highest yield values of the two studies fall close to the 850–900 kg per ha range, suggesting 850 kg per ha as a reasonable estimate of what a well-maintained field might have produced in prehistory. These upper and lower production boundaries in turn define a reasonable, if conservative, harvest yield range estimate of 350–850 kg per ha for the crop plant.

This achene harvest yield range estimate of 350–850 kg per ha is very similar to the 300–850 range estimate previously proposed (as shown in the upper harvest yield bar for *Iva annua* in Figure 7.11). This previous estimate for achene harvest yield, as well as the lower harvest yield bar of the marshelder pair, which factors out achene shell weight and represents kernel or true seed yields (Figure 8.9), was based on the work of Asch and Asch (1978). The Asches estimated that the ". . .tough fibrous indigestible shell is about 45% of the total achene weight" (Asch and Asch 1978:302), and that ". . .55% of the achene weight is kernel" (Asch and Asch 1978:312). No description of the data base for this estimate was given. Recent weighing of *Iva annua* samples indicates that kernels represent 71–73 percent of the total achene weight. An initial sample of 15 kernels weighed .0423 g (73 percent), while associated shells (pericarp) weighed .0158 g (27 percent). A second sample of 50 achenes yielded similar results, with the kernels weighing .1513 g (71 percent) and the shells .0618 g (29 percent). This new kernel-achene ratio changes the kernel harvest yield range value for *Iva annua* from 165–467 kg per ha to 255–620 kg per ha. This correction is shown by the shaded bar in Figure 7.11.

The harvest rate values obtained for 11 stands of marshelder during 1984 and 1985 showed considerable variation, ranging from a low of .60 kg per hour for stand 55, at 50 sq m the largest plot harvested, to 1.34 kg per hour for stand 60, a 25 sq m harvest plot. The average harvest rate was .90 kg per hour, somewhat higher than the .74 kg per hour average rate posted by the Asches for 20 stands (range .16–1.41 kg per hour).

Harvest Yield Comparisons

A fairly detailed comparison of the harvest yield values for different eastern North American premaize crop plants and other New and Old World species is presented in the discussion section of the preceding chapter. The correction of the modified range estimate for *Iva annua* from 165–467 kg per ha to 255–620 kg per ha (as shown by the shaded bar in Figure 7.11) brings marshelder more in line with the other eastern premaize crops, even without factoring in the larger size of the achenes of the prehistoric domesticate relative to those produced by modern wild stands of marshelder (Yarnell 1978; Asch and Asch 1978). *Iva annua's* harvest yield range brackets those of sunflower and erect knotweed and overlaps the lower end of the range estimate for *C. berlandieri,* as well as those of maize, quinoa, and wild emmer wheat.

Figure 8.9 Scanning electron micrograph of an *Iva annua* kernel, and its thin dry outer shell or pericarp (from stand 107, Tuscaloosa County, Alabama) (Scale bar, 2mm).

Marshelder as a Premaize Field Crop: Half-Hectare Fields of *Iva annua* and *C. berlandieri*

Given the harvest yield and harvest rate values for wild stands of *Iva annua,* as presented above, and those for *C. berlandieri,* outlined in the previous chapter, it is possible to offer reasonably informed speculation regarding the economic potential of these two domesticates as premaize field crops. More specifically, if what is known regarding the size and distribution of Hopewellian household units on the landscape (Chapter 9) is also taken into consideration, the relationship between field size, caloric contribution of field crops, and the labor requirements of premaize farming economies can be considered.

In those areas of eastern North America where adequate settlement information is available for Hopewellian societies (A.D. 0–200), individual households, representing the basic unit of production (with production in all likelihood controlled at the household level, Gregg 1988:131), were distributed along river and stream valley corridors in small single household settlements. Consisting of a small number of circular to oval house structures and associated pit clusters, these small single household settlements provide considerable evidence, in the form of carbonized plant remains, for the growing of indigenous seed-bearing plants, including *Iva annua* and *C. ber-*

Table 8.3 The Estimated Dietary Contribution of Different Sized Fields of *Iva annua* and *Chenopodium berlandieri* to a Hopewellian Household of Ten Individuals

	Chenopodium berlandieri				Iva annua			
Field size (sq.m)	Harvest yield (kg of seed)[a]	Percent household caloric requirement[b]	Harvest period (person hours)[c]	Required storage (liters)[d]	Harvest yield (kg kernels)[e]	Percent of household's caloric requirement[f]	Harvest period (person hrs)[g]	Required storage (liters)[h]
1000	130	6.1	87	285	60	3.8	46	190
2000	260	12.3	173	569	120	7.6	92	381
3000	390	18.5	260	854	180	11.5	138	572
4000	520	24.8	347	1139	240	15.2	185	763
5000	650	31.0	433	1424	300	19.1	200	953
6000	780	37.2	520	1708	360	22.9	240	1144
7000	910	43.4	607	1993	420	26.8	280	1335
8000	1040	49.6	693	2278	480	30.6	320	1525
9000	1170	55.7	780	2562	540	34.4	415	1716
1 Hectare	1300	61.9	867	2847	600	38.2	461	1906

[a] Assuming a harvest yield value of 1,300 kg per ha of complete *Chenopodium berlandieri* fruits.
[b] 1 kg of *Chenopodium berlandieri* fruits contains 4,000 calories.
[c] Assuming a harvest rate of 1.5 kg per hour (see Chapter 7).
[d] The volume of 1 kg of *Chenopodium berlandieri* is 2.19 liters.
[e] Assuming a harvest yield value of 600 kg per ha of *Iva annua* kernels.
[f] 1 kg of *Iva annua* contains 3,200 calories.
[g] Assuming a harvest rate of 1.3 kg per hour.
[h] The volume of 1 kg of *Iva annua* achenes is 2.32 liters (assuming whole achenes are stored).

landieri. As will be discussed in Chapter 9, however, the actual contribution of these and other crop plants to Hopewellian diets is still a matter of discussion. While I take the position that they were grown in large fields and made a substantial contribution to Hopewellian diets, others diminish their economic importance and argue that their cultivation was limited to small gardens. How can it be established whether they were major field crops or minor garden plants in the economy of Hopewellian populations?

One way of approaching this question is by comparing the projected caloric requirements of a Hopewellian household, on the one hand, with the harvest yield potential of field plots of different size, on the other, in order to estimate the area of land that household units would have needed to maintain under cultivation in order to generate different levels of agricultural production.

Although the size and composition of these small dispersed household units would certainly have varied across time and space, the following example is based on a Hopewellian family group of ten individuals, spanning three generations, and having an average daily per-capita caloric requirement of 2,300 calories. Given this estimate of household size and energy requirements, Table 8.3 presents the extent to which different sized fields of *Iva annua* and *C. berlandieri* would have satisfied the annual caloric needs of a Hopewellian family group. Gregg (1988) provides a similar, if far more detailed, consideration of early Neolithic farming economies in Europe. A 1,000 sq m field (32 m sq) of *Iva annua*, for example, could be reasonably expected to yield 60 kg of kernels, providing 3.8 percent of the reference household's total annual caloric budget. A field of this size could be harvested in one 9-hour day (46 person-hours) by five people. A five-person crew would need two 9-hour days to harvest of *C. berlandieri* of similar size, producing a yield of 130 kg of seed, and contributing an additional 6.1 percent of the annual household energy budget. The harvest from these two fields, covering a 45 m sq area, would have required 475 liters (123 gallons) of storage space. The harvest from 1 ha plots of *Iva annua* and *C. berlan-*

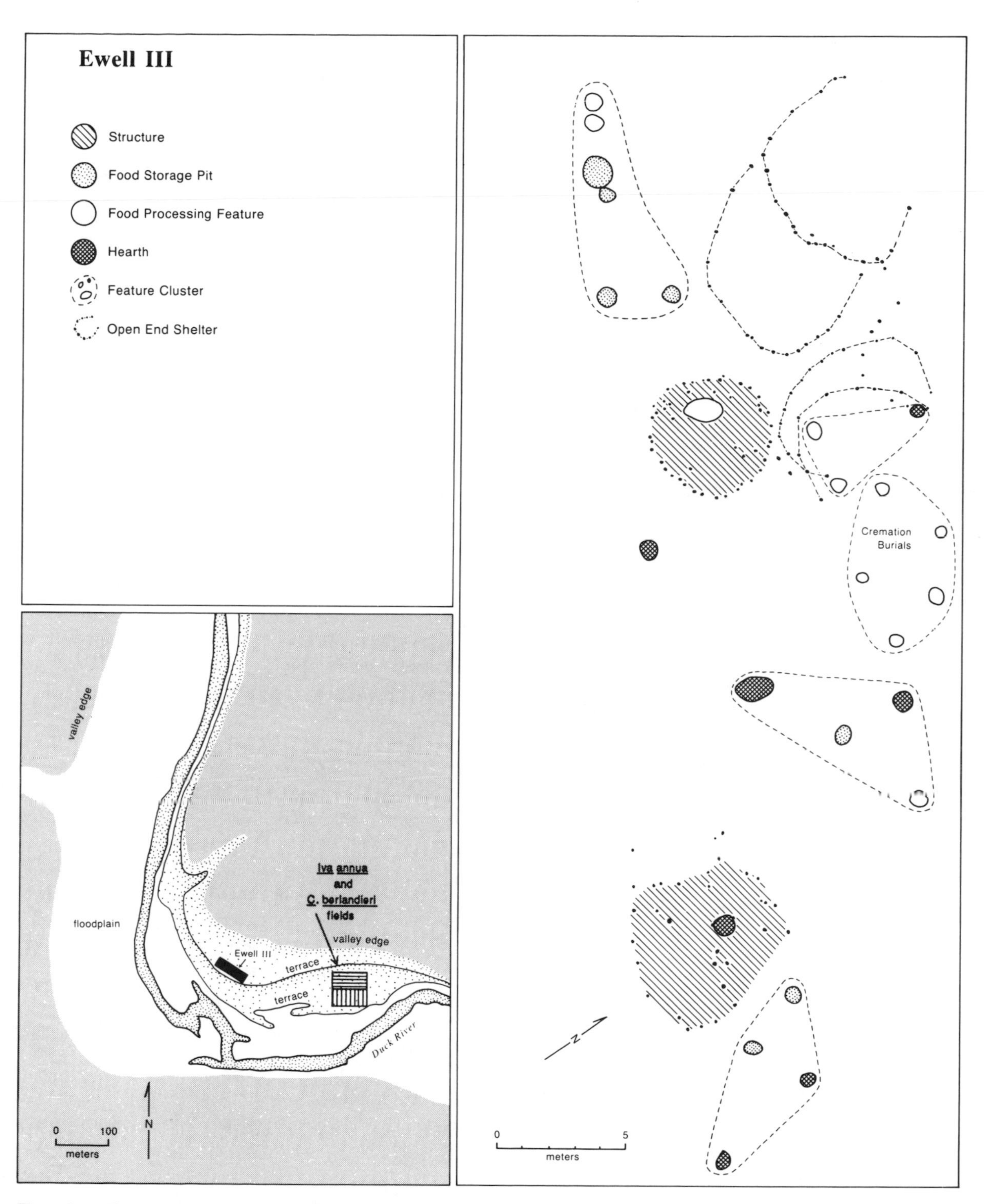

Figure 8.10 The Duck River Valley setting of the Ewell III site, showing the spatial context of a one-half hectare field of *Iva annua* and *C. berlandieri*, capable of providing 50 percent of the total caloric requirement of a ten person Hopewellian household over a period of six months.

dieri, which would take a crew of five, 30 9-hour days to collect, and which would require 4,753 liters (1,235 gallons) of storage, would meet the total annual caloric requirement of the reference household.

Obviously Hopewellian households did not rely exclusively on field crops such as *Iva annua* and *C. berlandieri* as food sources, any more than later Mississippian groups relied only on maize. Nor did Hopewellian households likely consume such crops uniformly through the year. If these crops can be viewed as partial year food sources, then the middle rows of Table 8.3 fall within the envelope of reasonable expectation regarding their economic role. A 2,500 sq m field of *Iva annua,* for example, and a similarly sized field of *C. berlandieri,* combined to form a 70 m sq area, could provide one-half of the total caloric requirement of the 10-person reference household over a 6-month period. The paired fields would take just over seven days to harvest by a five-person crew working 9-hour days, and would require 1,188 liters of storage space (a volume comparable to six 50-gallon drums). To place fields of this size in the spatial context of Hopewellian settlements and their river valley settings, a .5 ha sq (5,000 sq m) has been added, in Figure 8.10, to the locality map for the Ewell III site along the Duck River in central Tennessee (see Figure 9.17). With perhaps a few exceptions, all of the known localities of Hopewellian habitation sites provide sufficient high quality river valley alluvial soil in the immediate vicinity of the settlements to allow .5 ha fields (see Chapter 9). Such .5 ha fields, capable of yielding half the caloric budget of a household of 10 individuals over a 6-month period, would seem to have been well within the capability of such a family unit with only a week long effort by half the group required to bring in the harvest. Further, likely lengthy, processing of harvested seed would have been necessary, but could have been accomplished gradually, as stores were drawn down over the year. There would also have been considerable energy required for the preparation and maintenance of fields through the spring and summer growing season, as well as the work involved in clearing forest for new fields through a fallow cycle of unknown duration.

While certainly not demonstrating in any conclusive way that the premaize farming economies of Hopewellian households provided 50 percent or more of the caloric requirements of family groups for six months or more of each year, the foregoing example does serve to show that such a level of agricultural activity was well within the reach of such river valley farming societies.

Notes

1. The reproductive propagules of *Iva annua* are called achenes. Achenes consist of a single seed or kernel, enclosed within a thin dry outer shell (pericarp) (Figure 8.8).

Acknowledgments

The harvest yield research described in this article was made possible by Smithsonian Institution Research Opportunity Fund grants 1233F505 and 1233F611, awarded by David Challinor, Assistant Secretary for Research. C. Wesley Cowan, Cincinnati Museum of Natural History, accompanied the author during part of the 1985 harvest season, and helped with the harvest of stands 93, 99, 103, and 104.

Literature Cited

Asch, D. L., and N. B. Asch
1978 The Economic Potential of *Iva annua* and its Prehistoric Importance in the Lower Illinois Valley. In *The Nature and Status of Ethnobotany,* edited by R. I. Ford, pp. 301–341. Museum of Anthropology, Anthropological Papers No. 67. University of Michigan, Ann Arbor.

Gregg, S. A.
1988 *Foragers and Farmers.* University of Chicago Press, Chicago.

Steyermark, J. A.
1963 *Flora of Missouri.* Iowa State University Press, Ames, Iowa.

Yarnell, R. A.
1978 Domestication of Sunflower and Sumpweed in Eastern North America. In *The Nature and Status of Ethnobotany,* edited by R. I. Ford, pp. 289–300. Museum of Anthropology, Anthropological Paper No. 75. University of Michigan, Ann Arbor.

CHAPTER 9

Hopewellian Farmers of Eastern North America

Introduction

The appropriate point to begin a consideration of early farming settlements in the prehistoric Eastern Woodlands of North America is with the tropical cultigen maize *(Zea mays)*. The role of maize in forcing and fueling trajectories of prehistoric change in the Eastern Woodlands has been a popular topic of debate for over fifty years. This "maize debate" has centered on two related and rather basic questions. First, when was this tropical domesticate introduced into the temperate deciduous woodlands of the eastern United States? Second, what were the subsequent temporal and geographical dimensions of its transformation into a major agricultural crop?

The application of two technological innovations have recently provided solid and rather surprising answers to both of these questions, substantially resolving the maize debate. The first of these innovations is stable carbon isotope analysis. In contrast to the indigenous plants of economic importance in the prehistoric East, all of which utilize a C_3 photosynthetic pathway, maize is a C_4 pathway plant, which allows its relative dietary contribution to be accurately established by measuring the $^{13}C/^{12}C$ ratio in human bone collagen. Through the measurement of $^{13}C/^{12}C$ ratios in temporally ordered human skeletal series, regionally focused stable carbon isotope studies of the past decade have shown a consistent pattern of limited maize consumption until after A.D. 900–1000 (Vogel and Van der Merwe 1977; Van der Merwe and Vogel 1978; Bender et al. 1981; Broida 1983; Conrad 1985; Lynott et al. 1986) (Figure 9.1a). While stable carbon isotope studies have established a general post A.D. 900–1000 temporal context for the emergence of

This chapter will also appear in L'Homme et l'Habitat Pré- et Protohistoriques. Conférences generales du Congrès. *Monographien des Roëisch-Germanischen Zentralmuseums* 18:47–93

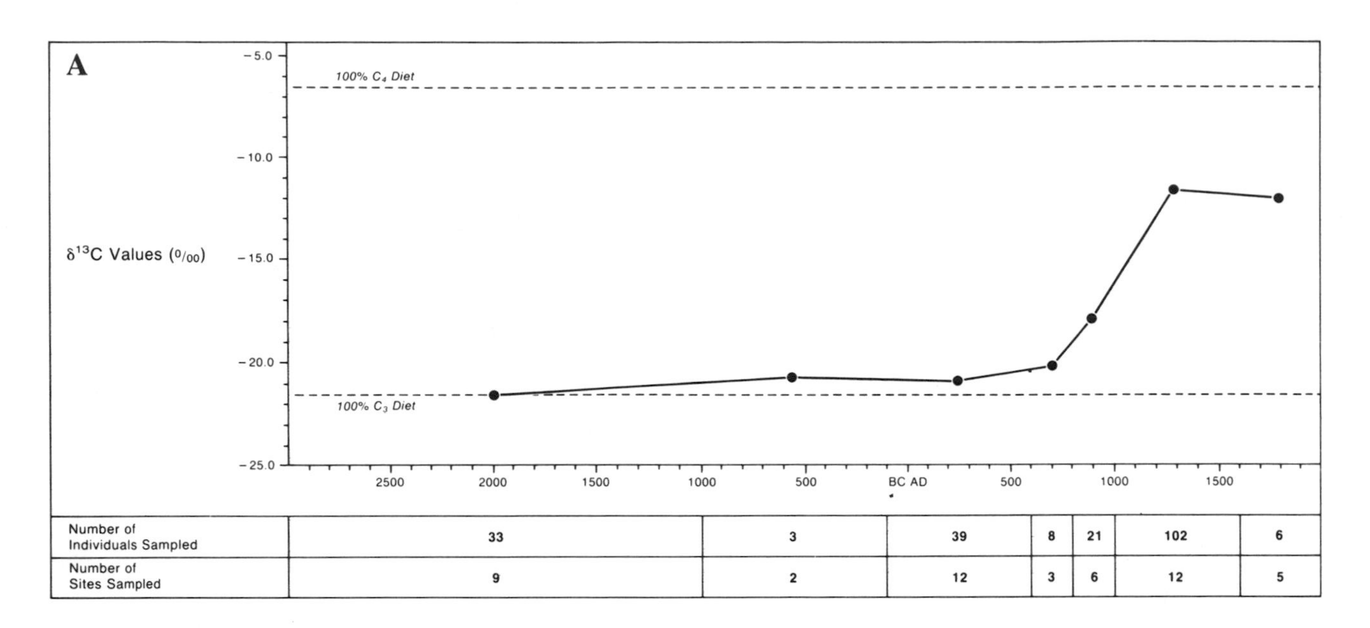

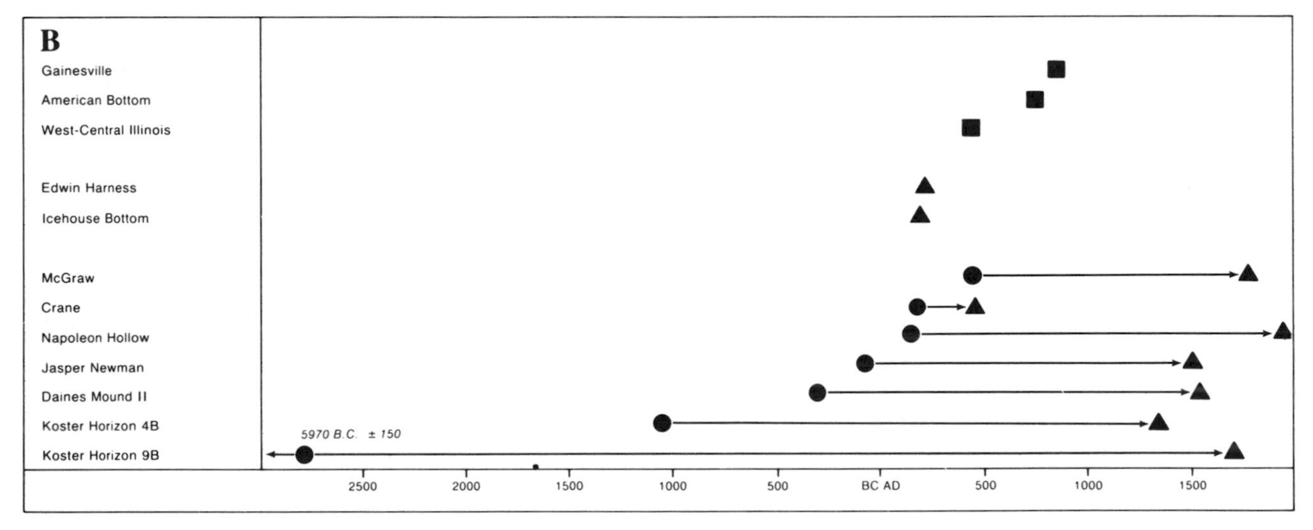

Figure 9.1 (a) Temporal change in the prehistoric human consumption of maize in eastern North America (modified from Wagner 1986). Points plotted represent the median values of separate skeletal series sampled within each time unit; (b) The initial introduction of maize into eastern North America, as indicated by direct AMS age determination of archaeological recovered maize kernels and cupules (see Table 9.1).

maize-centered field agriculture in the East, radiocarbon age determination of maize kernels and cupules by the accelerator mass spectrometer (AMS) method over the past five years has resolved the long-standing question of when this tropical cultigen was first introduced into the region. Although corn has been recovered from a number of well-sealed and dated cultural contexts dating back more than 2,900 years, direct AMS age determination of many of these early specimens has shown them to be intrusive contaminants (Table 9.1, Figure 9.1b). While some maize from purported earlier contexts has yet to be directly dated by the AMS method (for example, Meadowcroft), the recent AMS dates on maize from the Icehouse Bottom site in east Tennessee (A.D. 175 ± 110) and from the Edwin Harness site in Ohio (A.D. 220 ± 105)

Table 9.1. The Age of Early Context Maize in Eastern North America

Site/Project area	Radiocarbon age on associated materials (years B.P.)	Specimen age by direct AMS date (years B.P.)	Reference
Edwin Harness		1730 ± 85	
		1720 ± 105	Ford 1987
Icehouse Bottom	1511 ± 75	1775 ± 110	Chapman & Crites 1987
McGraw	1510 ± 80	170 ± 100	Ford 1987
Crane	1710 ± 70	1450 ± 350	Conard et al. 1984
Napoleon Hollow	1800 ± 70	0 ± 300	Conard et al. 1984
Jasper Newman	2030 ± 140	450 ± 500	Conard et al. 1984
Daines Mound II	2230 ± 140	385 ± 100	
		370 ± 90	Ford 1987
Koster Horizon 4B	2980 ± 70	600 ± 400	Conard et al. 1984
Koster Horizon 9B	7290 ± 150	250 ± 300	Conard et al. 1984

represent the most reasonable current estimate of the initial appearance of corn in the prehistoric East.

Since the Icehouse Bottom and Edwin Harness sites are located 400 km apart and along the eastern edge of the interior riverine area of eastern North America (Figure 9.2), one might expect that if this tropical cultigen entered the East from the Southwest or Mesoamerica, it should be present in river valley settings further west at an earlier point in time. Just the opposite is true, however. Maize does not even appear in the major river valley archaeobotanical sequences from west-central Illinois, the American Bottom, or west-central Alabama for another four to six centuries (Caddell 1981:32, 46; Asch and Asch 1982; Johannessen 1984).

In summary, based on stable carbon isotope analysis, AMS direct dating, and fine-grain archaeobotanical sequences, maize was initially introduced into eastern North America by A.D. 200–300, but was not present in some areas for another 400 to 600 years, and does not appear to have been anything more than a minor crop before A.D. 900–1000.

I have briefly traced the history of maize in the east, from first introduction to dietary dominance, because many researchers still tend to equate this tropical crop with an agricultural economy in North America. From such a perspective, since post A.D. 1000 populations in the prehistoric East were consuming a lot of maize, they therefore qualify as agriculturalists, while earlier "premaize" populations, by definition, are considered nonagricultural. What fate, then, awaits the premaize, "nonagriculturalists" of the prehistoric East? If they are not agriculturalists, because they don't grow maize, what are they?

How these premaize populations are perceived and characterized, is dependent, of course, upon the range of alternative categorical boxes available in the conceptual framework providing the context of consideration. Bender, for example, in a recent "social approach" that is geared to providing far greater illumination of premaize Hopewellian populations than that offered by a rigid and restricted scientific approach, employs a structural-dialectic framework of analysis (Bender 1985).

Within Bender's conceptual framework of dichotomous duality and opposition (cold and hot, changeless and changing, ecological and social), premaize Hopewellian populations are not maize agriculturalists, and therefore they must fall into the only other available category: hunter-gatherers. Similarly, in summarizing their stable carbon isotope research that documents changing $^{13}C/^{12}C$ ratios after A.D. 1000, Lynott et al. (1986:82) also characterize premaize populations as hunter-gatherers:

> This change in bone chemistry has been interpreted as a result of a shift from a foraging subsistence strategy, based upon exploitation of native plants and animals, to a subsistence strategy based upon the intensive cultivation of maize.

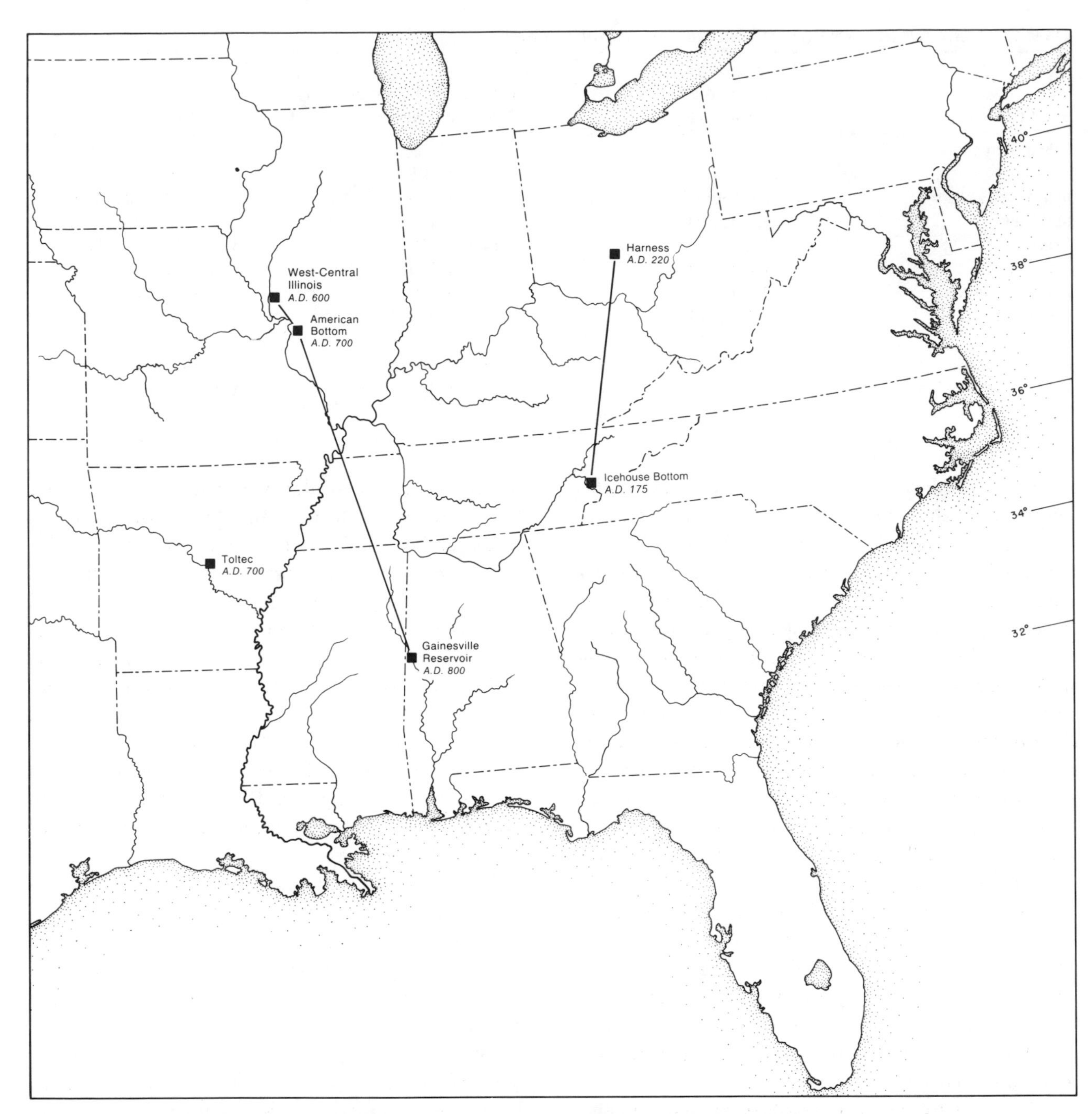

Figure 9.2 The location of the Harness and Icehouse Bottom sites, along with regions where maize has not been recovered in the archaeological record until after A.D. 500–700.

Having categorized premaize Middle Woodland Hopewellian populations as being gatherer-hunters, Bender then compares their richly elaborate mortuary programs and far-flung exchange networks to those of Neolithic societies of the Breton peninsula in France, and finds that the Old and New World societies are comparably complex, and that the Hopewellian hunter-gatherers "display social and ideological features usually attributed to farmers" (1985:21). Embedded in this case study comparison of Hopewell hunter-gatherer and Breton agricultural societies of apparently comparable social complexity is a larger theoretical issue. Bender hopes to demonstrate that social relations and social variation operate largely independently of the subsistence or economic realm and that cultural change can not be explained in terms of external (i.e., environmental or economic) causes, but only through consideration of the internal dynamics of specific social structures (Bender 1985:21–22).

Given an obvious interest in both building a case for the viability of a social approach to prehistory, and establishing the supremacy of internal explanations of social change, it is not really that surprising that Bender's "reading" of Hopewell groups as hunter-gatherers, which is critical to the case for a social approach, so lightly passes over considerable evidence of a contradictory nature. In one rather curious passage, Bender describes Hopewell groups as: ". . .still gatherer-hunters—but they behave like farmers, or perhaps they are farmers, but in a surreptitious and disappointingly invisible way" (1985:21).

In providing below a reading of Hopewell rather different from Bender's, I will argue that even though Hopewellian groups were not maize agriculturalists, they were farmers, and behaved like farmers in an obvious and encouraging manner. They planted, harvested, stored, and processed the seeds from a variety of cultivated and domesticated crops, and fortunately left behind richly detailed evidence of these farming activities in the pits, middens, and houses of their settlements. Similarly, Crawford and Yoshizaki (1987) have recently documented the prehistoric agricultural economy of the Ainu, long considered a prototypical "complex" hunter-gatherer society.

During the past thirty years a substantial number of the settlements of these premaize, Woodland period farmers have been excavated across a broad area of the Eastern Woodlands, documenting a complex spatial and temporal mosaic of variation in material culture, mortuary programs, settlement patterns, and subsistence strategies. Archaeobotanical assemblages recovered from these settlements, while exhibiting parallel patterns of regional variation in premaize farming systems, also document a broad-based increase in the relative importance of seed crops at around A.D. 0, two to three centuries prior to the initial introduction of maize into the East (Figure 9.3b). Since the two-century span from A.D. 0–200 brackets the initial emergence of premaize farming economies, as well as encompassing most of the Middle Woodland period regional manifestations of "Hopewellian" cultural interaction, it represents an appropriate temporal frame of reference for consideration of early farming settlements in the East. Before looking at patterns of similarity and variation in the Middle Woodland farming settlements of different regions of the East during this two-century span, I want to first briefly describe the food production economies of these Hopewellian populations.

The Nature and Development of Hopewellian Food Production Economies

Eastern North America was an independent center of plant domestication, with four indigenous seed bearing plants brought under domestication between 2000 and 1000 B.C.: squash *(Cucurbita pepo),* marshelder *(Iva annua),* goosefoot *(Chenopodium berlandieri),* and sunflower *(Helianthus annuus)* (Smith 1987a [Chapter 3]). Rind fragments of a small gourd assignable to the genus *Cucurbita* recovered from a number of Middle Holocene (6000–2000 B.C.) cultural contexts in the East and previously believed to document the initial introduction of a tropical domesticate from Mexico are now considered to represent an indigenous wild gourd (Decker and Wilson 1987). Domesticated varieties of squash *(Cucurbita pepo)* developed prehistorically in the East from this indigenous wild gourd (Decker 1988), with morphological changes associated with domestication (increase in rind thickness, changes in seed morphology) first appearing at ca. 1000–500 B.C. (Figure

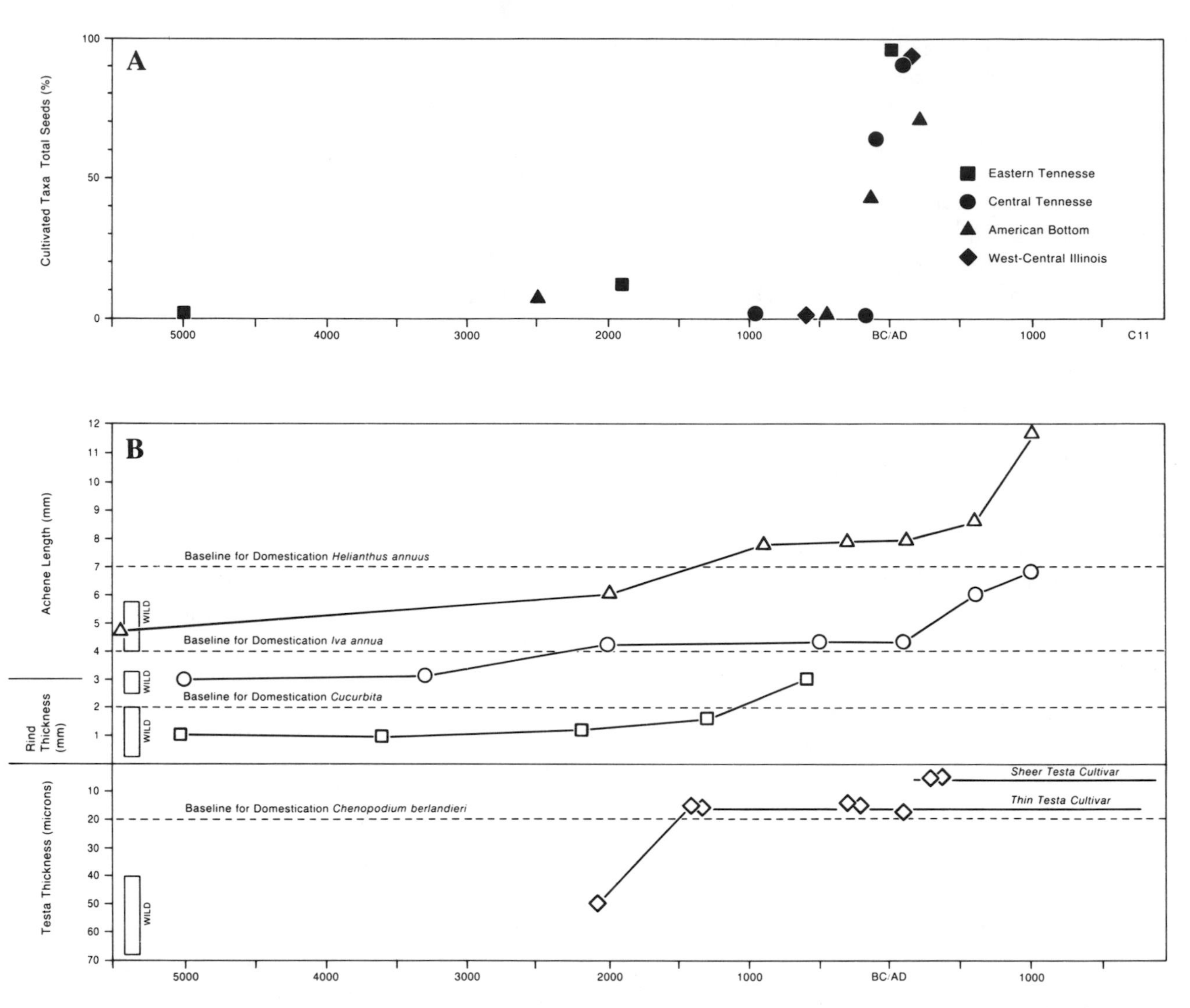

Figure 9.3 (a) The initial intensification of food production economies in eastern North America. The vertical scale indicates the relative abundance of seeds of cultivated taxa within total seed assemblages; (b) The temporal context of initial domestication of indigenous seed-bearing plants in eastern North America, based on morphological changes in seeds (see Figure 3.2 for a more detailed version of this graph).

9.3b) (Smith 1987a). Parallel morphological changes mark the transition to domesticated status in *Iva annua* by ca. 2000 B.C. (increase in achene size); *Chenopodium berlandieri* by ca. 1500 B.C. (reduction in testa thickness); and *Helianthus annuus* by ca. 1500–1000 B.C. (increase in achene size) (Figure 9.3b; Smith 1987a [Chapter 3]).

In addition to these four indigenous domesticates, premaize plant husbandry systems also included three other cultivated seed-crops for which a convincing case for domestication has yet to be made: erect knotweed *(Polygonum erectum)*, maygrass *(Phalaris caroliniana)*, and little barley *(Hordeum pusillum)*. Although all seven of these crop plants were probably under active cultivation by 1000 B.C., they are not, with a few notable exceptions (Salts Cave), very abundantly represented in archaeobotanical assemblages for another eight centuries.

By A.D. 0–200, archaeobotanical sequences over a broad area of the mid-latitude Eastern Woodlands register a dramatic increase in the representation of these seven crop plants in seed assemblages (Figure

Table 9.2 Regional Variability in Premaize Farming Economies

	West Central Illinois[a]	American Bottom[b]	Central Ohio[c]	Central Tennessee[d]	Eastern Tennessee[e]
Starchy-seeded annuals					
Fall Maturing					
Chenopod	8.6	26.4	13.6	48.5	88.0
Knotweed	31.8	0.6	32.8	8.9	
Spring Maturing					
Maygrass	37.5	29.1	20.7	35.3	
Little Barley	14.9				
TOTAL	92.8	56.1	67.2	92.7	88.0
Oily-seeded annuals					
Fall Maturing					
Sumpweed	0.3	0.1	5.5	0.1	1.6
Sunflower	0.3		.4	0.2	.9
Squash	.04			0.1	?
TOTAL	.64	.1	5.9	.4	2.5
TOTAL PERCENTAGE CULTIVATED TAXA	93.4	56.2	73.1	93.1	90.5
TOTAL NUMBER OF IDENTIFIED SEEDS	13,727	761	2879	5189	1627

[a]Seed count information for the Smiling Dan site (A.D. 50–150) (Asch and Asch 1985).
[b]Seed count information for Cement Hollow and Hill Lake phases (150 B.C.–A.D. 250) (Johannessen 1981, 1983).
[c]Seed count information for the Murphy and O.S.U. Newark Sites (ca. A.D. 200) (Wymer 1987).
[d]Seed count information for the Late McFarland phase, Normandy Reservoir (A.D. 0–200) (Crites 1985).
[e]Seed count information for Long Branch phase sites, Little Tennessee River Valley (Chapman and Shea 1981).

9.3a). This increase in turn marks the initial intensification of food production economies and the emergence of a Hopewellian farming adaptation. These premaize farming economics also appear to have differed considerably from region to region in terms of the presence and relative importance of different starchy-seeded and oily-seeded crop plants (Table 9.2; Smith 1985a, 1985b [Chapters 5, 6]).

Unfortunately there is as yet no straightforward chemical test such as stable carbon isotope analysis to convincingly quantify the relative economic importance of these premaize crops. In the absence of any such unequivocal measure of their dietary contribution, one might still attempt to minimize the importance of these seven food crops simply by characterizing the Hopewellian hunter-gatherer populations who were growing them as rather belatedly becoming involved in "small scale horticulture" (Bender 1985:25). This "horticulture" label, with its vague connotations of small vegetable gardens of a few squash plants and a sunflower or two, when applied to Hopewellian premaize farmers, conveniently relegates them conceptually to a shadowy and dimly defined purgatory of somewhat tainted hunter-gatherers who fall far short of an agricultural economy. If the simplistic semantics of a tripartite conceptual framework is set aside in favor of a consideration of both direct archaeobotanical and indirect present-day evidence regarding the economic potential of the crop plants in question, a far more useful characterization of this initial farming economy can be developed.

The only direct evidence regarding the dietary contribution of these seven premaize crops comes from the large sample of human paleofecal material recovered from Salts Cave, Kentucky, and dating to circa 650–250 B.C. Providing uniquely direct and unequivocal dietary evidence of 1–4 meals (Watson 1974), the Salts Cave human coprolites indicate that the four indigenous domesticates contributed a full

Table 9.3. The Nutritive Value of Seed of Hopewellian Crop Plants, Compared to Maize, Quinoa, and Bean (Percent dry basis).

Species	Protein	Fat	Carb.	Fiber	Ash
Starchy-seeded					
Goosefoot[a]					
C. berlandieri	19.12	1.82	47.55	28.01	3.50
Maygrass[b]					
P. caroliniana	23.7	6.4	54.3	3.0	2.14
Knotweed[a]					
P. erectum	16.88	2.41	65.24	13.33	2.34
Oily-seeded					
Marshelder[c]					
I. annua	32.25	44.47	10.96	1.46	5.80
Sunflower[d]					
H. annuus	24.00	47.30	16.10	3.80	4.00
Squash[c]					
C. pepo	29.0	46.7	13.10	1.9	4.9
Tropical Crops					
Maize[c]					
Z. Mays	8.9	3.9	70.20	2.0	1.2
Bean[e]					
P. vulgaris	22.0	1.6	60.8	4.3	3.6
Quinoa[f]					
C. quinoa	12.5	6.0	72.5	5.6	3.4

[a]Asch and Asch 1985:361.
[b]Crites and Terry 1984
[c]Asch and Asch 1978,
[d]Watt and Merrill 1963.
[e]Wu Leung 1961
[f]White et al. 1955.

two-thirds of the cavers' total food supply: *Chenopodium,* 25 percent; *Helianthus,* 25 percent; *Iva,* 14 percent; *Cucurbita,* 3 percent (Yarnell 1974). These cavers may have been subsisting on a special "trail mix," however, and their paleofeces may not be representative of the diet of subsequent Hopewell populations. Other, more indirect measures of the farming economy of Hopewellian populations include the frequent occurrence of chert hoes and hoe flakes at Hopewellian settlements (Farnsworth and Koski 1985), an increase in pollen and macrobotanical indicators of land clearing activities (Yarnell 1974; Johannessen 1984), and evidence of seed storage in a variety of containers and pit features (Smith 1985a, 1985b [Chapters 5, 6]). In addition to these various lines of archaeological evidence regarding the importance of these prehistoric crops (i.e., domesticated and cultivated status, abundance in archaeological assemblages, paleofecal analysis, hoe technology, evidence of land clearing, and deliberate large-scale storage), research on present-day descendant populations of these plants also demonstrates that they have both impressive nutritional profiles (Table 9.3) and substantial potential harvest yield values (Figure 9.4; Smith 1987b [Chapters 7, 8]).

When specifically compared to maize, all of the indigenous Hopewellian seed crops have a superior nutritional profile, and all overlap with maize in regard to economic potential. This potential harvest yield and nutritional composition information, when combined with the archaeological lines of evidence outlined above, fits neither the hunter-gatherer nor the small-scale horticultural profiles proposed for Middle Woodland Hopewellian populations. Contrary to what Bender concludes, the available evidence does *not* allow us to "abandon the notion that American Woodlanders are farmers and admit their basic dependence upon wild resources" (Bender 1985:21). They were farmers. They cleared forests, prepared fields, planted, cultivated, harvested, and stored at least seven different species of high yield, high nutritional profile seed crops. Before the introduction of maize, and more than seven centuries in advance of the transition to maize-centered field agriculture, farming economies and an agrarian way of life had been established in eastern North America.

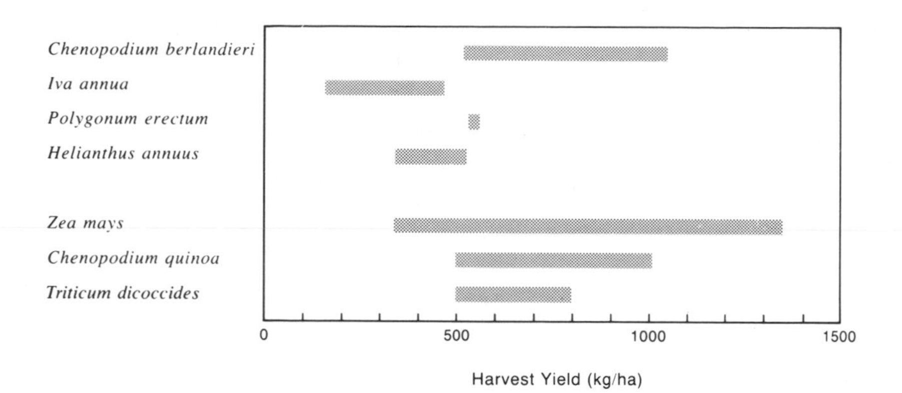

Figure 9.4 Potential harvest yield values for Hopewellian crop plants, compared to maize, quinoa, and wild emmer wheat (see Figure 7.11 for a more detailed version of this graph).

According to some models of subsistence change (Cleland 1976; Earle 1980; Christenson 1980), this initial development of food production economies and a substantial reliance on seed crops should be reflected in the archaeological record by a corresponding reduction in species diversity indices, as "high cost" wild plant and animal food sources are dropped off the menu. Within the context of these models, the absence of any noticeable narrowing of utilization of wild animal and plant species associated with the development of Middle Woodland Hopewellian farming economies has been interpreted as evidence for the continuation of a basic hunter-gatherer subsistence adaptation. But focalization, or narrowing of the exploited resource base of wild species of plants and animals, when defined in terms of a reduced species diversity index, can not be clearly recognized at any point in the archaeological record of the eastern United States. From their initial Middle Woodland period emergence in the East, through the subsequent Late Woodland and Mississippian periods of considerable intensification and elaboration, farming economies of the Eastern Woodlands were largely additive to pre-existing subsistence patterns, and their trajectories of development and increasing importance is effectively masked in many respects by long-term continuity in the exploitation of a wide variety of high-yield, low-cost wild species of plants and animals.

With the possible exception of the Middle Ohio River Valley Fort Ancient area, the post A.D. 1000 transition to maize-centered field agriculture was not accompanied by a reduction in the representation of wild species of plants and animals in archaeological assemblages, even though stable carbon isotope analysis indicates a dramatic rise in the dietary importance of maize. If this rather abrupt (ca. A.D. 700–1100) increase in the caloric contribution of maize from less than 10 percent to more than 50 percent is not reflected by any observable reduction in the number of species of wild plants or animals utilized by Mississippian populations, or the within-class representation of these species, a similar absence of evidence for focalization in terms of species diversity indices during the Middle Woodland period can hardly be considered as evidence of an unchanging hunter-gatherer way of life.

Having briefly considered the various lines of evidence that support the identification of Middle Woodland Hopewellian populations as being the first farmers of the prehistoric East, we can now turn to a consideration of their settlements.

Hopewellian Farming Communities

In the introductory section of this chapter an opening discussion of recent research regarding the tropical cultigen maize provided an essential background for consideration of nonmaize Hopewellian farming economies (Middle Woodland, A.D. 0–200). Similarly, Hopewellian settlements and the everyday domestic life of Hopewellian farmers can best be viewed against a background provided by an initial consideration of the general classes of corporate-ceremonial sphere activities recently documented within non-habitation precincts of Ohio Hopewell centers (Figure 9.5).

For a number of reasons it is appropriate that Ohio Hopewell corporate-ceremonial centers serve as a starting point for the present consideration of the habitation sites of Hopewellian communities. The

General Activity Class	Archaeological Manifestation
A. MORTUARY PROGRAMS 	• Crematory basins, processing areas • Burials and associated objects • Mortuary structures (e.g., Charnel houses, crypts)
B. CORPORATE LABOR 	• Geometeric earthworks • Burial mounds • Corporate structures (Mortuary and "workshop" structures)
C. PRODUCTION OF "EXOTIC" STATUS AND CEREMONIAL ITEMS FOR BURIAL, EXCHANGE, DISTRIBUTION 	• Nonmortuary, nondomestic, specialized workshop structures located in proximity to mortuary centers • Exotics, raw materials and manufacturing debitage • Utilized, broken bladelets
D. REDISTRIBUTION/FEASTING?	• Absence of domestic food processing and storage facilities • Limited food species representation • Limited meat cut representation • Located in proximity to mortuary centers

Figure 9.5 General classes of activities carried out within the corporate-ceremonial sphere of Hopewellian farming societies.

monumental geometric earthworks and mortuary mounds of Ohio Hopewell societies have captured and held the interest of archaeologists ever since they were initially documented in the first publication of the Smithsonian Institution, *Ancient Monuments of the Mississippi Valley,* by Squier and Davis, published in 1848. The corporate-ceremonial sphere of Ohio Hopewell also forms the core or center of the concept of "Hopewell" for most researchers. This is due both to the amount of research conducted on Ohio Hopewell earthworks and burial mounds over the past century, and their pre-eminent position among regional Hopewellian manifestations, based on the scale, richness, and complexity of their mortuary programs and earthworks.

There is a distinct, multifaceted boundary, however, between the intensively researched corporate-ceremonial sphere of the various regional variants of Hopewell, and the everyday domestic sphere of their farming communities. This distinct separation of the corporate-ceremonial and domestic spheres of Hopewellian life is clearly evident in the archaeological record, and is accordingly reflected in the remarkable imbalance in our knowledge, even today, regarding corporate centers vs. farming settlements. Nowhere is this imbalance more striking than in the south-central Ohio "heartland" of Hopewell, where only a single possible Middle Woodland habitation structure (Fischer 1971) has ever been excavated, and where the location, size, and degree of organizational complexity of Hopewell settlements remains largely undocumented. Fortunately, however, Hopewellian farming settlements and the domestic sphere are much better documented in a number of other areas of eastern North America, and when combined with the limited information available from Ohio, the habitation sites

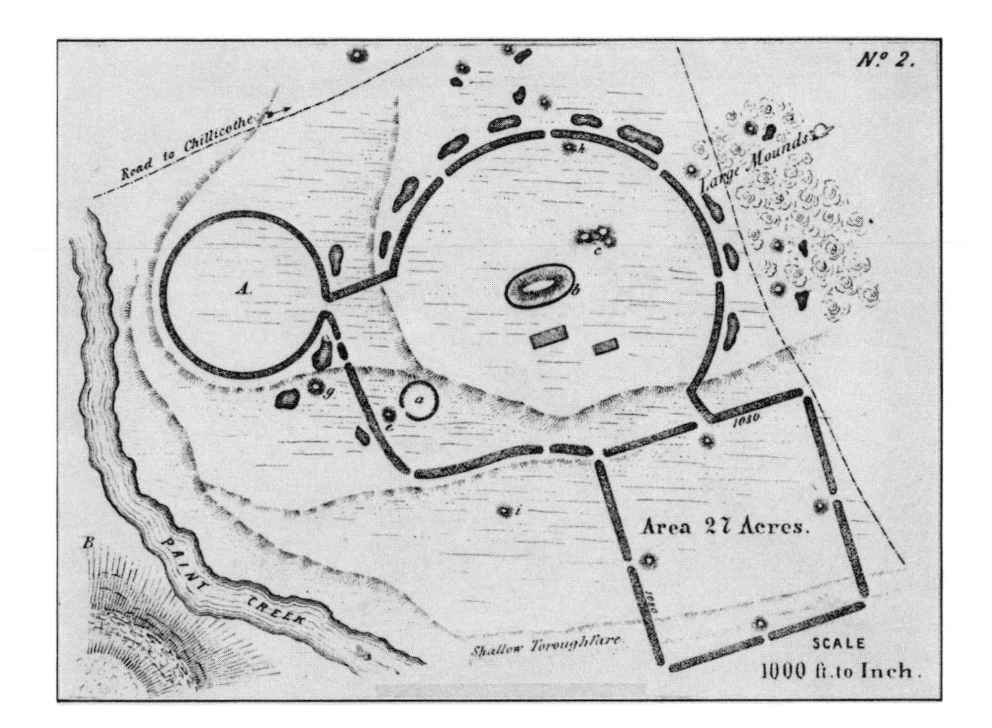

Figure 9.6 The Seip Mound site (Squier and Davis 1848, plate 21, original scale preserved).

of these other regions allow reasonable speculations to be offered regarding the setting and size of Ohio Hopewell settlements.

Working well within the corporate-ceremonial sphere of Ohio Hopewell, Raymond Baby has excavated a linear series of seven large rectangular structures at the Seip site (Figure 9.6; Baby and Langlois 1979). Situated halfway between the central mound and outer embankment (Figure 9.7), the structures have been interpreted as being specialized workshops for the production of ceremonial and ornamental objects, based on their size, spatial arrangement, and location within the Seip earthwork, an absence of hearths, a limited amount of food remains, and abundant evidence of the manufacture of status items.

Moving deeper within the corporate-ceremonial sphere of Ohio Hopewell, beneath the large earthen mounds at a number of sites, large rectangular structures, sometimes forming conjoined structural complexes (e.g., Harness, Figure 9.7) have been uncovered. These submound structures served as charnel houses, sheltering the dead, mortuary processing, and associated rituals and ceremonies (Brown 1979). In addition, these mortuary programs and structures were probably imbedded in broader contexts of corporate group integration such as redistribution and feasting (Seeman 1979), and likely served as the center and focus of the corporate-ceremonial sphere of Ohio Hopewell societies (Greber 1979, 1983).

At least four general classes of activity can be identified as having been carried out within the corporate-ceremonial sphere (Figure 9.5). While almost invariably separated, spatially and contextually, from the domestic sphere and habitation settlements, these corporate group activities are not always staged or carried out within imposing structures. The small circular single wall post structure components of the Harness "Big House" (Figure 9.7) are representative of the smaller scale corporate structures in Ohio and other regions as well, with the Russell Brown site structures, located just outside the Harness (Liberty) earthworks being an excellent case in point (Seeman and Soday 1980). The small circular structure components of the Harness "Big House" structural complex, and other low complexity corporate structures, are also similar in size and form to Hopewellian domestic structures across the East, often differing from them only in associated features and cultural assemblages. It is likely, therefore, that when Ohio Hope-

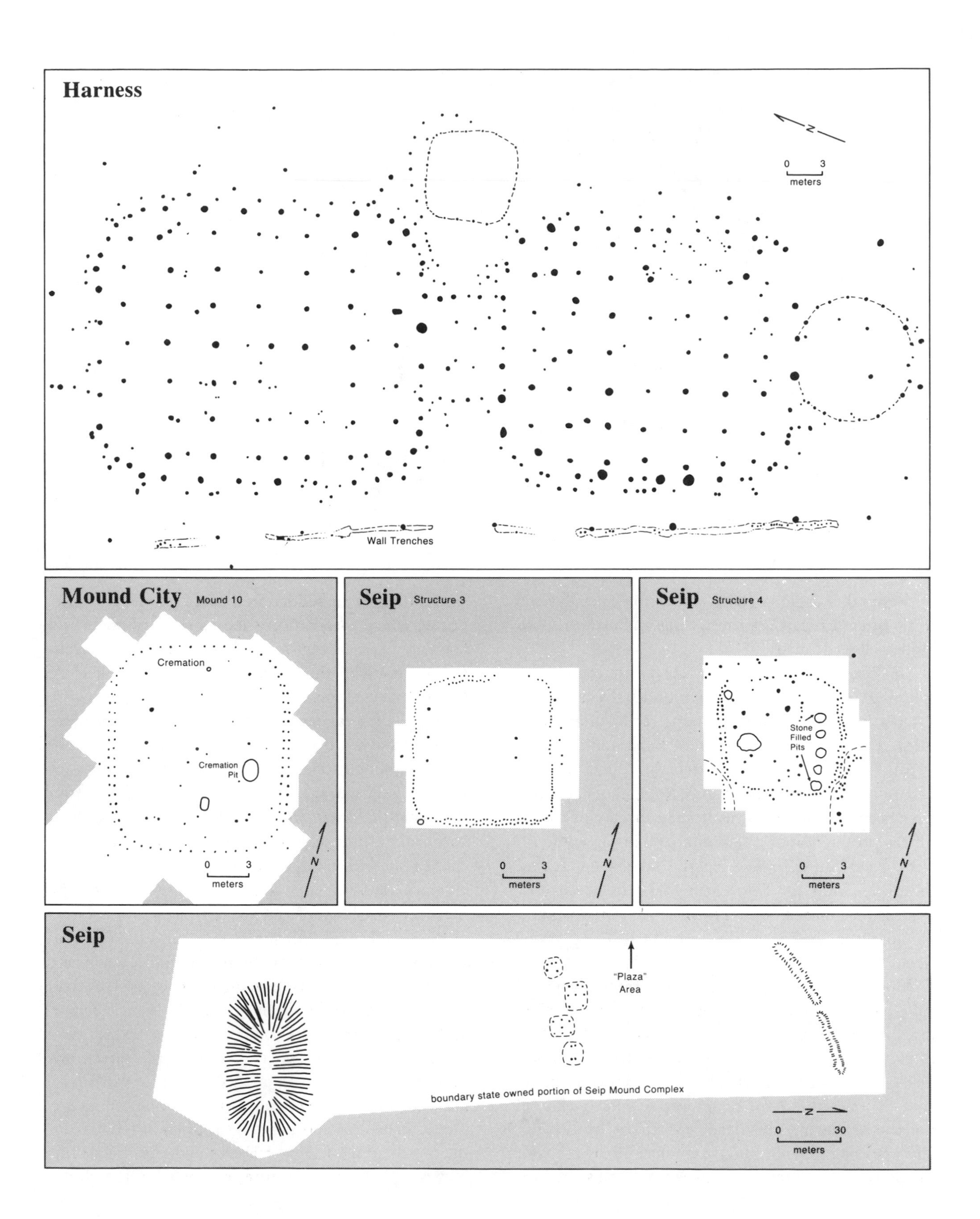

Harness
0 3
meters
Wall Trenches
Mound City Mound 10
Cremation
Cremation Pit
0 3
meters
Seip Structure 3
0 3
meters
Seip Structure 4
Stone Filled Pits
0 3
meters
Seip
"Plaza" Area
boundary state owned portion of Seip Mound Complex
0 30
meters

well domestic structures are eventually excavated, they will resemble, at least in size and general form, the peripheral circular structural elements at Harness. At the same time, however, a general dichotomy of scale exists between Hopewellian habitation structures and those of the corporate-ceremonial sphere, with most of the known habitation structures appearing to have been occupied by 5 to 13 individuals (Figure 9.8).

Hopewellian habitation structures, likely housing nuclear-extended family units, comprise the most obvious settlement element or component of what in the abstract can be called the "Hopewellian Household Unit" (Figure 9.9). Economically self-sufficient in large part, these household units were the basic socioeconomic building blocks of Hopewellian farming societies. In addition to single wall post houses, household units are reflected archaeologically by a number of other settlement components (Figure 9.9). Open ended arc or C-shaped post hole patterns, sometimes interpreted as wind-breaks, likely functioned as warm season open-sided shelters. A third, ubiquitous, settlement component of the Hopewell household unit is the food storage and processing pit cluster. Pit features remain the most common and most difficult class of Hopewellian archaeological features to characterize and interpret, as reflected in their all too common designation as nothing more than "refuse" pits. Based on both their diameter and depth, an initial storage function can be reasonably attributed to many Hopewellian pits, while comparably deep pits filled with fire-cracked rock are usually identified as earth ovens. While fire reddened and hardened earth, when preserved, allows the identification of surface hearth features, numerous shallow basins of variable plan and cross section frequently evade unequivocal functional attribution. They are often characterized as food processing facilities. Pit features of these four general classes, occurring in spatial-contextual clusters, provide a distinctive archaeological marker of the Hopewell household unit. Although these pit clusters exhibit considerable compositional and spatial variability, deep storage pits play a dominant role in the two most distinctive pit cluster patterns, as they both define within-structure food storage areas and often constitute a central element of the open-air food storage and processing pit clusters that were an important central element of Hopewellian domestic life. Four other settlement components of Hopewell household units can also be recognized: (1) scattered post hole patterns, likely marking the location of a variety of facilities—drying racks, fences, and wind breaks; (2) a general midden deposit, usually no more than 10–30 cm in depth, will often mark the location of Hopewell households if not eroded or plowed away; (3) in addition, overslope terrace edge and stream gully trash dumps a meter or more in thickness have occasionally been uncovered adjacent to household areas; (4) human burials, both isolated interments and burial clusters, are only rarely present in domestic household contexts. Taken together, or in various combinations, these settlement components constitute the archaeological signature of Hopewell household units.

How these household building blocks of Hopewellian society were spatially and socially organized

Figure 9.7 The relative size and placement of corporate structures associated with Hopewell mortuary mounds and earthworks in south-central Ohio. At Seip (Figure 9.6), excavations approximately halfway between the central mound and outer circular embankment have exposed a plazalike area lacking features (Greber 1983:86), and to the east of it, a linear arrangement of seven large rectangular "specialized workshop" structures that parallel the long axis of the central mound (Baby and Langlois 1979). Stratigraphic relationships and available radiocarbon dates indicate that the structures post-date the central mound (A.D. 55 ± 100) and were not in existence at the same time. Rather they were sequentially constructed, used, and capped with midden debris and gravel over a period of several centuries or more (Structure 7, A.D. 90 ± 85; Structures 1, 2, 3, A.D. 230 ± 80; Structure 5, A.D. 590 ± 105; Structure 4, A.D. 280 ± 55). In contrast to the Seip "specialized workshop" structures, submound structures of similar form excavated at a number of Ohio Hopewell sites, including Seip (Greber 1979), Harness (Liberty) (Greber 1983), and Mound City (Brown 1979) clearly functioned as charnel houses, sheltering both the dead and associated mortuary processing and ceremonial activities. Crematory basins, cremation burials, and associated artifacts, often also burned and fragmented, are unequivocal diagnostic features of such charnel houses. Seeman (1979) has proposed that Ohio Hopewell charnel houses and mortuary programs also provided the spatial-ritual context for redistribution of meat through feasting. Similarly, Greber (1983:92) emphasizes the broader civic, ceremonial, and religious role of such central corporate structures.

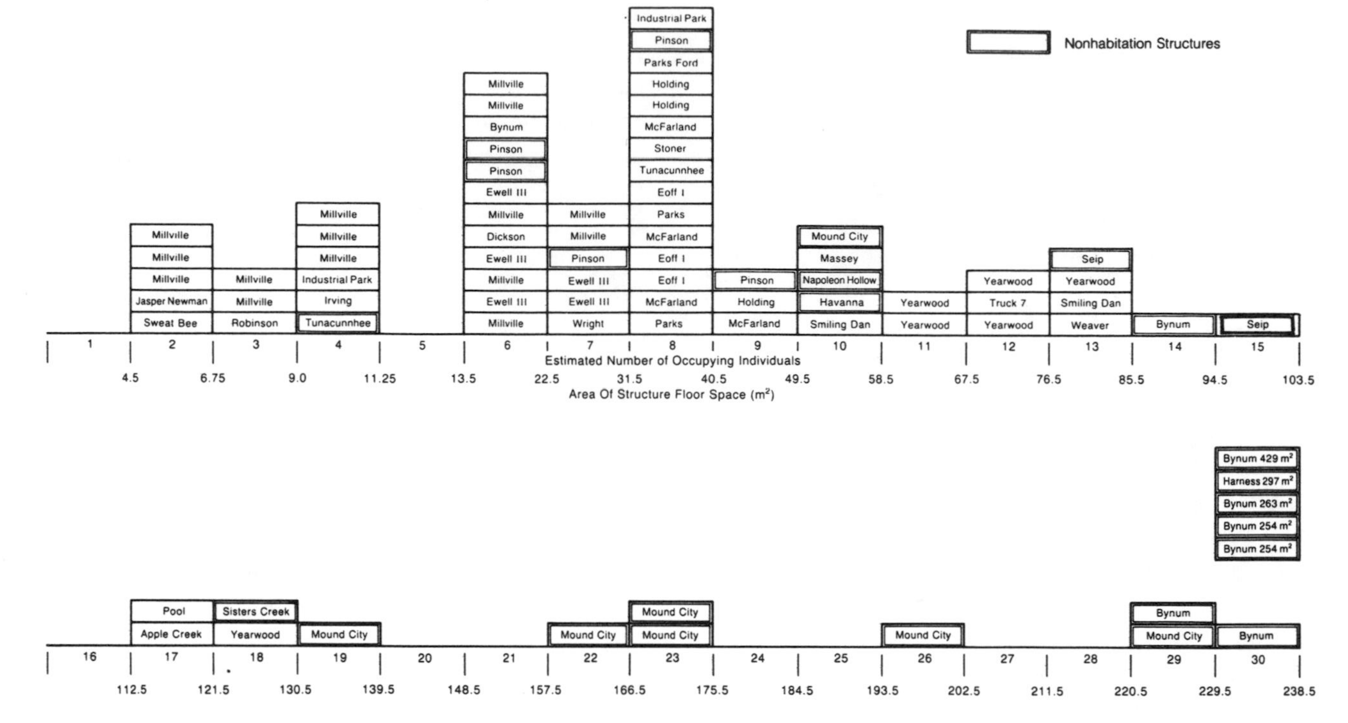

Figure 9.8 Floor area in square meters of 83 Middle Woodland habitation and corporate-ceremonial structures in eastern North America. The number of occupants of structures is also estimated, based on Cook's "Rule of thumb formula": "For measuring space a fair rule of thumb is to count 25 square feet for each of the first six persons and then 100 square feet for each additional individual" (Cook 1972:16). Site names are listed, with nonhabitation corporate ceremonial structures indicated by a double line.

and integrated is still a matter of considerable speculation, with a number of different annual cycle settlement pattern models having been proposed (Figure 9.10). The most frequently mentioned habitation settlement category for Hopewellian groups is the permanent year-round village, comprised of an unspecified number of household units, and exhibiting unspecified indications of supra-household sociopolitical integration. At various times through the annual cycle small, task specific groups would leave the village for short periods of time, establishing short-term limited activity camps (for example, hunting and plant harvesting camps, floodplain agricultural camps, and raw material procurement camps).

Such seasonal, short-term specialized camps are also components in an alternative settlement pattern model that has small numbers of Hopewell household units, perhaps three to five, forming growing season farming hamlets in floodplain settings, and then fissioning into higher (drier) elevation cold season homesteads (Fortier et al. 1989).

A third alternative model has Hopewell household units distributed along river and stream valley corridors, in small spatially isolated year-round settlements.

Bracketing the logical spectrum of settlement pattern alternatives, from year-round nucleated through seasonally shifting to year-round dispersed, these models provide a theoretical abstract background of expectation for consideration of Middle Woodland (A.D. 0–200) farming settlements recently excavated within the context of long-term regional research projects in three different regions of the eastern United States: (1) the Upper Duck River Valley of central Tennessee; (2) the Lower Illinois River Valley of west-central Illinois; and (3) the American Bottom area of the Central Mississippi River Valley (Figure 9.11). In addition to providing the best available regional data sets regarding Hopewellian farming econ-

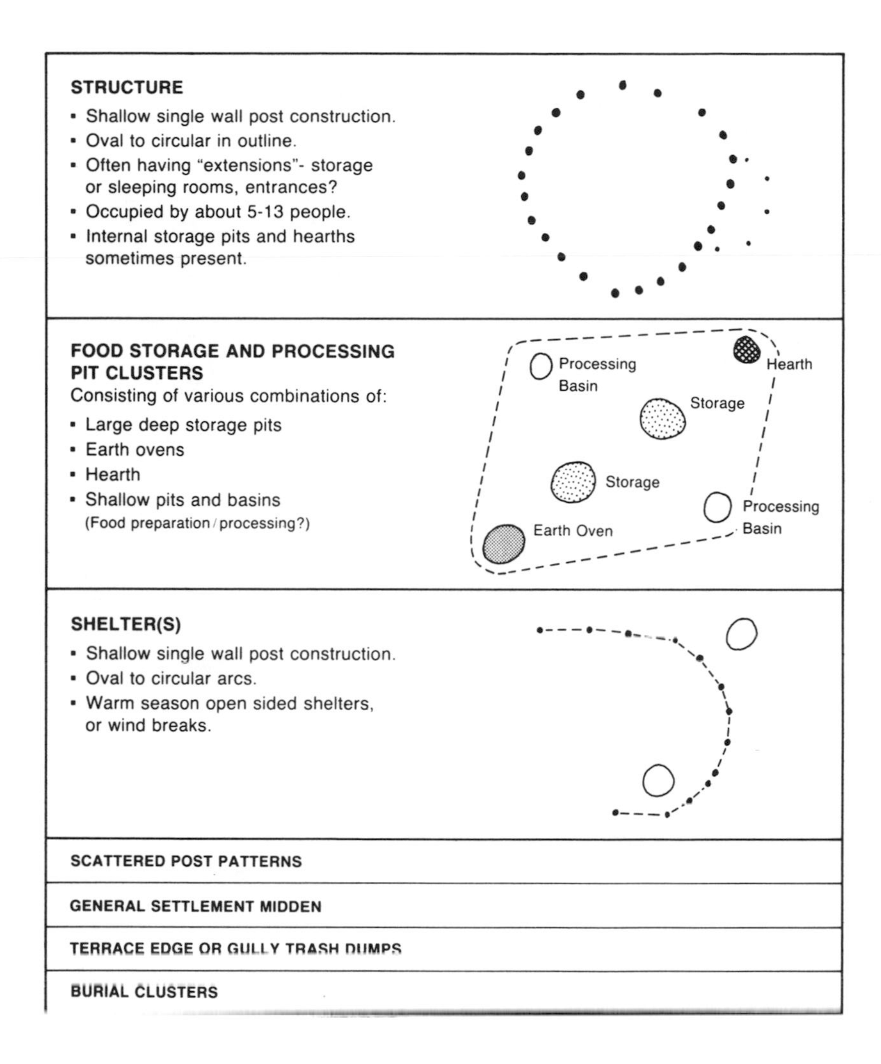

Figure 9.9 Settlement components of the Hopewellian household unit.

omies and settlements, these three river valley segments also present an interesting opportunity for comparison of Middle Woodland communities occupying riverine systems of quite different scale. While the floodplain of the 25-km-long section of the Upper Duck River Valley under consideration does not exceed 1 km in width, the floodplain corridor of the Lower Illinois River is almost 7 km wide, and is fed by a series of four large streams, which are roughly comparable in size to the Upper Duck River. Finally, within the American Bottom, a 500 sq km expansion of the Mississippi River Valley just downstream from the confluence of the Mississippi and Missouri rivers, the floodplain reaches a width of over 16 km.

Middle Woodland (A.D. 0–200) farming settlements from other areas of the eastern United States will also be brought into the discussion when appropriate, as we move from the Upper Duck River Valley to the Lower Illinois River Valley, and finally to the American Bottom.

The Upper Duck River Valley of Central Tennessee

As a result of initial surveys and surface collections, followed by testing and large-scale excavation of selected sites, more than 35 Middle Woodland McFarland phase sites were located by University of Tennessee archaeologists within the Normandy Reservoir area of the Upper Duck River Valley (Figure 9.12). Although no McFarland phase mortuary mounds have been excavated to date, the Old Stone Fort, a 20 ha geometric earthwork placed in an upland setting between two forks of the headwaters of the Duck

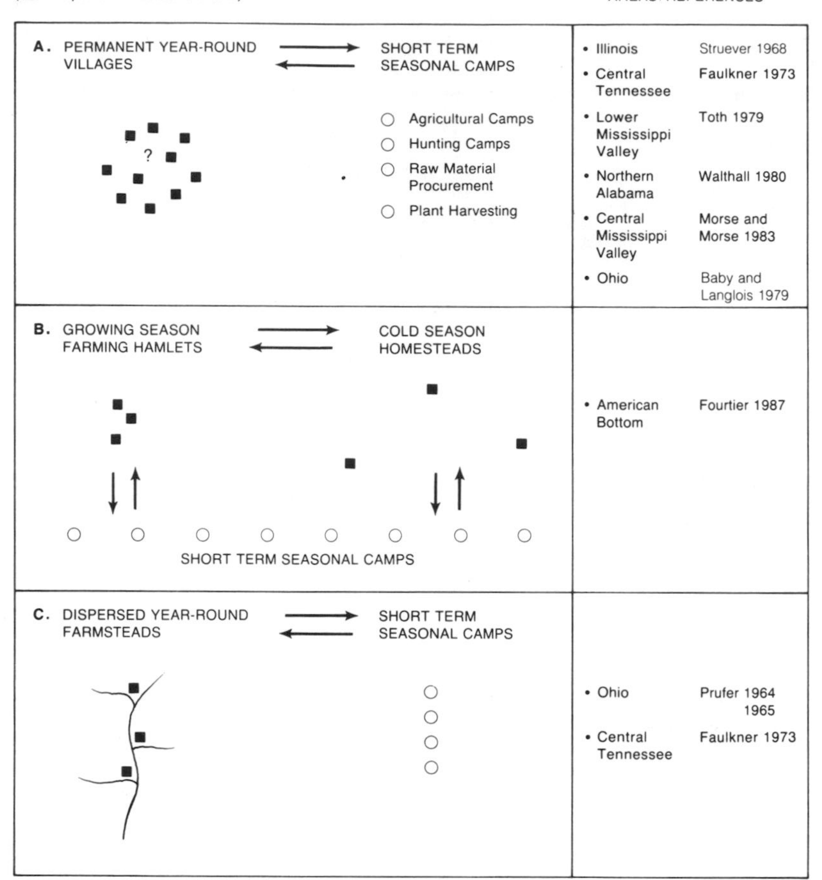

Figure 9.10 Alternative Hopewellian settlement pattern models.

River (Figures 9.12, 9.13), and near a major interregional trail, has been identified as the corporate-ceremonial center of the McFarland phase (Faulkner 1968). Of the McFarland phase domestic sphere components excavated as part of the Normandy Project, two (Aaron Shelton, Wiser-Stephens) have been identified as short-term seasonal camps, while five were found to be more substantial habitation settlements (Banks, Parks, Eoff I, Ewell III, and McFarland; Figure 9.12). These excavated habitation sites form a fairly evenly spaced linear sample of Hopewellian farming settlements, and after first considering an excavated McFarland phase habitation site located in the Upper Elk River drainage 30 km to the south, each will be briefly considered in turn, moving upstream in the Duck River Valley toward the Old Stone Fort.

Since a number of proposed Hopewellian settlement pattern models place nucleated village settlements adjacent to corporate-ceremonial centers, it will be particularly interesting to see to what degree changes in the size and complexity of habitation sites is correlated with proximity to the Old Stone Fort. Do Hopewellian farming settlements become larger and exhibit greater spatial and social organization as one moves from the edge, the periphery, toward the corporate center?

PARK'S FORD. The Park's Ford site (40Fr47) is situated on a small topographically restricted terrace spur in the Upper Elk River Valley, about 30 km south of the Duck River drainage and the Old Stone Fort. Overlooking a "pocket" enlargement of the floodplain resulting from a tributary

Figure 9.11 Map of the eastern United States showing the location of the sites and river valley research areas that have yielded information regarding the Middle Woodland farming settlements of Hopewellian populations.

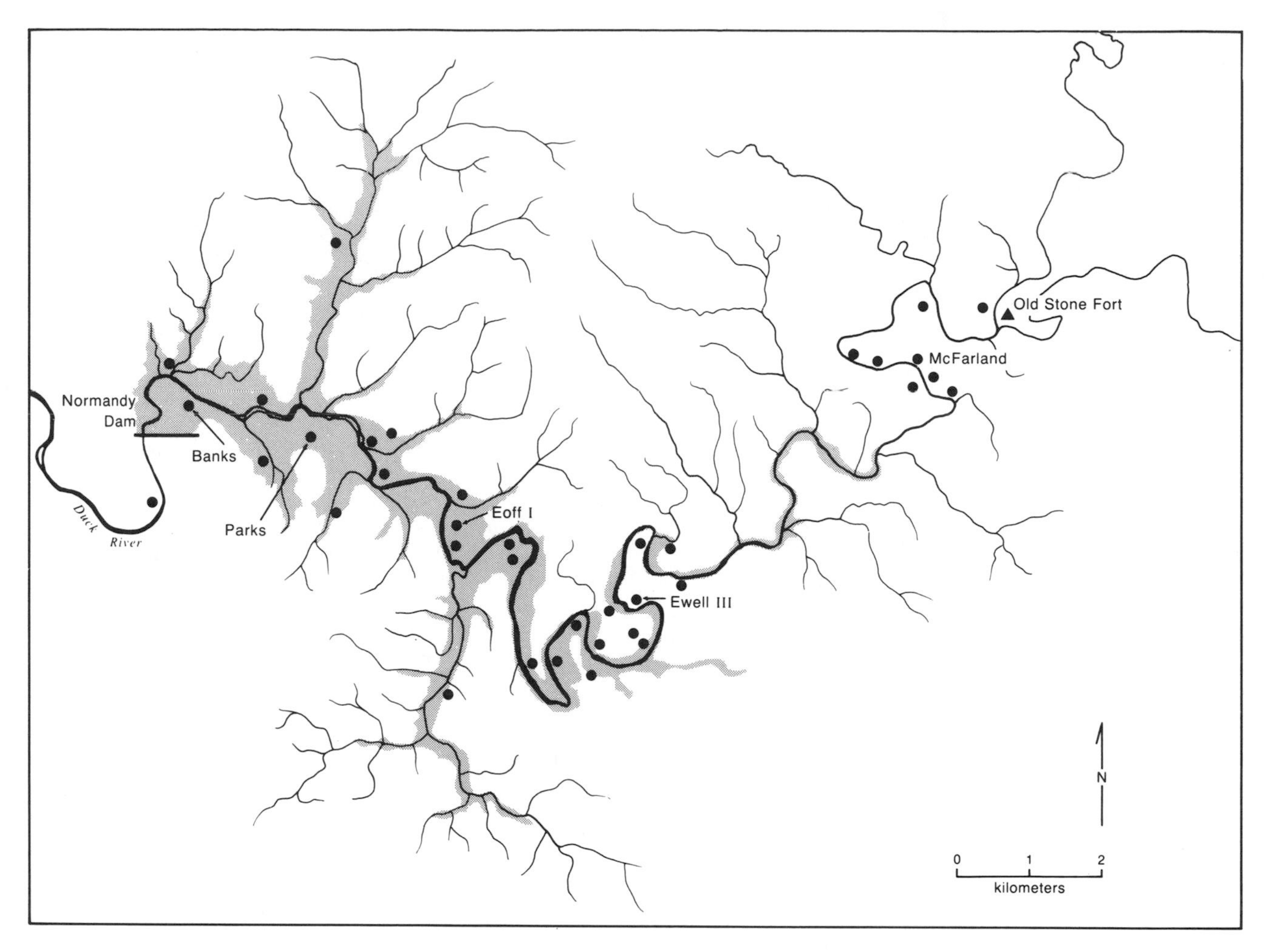

Figure 9.12 The location of McFarland phase settlements within the Duck River drainage (after Kline et al. 1982, fig. 2).

stream entering the Elk River Valley from the north, this small (400 sq m) habitation site contained three settlement components of the Hopewellian household unit. Two food storage and processing feature clusters, comprising a total of 14 pits and hearths, were situated just to the west of an open-end shelter, between it and the terrace edge. A number of scattered post molds and a single burial were also excavated (Figure 9.14). The absence of any more substantial structure than the C-shaped shelter, or any evidence of rebuilding, led to Park's Ford being characterized as a late summer-fall habitation site (Bacon and Merryman 1973). I think, however, that the only major component of the Hopewellian household unit missing at Park's Ford, a circular habitation structure, likely went undiscovered and unexcavated in a nearby location. Such a pattern of spatial separation of settlement components is evident in most of the McFarland phase farming settlements that have been excavated in the Upper Duck River Valley.

PARKS. At the 10 ha multicomponent Parks site, for example (Figure 9.15), extensive testing and six block-excavation units exposed only a single McFarland phase structure, located adjacent to a pit feature cluster and within a shallow 900 sq m midden area (Faulkner and McCollough 1982). In addition to this circular single wall post structure, a number of other settlement components were scattered some distance away across the terrace. A group of open-

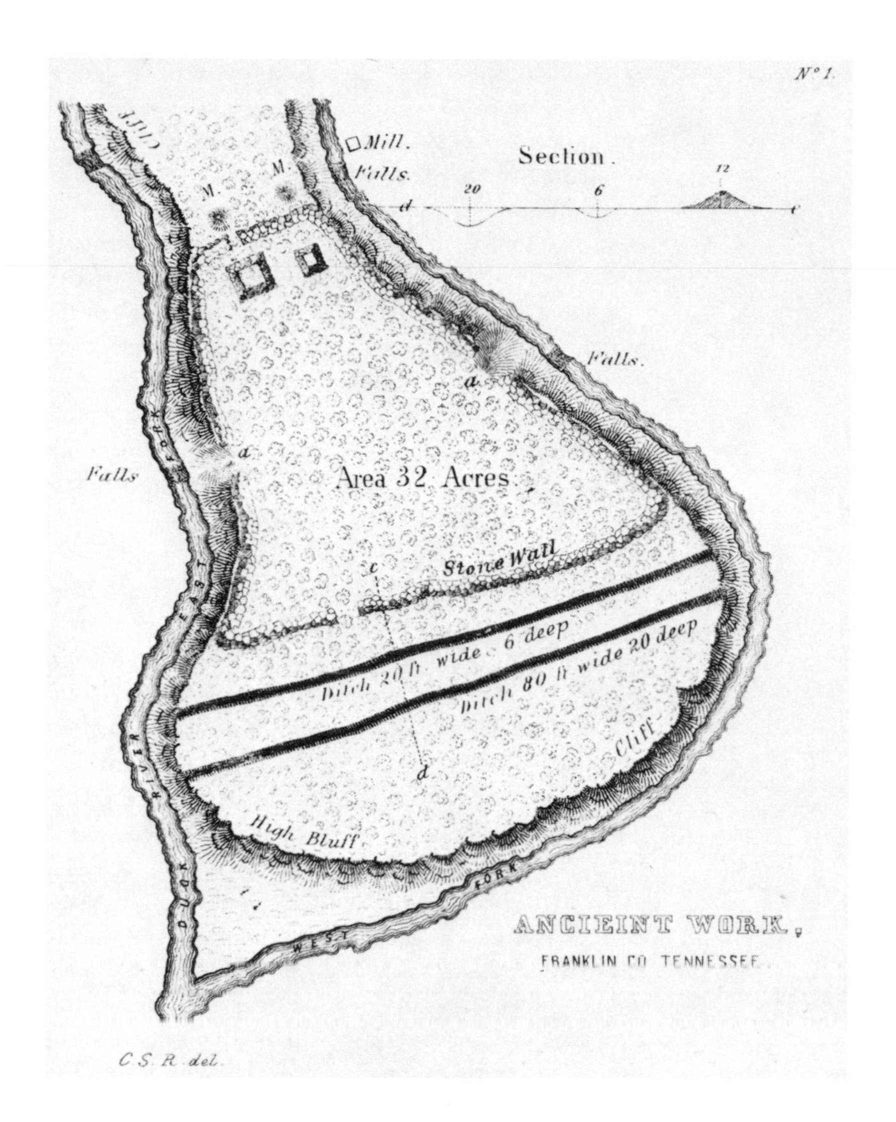

Figure 9.13 The Old Stone Fort, corporate-ceremonial center of the McFarland phase (Squier and Davis 1848, plate XII).

end shelters and pit clusters were located over 50 m away from the circular structure, and two isolated cremation burial clusters were placed 90 m and 200 m away in excavation units 6 and 7.

EOFF I. Moving upstream several kilometers, the 7.5 ha multicomponent Eoff I site, like the Parks site, is situated on a large terrace across the valley floor from a tributary stream, and exhibits considerable dispersal of settlement components (Figure 9.16). Following two block-unit excavations and the mechanical stripping of overburden from the entire southern portion of the site, three circular single wall post structures, spaced from 60 m to 160 m apart, were found to occupy the central portion of the terrace (Figure 9.16:4, 5, 6), while three food storage and processing pit cluster areas, one with an accompanying C-shelter (1), and another consisting of three pit clusters (3), were located closer to the terrace edge, at the site periphery (Figure 9.16; Faulkner 1977, 1982). It is tempting to pair these three structures and three food storage and processing feature cluster areas, and even to suggest that their spacing reflects intervening terrace field areas, like those documented for Neolithic settlements in Europe (Whittle 1987).

EWELL III. At the Ewell III site, situated on a smaller river terrace 6 km upstream toward the Old Stone Fort, testing and a large block-excavation unit

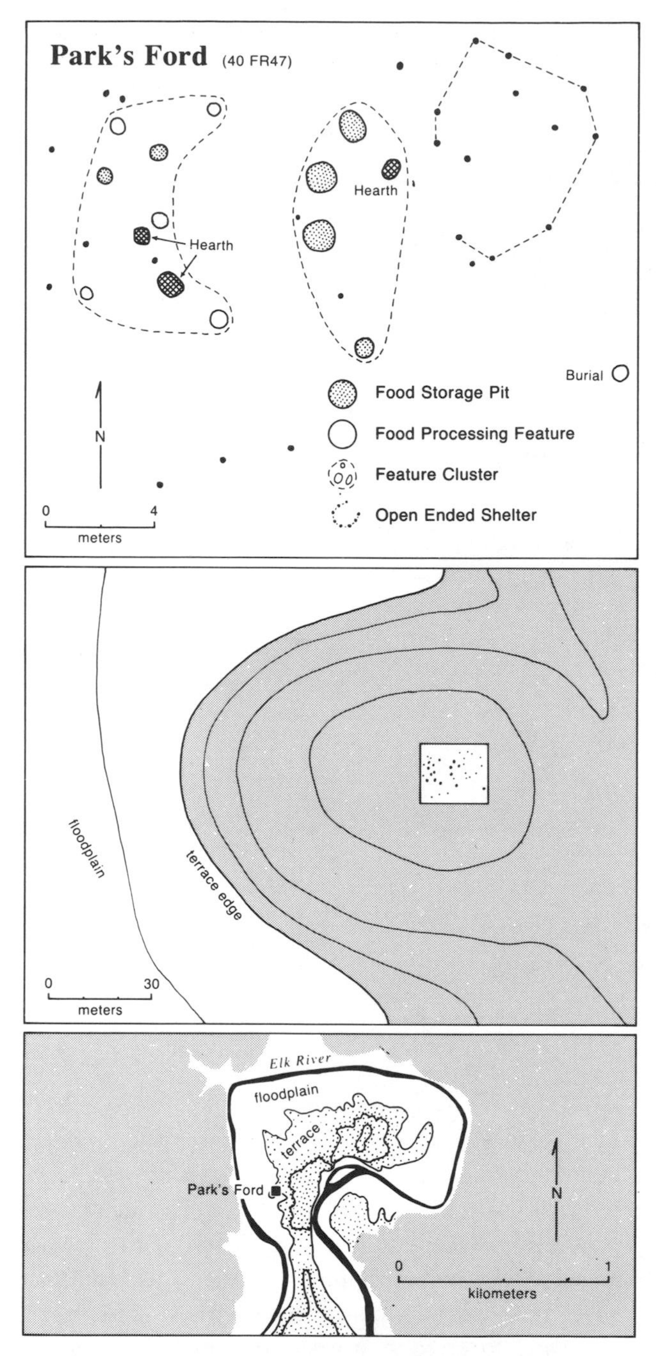

Figure 9.14 The 400 sq m McFarland phase settlement at the Park's Ford site, located on a terrace overlooking the Upper Elk River, about 30 km south of the Duck River. Two distinct food storage and processing feature clusters were situated between an open-end C-shelter and the terrace edge (Bacon and Merryman 1973). It is possible that the C-shelter post hole pattern may represent a circular structure.

exposed a much less dispersed pattern of Hopewellian household components. A total of seven food storage and processing feature clusters were spaced along the terrace edge, with the four southernmost of these (Figure 9.17) in close proximity to two circular single wall post structures, four open-end C-shelters, and a centrally placed cremation burial cluster (Duvall 1982). This "central" burial cluster at Ewell III, along with the suggestion of a possible open area to the west of it that is flanked by two structures, provides the first, if admittedly weak, indication of any spatial or social integration above the household level within the domestic sphere of McFarland phase society.

MCFARLAND. Situated 7 km upstream from Ewell III and within 1.5 km of the corporate-ceremonial center of the McFarland phase, the 2 ha McFarland site is situated on a large terrace in a bend of the Duck River. Identified on the basis of surface collections as the largest, most intensively occupied Middle Woodland habitation site in the region, and frequently characterized as a substantial year-round farming village and a center for specialized ceremonial activities associated with the nearby Old Stone Fort, the McFarland site was extensively excavated in 1979 (Kline et al. 1982). In order to identify the location of structures and feature clusters, 26 parallel 1-m-wide trenches were spaced every 5 m across the site, exposing three separate areas of habitation (Figure 9.18). A total of 92 pit features were excavated in the initial trenches and subsequent excavation units (basin-shaped processing pits, storage pits, earth ovens), found both in association with the five circular structures that were uncovered, and in spatially separate feature clusters. Single structures, accompanied by feature clusters and possible C-shelters, were found in two of the habitation areas defined, while the third area yielded three conjoined structures. These three apparently connected structures and a possible adjacent open area would appear to represent some minimal level of social integration above the household level. But even if all five structures were contemporaneous, rather than reflecting a temporal sequence of single household unit occupational episodes, this farming "village" would have been in-

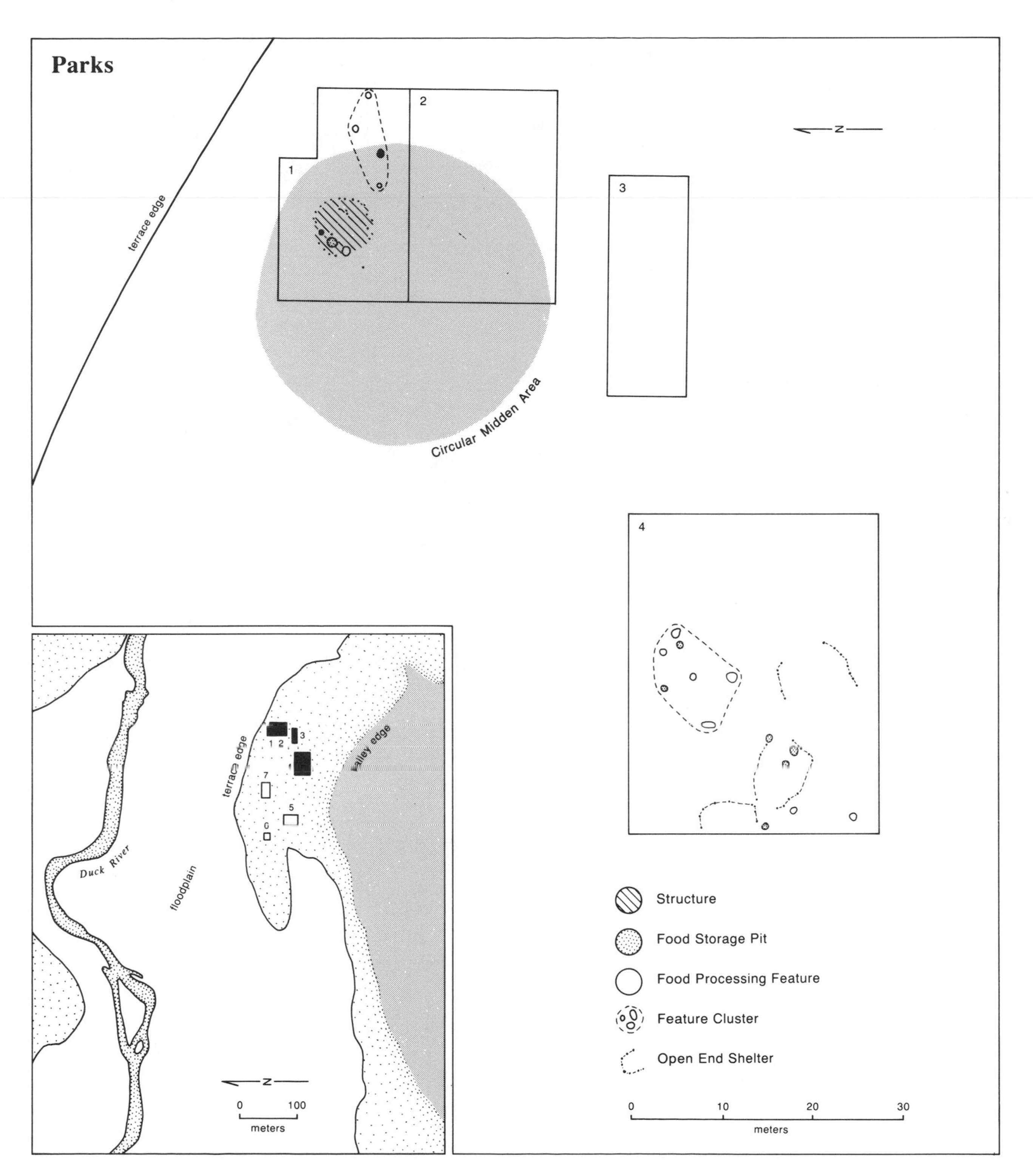

Figure 9.15 The McFarland component settlement plan at the Parks site.

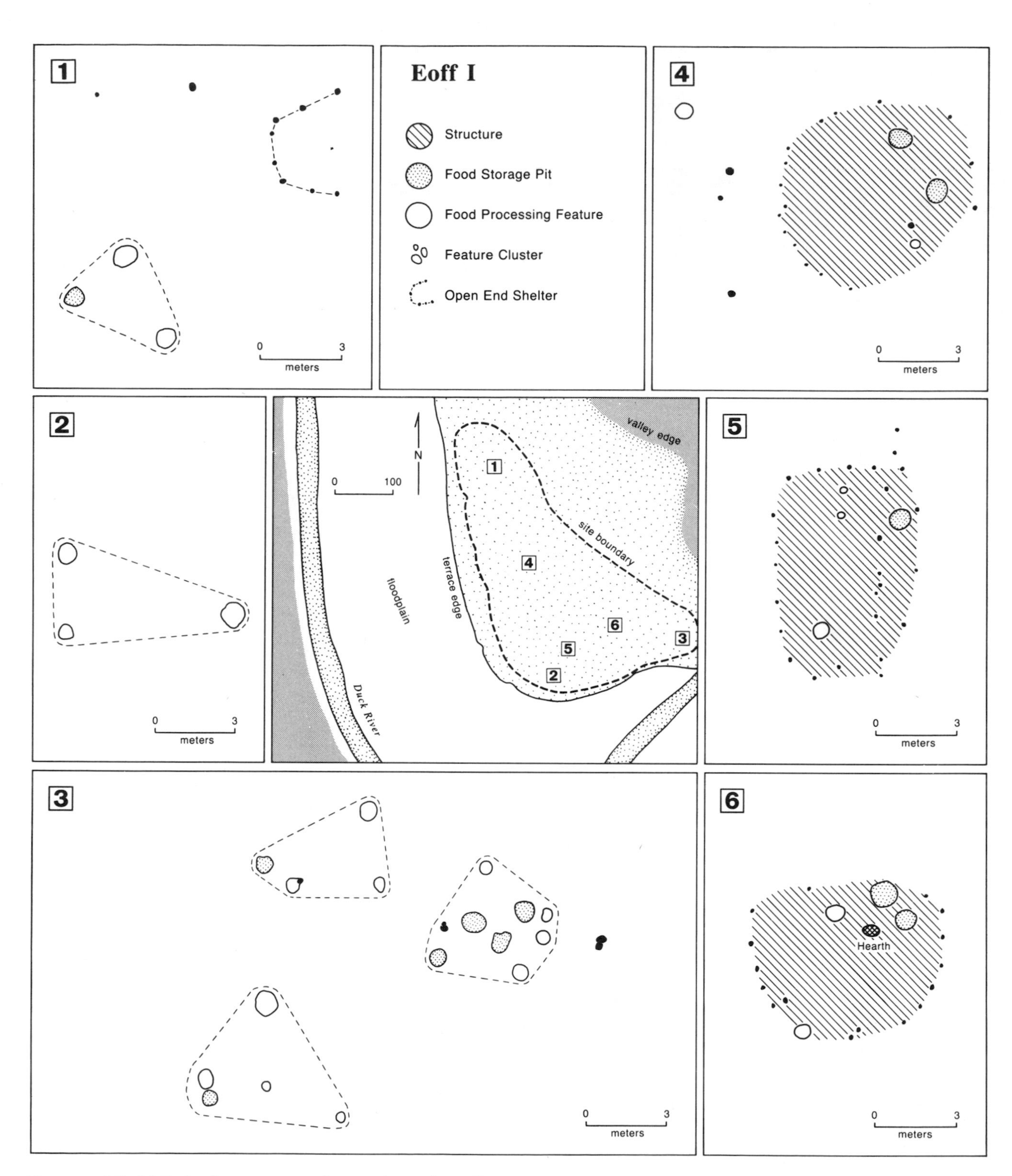

Figure 9.16 The McFarland component settlement plan at the Eoff I site.

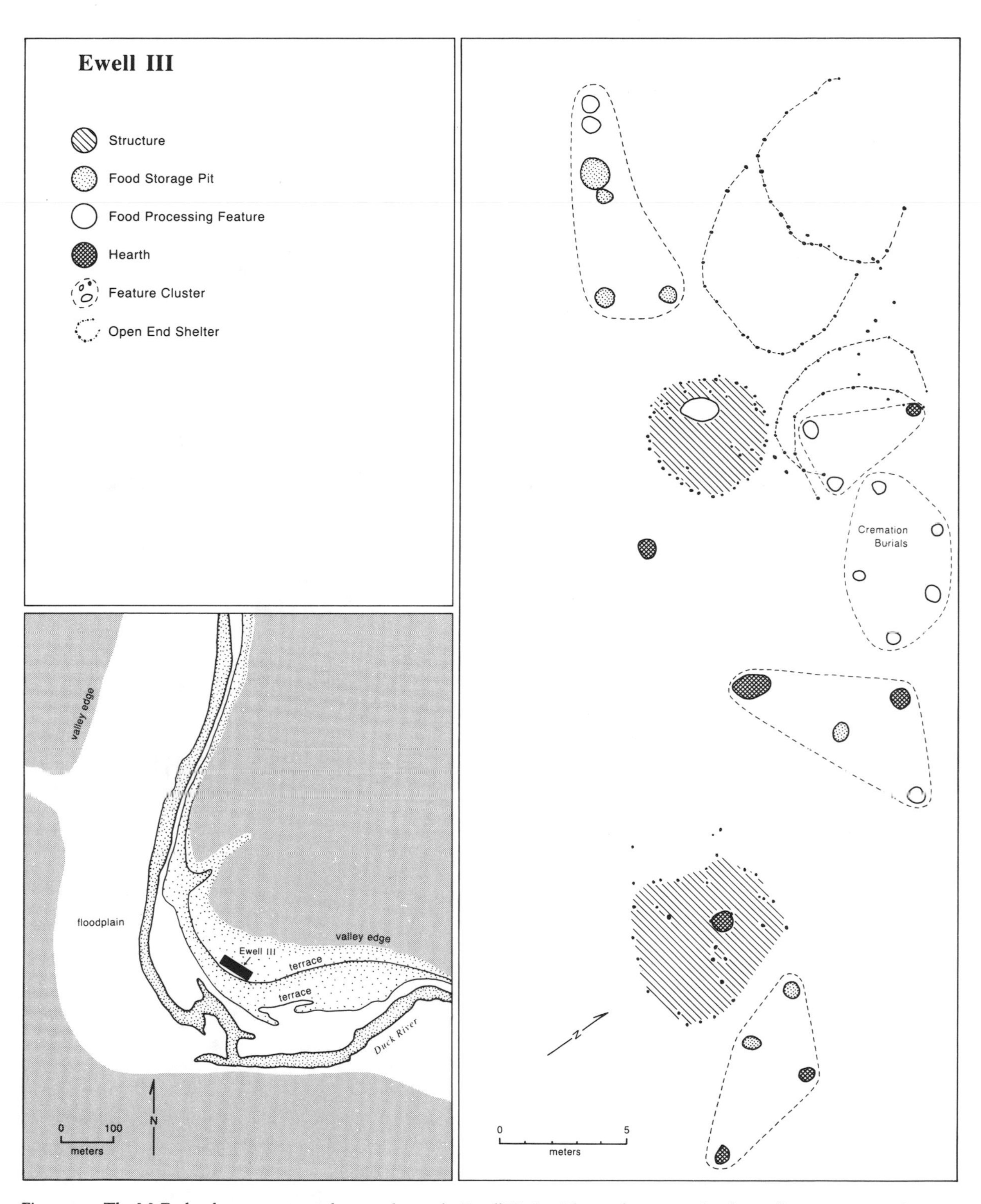

Figure 9.17 The McFarland component settlement plan at the Ewell III site. The northernmost circular wall post structure shown as belonging to the McFarland phase occupation of the site was originally assigned to the Terminal Archaic Wade phase component at the site (DuVall 1982), and its exact temporal placement remains in question.

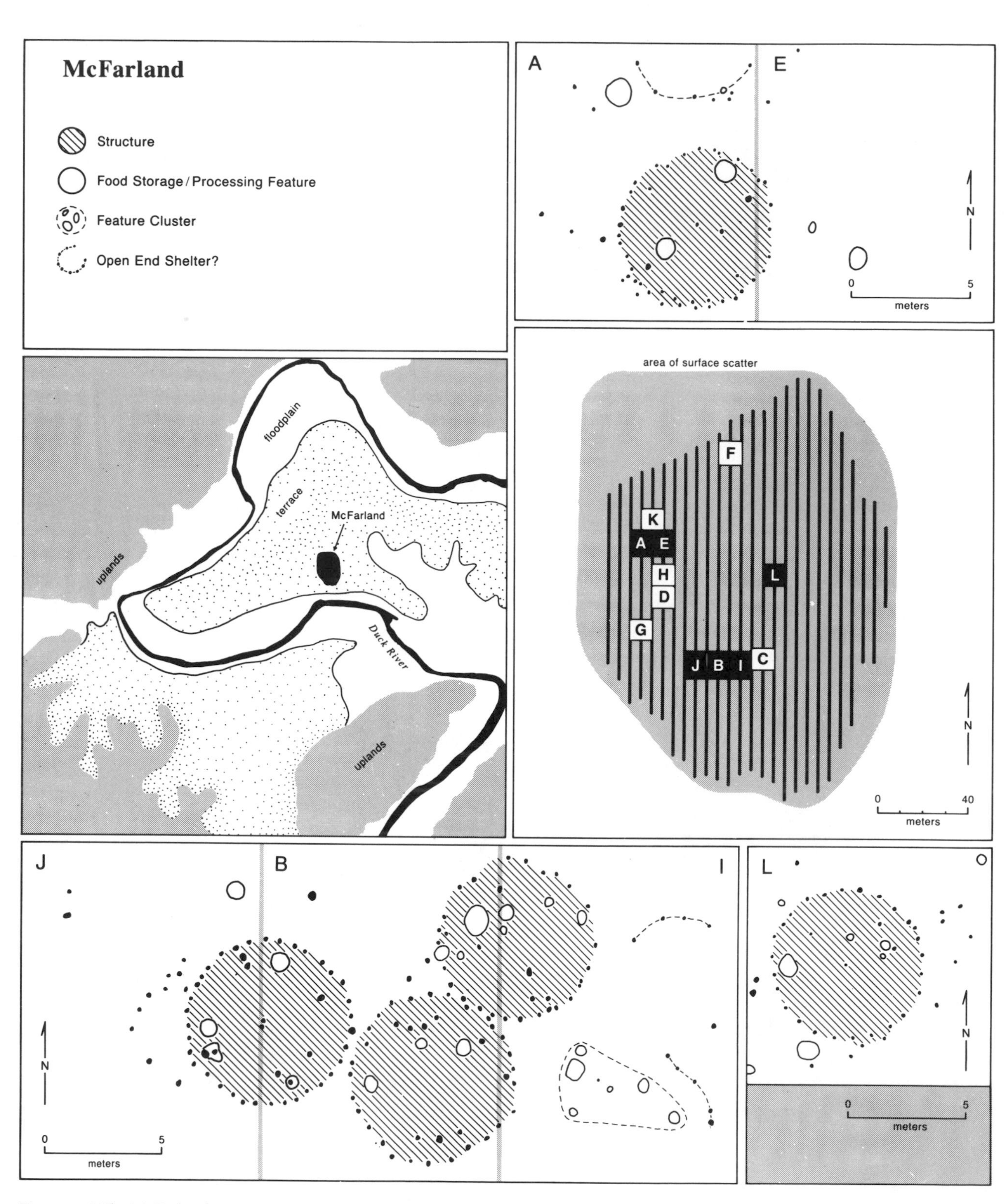

Figure 9.18 The McFarland component settlement plan at the McFarland site.

habited at most by only about 40–50 individuals. No evidence of corporate-sphere ceremonial activities was recovered.

In summary, there is very little evidence in the Upper Duck River Valley for social or spatial integration very far beyond the basic household unit. This central Tennessee Hopewellian society appears to have been distributed along the terraces of the Upper Duck River Valley in small farming settlements varying in size from one to perhaps three households. The spatially dispersed nature of these small one to three household occupations, along with the often intensive subsequent prehistoric occupation of the terraces on which they are located, makes McFarland phase settlements appear larger on the surface than they really are, while at the same time making them difficult to fully recognize, even when extensively excavated.

Before comparing this small drainage McFarland phase pattern of dispersed one to three household terrace settlements with the Hopewellian settlements documented in the higher-order drainage system of the Lower Illinois River Valley, two Hopewellian corporate-ceremonial centers of the Midsouth region should be briefly considered, since large village settlements have been recognized in association with each of them.

Bynum Mounds

Situated on a flat ridge between two forks of a small tributary of the Tombigbee River in northeast Mississippi (Figures 9.11, 9.19), the 2.8 ha Bynum Mounds site contained six mortuary mounds and an associated "village" area (Cotter and Corbett 1951). Each of the three mortuary mounds excavated yielded evidence of charnel houses and complex mortuary processing activities, while extensive trenching and block excavation of the "village" area uncovered thousands of post holes, and seven circular post hole patterns (Figure 9.19). While one of these circular post hole patterns falls within the size range of Hopewellian habitation structures, five of the remaining six form a linear arrangement in association with Mound D, and are at the upper end of the size range curve of Middle Woodland structures (Figure 9.8), suggesting a corporate-ceremonial function rather than a habitation role. In addition, the almost total lack of evidence for internal support posts within these large circular post hole patterns suggests that they were unroofed corporate enclosures rather than roofed habitation structures. Finally, pit features of any kind were rare in the "village" precinct of the Bynum site, underscoring its appropriate placement within the corporate rather than the domestic sphere of Hopewellian society. A similar reassignment of a proposed mound center Hopewellian village to the corporate sphere has taken place at the Pinson Mounds site in west Tennessee.

Pinson Mounds

As a result of recent research by Robert Mainfort (1986), the 160 ha Pinson Mounds site, situated adjacent to the floodplain of the Forked Deer River in west Tennessee (Figures 9.11, 9.20), has been shown to be the largest Middle Woodland period mound center in eastern North America in terms of both site area and mound mass, with its 12 mounds and geometric earthwork containing more than 100,000 cu m of fill.

On the basis of surface surveys in the 1970s, what appeared to be a large village area was identified at the western edge of the corporate-ceremonial center and named the Cochran site (Mainfort 1986). Subsequent testing of the south, central, and north sectors of the Cochran site, however, yielded little other than a few scattered post molds, except in one area of abundant surface scatter. A block-excavation unit at this location (Figure 9.20b) uncovered evidence of three single wall post structures, all within the size range of domestic structures (Figure 9.8). The absence of food storage and preparation features, the presence of crematory pits, and the recovery of exotic raw materials (copper, quartz, mica), strongly suggested that they were corporate-sphere specialized workshops similar to those at Seip (Figure 9.7) in general function if not in size (Mainfort 1986). Block excavations in three other widely scattered areas within the corporate-ceremonial precincts at Pinson (Figure 9.20a, c, d) uncovered four other oval to circular structures, with three of these clearly reflecting mortuary programs and burial preparation. Thus a number of scattered "habitation" areas at Pinson, upon excavation, have been found to be the location of specialized workshop and mortuary structures,

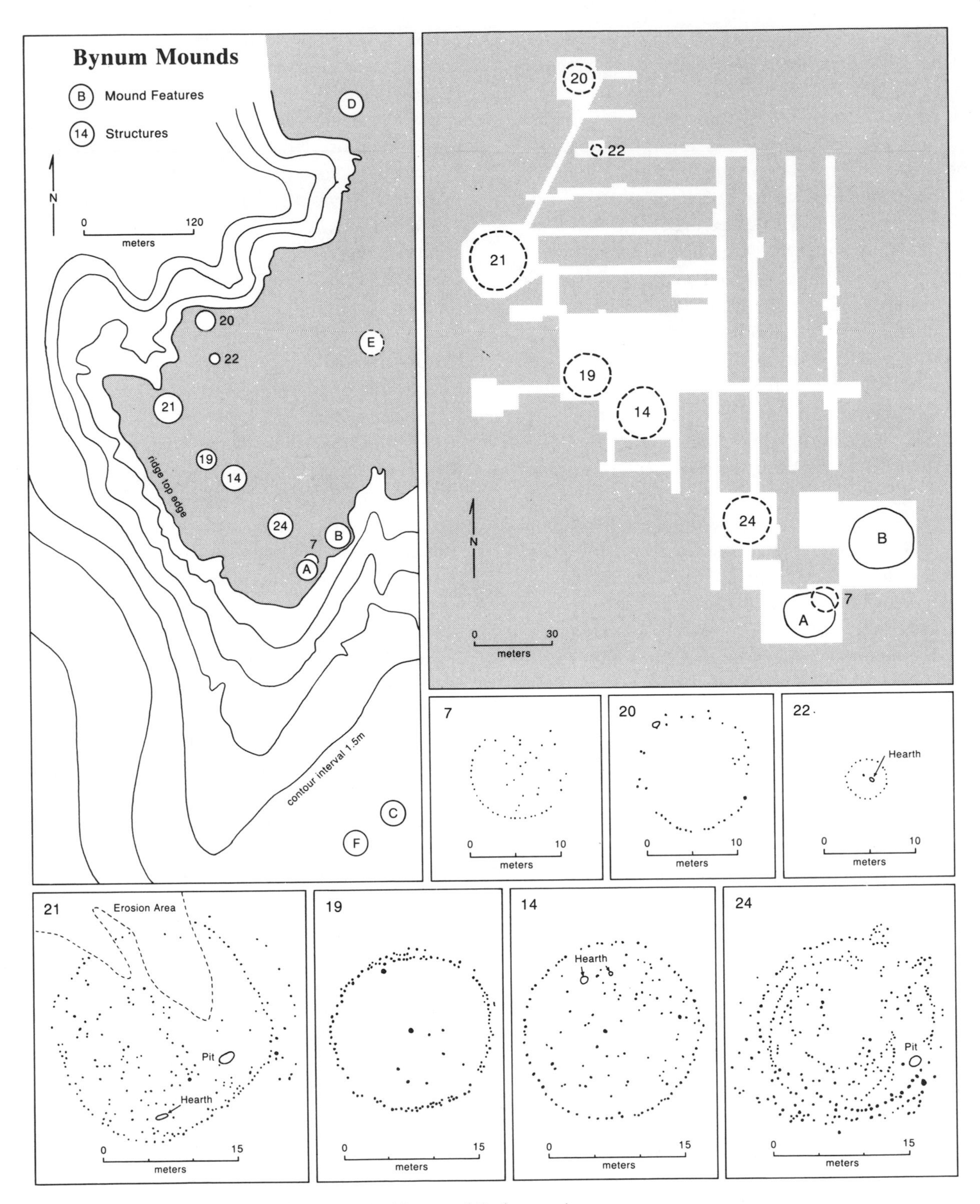

Figure 9.19 The settlement plan of the Bynum site (Cotter and Corbett 1951).

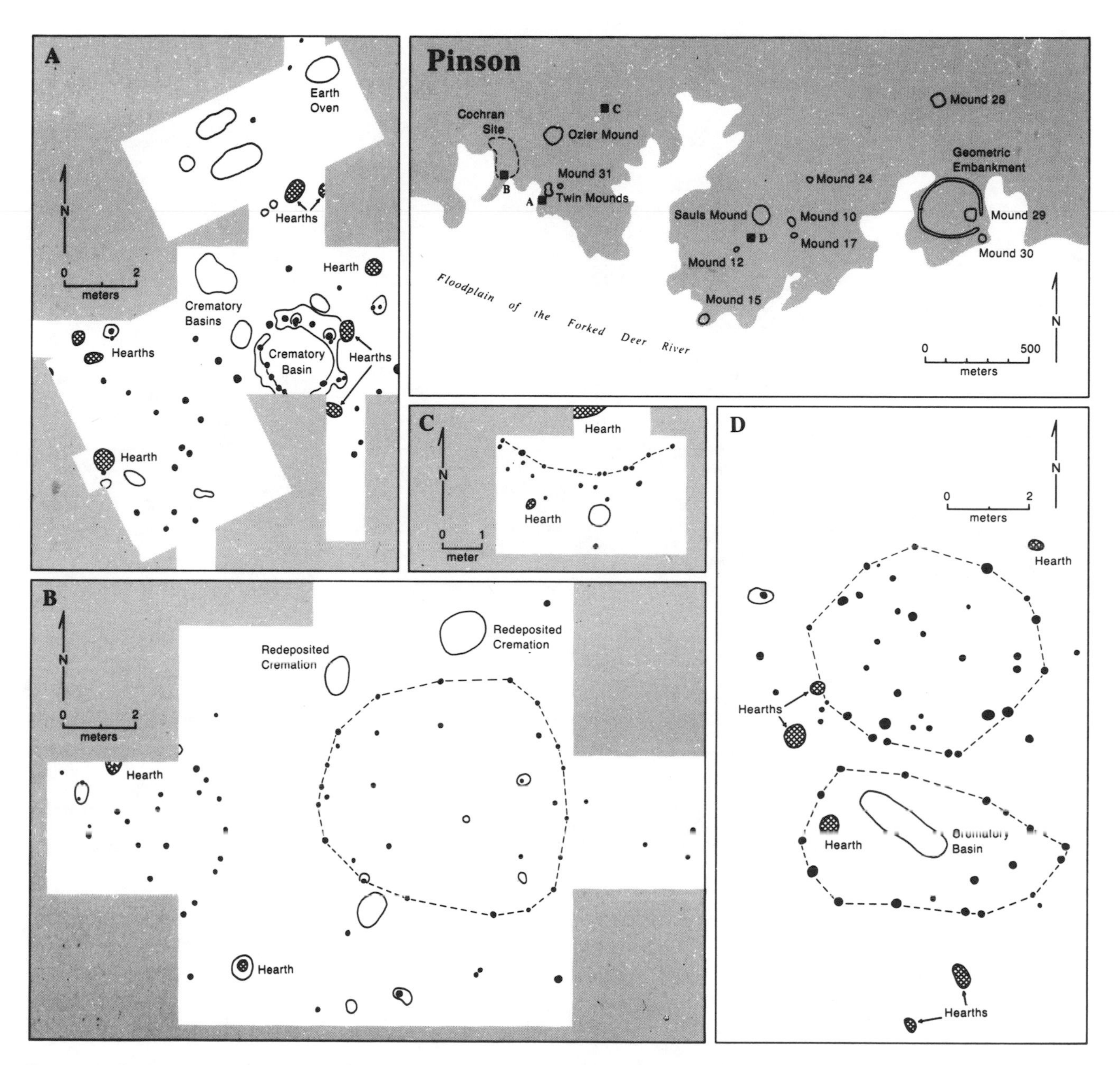

Figure 9.20 Corporate structures excavated at the Pinson Mounds site in western Tennessee. The largest mound at Pinson (Sauls Mound) is situated at the center of the site, measuring 22 m in height and containing 60,500 cu m of earth. Three other large platform mounds (Ozier Mound, Mound 28, Mound 29) are placed 1,000–1,150 m from the central mound, marking the northwest, northeast, and southeast corners of the ceremonial precinct. Breaking this pattern in both size and placement, Mound 15 nonetheless marks the topographic southwest corner of the site. Small block excavation units in four separate areas (A–D) exposed ovoid single wall post structures and associated features. Interpreted as representing small short-term ceremonial areas associated with mortuary processing and the manufacture of objects for interment and exchange, these mortuary encampments lacked midden deposits and much evidence of food storage or preparation, while yielding exotic raw materials (copper, quartz, mica) and crematory facilities (Mainfort 1986).

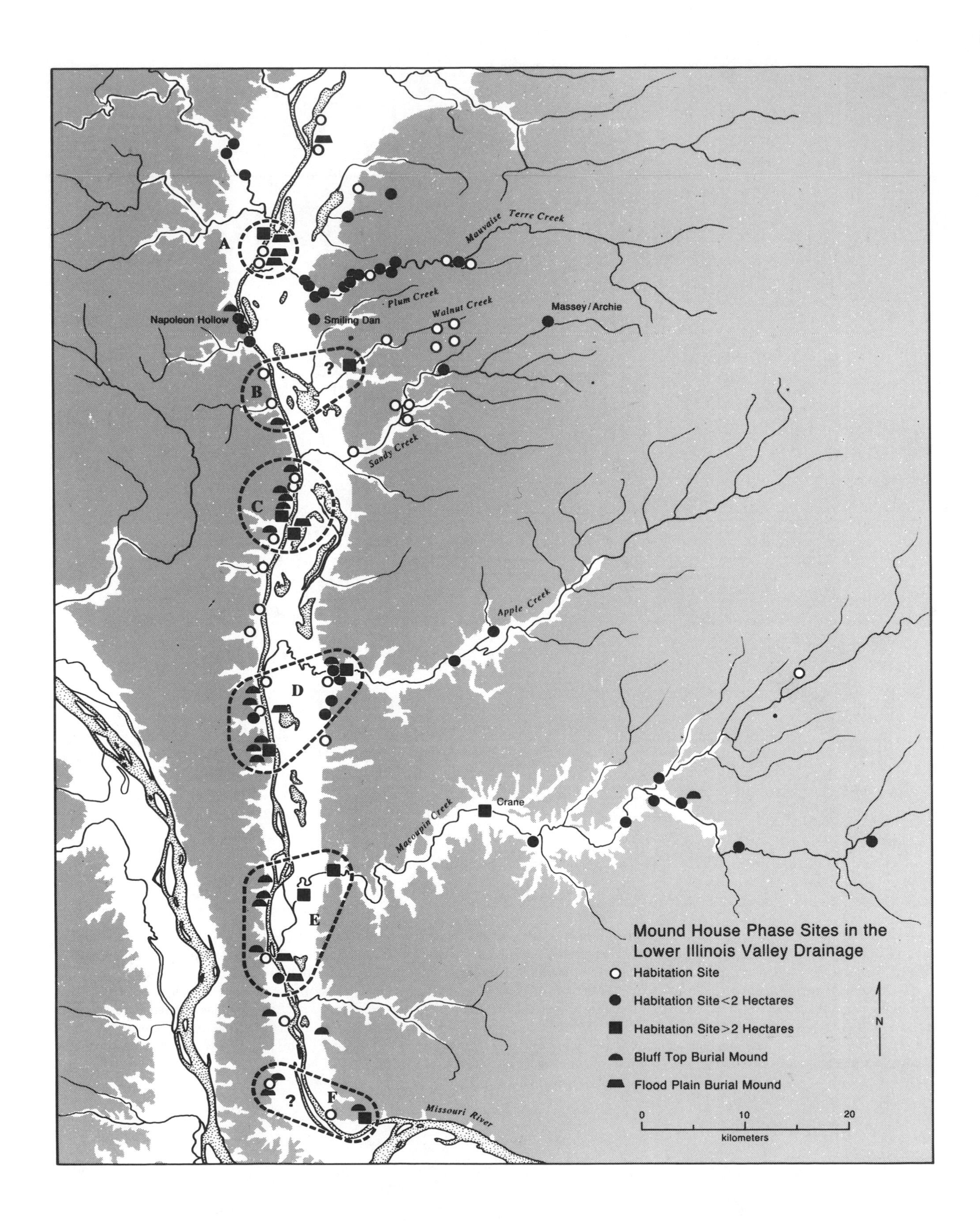
Mound House Phase Sites in the Lower Illinois Valley Drainage
Habitation Site
Habitation Site<2 Hectares
Habitation Site>2 Hectares
Bluff Top Burial Mound
Flood Plain Burial Mound
0
10
20
kilometers
N
A
B
C
D
E
F
?
Mauvaise Terre Creek
Plum Creek
Walnut Creek
Massey/Archie
Napoleon Hollow
Smiling Dan
Sandy Creek
Apple Creek
Macoupin Creek
Crane
Missouri River

and the large Cochran site "village" adjacent to the Pinson site, like that at Bynum, clearly belongs within the corporate rather than the domestic sphere. While village-sized settlements have not been identified in association with the Pinson site, surveys of the surrounding drainage systems have recorded numerous small and as yet unexcavated Middle Woodland sites in stream terrace and valley edge settings, suggesting a pattern of dispersed single household farming settlements similar to that documented in the Upper Duck River drainage.

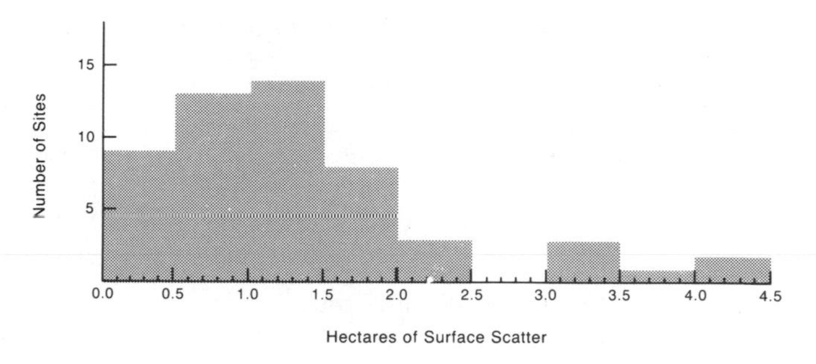

Figure 9.22 The projected size of Mound House phase settlements in the Lower Illinois Valley drainage area, based on area of surface scatter (information from Asch 1976; Asch and Asch 1985).

The Lower Illinois River Valley

For over two decades the Lower Illinois River Valley has been the focus of a long-term regional research program having a particular emphasis on Middle Woodland Havana-Hopewell (Mound House phase) habitation and mortuary mound sites (Figure 9.21).

Hopewellian mortuary programs in the Lower Illinois Valley centered on burial crypts rather than charnel houses (Brown 1979), with the earthen mounds covering these mortuary structures placed, with a few notable exceptions, in bluff top locations along the western edge of the valley (Buikstra 1976). The most obvious exception to the bluff top placement of burial crypts involves the large keel or loaf-shaped burial mounds situated in eight separate locations on the valley floor in association with what have been characterized as large permanent village settlements.

Of the more than 80 Mound House phase sites identified to date, size estimates based on area of surface scatter are available for 54, and of these 44 are less than 2 ha in size (Figure 9.22). These small habitation sites, located both along the main valley and within the four major and numerous minor tributary streams of the Lower Illinois River, are quite probably small one to three household Hopewellian farming settlements.

With the exception of the Crane site, which is centrally located in the largest tributary valley, all of the large habitation sites (> 2 ha), frequently characterized as "permanent village settlements," are situated close to the main valley, usually in association with one of the floodplain burial mound groups (Figure 9.21). At least four spatially distinct floodplain mortuary mound-permanent village concentrations or "complexes" have been recognized, spaced 15–20 km apart and associated with four large tributary streams entering the main valley from the east (Struever 1968; Struever and Houart 1972; Asch et al. 1979; Table 9.4, Figure 9.21a, c, d, e). In addition to these four spatially distinct "complexes," two other apparent areas of mortuary-habitation concentration can be proposed (Figure 9.21b, f), filling gaps in the 15–20 km spacing of Lower Illinois River Valley Hopewellian cultural units. Although relatively little research has been carried out to date regarding the degree to which this pattern of spatial separation may reflect cultural boundaries within Havana-Hopewell

Figure 9.21 The location of known Mound House phase Havana-Hopewell sites in the Lower Illinois Valley and its tributaries. Based on area of surface scatter, ten habitation sites larger than 2 ha have been identified (Asch 1976:66; Asch and Asch 1985). In addition, eight floodplain (as opposed to bluff top) burial mound groups have been recorded (Buikstra 1976). Seven of the ten greater than 2 ha habitation areas and seven of the eight floodplain mound groups are distributed in four distinct spatial concentrations spaced 15 to 20 km apart and associated with the four large tributary stream systems entering the valley from the east. Two smaller tributaries (Plum Creek and Walnut Creek) break this pattern of spatial concentration of floodplain mound groups and large habitation areas in that they apparently lacked a main valley floodplain mound group. In addition, Macoupin Creek, the largest of the tributaries, supported the mid-valley Crane settlement, as well as an upstream burial mound group.

Table 9.4. Floodplain Burial Mound Groups and > 2 ha Habitation Areas Associated with Tributary Streams in the Lower Illinois Valley

Tributary stream	Floodplain mound	Habitation area
A Mouth of Mauvaise Terre Creek	Naples Abbot Naples Chambers Robertson Naples Castle	Naples Abbot 3.5 ha
B Mouth of Plum and Walnut Creeks	?	Plum Creek 3.1 ha
C Mouth of Sandy Creek	Mound House	Mound House 4.1 ha Springer 3.4 ha
D Mouth of Apple Creek	Kamp	Audrey 4.1 ha Gardens of Kampsville 4.0 ha
E Mouth of Macoupin Creek	Merrigan Peisker	Titus 2.3 ha Macoupin 2.4 ha
F Mouth of the Illinois River	?	Duncan Farm 3.1 ha

society, recent ceramic studies (Morgan 1986:424) suggest that there may be discernible differences between material culture assemblages. Similarly, Buikstra (1976) has documented biological variability between populations along the Lower Illinois Valley corridor.

In addition to the information regarding the estimated size and spatial distribution of Mound House phase farming settlements recovered through surface surveys, a number of habitation sites have been excavated. The construction of the Central Illinois Expressway (FAP 408) recently resulted in the excavation of four sites of particular importance in gaining a better understanding of Hopewellian farming settlements in the region.

Forming both an environmental and cultural gradient, the four FAP 408 sites in question are respectively located along a small tributary of Sandy Creek 25 km east of the main valley (Archie and Massey), in a tributary valley only 700 m from the main valley (Smiling Dan), and on the floodplain of the Illinois River directly below a bluff top burial mound group (Napoleon Hollow; Figure 9.21). In addition to being located in different riverine settings and at different spatial and social distances from a corporate mortuary center, these sites also differ considerably in estimated size (area of surface scatter) prior to excavation. Massey and Archie were both quite small, about .2 ha in size, while Smiling Dan was thought to cover about .7 ha, and Napoleon Hollow, at 15 ha, was considered to be one of the largest Middle Woodland sites in the main valley.

MASSEY AND ARCHIE. Because of its upland location 25 km east of the Illinois River Valley, the Massey site was initially thought to be a short-term hunting camp. A block excavation unit, however, exposed an oval (rectangular?) single wall post structure and associated food storage and processing pits (Figure 9.23), leading to its identification as the year-round rural farmstead settlement of a largely independent extended family group (Farnsworth and Koski 1985). An arc of four storage pits excavated at the largely destroyed Archie site, situated in a bluff top setting less than a kilometer from Massey, suggests that it too may have been an isolated year-round farming settlement occupied by a single Hopewellian household.

Although the household units occupying Massey and Archie are not thought to have been that closely linked to main valley groups, based on the composition of the recovered ceramic assemblages, these two sites are likely very similar in size and complexity to the numerous small (< 2 ha) settlements located along the tributary streams of the Lower Illinois River Valley.

SMILING DAN. Located in Campbell Hollow, less than 700 m east of the floodplain of the Lower Illinois River and adjacent to the Hog Bluff

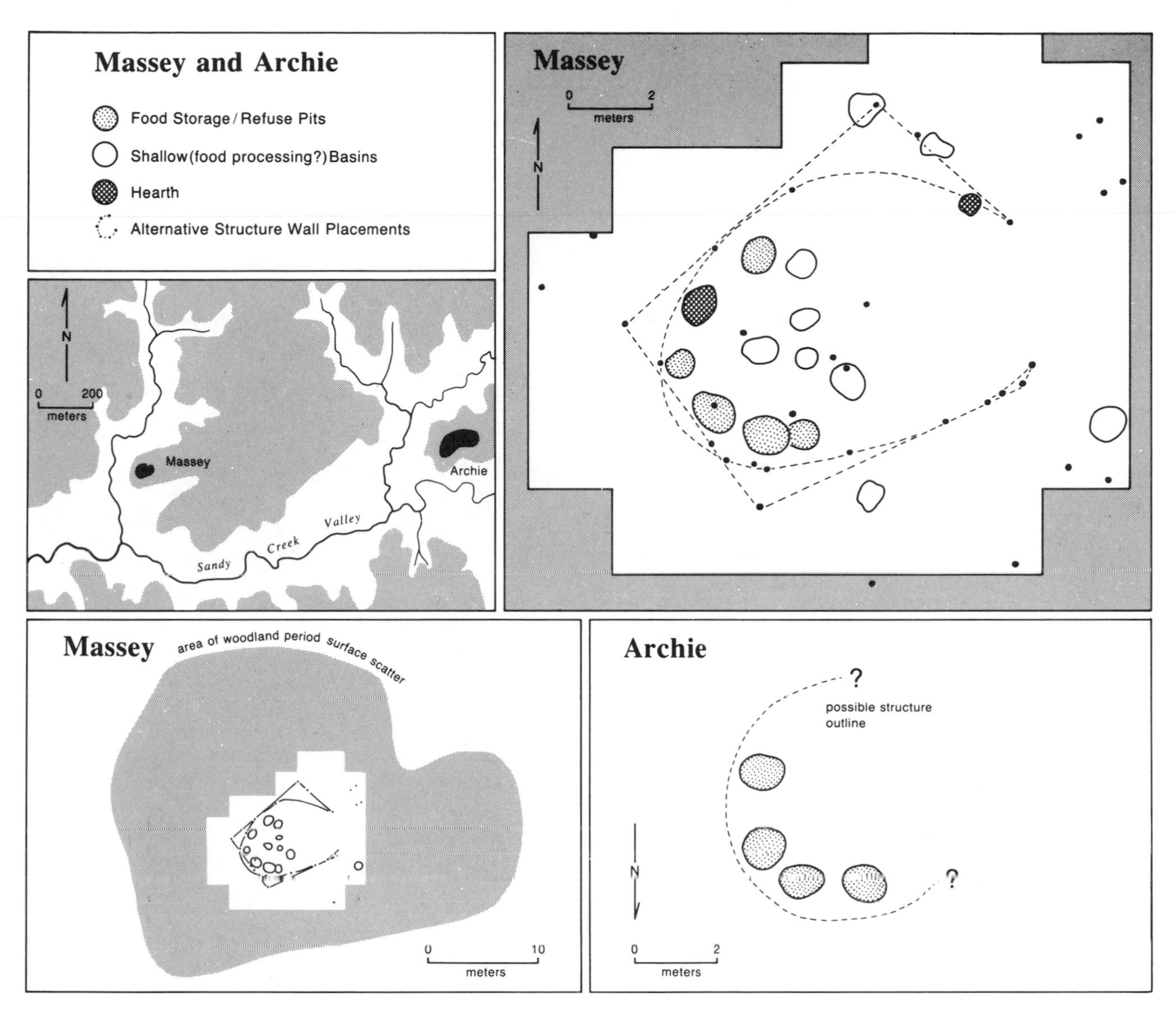

Figure 9.23 The settlement plan of the Massey phase components at the Massey and Archie sites. The two sites are situated less than a kilometer apart in bluff top settings where intermittent tributary stream valleys join Sandy Creek, serving to widen the floodplain. The Archie site was discovered only after highway construction had destroyed most subsurface features, and only four deep storage pits were recorded. At the nearby Massey site, located prior to highway construction, a block excavation unit centered on a .2 ha area of surface scatter of Middle Woodland material uncovered a single structure containing two hearths and a series of five deep storage pits, along with a series of five shallow food processing(?) basins. The similar grouping of deep storage pits at the Archie site may reflect the presence of a similar structure (Farnsworth and Koski 1985).

site, a short-term encampment, the Smiling Dan site contained a number of different settlement components of the Hopewellian household unit (Figure 9.24; Stafford and Sant 1985). A general midden deposit 50–60 cm thick extended over an area of .5 ha, reaching a depth of 2 m where a small tributary channel of Campbell Hollow Creek that bisected the site had served as a trash dump. Removal of the midden exposed three and possibly four subrectangular single wall post structures spaced from 15 to 30 m apart, along with a total of 182 pit features distributed in at least four spatial concentrations (Figure 9.24; Stafford and Sant 1985). While the depth and size of the midden, combined with the abundance of pit fea-

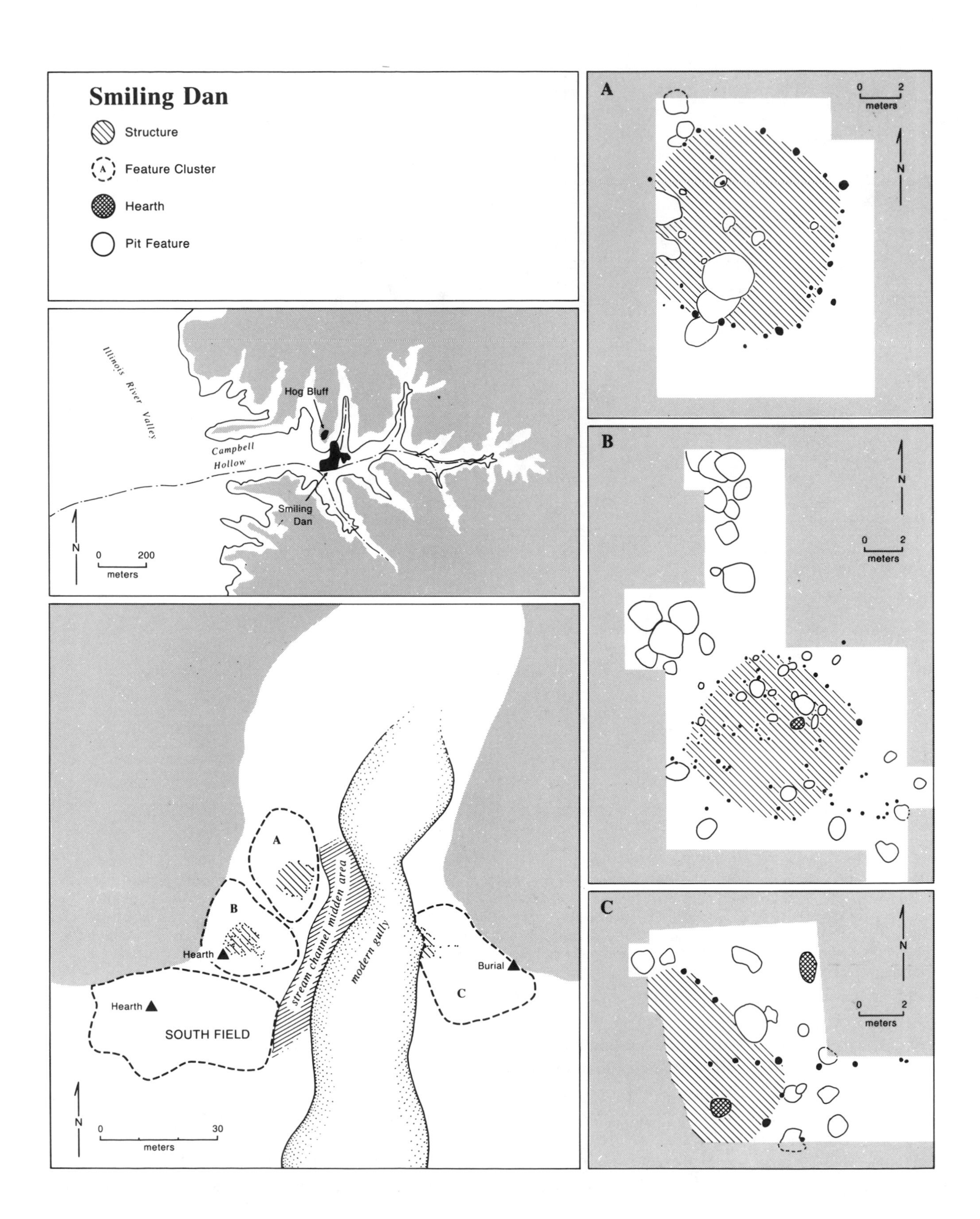
Smiling Dan
Structure
Feature Cluster
Hearth
Pit Feature
Illinois River Valley
Hog Bluff
Campbell Hollow
Smiling Dan
N
0 200
meters
A
B
C
Hearth
Burial
SOUTH FIELD
stream channel midden area
modern gully
N
0 30
meters
0 2
meters
N

tures and apparent temporal range of the recovered ceramic assemblage suggests a fairly long span of habitation at Smiling Dan, it is difficult to establish which, if any, of the structures were contemporaneously occupied. While not overlapping spatially, the three to four structure-pit feature cluster complexes do not exhibit any indications of social or spatial integration above the basic household unit level. It is quite possible that the Smiling Dan site represents a sequence of single household unit farming settlements (with the sequence of structure occupation being C, A, B; Stafford and Sant 1985). At most this Hopewellian farming settlement would have been occupied by three to four households, perhaps 30 individuals, leading Farnsworth (1985:vii) to conclude:

> The Smiling Dan study suggests that many sites considered to be Middle Woodland base-camp villages in the Lower Illinois Valley may actually have housed very small resident populations—perhaps no more than a single extended family.

Investigation of the 15 ha Napoleon Hollow site, situated on the Illinois River floodplain 9 km west of Smiling Dan (Figure 9.21), served to further underscore the dangers inherent in characterizing Hopewellian settlements as "villages" on the basis of area of surface scatter.

NAPOLEON HOLLOW. Situated at the mouth of Napoleon Hollow, where Napoleon Creek enters the floodplain of the Lower Illinois River, the Napoleon Hollow site is just below the Elizabeth Mounds bluff top burial mound group (Figure 9.25; Wiant and McGimsey 1986). Buried by up to 1.5 m of colluvial deposits, the Mound House phase Havana-Hopewell component at Napoleon Hollow was estimated to cover 15 ha. Because of its large size, main valley tributary mouth location, and its close proximity to a mortuary mound complex, the Napoleon Hollow site fit closely the profile of a village level Hopewellian farming settlement.

In contrast to other large sites in the main valley, however, the Napoleon Hollow site yielded limited surface material, and its "village" status was quickly questioned. Extensive testing of the site indicated that occupational evidence, while extensive, was light except in two locations. Midway up the bluff toward the Elizabeth Mounds mortuary precinct, the Block I excavation unit exposed a small 15–20-cm-thick refuse midden that likely resulted from bluff top mortuary program activities. The Block IV excavation unit, placed on the crest of a former natural levee of the Lower Illinois River, exposed, at a depth of 1.5 m, a circular wall post structure 8 m in diameter, along with 12 associated hearths and pit features (Figure 9.25). Although 60 test units, 11 backhoe trenches, and over 100 cores were placed in other areas across the site, only four other pit features were discovered outside of the Block IV excavation unit. This absence of pit features at Napoleon Hollow, particularly storage pits and earth ovens, along with a low diversity of subsistence remains and tool types and a high proportion of exotic raw materials and artifacts, indicated that the single structure uncovered at the site, like the bluff side midden, was closely related to ritual activities of the bluff top mortuary mound group. Rather than being a large permanent village, or even the isolated settlement of a single household unit, Napoleon Hollow, once excavated, was found to clearly fall within the corporate-ceremonial sphere of Hopewellian society.

While Napoleon Hollow yielded few indications of domestic sphere activities, limited excavations at a number of other large multicomponent sites in the Lower Illinois River Valley with Mound House phase components, have provided, in contrast, substantial evidence of habitation by Middle Woodland and subsequent Late Woodland groups. The Apple Creek site (Figure 9.21) provides a good example of the diffi-

Figure 9.24 The Mound House phase settlement at the Smiling Dan site. A linear arrangement of five rock-lined post molds similar to those identified for structures A, B, and C was uncovered in the South Field area of the site, 20–30 m west of a fourth feature cluster, and may represent a fourth structure. The bluff top Hog Bluff site, consisting of a single pit feature and a small cultural assemblage, has been interpreted as a short-term encampment (Stafford and Sant 1985).

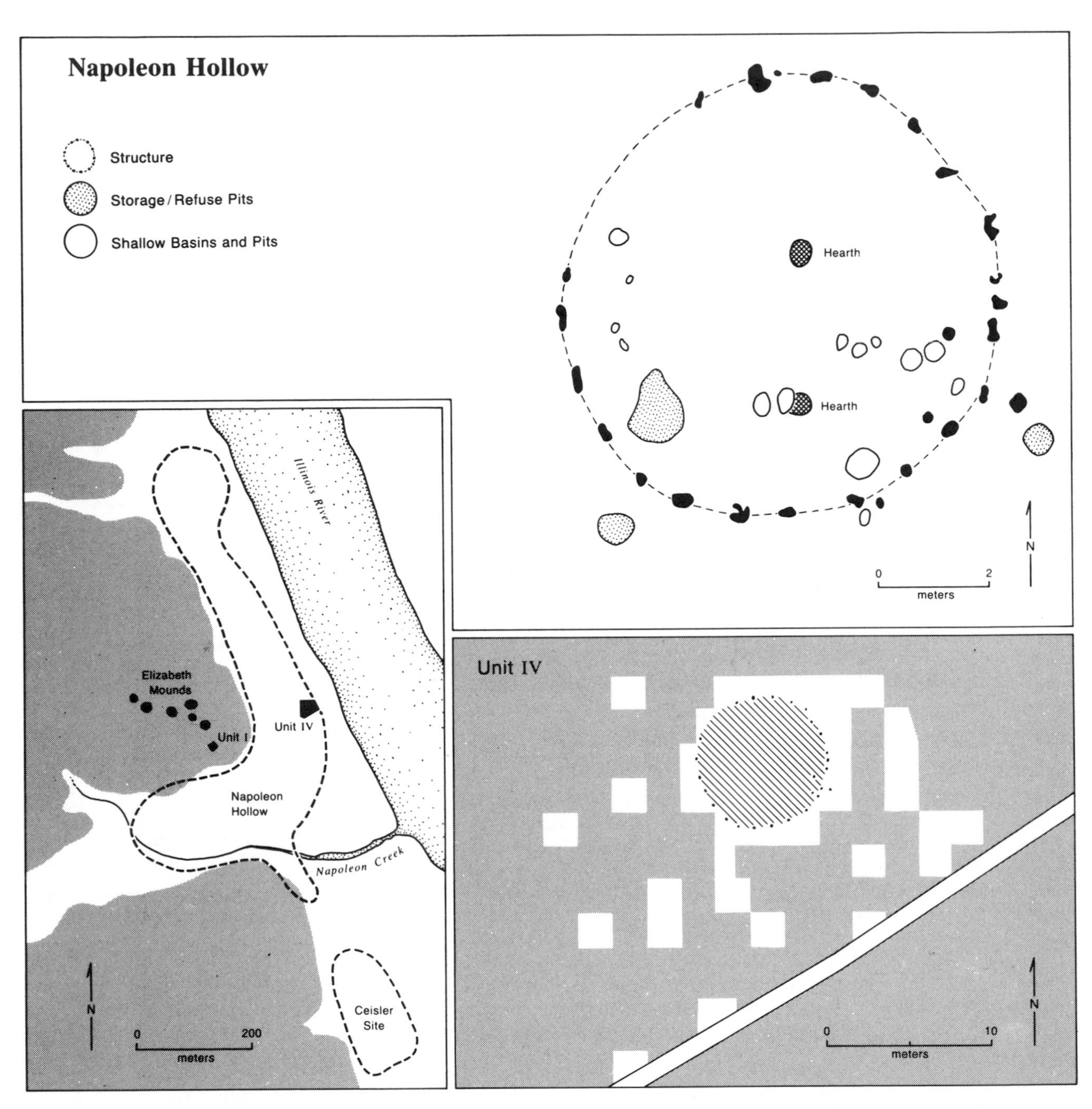

Figure 9.25 The settlement plan of the Mound House phase component at the Napoleon Hollow site. Situated midway up the bluff face, the Block I excavation unit uncovered a small, discrete midden dump apparently associated with the bluff top Elizabeth Mound group. The Block VI excavation in the floodplain exposed a circular wall post structure, associated pit features, and a riverbank refuse dump (Wiant and McGimsey 1986).

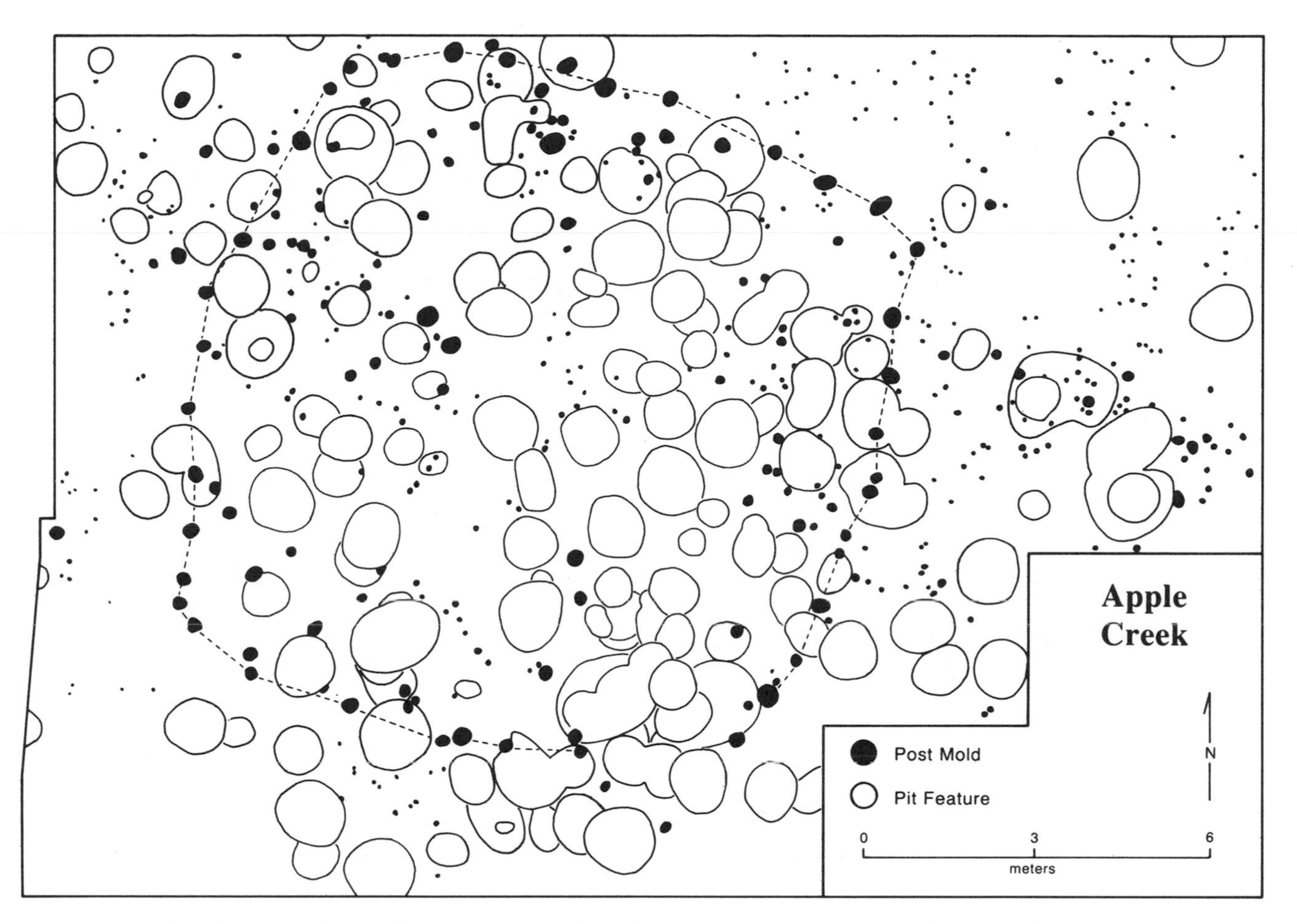

Figure 9.26 Middle and Late Woodland pit features and a possible Middle Woodland structure excavated at the 5.6 ha multicomponent Apple Creek site in the Lower Illinois River Valley (after Marshall 1969).

culties often inherent in investigating the large (> 2 ha) Hopewell "villages" in this region.

APPLE CREEK. Although a site report has yet to be published, an analysis of the Apple Creek faunal assemblage (Parmalee et al. 1972), and an engineering study of a possible Hopewell house structure (Marshall 1969) provide some information regarding the site (Figure 9.26). Reflecting apparent intensive occupation over perhaps a five-century span, this 1.4 ha habitation site exhibits considerable evidence of disturbance of Hopewellian features by subsequent Late Woodland pit digging activities, with a resultant mixing of cultural components. Of the 143 pit features shown in Marshall's illustration, for example, more than 80 overlapped other pit features (Marshall 1969, fig. 1). Similarly, Parmalee (Parmalee et al. 1972:57) describes "a large quantity of superimposed structures and a general mixing of cultural debris. Features and other excavation units containing materials from a pure or single cultural component were few." While such block excavations often provide only limited information regarding the spatial extent or organizational complexity of Hopewellian settlements, they do indicate the problems involved in sorting out the various overlapping Middle and Late Woodland occupational episodes comprising such multicomponent sites of the Lower Illinois River Valley.

In summary, recent excavation of four sites (Massey, Archie, Smiling Dan, Napoleon Hollow) document both the presence of discrete single household unit farming settlements as a basic component of Mound House phase Hopewellian communities, and the existence of small specialized corporate-ceremonial sphere sites in proximity to mortuary mound precincts. When used to interpret existing surface survey information (Figure 9.21), the Smiling Dan and Massey sites strongly suggest that many of the small (< 2 ha) Mound House phase sites located in tributary stream valleys and within the Lower Illinois River Valley were likely one to perhaps three household unit settlements. Similarly, the Napoleon Hollow site indicates that a possible specialized corporate-ceremonial sphere role should be considered for those sites situated in close proximity to burial mound groups. Such sites adjacent to mortuary mound corporate-ceremonial sphere precincts, particularly those associated with the floodplain mound complexes, might be more substantial than Napoleon Hollow, and could incorporate larger corporate structures, perhaps exhibiting some spatial organization similar to that seen at Seip and Bynum. Alternatively, they might exhibit a fairly dispersed pattern of smaller corporate structures such as that observed at Pinson.

Similarly, the main channel domestic sphere habitation sites with areas of surface scatter greater than 2 ha (or even in the 1–2 ha range) might on the one hand turn out to be Hopewellian farming settlements of more than three households, with some greater degree of social and spatial integration than previously documented. Concentrated within distinct river valley segments of apparent higher settlement and population density and centered on a floodplain mortuary mound group (Figure 9.21), these > 2 ha sites certainly hold the greatest potential for illuminating the full range of variation in Mound House phase domestic sphere settlements.

Situated at the mouth of tributary stream valleys or adjacent to protein rich floodplain lakes and marshes (Asch et al. 1979), and often disturbed and obscured by subsequent Late Woodland settlements, these main valley Hopewellian farming "villages" on the other hand may sometimes turn out, upon excavation, to consist of a relatively diffuse spatial agglomeration of household units showing few indications of integration or organization. Like the Smiling Dan site, the household units comprising such sites may be difficult to link together in a single contemporaneous settlement, and might represent a temporal sequence of staggered, overlapping, and uncoordinated occupational episodes by individual households. If this is in fact the case, and these sites, when excavated, appear to lack an overall community plan, or any indications of spatial or social integration above the household level, they could perhaps best be viewed as a concentration or spatial compression of the tributary stream settlement pattern of dispersed single households that was facilitated by rich localized resource zones.

The American Bottom

Until a large-scale archaeological research program was undertaken in association with the construction of a federal interstate expressway (I-270) through the American Bottom (Bareis and Porter 1984), almost nothing was known regarding Middle Woodland Hopewellian occupation of the region. As a result of the I-270 project, two Hopewellian habitation settlements have now been excavated, and more than 100 other Middle Woodland sites have been located within this broad expansion of the Central Mississippi River Valley, providing an interesting case study comparison for the Duck and Lower Illinois River Valleys.

Interestingly, many of the Middle Woodland sites located to date in the American Bottom occur within 2 km of the eastern bluff line of the valley edge, with many of these concentrated in those areas where shallow oxbow lakes and marshes extend eastward close to the eastern bluff line. In these areas along the eastern edge of the American Bottom, where shallow lake and marsh areas approach the bluff line from the west, and tributary streams enter from the east, forming alluvial fans, the general habitat configuration resembles that of the Lower Illinois Valley, and to a lesser degree, the Upper Duck River Valley. Similarly, in terms of location, size, and internal organization, the two Hopewellian farming settlements that have been excavated within these eastern edge habitat areas of the American Bottom also share certain similarities with those of the Lower Illinois and Upper Duck River Valleys.

TRUCK 7. In the Hill Lake Meander locality of the American Bottom, Hill Lake, a former channel of the Mississippi River, comes within 500 m of the valley edge bluff line, paralleling it for over 3 km (Figure 9.27). The multicomponent Truck 7 site was located on a sand bar deposit within this partially filled river channel, which at the time of occupation formed a shallow lake and marsh habitat (Fortier 1985). The dynamic alluvial fan of Hill Lake Creek, located just to the east of the Truck 7 site, would eventually expand to cover it with more than a foot of silty loess.

Within a 1.8 ha area of surface scatter (including the contiguous Go Kart South site), the Middle Woodland occupation of this multicomponent site was indicated by a .16 ha area of dense ceramic and chert debris. Machine stripping of a .23 ha area within the right of way exposed a total of 25 features comprising a number of settlement components of a Hopewellian household unit. A circular wall post structure containing 10 pit features (5 storage pits) had a small chambered area (a covered entrance?) along its east wall. A feature cluster consisting of four surface hearths and two processing pits was located 10 m east of the structure, while two other processing pit clusters were placed to the south and southwest of the structure (Figure 9.27). The absence of a general midden deposit and the recovery of a fairly limited material culture assemblage (only 10 ceramic vessels were represented), primarily from within the structure, implied a relatively limited temporal span of occupation, and the site was interpreted as representing the cold season (late fall–winter) habitation unit of a limited number of family units (Fortier 1985:275–277). Based on a continuing surface scatter of Middle Woodland materials, the settlement appears to extend 25 m beyond the right of way excavation units to the northwest, and additional structures or feature clusters may have been present in this area. There are no indications, however, that this Hopewellian farming settlement was occupied by more than a single household unit.

HOLDING. Situated 23 km north of the Truck 7 site in the Edelhardt Lake locality, the Holding site occupies a very similar habitat setting. A vast complex of channel remnant lakes and marshes extends away from the Holding site to the north, west, and southwest, while nearby to the east is the alluvial fan deposits of School House Branch Creek (Figure 9.28).

Within this 5.6 ha multicomponent site, a large block excavation uncovered most of a .6 ha Hopewellian midden deposit. Removal of the midden in turn exposed a total of 163 Middle Woodland features, including structures, possible structures, a C-shelter, midden stain areas, and 143 pit features arranged in seven feature clusters (Figure 9.28; Fortier et al. 1989). An additional 74 post molds not clearly associated with structures or shelters were also recorded. Circular post hole patterns defined two definite structures, while three adjacent circular midden stain areas, each with ambiguous post hole patterns, might represent additional structures or C-shelters. Another circular stain area with post molds located in the southeast corner of the excavation marks the location of a fourth possible structure or shelter. An open-end C-shelter and two of the seven pit clusters (Figure 9.28f–g) are located north of the two structures, while the remaining five food storage and processing pit clusters (Figure 9.28a–e) are grouped to the south of the circular houses and their associated midden stains and post hole patterns.

These five feature clusters, along with the two known structures and three midden stain-post hole areas (possible structures) appear to encircle and define a central open area or courtyard containing a large (communal?) storage pit (Fortier et al. 1989). This apparent spatial organization of settlement components, generally comparable to that exhibited at the Ewell III and McFarland sites in the Upper Duck River Valley, suggests at least some minimal social integration above the household level, involving two and perhaps as many as five households. Fortier (Fortier et al. 1989) considers the Holding site to be a growing season (spring–summer) farming hamlet occupied by closely knit family units, with family work spaces organized around a common central courtyard.

In summary, two recently excavated Middle Woodland (A.D. 0–200) sites, both situated in proximity to tributary mouth alluvial fans and shallow lake and marsh habitats along the eastern edge of the

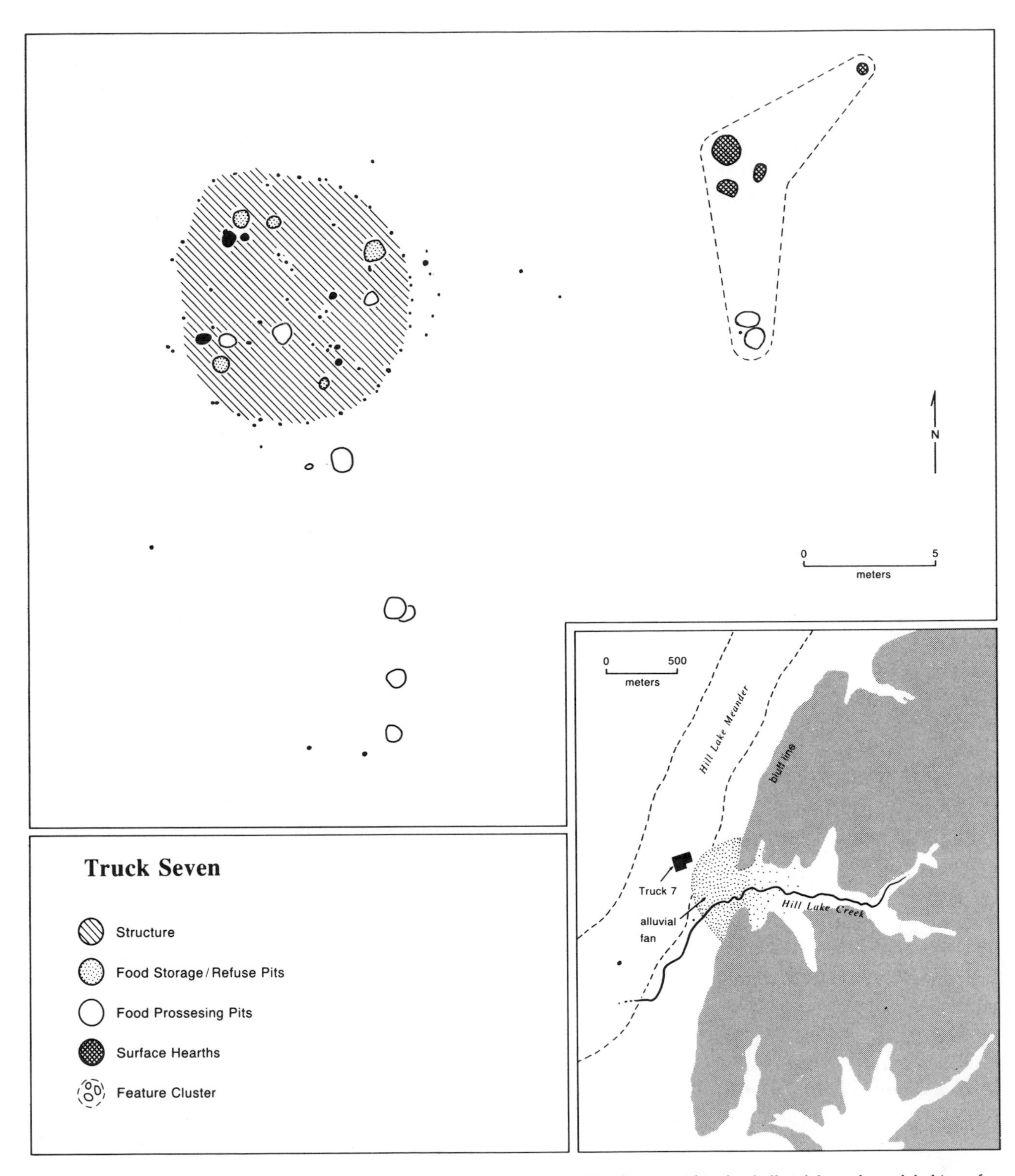

Figure 9.27 The Hill Lake phase settlement at the Truck 7 site. Situated on a sand bar feature within the shallow lake and marsh habitat of the largely filled in Hill Lake Meander, the Truck 7 settlement was also adjacent to the Hill Lake Creek alluvial fan.

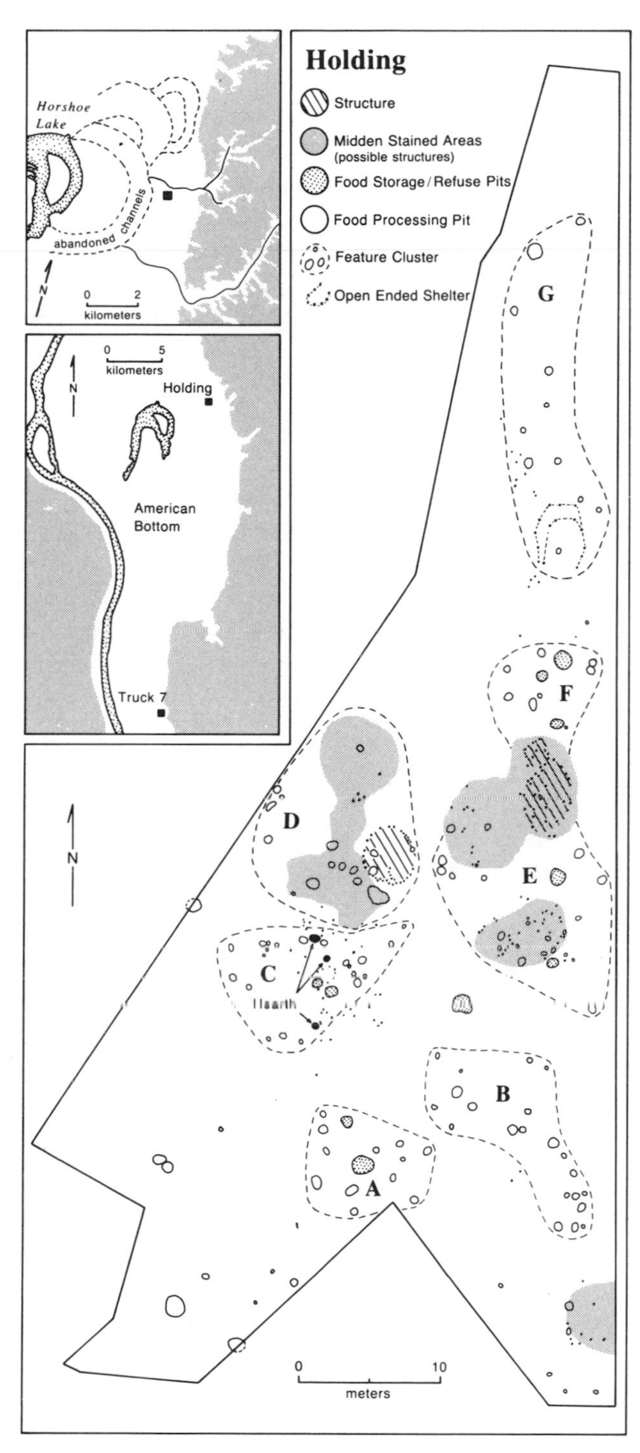

Figure 9.28 The Holding phase settlement plan at the Holding site, located adjacent to oxbow lake and marsh habitats in the American Bottom area of the Central Mississippi Valley.

American Bottom, represent small Hopewellian farming settlements. Truck 7, the smaller of the two, was occupied by a single household unit, while at the Holding site two and perhaps as many as five households were present. Fortier (Fortier et al. 1989) considers these two sites to be representative of a general annual cycle of relocation of Hopewellian social groups linked to the growing season, with single households dispersed through the late fall and winter (Truck 7), and coalescing into small multifamily farming hamlets (Holding) during the spring–summer growing season. Alternatively, the Truck 7 and Holding sites might both have been permanent year-round settlements.

Employing surface collection data regarding as yet unexcavated sites, as well as unpublished information concerning recently excavated ones, Fortier (Fortier et al. 1989) provides a larger Hopewellian community context for the Holding site. With a possible Middle Woodland mound, the McDonough Lake site, located 2 km north of Holding on the south shore of a channel remnant lake, is identified as a possible larger Hopewellian village habitation site and "transaction center," while two nearby bluff top mounds (Fox Hill and Sugar Loaf) may have been part of the mortuary program of Hopewellian groups in the Edelhardt Lake locality. The nearby Nochta site, along with a number of other sites similar in size to Holding (Esterlein, Meridian Hills, Creve Coeur) are proposed as likely farming hamlets, while a number of sites that may be short-term limited activity "extractive camps" are also mentioned (for example, Willoughby and Widman) (Fortier et al. 1989:568).

Conclusions

By A.D. 0–200 and perhaps earlier, before the introduction of any tropical food crops from Mesoamerica or the Southwest, and a full eight centuries prior to the emergence of maize-dominated agricultural systems, food production economies based on at least seven different indigenous domesticated and cultivated seed plants were well established in a number of different regions of the prehistoric Eastern Woodlands of North America. During the past ten to twenty years long-term regionally focused archaeological research programs carried out within drainage systems of different scale located in three different regions (Upper Duck River, Lower Illinois River, American Bottom) have resulted in the excavation of more than a dozen habitation settlements of these first farmers of the prehistoric East. Offering the first detailed documentation of the size and internal organization of the initial farming settlements of A.D. 0–200, these excavated sites, when combined with surface survey information from the three regions discussed, provide the first good look within the domestic sphere of Hopewellian society. These three regional data sets in fact constitute almost the complete currently available data base regarding Hopewellian farming settlements of the critical two-century span being considered.

Information regarding habitation settlements of this time period is certainly available from other regions of the East, and descriptions of structures or other settlement components are not entirely absent in the literature, but with a few notable exceptions, to be discussed below, none of the partial and fragmentary descriptions of Hopewellian farming settlements of other regions either conflicts with or adds much to what has been briefly summarized so far in this chapter. A brief consideration of the limited settlement information available for three additional river valleys in the Eastern Woodlands will help to underscore this point.

Based on available surface collection information, the Marksville Hopewellian settlements of the Lower Mississippi Valley appear to have been less than 1 ha in size, and located along the natural levees bordering active streams and floodplain lakes and marshes (Toth 1979:197).

Similarly, Copena Hopewellian settlements along the Middle Tennessee River in northern Alabama are described as small "villages" located on bottomland ridges, often near the valley edge. Located 180 m from a Copena burial mound group on a ridge near the valley edge, and having a .6 ha midden up to 30 cm in thickness, the Wright site is the only excavated Copena habitation site reported in the literature (Walthall 1980). A 190 sq m block-excavation unit placed in the center of the midden exposed a single circular wall post structure 3.6 m in diameter, along with a total of 66 pit features.

Within the Hopewellian habitation area situated at a distance of 200 m from the Tunacunnhee mound group in northwest Georgia, a 200 sq m block-excavation unit exposed two circular wall post structures (Figure 9.8) and associated pit features (Jefferies 1976).

Within the Little Tennessee River Valley of east Tennessee block excavation units at the Icehouse Bottom site exposed a Middle Woodland midden, pit features, and post holes, suggesting the likely presence of a small single household Hopewellian farming settlement (Chapman 1985).

Like the Duck, Illinois, and Central Mississippi Valley, the Lower Mississippi, Middle Tennessee, and Little Tennessee River Valley, along with numerous other, rather poorly known areas in the East, provide very little information indicating the presence of Hopewellian farming settlements that included more than three household units. The consistent, redundant pattern in the archaeological record is of spatially discrete and dispersed single household settlements in valley edge tributary mouth settings or adjacent to floodplain lakes and marshes. And in those documented settlements that might have involved two or three household units (Ewell III, McFarland, Smiling Dan, Holding), there is no dramatic evidence, other than simple spatial proximity, to suggest a very strong organization or integration of the household units that were present. Almost all of the available evidence regarding Hopewellian farming settlements seems to indicate a largely independent and autonomous status for individual household units, and their consistent dispersal across the Middle Woodland landscape of the prehistoric East.

This consistent pattern of small one to three household Hopewellian farming settlements clearly calls into question the concept of Hopewellian "villages." Never specifically defined or described in terms of site size, number of structures or inhabitants, range of activities, or settlement plan, the term "village" was initially used in the last century to differentiate habitation sites from earthworks and mound groups. Since this initial recognition and demarcation of the corporate and domestic spheres, any sizable surface scatter or apparent midden yielding Middle Woodland cultural materials has had a good chance of being called a "village." But where are these "villages"? The one to three household settlements described above certainly do not qualify for "village" status. Upon excavation, the "villages" associated with corporate-ceremonial centers at Seip, McFarland, Bynum, Pinson, and Napoleon Hollow turned out to either be small habitation sites (McFarland) or to clearly fall within the corporate-ceremonial sphere.

The evidence for permanent Hopewellian farming settlements comprised of more than three households is admittedly meager at present, but rather than leading to the conclusion that they either did not exist, or were an infrequent occurrence, this absence or information will, it is hoped, serve to focus research interest and attention in their direction. Rather than simply applying or removing a "village" label from the large multicomponent sites in the Lower Illinois Valley and elsewhere that seem to hold the best promise of filling the empty category of "larger than three household" Hopewell farming settlement, these complex sites should be investigated with the specific problem orientation of identifying the composition and internal organization of individual communities (Kelly et al. 1987). In addition to simply documenting the number of structures and other settlement components present, the spatial organization of the household units comprising such settlements, and the inherent indication of supra-household social integration, would be of particular importance.

In terms of internal organization two excavated Middle Woodland settlements located over 1,000 km apart provide alternative examples of what such Hopewellian farming settlements may look like.

Excavation in 1969 of the Industrial Park site, located on a 1.2 ha terrace of the Chattahoochee River in southwest Atlanta, Georgia (Figure 9.11), exposed "some 30 randomly distributed round to broad oval" double wall post structures (Kelly 1973). The absence of a published site report makes it impossible to assess how many of these structures may have been occupied contemporaneously, as opposed to representing individual, temporally discrete occupational episodes. This loose and unorganized spatial agglomeration of structures, however, might perhaps indicate that at some point during its span of occupation the Industrial Park site was occupied by more than three household units.

In distinct contrast to the random and somewhat dispersed distribution of structures at the Industrial Park site, excavation of the Millville site in southwest Wisconsin (Figure 9.11), exposed a compact and organized settlement plan (Figure 9.29). The .6 ha site was located in a terrace edge setting at the mouth of Dutch Hollow, where a tributary stream enters the floodplain of the Wisconsin River. Almost total excavation of the partially eroded site exposed a compact settlement consisting of 14 separate house structures and 176 associated features, including 4 burials, 139 storage pits, 40 hearths/firepits (only those within structures are shown), and three earth ovens. Six of the 14 structures had additional structural elements and with the exception of structure 12, all contained a shallow house basin midden fill zone. The numerous additional recorded posts often appear to form connecting walls between structures, suggesting a possible compound configuration around a small central courtyard (Freeman 1969; Stoltman 1979). While the absence of archaeobotanical assemblages from Millville or related sites in southwest Wisconsin makes it impossible to determine whether or not farming economies had been established this far north by the Middle Woodland period, the site does provide an indication of what more compact and organized "large" Hopewellian farming settlements further south may have looked like.

Returning to the south-central Ohio "heartland" of Hopewell, what can be said regarding the likely size, complexity, and spatial distribution of Ohio Hopewell farming settlements? In the two decades that have passed since Prufer (1964, 1965) suggested that Ohio Hopewell populations were distributed across the landscape in small settlements rather than large villages, a number of small Ohio Hopewell habitation sites have been documented that fit his settlement pattern model and the profile of a discrete single family household settlement quite well. Seeman (personal communication, 1985) and Prufer (1975:311–316) both note the occurrence of small Hopewell habitation site surface scatters along the

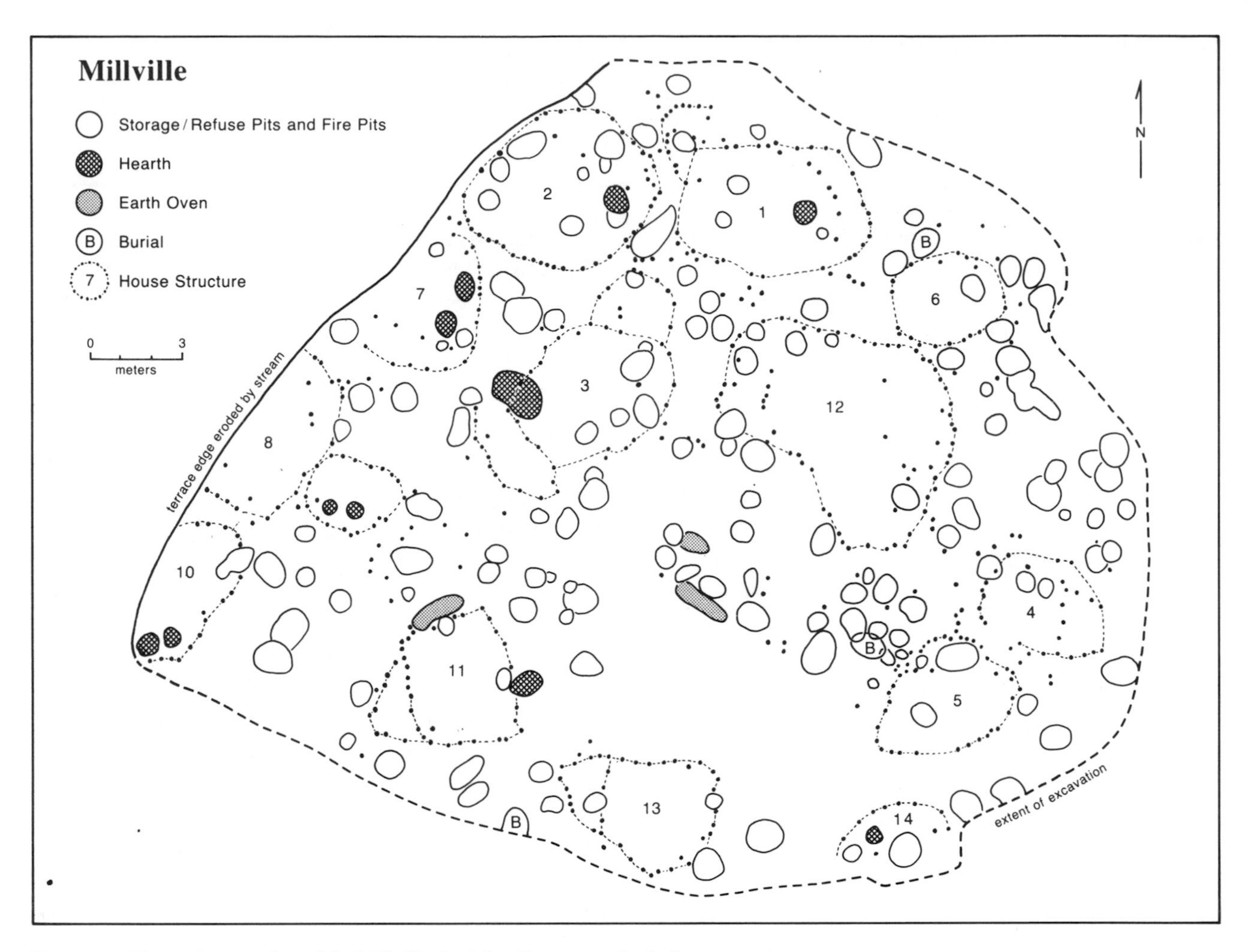

Figure 9.29 The settlement plan of the Millville site (after Freeman 1969; Stoltman 1979).

Scioto River and its tributaries, and Prufer (1965) has excavated a portion of the midden or overslope dump of one such site (McGraw) in the lower Scioto drainage. In addition, several similar small habitation sites have been at least partially excavated in other regions of Ohio (Twin Mounds, Fischer 1971; Converse 1984).

Prufer (1975:316) also suggests that loose concentrations of these small Hopewellian habitation sites can be found in the vicinity of corporate earthwork centers, and Shane (1970:145) describes a total of 21 areas of surface scatter, each from 15 to 30 m in diameter, occurring within a 62 ha floodplain segment of the lower Scioto River Valley in the vicinity of the High Banks Hopewell earthworks. As Brose has noted (1970:146), it is again difficult to establish the degree to which such apparent spatial concentrations of small Ohio Hopewell habitation sites as that described by Shane represent a loosely concentrated "large community" as opposed to a series of temporally discrete occupational episodes by single family households in an attractive main valley habitat setting.

While the published information available regarding habitation sites of Hopewell populations in Ohio is admittedly meager, it does suggest a Middle Woodland settlement pattern that is generally similar to that outlined above for other regions of eastern North America. Small farming settlements of one to three households were likely distributed along the valleys of smaller tributary streams, while loose spatial concentrations of such settlements were located in the richer habitat zones of main valley settings, in the general proximity of corporate-ceremonial centers. As is the case in the Lower Illinois Valley area, it will perhaps be quite difficult to establish either the degree of contemporaneity of occupation, or the level of social and spatial integration that existed for the individual household units that were present in these main valley habitation zones.

Agricultural communities consisting of individual single family farmsteads scattered along river valley segments in a temporally shifting mosaic of relatively short-term occupational episodes are well documented for the late prehistoric and early historic periods (Smith 1978). Such a community pattern of small, dispersed, river valley farming settlements may well be a reasonable analog for Hopewellian farming societies in general, and Ohio Hopewell groups in particular.

Thus even within the south-central Ohio "heartland" of Hopewell, there is a marked, and to some an unsettling, incongruity between the corporate-ceremonial and domestic spheres of Hopewellian society. The almost total absence of evidence from habitation sites for spatial organization or social integration of Hopewellian society above the household level, when combined with the lack of any evidence within the domestic sphere for differential status or social ranking, stands in stark contrast with information recovered from corporate-ceremonial contexts. While Hopewellian farming settlements were small, dispersed, apparently independent, and redundantly egalitarian in appearance, excavations in Hopewellian corporate-ceremonial precincts has often illuminated rich, elaborately expressed, and regionally quite variable programs of social ranking (Braun 1979; Greber 1979, 1983), along with considerable evidence of strong community-scale social integration in the form of corporate labor efforts (corporate-ceremonial structures, mounds, geometric earthworks).

Rather than representing an inherent contradiction that requires resolution, the quite different perspectives offered by a balanced consideration of the corporate and domestic spheres of Hopewellian society hold the promise of a better understanding of these early farmers of the prehistoric East. To achieve such a balanced consideration of Hopewell society, however, main channel Hopewellian habitation zones in a number of different regions require extensive excavation in order to establish the level of social and spatial organization inherent in these poorly known settlements.

Acknowledgments

A number of individuals read and commented on early drafts of this chapter, including David Brose, C. Wesley Cowan, Charles Faulkner, Andrew Fortier, Gayle J. Fritz, James B. Griffin, and Richard Yarnell. Their corrections and suggestions are gratefully appreciated.

Literature Cited

Asch, D.

1976 *The Middle Woodland Population of the Lower Illinois Valley: A Study in Paleodemographic Methods*. Northwestern University Archeological Program, Scientific Papers 1. Evanston, Illinois.

Asch, D., and N. Asch

1978 The Economic Potential of *Iva annua* and its Prehistoric Importance in the Lower Illinois Valley. In *The Nature and Status of Ethnobotany*, edited by R. I. Ford, pp. 301–341. Museum of Anthropology, Anthropological Paper No. 67. University of Michigan, Ann Arbor.

1982 A Chronology for the Development of Prehistoric Horticulture in West-Central Illinois. Paper presented at the 47th annual meeting of the Society for American Archaeology, Minneapolis.

1985 Archaeobotany. In *Smiling Dan,* edited by B. D. Stafford and M. B. Sant, pp. 327–399. Kampsville Archaeological Center Research.

Asch, D. L., K. B. Farnsworth, and N. B. Asch

1979 Woodland Subsistence and Settlement in West-Central Illinois. In *Hopewell Archaeology,* edited by D. S. Brose and N. Greber, pp. 80–85. Kent State University Press, Kent, Ohio.

Baby, R. S., and S. M. Langlois

1979 Seip Mound State Memorial: Nonmortuary Aspects of Hopewell. In *Hopewell Archaeology,* edited by D. Brose and N. Greber, pp. 16–18. Kent State University Press, Kent, Ohio.

Bacon, W. S., and H. L. Merryman
1973 *Salvage Archaeology at 40FR47*. Tennessee Archaeological Society Miscellaneous Paper 11.

Bareis, C., and J. Porter (editors)
1984 *American Bottom Archaeology.* University of Illinois Press, Urbana.

Bender, B.
1985 Prehistoric Developments in the American Midcontinent and in Brittany, Northwest France. In *Prehistoric Hunter-Gatherers the Emergence of Cultural Complexity,* edited by T. D. Price and J. A. Brown, pp. 205–227. Academic Press, Orlando, Florida.

Bender, M. M., D. A. Baerreis, and R. L. Steventon
1981 Further Light on Carbon Isotopes and Hopewell Agriculture. *American Antiquity* 46:346–353.

Braun, D. P.
1979 Illinois Hopewell Burial Practices and Social Organizations: A Re-examination of the Klunk-Gibson Mound Group. In *Hopewell Archaeology,* edited by D. Brose and N. Greber, pp. 66–80. Kent State University Press, Kent, Ohio.

Broida, M. O.
1983 Maize in Kentucky Fort Ancient Diets: An Analysis of Carbon Isotope Ratios in Human Bone. M.A. thesis, Department of Anthropology, University of Kentucky, Lexington.

Brose, D.
1970 Comment in the Discussion to Orrin Shane's Chapter. In *Adena: The Seeking of an Identity,* edited by B. K. Swartz, p. 146. Ball State University, Muncie, Indiana.

Brown, J. A.
1979 Charnel Houses and Mortuary Crypts: Disposal of the Dead in the Middle Woodland Period. In *Hopewell Archaeology,* edited by D. S. Brose and N. Greber, pp. 211–219. Kent State University Press, Kent, Ohio.

Buikstra, J.
1976 *Hopewell in the Lower Illinois Valley.* Northwestern University Archeological Program, Scientific Papers 2. Evanston, Illinois.

Caddell, G. M.
1981 Plant Resources, Archaeological Plant Remains, and Prehistoric Plant-Use Patterns in the Central Tombigbee River Valley. In *Biocultural Studies in the Gainesville Lake Area,* vol. 4, *Archaeological Investigations in the Gainesville Lake Area of the Tennessee-Tombigbee Waterway,* edited by G. Cadell, A. Woodrick, and M. C. Hill, pp. 1–86. Office of Archaeological Research, Report of Investigations 14. University of Alabama. Report submitted to the National Park Service, Southeastern Region.

Chapman, J.
1985 *Tellico Archaeology.* Report of Investigations 43, Department of Anthropology, University of Tennessee, Knoxville.

Chapman, J., and G. D. Crites
1987 Evidence for Early Maize *(Zea mays)* from the Icehouse Bottom Site, Tennessee. *American Antiquity* 52:352–354.

Chapman, J., and A. B. Shea
1981 The Archaeobotanical Record: Early Archaic Period to Contact in the Lower Little Tennessee River Valley. *Tennessee Anthropologist* 6:61–84.

Christenson, A.
1980 Change in the Human Food Niche in Response to Population Growth. In *Modeling Change in Prehistoric Subsistence Economies,* edited by T. Earle and A. Christenson, pp. 31–72. Academic Press, Orlando, Florida.

Cleland, C. E.
1976 The Focal-Diffuse Model: An Evolutionary Perspective on the Prehistoric Cultural Adaptations of the Eastern United States. *Midcontinental Journal of Archaeology* 1:59–76.

Conard, N., D. Asch, N. Asch, D. Elmore, H. Gove, M. Rubin, J. Brown, M. Wiant, K. B. Farnsworth, and T. Cook
1984 Accelerator Radiocarbon Dating of Evidence for Prehistoric Horticulture in Illinois. *Nature* 308:443–446.

Conrad, A. R.
1985 A Preliminary Report on the Incinerator Site (33My57) Stable Carbon Isotope Ratios Used in Dietary Reconstruction. M.A. thesis, Department of Anthropology, University of Cincinnati, Ohio.

Converse, R.
1984 An Open Hopewell Site in Ross County, Ohio. *Ohio Archaeologist* 34:23–35.

Cook, S. F.
1972 *Prehistoric Demography.* McCaleb Module in Anthropology, Module No. 16. Addison Wesley, Reading, Massachusetts.

Cotter, J. L., and J. M. Corbett
1951 *Archaeology of the Bynum Mounds, Mississippi.* National Park Service Archaeological Research Series 1. Washington, D.C.

Crawford, G., and M. Yoshizaki
1987 Ainu Ancestors and Prehistoric Asian Agriculture. *Journal of Archaeological Science* 14:201–213.

Crites, G.
1985 Middle Woodland Paleoethnobotany of the Eastern Highland Rim of Tennessee: An Evolutionary Perspective on Change in Human-Plant Interaction. Ph.D. dissertation, Department of Anthropology, University of Tennessee, Knoxville.

Crites, G. D., and R. D. Terry
1984 Nutritive Value of Maygrass. *Economic Botany* 38:114–120.

Decker, D.
1988 Origin(s), Evolution, and Systematics of *Cucurbita pepo. Economic Botany* 42:4–15.

Decker, D., and H. D. Wilson
1987 Allozyme Variation in the *Cucurbita pepo* Complex: *C. pepo* var. *ovifera* vs. *C. texana. Systematic Botany* 12:263–273.

DuVall, G. D.
1982 The Ewell III site (40CF118). In *Seventh Report of the Normandy Archaeological Project,* edited by C. Faulkner and M. McCollough, pp. 8–152. Report of Investigations 32. Department of Anthropology, University of Tennessee, Knoxville.

Earle, T.
1980 A Model of Subsistence Change. In *Modeling Change in Prehistoric Subsistence Economies,* edited by T. Earle and A. Christenson, pp. 1–29. Academic Press, Orlando, Florida.

Farnsworth, K. B.
1985 Preface. In *Smiling Dan,* edited by B. Stafford and M. Sant, p. vii. Kampsville Archaeological Center, Research Series 2.

Farnsworth, K. B., and A. Koski
1985 *Massey and Archie.* Kampsville Archaeological Center, Research Series 3.

Faulkner, C.
1968 *The Old Stone Fort.* University of Tennessee Press, Knoxville.

1973 Middle Woodland Subsistence-Settlement Systems in the Highland Rim: A Commentary. In *Salvage Archaeology at 40FR47,* by Willard S. Bacon and N. L. Merryman, pp. 35–45. Tennessee Archaeological Society Miscellaneous Paper 11.

1977 Eoff III Site (40OF107). In *Fourth Report of the Normandy Archaeological Project,* edited by C. Faulkner and M. McCollough. Report of Investigations 19, Department of Anthropology, University of Tennessee, Knoxville.

1982 The McFarland Occupation at 40CF32: Interpretations of the 1975 Field Season. In *Eighth Report of the Normandy Archaeological Project,* edited by C. Faulkner and M. McCollough, pp. 303–388. Report of Investigations 33, Department of Anthropology, University of Tennessee, Knoxville.

Faulkner, C., and M. McCollough
1982 The Investigation of the Parks Site (40CF5). In *Seventh Report of the Normandy Archaeological Project,* edited by C. Faulkner and M. McCollough, pp. 313–352. Report of Investigations 32, Department of Anthropology, University of Tennessee, Knoxville.

Fischer, F.
1971 Preliminary Report on the University of Cincinnati Archaeological Investigations, 1970. Typewritten ms. in possession of author.

Ford, Richard I.
1987 Dating Early Maize in the Eastern United States. Paper presented at the annual meeting of the American Association for the Advancement of Science, February 14–18, Chicago.

Fortier, A.
1985 Middle Woodland Occupations at the Truck 7 and Go-Kart South Sites. In *Selected Sites in the Hill Lake Locality,* edited by Andrew Fortier, pp. 163–281. American Bottom Archaeology FAI-270 Site Reports 13. University of Illinois Press, Urbana.

Fortier, A., T. Maher, J. Williams, M. Meinkoth, K. Parker, and L. Kelly
1989 *The Holding Site: A Hopewell Community in the American Bottom.* American Bottom Archaeology

FAI-270 Site Reports 19. University of Illinois Press, Urbana.

Freeman, J. E.
1969 The Millville Site, a Middle Woodland village in Grant County, Wisconsin. *Wisconsin Archaeologist* 50:37–88.

Greber, N.
1979 A Comparative Study of Site Morphology and Burial Patterns at Edwin Harness Mound and Seip Mounds 1 and 2. In *Hopewell Archaeology,* edited by D. Brose and N. Greber, pp. 27–38. Kent State University Press, Kent, Ohio.

1983 Recent Excavations at the Edwin Harness Mound, Liberty Works, Ross County, Ohio. *Kirtlandia* 39.

Jefferies, R.
1976 *The Tunacunnhee Site.* Anthropological Papers of the University of Georgia No. 1. Athens.

Johannessen, S.
1981 Plant Remains from the Truck 7 Site. In *Archaeological Investigations of the Middle Woodland Occupation at the Truck* 7 and Go-Kart South Sites, by A. C. Fortier, pp. 116–130. American Bottom Archaeology FAI-270 Archaeological Mitigation Project Report 30. University of Illinois Press, Urbana.

1983 Plant Remains from the Cement Hollow Phase. In *The Mund Site,* by A. C. Fortier, F. A. Finney, and R. B. LaCampagne, pp. 94–103. American Bottom Archaeology FAI-270 Site Reports 5. University of Illinois Press, Urbana.

1984 Paleoethnobotany. In *American Bottom Archaeology,* edited by C. Bareis and J. Porter, pp. 197–214. University of Illinois Press, Urbana.

Kelly, A. R.
1973 Early Villages on the Chattahoochee River Georgia. *Archaeology* 26:32–37.

Kelly, J., A. Fortier, S. J. Ozuk, and J. A. Williams
1987 *The Range Site: Archaic through Late Woodland Occupations.* American Bottom Archaeology FAI-270 Site Reports 16. University of Illinois Press, Urbana.

Kline, G. W., G. Crites, and C. Faulkner
1982 *The McFarland Project.* The Tennessee Anthropological Association Miscellaneous Paper 8.

Lynott, M. J., T. W. Boutton, J. E. Price, and D. E. Nelson
1986 Stable Carbon Isotope Evidence for Maize Agriculture in Southeast Missouri and Northeast Arkansas. *American Antiquity* 51:51–65.

Mainfort, R. C.
1986 *Pinson Mounds: A Middle Woodland Ceremonial Center.* Research Series 7. Division of Archaeology, Tennessee Department of Conservation, Nashville.

Marshall, J. A.
1969 Engineering Principles and the Study of Prehistoric Structures: A Substantive Example. *American Antiquity* 34:166–171.

Morgan, D. T.
1986 Ceramics. In *Woodland Period Occupations of the Napoleon Hollow Site in the Lower Illinois Valley,* edited by M. D. Wiant and C. R. McGimsey, pp. 364–426. Kampsville Archaeological Center, Research Series 6.

Morse, D. F., and P. A. Morse
1983 *Archaeology of the Central Mississippi Valley.* Academic Press, Orlando, Florida.

Parmalee, P., A. Paloumpis, and N. Wilson
1972 *Animals Utilized by Woodland Peoples Occupying the Apple Creek Site, Illinois.* Illinois State Museum, Reports of Investigations 23. Springfield.

Prufer, O. H.
1964 The Hopewell Complex of Ohio. In *Hopewellian Studies,* edited by J. Caldwell and R. Hall, pp. 35–84. Illinois State Museum, Scientific Papers 22. Springfield.

1965 *The McGraw Site: A Study in Hopewellian Dynamics.* Cleveland Museum of Natural History, Scientific Publications, new series, 4(1). Ohio.

1975 The Scioto Valley Archaeological Survey. In *Studies in Ohio Archaeology,* edited by O. Prufer and D. McKenzie, pp. 267–328. Kent State University Press, Kent, Ohio.

Seeman, M. F.
1979 Feasting with the Dead: Ohio Hopewell Charnel House Ritual as a Context for Redistribution. In *Hopewell Archaeology,* edited by D. S. Brose and N. Greber, pp. 39–48. Kent State University Press, Kent, Ohio.

Seeman, M. F., and F. Soday
1980 The Russell Brown Mounds: Three Hopewell Mounds in Ross County, Ohio. *Midcontinental Journal of Archaeology* 5:73–116.

Shane, O. C.
1970 The Scioto Hopewell. In *Adena: The Seeking of an Identity,* edited by B. K. Swartz, Jr., pp. 142–145. Ball State University Press, Muncie, Indiana.

Smith, Bruce D.
1978 Variation in Mississippian Settlement Patterns. In *Mississippian Settlement Patterns,* edited by B. Smith, pp. 479–503. Academic Press, Orlando, Florida.

1985a The Role of *Chenopodium* as a Domesticate in Pre-Maize Garden Systems of the Eastern United States. *Southeastern Archaeology* 4:51–72.

1985b *Chenopodium berlandieri* ssp. *jonesianum:* Evidence for a Hopewellian Domesticate from Ash Cave, Ohio. *Southeastern Archaeology* 4:107–133.

1987a The Independent Domestication of Indigenous Seed-Bearing Plants in Eastern North America. In *Emergent Horticultural Economies of the Eastern Woodlands,* edited by W. Keegan, pp. 3–48. Center for Archaeological Investigations Occasional Paper 7. Southern Illinois University, Carbondale.

1987b The Economic Potential of *Chenopodium berlandieri* in Prehistoric Eastern North America. *Ethnobiology* 7(1):29–54.

Squier, G. E., and E. H. Davis
1848 *Ancient Monuments of the Mississippi Valley.* Smithsonian Contributions to Knowledge 1. Washington, D.C.

Stafford, B., and M. B. Sant (editors)
1985 *Smiling Dan.* Kampsville Archaeological Center, Research Series 2.

Stoltman, J. B.
1979 Middle Woodland Stage Communities of Southwestern Wisconsin. In *Hopewell Archaeology,* edited by D. S. Brose and N. Greber, pp. 122–140. Kent State University Press, Kent, Ohio.

Struever, S.
1968 A Re-examination of Hopewell in Eastern North America. Ph.D. dissertation, Department of Anthropology, University of Chicago, Illinois.

Struever, S., and G. L. Houart
1972 An Analysis of the Hopewell Interaction Sphere. In *Social Exchange and Interaction,* edited by E. N. Wilmsen, pp. 47–79. Museum of Anthropology, Anthropological Papers No. 46. University of Michigan, Ann Arbor.

Toth, A.
1979 The Marksville Connection. In *Hopewell Archaeology,* edited by D. Brose and N. Greber, pp. 188–199. Kent State University Press, Kent, Ohio.

Van der Merwe, N. J., and J. C. Vogel
1978 $^{12}C/^{13}C$ Content of Human Collagen as a Measure of Prehistoric Diet in Woodland North America. *Nature* 276:815–816.

Vogel, J. C., and N. J. Van der Merwe
1977 Isotopic Evidence for Early Maize Cultivation in New York State. *American Antiquity* 42:238–242.

Wagner, G.
1986 The Corn and Cultivated Beans of the Fort Ancient Indians. *The Missouri Archaeologist* 47:107–137.

Watson, P.
1974 Theoretical and Methodological Difficulties Encountered in Dealing with Paleofecal Material. In *Archaeology of the Mammoth Cave Area,* edited by P. Watson, pp. 239–242. Academic Press, Orlando, Florida.

Walthall, J.
1980 *Prehistoric Indians of the Southeast.* University of Alabama Press.

Watt, B. K., and A. L. Merrill
1963 *Composition of Foods.* Agriculture Handbook 8. U.S. Department of Agriculture, Washington, D.C.

White, P. L., E. Alvistor, C. Dias, E. Vinas, H. S. White, and C. Collazos
1955 Nutrient Content and Protein Quality of Quinoa and Canihua: Edible Seed Products of the Andes Mountains. *Journal of Agriculture and Food Chemistry* 3:531–534.

Whittle, A.
1987 Neolithic Settlement Patterns in Temperate Europe: Progress and Problems. *Journal of World Prehistory* 1:5–52.

Wiant, M., and C. R. McGimsey (editors)
1986 *Woodland Period Occupations of the Napoleon Hollow Site in the Lower Illinois Valley.* Kamps-

ville Archaeological Center, Research Report 6, Illinois.

Wu Leung, W.
1961 *Food Composition Table for Use in Latin America.* A research project sponsored jointly by INCAP-ICNND and National Institute of Health, Bethesda, Maryland.

Wymer, D. A.
1987 The Middle Woodland-Late Woodland Interface in Central Ohio: Subsistence Continuity and Cultural Change. In *Emergent Horticultural Economies of the Eastern Woodlands,* edited by W. F. Keegan, pp. 201–216. Center for Archaeological Investigations, Occasional Paper 7. Southern Illinois University, Carbondale.

Yarnell, R. A.
1974 Plant Food and Cultivation of the Salts Cave. In *Archaeology of the Mammoth Cave Area,* edited by P. Watson, pp. 113–122. Academic Press, Orlando, Florida.

CHAPTER 10

❑ ❑ ❑

In Search of *Choupichoul*, the Mystery Grain of the Natchez

Introduction

The agricultural economies and cultivated crops of southeastern Indian groups were a frequent focus of interest for the Europeans who explored and settled the region during the fifteenth, sixteenth, and seventeenth centuries. Exhibiting considerable variation in their focus, level of detail, and degree of cultural distortion, the surviving early European accounts provide a vivid, if frustratingly brief, glimpse of indigenous southeastern agriculture.

Although the tropical plant trinity of corn, beans, and squash almost invariably dominates early written descriptions of Indian agriculture in the Southeast, other crops are sometimes mentioned. These all too fleeting references of obscure, long-forgotten cultigens often attract the attention and interest of those archaeobotanists interested in the first beginnings of plant husbandry in the Eastern Woodlands of North America. Why such curiosity concerning the minor crops of contact-period maize-dominated agricultural systems? Because of the tantalizing possibility that these minor crops might represent remnant cultivar survivors from early plant husbandry systems that flourished long before the emergence of maize agriculture. In 1924 Ralph Linton first suggested the possible existence of such premaize plant husbandry systems, which he thought might have been based on various small grains of the Southeast (Linton 1924).

During their respective analyses of archaeobotanical assemblages from Ozark and eastern Kentucky rock shelters in the 1930s, Melvin Gilmore (Gilmore 1931) and Volney Jones (Jones 1936) found both morphological and contextual evidence for assigning domesticated status to several indigenous species of plants, and Jones tentatively speculated that these native domesticates may have been grown in the East prior to the introduction of corn.

Thanks to cultural resource protection legislation and the development of flotation recovery methods, the past quarter century has witnessed a remarkable increase in both the magnitude and quality of the archaeobotanical data base for the prehistoric East (Yarnell and Black 1985). This enriched data base in turn has allowed a number of researchers to carefully document, in increasing detail, the developmental history and economic importance of the oily- and starchy-seed plants that comprised these regionally variable premaize plant husbandry systems. When maize *(Zea mays)* did arrive in the East, apparently no earlier than about A.D. 200, it was adopted as just another starchy-seed crop within already established garden or field plots. At least four premaize seed crops—marshelder *(Iva annua),* lambsquarter *(Chenopodium berlandieri),* sunflower *(Helianthus annuus),* and squash *(Cucurbita pepo)*—were by then long-standing domesticates in the East. In addition to these long-standing domesticates, little barley *(Hordeum pusillum),* erect knotweed *(Polygonum erectum),* and maygrass *(Phalaris caroliniana)* were also important in premaize food production economies. Although maize was added to at least some eastern cultivation plots by around A.D. 200, this tropical plant remained below, or just at, the level of archaeological visibility until the Woodland to Mississippian period transition around A.D. 750–800, when it abruptly becomes abundantly present in archaeobotanical assemblages over a wide area of the Southeast and Midwest. Judging from stable carbon isotope studies, however, it was not for another two to four centuries, until around A.D. 1000–1200, that maize became the dominant, central crop in Mississippian field agriculture systems.

Against this backdrop of Woodland to Mississippian period cultural change, and the evidence for an increasingly important role for maize, the general question of what happened to the constituent crops of the premaize plant husbandry systems is increasingly attracting research interest. Although there is apparently no broad and uniform pattern, and good evidence of regional variation, the seven cultigens all continued under cultivation in a number of areas of the East through the A.D. 800–1100 Late Woodland to Mississippian transition, and were still present in the post A.D. 1200 maize-dominated food production economies. But they seem to have all but disappeared from view by the time of European contact. With the possible exception of Harriot's 1586 account from the Carolinas of sunflower and melden (Sturtevant 1965), and that of Antoine Simon Le Page du Pratz from the early eighteenth-century Natchez area of the Lower Mississippi Valley (Le Page du Pratz 1758), the ethnohistorical record is silent in regard to possible remnant premaize cultigens of the Southeast.

Although Le Page's account has generally been considered more ambiguous and less valuable than Harriot's, it is his description of *Choupichoul,* the mysterious Mississippi River grain of the Natchez, that will be pursued in this chapter. I will be arguing that this mysterious grain of the Natchez was in fact a domesticated variety of *Chenopodium berlandieri*—a holdover from premaize plant husbandry systems—an ancient crop that had been under continuous cultivation in the East for more than 3,000 years. To build the case for *Choupichoul* as *Chenopodium* it will be necessary to follow divergent paths of supporting evidence, which lead into three quite distinct disciplines, each with its own pitfalls and problems.

Le Page, the Natchez, and Choupichoul

In turning first to the ethnohistorical record, Le Page's written account of *Choupichoul,* it will be necessary to view it critically, against a backdrop of relevant concerns. How reliable are Le Page's observations, particularly in regard to Natchez use of plants, for example? How reliable are the French to English translations of Le Page's accounts? Which of the various botanical and taxonomic interpretations that have been made of Le Page's plant names are accurate? Once Le Page's circa 1720 descriptions of *Choupichoul* have been considered, they can be framed in a broader interpretive context by looking at recent discoveries regarding the history of *Chenopodium berlandieri* as a prehistoric domesticated crop in the Eastern Woodlands on the one hand, and recent studies of present-day *Chenopodium berlandieri* populations on the other. When *Choupichoul,* as described by Le Page, is viewed within such a developmental context, in terms of both its possible ancestry and likely present-day weedy descendants, its identification as a domesticated variety of *Chenopodium berlandieri* becomes much more compelling.

I don't mean to imply that no one has previously considered Le Page's account of *Choupichoul,* or that others haven't suggested that it might be *Che-*

nopodium. But while *Choupichoul* has attracted attention on a number of occasions over the past 50 years, this attention has always been rather limited, has focused on the common names used by Le Page to identify *Choupichoul*, and did not have the benefit of either recent archaeological research or modern field studies of *Chenopodium berlandieri* stands. In his 1931 publication on Ozark bluff shelter plant remains, Gilmore made a brief reference to *Choupichoul*, and concluded that it was a cultivated form of *Chenopodium* because Le Page had called it *belle dame sauvage*, and *belle dame* was, according to Gilmore, the French common name for a European species of *Chenopodium* (Gilmore 1931:97–98). Opinions differ, however, as to what Le Page meant when he used the term *belle dame sauvage*. Some agree with Gilmore's identification of *belle dame* as a European chenopod (Seeman and Wilson 1984:30), while others identify *belle dame* as buckwheat (see the 1774 English translation of Le Page) or a "common French name for *Atriplex*" (Asch and Asch 1977:97). A distinct genus from *Chenopodium*, *Atriplex* is within the large and diverse family *Chenopodiaceae*, which includes about one hundred genera and fourteen hundred species worldwide (Wilson and Walters n.d.).

In addition, because Le Page also describes *Choupichoul* as being a kind of millet, a number of others have concluded that it was not *Chenopodium* at all, but a species of grass (Asch and Asch 1977:9), perhaps cockspur *(Echinochloa crusgalli)* (Swanton 1911:76; 1946:291) or maygrass *(Phalaris caroliniana)* (Yarnell 1972:10).

I will return to Le Page's use of the terms *belle dame* and millet, but before considering at what level of botanical and taxonomic specificity they should be interpreted, it is necessary to first briefly consider Le Page, the Natchez, and his description of the plant itself: what it looked like, and where and how it was grown.

After arriving in Louisiana in 1718 Le Page spent about a year in the New Orleans area before traveling up the Mississippi Valley to the vicinity of the present-day city of Natchez, Mississippi. In January of 1720 he established himself as a neighbor of the Natchez Indians, whose communities of widely scattered homesteads and mound centers were located along St. Catherines Creek, a tributary of the Mississippi. Le Page lived among the Natchez for eight years before returning to New Orleans in 1728, and eventually to France in 1734.

His personal observations of Louisiana, first published as a series of articles beginning in 1751, and then expanded into the 1758 memoir *Histoire de la Louisiane*, were intended to correct false impressions and deliberate misrepresentations of the French colony, as well as to serve as a practical handbook for subsequent settlers along the Mississippi. While Le Page's 1758 memoir provides ample proof of his gullibility in reliance on the observations of others, his first-hand descriptions of wildlife and vegetation are considered accurate, and his description of Natchez lifeways is acknowledged as the "unrivaled chief source of modern knowledge concerning the Natchez" (Tregle 1975:xl). While unrivaled as a source on the Natchez, Le Page has not, however, gone uncriticized. His interpretations of Natchez culture, particularly his perceptions of Natchez social organization, as viewed through the French monarchy "Sun King" cultural filter, have been justifiably questioned, while at the same time providing a fascinating challenge of interpretation for generations of scholars (Knight 1990).

His straightforward first-hand descriptions of Natchez crops and cultivation have not attracted any criticism, however, although some of the illustrations of animals and plants of the region published in his memoir have been characterized as "bizarre misconceptions" (for example, the vertically striped skunk) (Tregle 1975:xxxvii). Many of the illustrations of plants and animals are fairly accurate, however (Figure 10.1), and it is certainly unfortunate that *Choupichoul* was not illustrated.

Le Page discusses the cultivation and preparation of a broad spectrum of cultivated plants including flour maize, white, yellow, red, and blue varieties of grits maize, little maize, red and black beans, two varieties of squash and tobacco, along with several more recent introductions, including figs, watermelon, and clingstone peaches. In addition he mentions the harvesting of a variety of wild plant foods including cane seeds, nuts, persimmon, and mushrooms. Le Page also describes responding to a request for Natchez "simples" or medicinal plants by collect-

Figure 10.1 The alligator gar *(Lepisosteus spatula),* as depicted in *Historie de la Louisiane,* by Le Page du Pratz (1758).

ing more than 300 samples and sending them, in cane baskets, and accompanied by descriptions of their uses and preparation, south to New Orleans. A botanic garden was apparently expressly made for their cultivation.

Le Page apparently had not had much, if any, formal training in botany. He described himself as a professional architect, received his degree in mathematics, had published in the field of astronomy, and was considering applying his skills as a hydraulic engineer to solve the silting problem of the Mississippi River, all before turning to anthropological observation and ethnobotany.

The Passages that Refer to Choupichoul

Having briefly discussed Antoine Simon Le Page du Pratz's standing as an observer of the customs and crops of the Natchez, I can turn now to the three references to *Choupichoul* that are scattered through his 1758 publication.

The shortest of these passages is hidden in Book 2 (Tome Second), Chapitre XIII *(Travaux des Naturels de la Louisiane: Construction de leurs Cabannes),* under the margin heading "Cultivation of the earth" *(Culture de la terre).* This passage was missed by John R. Swanton and not included in his 1911 publication on the Indian tribes of the Lower Mississippi Valley. As a result it is not mentioned by any subsequent researchers. It pairs the crop with maize as being both delicious and nutritious:

> Il y a apparence que ces Peuples rassemblés, & composant une Ville & ou un Village, devinrent plus sédentaires, ne pouvant comme auparavant, emporter leurs demeures qu'ils avoient rendues stables en les bâtissant. Ils cultiverent la terre, afin qu'elle pourvût à leur nourriture; ils s'adonnerent à la culture du Mahiz, soit qu'ils

> l'eussent trouvé en Amérique, soit qu'ils l'eussent apporté de la Scythie ou de la Tartarie qui en produisent. Ce grain est très-bon & très-nourrissant, de méme que le Choupichoul qui vient sans qu'on le cultive. (Le Page du Pratz 1758, Tome Second:176)

In addition to providing information about *Choupichoul,* this passage also serves to demonstrate the often rather loose and sometimes misleading nature of the first English translation of Le Page. Presented in two volumes in 1763 and then offered in a single volume in 1774, this first English translation was published by T. Becket, London. The 1774 single-volume version of this translation has been reproduced in facsimile (Le Page du Pratz 1975). In his introduction to this facsimile edition, Joseph Tregle describes the first English translation of Le Page as "a total reordering of the work's content and plan of organization" (Tregle 1975:xxviii). Tregle comments on the "wretchedness of the translator's 'improvement' " (Tregle 1975:xxix).

In this first translation of Le Page into English, the above passage appears in Book IV (Of the Natives of Louisiana), Chapter III (A Description of the Natives of Louisiana), Section V (Of the Arts and Manufactures of the Natives) (Le Page 1975:360):

> The natives having once built for themselves fixed habitations, would next apply themselves to the cultivation of the ground. Accordingly, near all their habitations, they have fields of maiz, and of another nourishing grain called Choupichoul, which grows without culture.

A closer translation of the original French is offered by Valérie Chaussonnet:

> It appears that these people assembled, and forming a town or a village, became more sedentary, and now unable to carry their habitations, which they had rendered stable by building them. They cultivated the earth, so that it would give them food; they started to cultivate Maize, which they either found in America or brought from Scythia or from Tartary that produce it. This grain is very delicious and very nourishing, in the same way as Choupichoul that grows without being cultivated.

Notice that this second translation does not place maize and *Choupichoul* in fields near habitation sites, but rather is limited to characterizing the taste and nutritional attributes of the two crops. This brief passage provides a good start toward building a profile of *Choupichoul* in that it indicates the following attributes:

1. *Choupichoul* was delicious and nourishing,
2. It was a grain, harvested for its seeds,
3. It grew without culture (attention during the growing season).

It is also interesting to speculate that this passage might suggest that *Choupichoul* was grown, without much attention or cultivation, along with maize, in the fields that were an integral part of the Natchez agricultural landscape of scattered farmsteads and intervening field and garden plots. While the original passage (as opposed to the 1774 English translation) does not provide much support for this speculation, other than mentioning the two crops together, it would, if correct, expand the context of cultivation for *Choupichoul.* The dispersed farmsteads and centers that comprised Natchez communities were situated along St. Catherines Creek, on top of the dramatic 100 ft high loess bluff line, which defines the eastern edge of the Mississippi Valley in the Natchez area (Brown 1984; Figure 10.2). These farmsteads and their nearby fields were far removed from the sandbanks of the main channel of the Mississippi River, and it is these sand banks that were the setting of Le Page's better known account of *Choupichoul* cultivation.

Before turning to this description of sand-bank cultivation, however, a second passage needs to be briefly considered. Appearing in Book 3 *(Troisieme partie),* Chapter 1 *(Chapitre Premier),* entitled *Suite des Meurs; Jeux des hommes, des femmes & des garçcons: Conversations, nourritures, repas & jeunes des Naturels,* and with the margin caption "Grains that were employed for nourishment" *(Grains qu'ils employent pour leur nourriture).* This famous "millet" passage has misled scholars for over 75 years, serving as a red herring in regard to the identity of *Choupichoul*:

> Ils font encore du manger avec deux graines dont l'une se nomme *Choupichoul,* (2) qu'ils cultivent sans peine, l'autre est le *Widlogouill,* qui vient naturellement & sans aucune culture; ce sont deux espéces de millet qu'ils écalent de même que le riz. (Le Page du Pratz 1758, Troisieme partie:9)

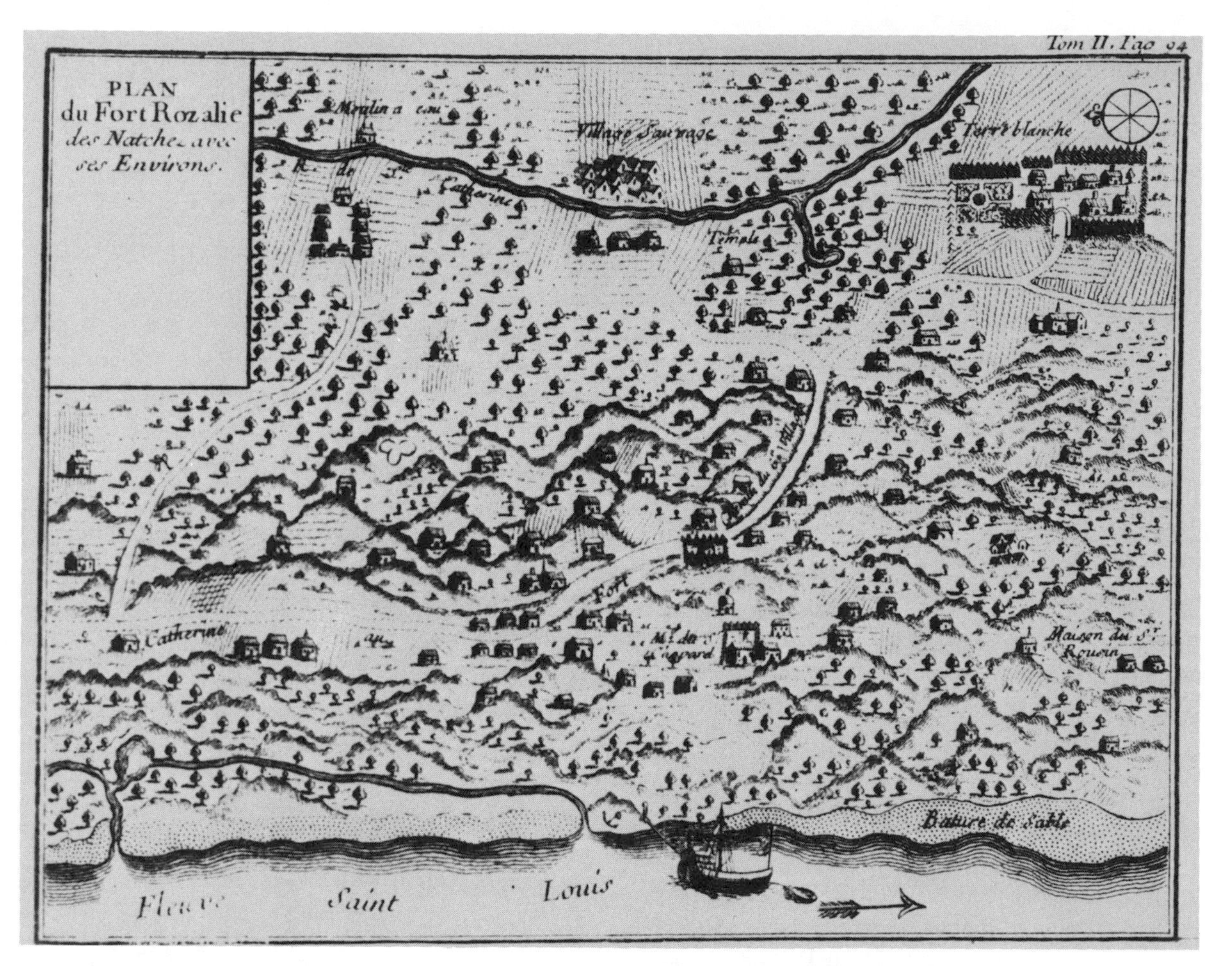

Figure 10.2 Plan of Fort Rosalia, the concessions of St. Catherine and White Earth, showing the distribution of Natchez settlements and fields. (From Dumont de Montigny 1753, II, 94e; National Anthropological Archives negative number 1168-b-7.)

In the 1774 English translation this passage is relocated to Book IV (Of the Natives of Louisiana), Chapter III (A Description of the Natives of Louisiana), Section VI (Of the Attire and Diversions of the Natives: Of their Meals and Fastings) (Le Page du Pratz 1975:368).

> They likewise use two kinds of millet, which they shell in the manner of rice; one of these is called *Choupichoul,* and the other *Widlogouil,* and they both grow almost without any cultivation.

Compare the translation by John R. Swanton:

> They also make food of two grains, of which one is called *Choupichoul* (d) which they cultivate without difficulty, and the other is the *Widlogouil,* which grows naturally and without any cultivation. These are two kinds of millet which they hull in the same way as rice.
>
> (d) Perhaps cockspur grass *(Echinochloa crusgalli)*
>
> (Swanton 1911:76)

The key phrase in the passage is: "these are two kinds of millet," or "they likewise use two kinds of millet." This phrase has caused researchers to categorize *Choupichoul* as a milletlike plant. Swanton suggests that it was a kind of grass, perhaps cockspur *(Echinochloa crusgalli),* and Yarnell offers maygrass as a likely identification (Yarnell 1972:10).

Le Page's subject headings "of the nourishment," and "meals of the natives," as well as the food preparation narrative of this section of his memoir, however, provide the clarifying context for this millet reference. When Le Page refers to the grain *Choupichoul* as a kind of millet in this passage, he is referring not to the entire plant, but just to the seeds of the plant, and within the context of food preparation. Within such a food preparation context, millet constitutes an appropriate European analog for a grain or seed presenting similar processing requirements. Proof that this is the correct reading of this passage—that millet is a European analog for *Choupichoul* grain (seeds) within a food preparation context, but not an analog for the *Choupichoul* crop plant itself, is fortunately provided by Le Page himself in his third passage describing *Choupichoul.* Before considering this third and final passage in full, it is worthwhile to first pull out those parts of it that are pertinent in reference to whole plant vs. seed analogs.

On the one hand, Le Page provides an unequivocal European plant species analog for *Choupichoul*: ". . .this *plant* is that which is called 'beautiful savage lady' and which grows in all countries. . ." (Swanton 1911:76). In this same passage, Le Page also provides a clear European food preparation analog for the seeds or grain of *Choupichoul*: ". . .the women and children cover the *grain,* with a great deal of indifference, with their feet, almost without looking at it. After this sowing and this kind of cultivation they wait until autumn and then gather a great quantity of this *grain* [the seeds]. They *prepare it like millet* and it is very good eating" (Swanton 1911:76; emphasis added).

So in describing *Choupichoul* for a European audience, under a number of different topical headings, Le Page provides his readers with two distinct analogs within a European context of reference: (1) The *Choupichoul* plant is that which is called *belle dame sauvage;* (2) The product of the *Choupichoul* plant is a kind of millet, and is prepared like millet. Having considered Le Page's two brief references to the mystery grain of the Natchez and parts of the final and longest passage, let's look at this longer passage in full, to see what other attributes can be added to the descriptive profile of *Choupichoul.* Located in Book 1 *(Tome Premier),* Chapter 23 *(Chapitre XXIII),* and having the margin heading "Sand Banks of the S. Louis River" *(Battures du Fleuve S. Louis):*

> Je ne dois point omettre ici que depuis les terres basses de la Louisiane, le Fleuve S. Louis a beaucoup de battures de sable en le remontant, qui paroît très-sec, après que les eaux se sont retirées à la fin de son débordement: ces battures sont plus ou moins longues, il y en d'une demie lieue de long, qui ne laissent pas d'avoir une bonne largeur. J'ai vû les Natchez & autres Naturels sémer une graine qu'ils nommoient *Choupichoul,* sur les battures; ce sable n'étoit nullement cultivé, & les femmes & les enfans avec leur pieds couvroient tellement quellement cette graine sans y regarder de près. Après cette sémaille, & cette espéce de culture, ils attendoient l' Automne, & recueilloient pour lors une grande quantité de cette graine: ils la préparoient comme du millet, & elle étoit très-bonne à manger. Cette plante est ce que l'on nomme *Belle Dame sauvage,* qui vient en tout pays, mais il lui faut une bonne terre; & quelque bonne qualité qu'ait une terre en Europe, elle ne vient que d'un pied & demi de haut; & sur ce sable du Fleuve, sans culture elle s'éléve jusqu'à trois pieds & demi & quatre pieds. (Le Page du Pratz 1758, Tome Premier:316–317).

In the 1774 English translation (Le Page du Pratz 1975:175) this long passage appears in Book II (Of the Country and its Products), Chapter VI (A Brook of Salt Water: Salt Lakes. Lands of the River of the Arkansas. Red-veined Marble: Slate Plaster. Hunting the Buffalo. The Dry Sand-banks in the Mississippi). The passage is translated as follows:

> I ought not to omit mentioning here, that from the low lands of Louisiana, the Mississippi has several shoal banks of sand in it, which appear very dry upon the falling of the waters, after the inundations. These banks extend more or less in length; some of them half a league, and not without a considerable breadth. I have seen the Natchez, and other Indians, sow a sort of grain, which they called Choupichoul, on these dry sand-banks. This sand received no manner of culture; and the women and children covered the grain any how with their feet, without taking any great pains about it. After this sowing, and manner of culture, they waited till autumn, when they gathered a great quantity of the grain. It was prepared like millet, and very good to eat. This plant is what is called Belle Dame Sauvage,* which thrives in all countries, but requires a good soil: and whatever good quality the soil in Europe may have, it shoots but a foot and a half high; and yet, on this land of the Mississippi, it rises, without any culture, three feet and a half, and four feet high.

*He seems to mean Buck-wheat.

Compare the translation by John R. Swanton:

> I ought not to omit here that from the lowlands of Louisiana upward the river St. Louis [Mississippi] has many sand banks, which become entirely dry after the waters have gone down at the end of the flood. These sand banks vary in length. There are some half a league long which do not lack a good breadth. I have seen the Natchez and other natives sow a grain which they called *choupichoul* on these sand banks. This sand is never cultivated and the women and children cover the grain, with a great deal of indifference, with their feet, almost without looking at it. After this sowing and this kind of cultivation they wait until autumn and then gather a great quantity of this grain. They prepare it like millet and it is very good eating. This plant is that which is called "beautiful savage lady"(f) and which grows in all countries, but it needs a good soil, and however, good is the quality of any European soil it there reaches a height of only 1½ feet, while on this river sand without cultivation it reaches a height of 3½ to 4 feet.
>
> (f) *Belle dame sauvage*
>
> (Swanton 1911:76)

The information provided in this long account of sand-bank cultivation considerably expands the descriptive profile of *Choupichoul,* which can be considered to have eight distinctive attributes based on Le Page's descriptions:

1. Its closest European counterpart, a plant that grew in all countries, and in good soils, to a height of 1½ ft, went by the common name *belle dame sauvage;*
2. It was a grain (seed) crop, harvested for its seeds;
3. It was very good eating, and very nourishing;
4. It was planted along the exposed sand banks of the Mississippi River and perhaps in fields close to Natchez settlements;
5. It grew without difficulty, and without attention during the growing season;
6. It reached a height of 3½–4 ft, some 2 ft more than its European analog, *belle dame sauvage.*
7. It was harvested in the fall, and
8. It yielded a great quantity of grain.

Armed with this fairly detailed description, we can go in search of the beautiful savage lady, first between the covers of French dictionaries of the 1700s, and then along the exposed sand banks of the Mississippi River.

In Search of Belle Dame Sauvage

While *belle dame sauvage* is not to be found in eighteenth-century French dictionaries, *belle dame* is commonly listed as referring to two quite different plants: (1) belladonna or deadly nightshade *(Atropa belladonna)* and (2) garden orach *(arroch a des jardins)* (Richelet 1759:190). The 1972 edition of *Harrap's New Standard French and English Dictionary* also assigns the *belle dame* label to belladonna and garden orach. Devries's 1968 *French-English Science Dictionary* identifies *bonne dame* as mountain spinach, garden orach *(Atriplex hortensis).* Assuming that Le Page was not drawing an analogy between *Choupichoul* and belladonna, which could hardly be considered an appropriate analog for a food crop, he was quite probably identifying garden orach as the Old World plant most similar to *Choupichoul.* Orach, in turn, is the common name for the genus *Atriplex,* and garden orach is listed in Usher's *Dictionary of Plants Used by Man* (1974) as the common name for the species *Atriplex hortensis.* While the genus *Atriplex,* and perhaps the species *Atriplex hortensis* might have been the plant taxon identified by Le Page as the European analog for *Choupichoul, Atriplex hortensis* could not have been the mystery grain of the Natchez, and it is highly doubtful that *Choupichoul* belonged to the genus *Atriplex.* The garden orach *(Atriplex hortensis)* is an Old World plant, cultivated not for its seeds but for its leaves, which are cooked as a vegetable (Usher 1974:69). It apparently was not introduced into eastern North America until the mid-nineteenth century, and is rare even today. Its first record in Missouri was 1839 (Steyermark 1963:615), and of the 17 specimens in the National Herbarium, National Museum of Natural History, the earliest was collected in 1895. In addition, there is no record of *Atriplex hortensis* in Louisiana or Mississippi, and it is not listed in the recent index *Vascular Flora of the Southeastern United States* (Wilson and Walters n.d.). In addition, neither of the two species of *Atriplex* that are listed in this index, *A.*

pentandra and *A. patula,* seem to be very promising *Choupichoul* candidates.

Atriplex pentandra, on the one hand, can be ruled out quickly since it occurs only in coastal salt marsh and sand dune habitat settings (Lowe 1921:139; Eleuterius 1980; Wilson and Walters n.d.:10–11). The most common species of *Atriplex* in eastern North America, *Atriplex patula,* is a more viable candidate, growing in salt marshes but also ranging inland in saline soils and waste places. It grows to 90 cm in height, however, somewhat short of the 105–120 cm cited for *Choupichoul,* and is not found in undisturbed floodplain habitats. Additionally, while sometimes cooked for its greens, I could find no mention of its seeds having being used as a food source in the East. As the most viable *Choupichoul* candidate within the genus *Atriplex, Atriplex patula* can't be rejected out of hand, however, and it should be kept in mind as we turn to a consideration of the candidacy of *Chenopodium.* We can begin by returning to the modifier *savage,* which translates as wild or feral, and which Le Page attaches to the common name *belle dame,* in identifying his European analog for *Choupichoul.* Perhaps this modifier was added to indicate to his readers that he was referring not to the cultivated *belle dame,* garden orach, but to its uncultivated weedy look-alike relative—*Chenopodium* (Asch and Asch 1977:9). The two genera *Chenopodium* and *Atriplex* both belong to the family Chenopodiaceae, and some species belonging to each look quite similar. In describing the eastern North American species of *Atriplex,* Fernald and Kinsey remark that they resemble *Chenopodium,* "differing only in technical characters of the flowers and fruits" (1958:180). In addition, several of the garden orach *(A. hortensis)* specimens in the National Herbarium, National Museum of Natural History, exhibited leaf shapes quite similar to those of *Chenopodium* (Figure 10.3).

Figure 10.3 Leaf shape of *Atriplex hortensis,* from a specimen in the National Herbarium, NMNH.

This, I think, is about as far as it is possible to pursue *Choupichoul* and what Le Page may have meant by *belle dame sauvage* down the tangled path of French dictionaries. The identity of the mystery grain of the Natchez is not yet firmly in hand as we turn from a consideration of orachs, mountain spinach, and beautiful savage ladies. But the field has been narrowed considerably in developing a short list of *Choupichoul* candidates. In Europe of the 1700s, *belle dame (Atriplex hortensis),* the garden orach, is the leading candidate for Le Page's European analog for *Choupichoul,* with other species of *Atriplex* and their look-alike cousins—the weedy chenopods—also likely possibilities. On this side of the Atlantic on the other hand, the genus *Atriplex* provides a single, lonely, and just barely viable *Choupichoul* candidate—*Atriplex patula,* while the richly diverse genus *Chenopodium* is crowded with look-alike candidates for the *Choupichoul* crown.

The Case for Chenopodium berlandieri

The richly diverse genus *Chenopodium,* however, also presents a darkly tangled taxonomic thicket that has claimed many an unwary student of goosefoot. Fortunately it is not necessary to wander unarmed through the dark forest of the genus *Chenopodium* in search of *Choupichoul* candidates. A number of clues have survived down through the centuries that point directly at a particular species of chenopod as having provided the mystery grain of the Natchez. The clues in question are not hidden in French manuscripts or documents, but rather are to be found in present-day stands of *Chenopodium*. We know from Le Page's account that the grain *Choupichoul* was a domesticated plant in the Lower Mississippi Valley in the 1720s, deliberately planted and harvested, with storage of seed stock. We could also expect that as a result of this degree of human intervention in the life cycle of the plant that it would have responded in a number of well-documented and well-understood ways under the general heading of the adaptive syndrome of domestication.

At the same time, Le Page also repeatedly mentions the undemanding nature of *Choupichoul* and its weedy enthusiasm in sand bank situations. Given its weedy and undemanding nature and its ability to thrive with little human attention during the growing season, it is not unreasonable to expect that *Choupichoul* has survived down to the present day while retaining at least some of its characteristics of domestication. In order to pursue this possibility—that the present-day weedy descendant of *Choupichoul* might be identified by looking for a species still exhibiting characteristics of prior domestication—it is only necessary to turn to recent studies of *Chenopodium*. Two such lingering indications of possible prior domesticated status have been recognized as still present in modern plants belonging to the species *C. berlandieri*. In dramatic contrast to the sequential flowering and resultant staggered maturation of fruit typical of wild chenopods, *Chenopodium berlandieri* exhibits a relatively uniform and precise photoperiodic flowering response, with plants flowering extremely late in the year, generally during the last week in August (Wilson 1981:238). The resultant uniform maturation of fruit in *Chenopodium berlandieri*, so uncharacteristic of uncultivated species, is common in domesticated crops (Wilson 1981:234), where harvesting by human populations strongly selects against maturation of fruits over a long period of time (Harlan et al. 1973:316). Within garden or field plots, plants that have simultaneous flowering and uniform maturation of fruit synchronized to an annual harvest by humans will make a greater contribution to the gene pool of the next generation. A second rapid and distinctive response by domesticated plants to human harvesting is the reduction or loss entirely of natural mechanisms of seed dispersal. This most diagnostic of the morphological characteristics that separate domesticated and uncultivated forms of a seed-bearing species (Harlan et al. 1973:314–315; Wilson 1981:234) is also strongly and automatically selected for during harvest. Plants that retain their seeds through the harvest season are more likely to contribute to the next generation than are those that continue to disperse seeds through natural shatter mechanisms.

Although not yet documented or quantified in detail, there is no question that *C. berlandieri* exhibits a considerable reduction in natural shatter/seed dispersal mechanisms relative to other chenopods of the East (Seeman and Wilson 1984:303). Munson (1984:309, 463) has recently described *C. berlandieri* stands in Indiana that retained a substantial percentage (90 percent) of their fruits/seeds into late December. For a number of years, I followed an extensive mixed stand of *Chenopodium berlandieri* and *Chenopodium missouriense* located in several vacant lots in Detroit, and photographs taken in January of 1985 (Figure 10.4) provide a good example of the loss of shatter mechanisms in *Chenopodium berlandieri*. Even at a distance the *C. berlandieri* plants could be differentiated from *C. missouriense* plants based on relative fruit retention. The *C. missouriense* plants were consistently devoid of fruit, having dispersed them long before. *C. berlandieri* plants, in contrast, retained their fruits, making it difficult not to notice this distinguishing characteristic. Because uncultivated present-day populations of *C. berlandieri* exhibit these two characteristics of domestication—uniform maturation of fruit and a reduction in natural dispersal—Wilson and Seeman have postulated that *C. berlandieri* is the weedy descendent of a prehistoric domesticate. These characteristics also significantly strengthen the case of *C. berlandieri* actually being the mysterious lost grain of the Natchez.

Figure 10.4 The relative seed retention of *Chenopodium berlandieri* (left) and *Chenopodium missouriense* (right) against a background of snow, in January 1985, Wayne County, Michigan.

Perhaps small stands of this plant have been waiting patiently for over 250 years, for the partnership with humans, the contract, to be resumed. Through the endless cycle of seasons, across the decades, *Chenopodium berlandieri* has been flowering, in unison, in the heat of late August, and retaining its seed through the long dark nights of winter, waiting to be harvested. Interestingly, the present-day harvesting of *C. berlandieri* offers another pathway along which to search for the identity of *Choupichoul.*

Along the Sand Banks of the Mississippi

If Le Page's *Choupichoul* grew so prolifically, with so little human attention, along the exposed sand banks of the main channel of the Mississippi River in 1720, perhaps its weedy descendants still colonize those sandy shores. Perhaps *Choupichoul* grew so well along the Mississippi with so little attention from the Natchez because that sand-bank habitat was very similar to the natural habitat setting of the wild ancestral populations of the mystery grain of the Natchez. The present-day meander belt of the Mississippi River might in fact still harbor two-thirds of a postulated prehistoric trinity of wild, weedy, and domesticated forms, or personas, of the same species. Perhaps the species of plant from which *Choupichoul* initially developed is still present, as an integral component in river margin floodplain vegetation communities, co-existing along with more weedy forms of the species, which colonize both anthropogenic and naturally disturbed natural levee soil situations.

Some initial indications that *Chenopodium berlandieri* might well fit the profile of a colonizing weed particularly adapted to the sandy floodplain soils of meander belt natural levee settings can be found scattered through the literature. Wahl (1954:44) states that "*C. bushianum* occurs most often as a weed of cultivated places but is found also in alluvium along streams and in waste places." Steyermark (1963:614) states that it "occurs in sandy fields and alluvial ground along rivers, waste places, wooded slopes and dry open or shaded ground." More recently, Wilson has described its habitat as "disturbed ground, especially alluvial soils of agricultural areas" (Wilson and Walters n.d.). While these few brief published habitat descriptions for *Chenopodium berlandieri* are encouraging, this species is also generally acknowledged to be very elusive and difficult to study in the field, being described as occurring "sporadically throughout northeastern North America" (Wilson 1981:238), and rarely forming large stands (Seeman and Wilson 1984:303, 304).

In searching for stands of *C. berlandieri* over the past five years, I independently reached the same sporadic small stand conclusion and characterization. It was easy enough to find large stands of *Chenopodium* in gardens, fields, construction sites—anywhere the soil had been recently disturbed—but *Chenopodium berlandieri* was rarely present in these stands, which were almost invariably dominated by much more common species—*Chenopodium missouriense*. In those rare situations where *Chenopodium berlandieri* was present, it was in small isolated pockets surrounded by a large stand of *C. missouriense*. But not having paid much attention to the few and scattered habitat clues for *C. berlandieri* in the literature, I had been looking in the wrong habitat settings. In November of 1985, however, my search for *Chenopodium berlandieri* led me to a sandy floodplain setting, and in an overgrown vegetable garden in Pike County, Ohio, and I finally encountered *Chenopodium berlandieri* on its home ground. A leafless, visually arresting yellow plant, 5 ft in height and occupying 1.5 sq m of ground, had numerous lateral branches with black terminal infructescences (see Figure 7.2). After hand-stripping the fruit, the plant was felled, later yielding a dry stalk that weighed 497 g. In comparison, the plant produced 428 g of clean fruit or seeds. This translates into a remarkable estimated harvest yield value of 2,854 kg per ha (see Chapter 7). Clearly a plant to be reckoned with in alluvial sandy soil settings, *C. berlandieri* was fairly abundant in this floodplain garden, while *C. missouriense,* the king of upland settings, was represented by only a few plants. Any lingering doubts that this plant was an anomaly, a floodplain freak, were erased three days later along the natural levee sand ridges of the main channel of the Mississippi River near the Dorena Ferry Landing in Mississippi County, Missouri. Although located 200 miles upriver from Natchez, this floodplain setting was essentially the same habitat in which the Natchez planted and harvested *Choupichoul.* The river valley location (Figure 7.4) contained a series of adjacent linear habitat zones that parallel the river. Exposed low water sand banks can be seen along the near and far shore. Moving away from these sand banks on the near shore there is a narrow 20-m-wide black willow vegetation zone and then a 70-m-wide overgrown and abandoned soybean field. A 100-m-wide swale of wet clay soil separates this field from other soybean fields. Crossing that swale to the soybean field, one enters the land of *Choupichoul,* and its present-day wild and weedy personas. Within this neglected field of soybeans, a rather scattered linear stand of 46 *Chenopodium berlandieri* plants ranging in size from 2 to 6 ft was located and harvested by hand-stripping over a period of four hours (see Figure 7.5). In spite of being seriously crowded by soybeans and other weeds, including only a few *C. missouriense* plants, these *C. berlandieri* plants had harvest yield values in the 400–1,800 kg per ha range, comparable to reported yields of *C. quinoa,* the South American chenopod domesticate. Growing in an anthropogenic disturbed soil situation along a natural levee of the Mississippi River, less than 50 m from the river's edge, these plants fit the *Choupichoul* profile quite closely in terms of growing without attention in sandy alluvial soils to a height of 3 ft or more, and when harvested in the fall yielding a great quantity of grain.

Moving toward the Mississippi River (and 50 m downstream), out of the bright sun of a human-created habitat into the understory shade of the black willow vegetation zone, a very different persona of *C. berlandieri* is encountered. In this undisturbed natural vegetation association, *C. berlandieri* thrives as a light shade understory dominant, often occurring in long linear stands of hundreds of plants. In this light shade setting, however, the plants are slender and delicate, often reaching a height of 7 or 8 ft, have fewer lateral branches, small diffuse infructescences, and have considerably reduced harvest potential (see Figure 7.7). A stand of 27 plants, occupying 3.5 sq m of ground, yielded about one-fifth the seed produced by the Pike County plant. I think that this delicate light shade understory persona of *C. berlandieri* is the present-day descendent of the wild ancestor of *Choupichoul,* and that it has been an integral component of such main channel vegetation communities throughout the Holocene.

Two distinctly different personas of *Chenopodium berlandieri* thus coexist today, growing only meters apart in two adjacent but quite different habitat zones that parallel the river. A tall delicate form having few and scattered infructescences grows abundantly in the light shade understory of undisturbed black willow vegetation communities, while a few steps away, in the bright sun of a cultivated field, a robust form adapted to anthropogenic disturbance opportunities along the natural levee produced five times the seed of the understory form. While this black willow vegetation community is bordered on one side by a humanly disturbed habitat setting, it is bordered on the other, river side, by a naturally disturbed habitat—the exposed sand banks or beaches of the Mississippi river.

Stepping out of the light shade of the willows into the bright sunlight of the river's edge, the weedy persona of *C. berlandieri* is again encountered, this time growing abundantly, within the exact habitat setting where *Choupichoul* was cultivated by the Natchez. Often supporting wild bean *(Strophtostyles helvola)* runners, these sand-bank plants looked just like the soybean field specimens—large, robust, with numerous lateral branches and terminal infructescences (Figure 7.6). With the exception of a few *C. missouriense* and *C. ambrosioides* plants, no other chenopods were observed, and *Atriplex* was not present.

So in searching for the modern wild and weedy descendants of the *Choupichoul* weed-crop complex within the described context of *Choupichoul* cultivation, we find a single species of chenopod, *Chenopodium berlandieri,* generally dominating the sandy soils of main channel natural levee ridges to the exclusion of other chenopods, ready to expand out of the willow understory into the disturbed soil opportunities created both by the river and by humans (see Chapter 7 for a detailed discussion of modern harvest research on *C. berlandieri*).

Manifesting several characteristics of domestication, this species also fits Le Page's profile of *Choupichoul* quite well. But while modern field studies certainly provide strong evidence of weedy descendants and support for the *Choupichoul* candidacy of *Chenopodium berlandieri,* we need to very briefly turn to the archaeobotanical record to find the domesticated persona of *Choupichoul.*

Chenopodium berlandieri in Prehistory

As a result primarily of accelerator mass spectrometer (AMS) and scanning electron microscopy (SEM) based research carried out on dry cave and rock shelter collection in the past seven years, there now exists a basic temporal-developmental framework of understanding of the prehistory of domesticated *Chenopodium* in the East. At least two distinct cultivar varieties of domesticated *Chenopodium* were present in the East prehistorically. A thin-testa variety, *C. berlandieri* ssp. *jonesianum,* was being cultivated over a broad mid-latitude area of the East by the birth of Christ, and has been recovered in abundance from dry cave and rock shelter storage contexts of the Middle Woodland time period (see Chapters 5 and 6). Widely cultivated by 2,000 B.P., this thin-testa cultivar had been brought under domestication by 3,500 B.P. based on recent accelerator dates on specimens recovered from eastern Kentucky rock shelters (Smith and Cowan 1987). A pale-seeded variety of *Chenopodium berlandieri* was also present in the East by 1000 B.C., and perhaps earlier. Both of these varieties of *C. berlandieri* were cultivated into the post A.D. 1200 maize-centered agricultural systems of Mississippian populations. Both pale-colored and thin-testa cultivar chenopods have been documented in post A.D. 1000 Ozark Bluff dweller contexts, and both have been recovered from the Gypsy Joint site, a Mississippian farmstead dating to around A.D. 1300 and located less than 160 km (100 miles) from the Dorena Ferry modern stand mentioned above. In contrast, as far as I can determine, the genus *Atriplex* is not represented in the archaeobotanical record for the eastern United States.

Unfortunately, even though these thin-testa and pale-colored (extremely thin-testa) domesticated varieties of *Chenopodium berlandieri* can be tracked from 1500 B.C. up to A.D. 1300, neither has yet been documented in post A.D. 1300 contexts. But I have no doubts that their record of cultivation will be extended up into the historic period as good quality archaeobotanical assemblages are recovered from late prehistoric and early historic contexts in the Lower Mississippi Valley. Similarly, I am convinced that when known Natchez sites of the 1700s are excavated with an emphasis on the recovery of plant remains, domesticated forms of *Chenopodium berlandieri* will be represented.

Conclusion

The various lines of inquiry outlined in this chapter all converge on *Chenopodium berlandieri* as being *Choupichoul,* the mystery crop of the Natchez. *C. berlandieri* was grown as a domesticated crop in river valleys of eastern North America beginning as early as 3,500 years ago, and its continued role as a field crop is documented in fourteenth-century sites located in the Lower Mississippi Valley. While retaining several attributes of its prior life as a domesticated crop (uniform maturation of fruit and a reduction of natural seed dispersal), *Chenopodium berlandieri* grows today in the exact sand bank setting in which Natchez women and children planted *Choupichoul* in the 1720s. In addition to filling the habitat of *Choupichoul, C. berlandieri* also matches Le Page's description of *Choupichoul* in terms of size and producing a "grain." Finally, in his efforts to characterize *Choupichoul* in terms of a European analog that his readers could identify, he chose *belle dame sauvage.* The common name *belle dame* in turn quite likely refers to *Atriplex hortensis,* a European garden plant that closely resembles *Chenopodium.* The addition of *sauvage,* indicating a wild plant, strengthens the likelihood that du Pratz was indicating *Chenopodium,* the wild look-alike relative of *Atriplex,* as his European analog for *Choupichoul.*

It would appear, therefore, that in the 1720s, on the broad battures of the Mississippi River, Le Page du Pratz witnessed and described the closing chapter of an agricultural economy that had existed in eastern North America across 4,000 years. This last brief glimpse, through a European's eyes, of the only remaining seed crop component of a premaize economy that had flourished for so long, provides an appropriate closure. It returns us full circle, in a way, back to the initial emergence of farming economies 4,000 years earlier, when, in all likelihood, women first planted seeds, returning in the fall for the harvest.

Acknowledgments

William Sturtevant offered comments on an early draft of this chapter, and Valérie Chaussonnet very generously provided careful translation and copy edited of a number of original passages in *Histoire de la Louisiane.*

Literature Cited

Asch, D., and N. Asch
1977 Chenopod as Cultigen: A Re-evaluation of Some Prehistoric Collections from Eastern North America. *Midcontinental Journal of Archaeology* 2:3–45.

Brown, I.
1984 *Natchez Indian Archaeology.* Lower Mississippi Valley Archaeological Survey Report. Peabody Museum, Harvard University, Cambridge, Massachusetts.

Devries, Louis
1968 *French-English Science Dictionary.* McGraw Hill, New York.

Eleuterius, L. N.
1980 *Tidal Marsh Plants of Mississippi and Adjacent States.* Mississippi-Alabama Sea Grant Program Publication 77–039.

Fernald, M. L., and A. C. Kinsey
1958 *Edible Wild Plants of Eastern North America.* Idlewild Press, New York.

Gilmore, M. R.
1931 Vegetal Remains of the Ozark Bluffdweller Culture. *Papers of the Michigan Academy of Science, Arts, and Letters* 14:83–105.

Harlan, J. R., J. M. J. de Wet, and E. G. Price
1973 Comparative Evolution of Cereals. *Evolution* 27:311–325.

Harrap, A.
1972 *Harrap's New Standard French and English Dictionary.* Scribners, New York.

Heiser, C.
1985 Some Botanical Considerations of the Early Domesticated Plants North of Mexico. In *Prehistoric Food Production in North America,* edited by R. I. Ford, pp. 57–72. Museum of Anthropology, Anthropological Papers No. 75. University of Michigan, Ann Arbor.

Jones, V.
1936 The Vegetal Remains of Newt Kash Hollow Shelter. In *Rock Shelters in Menifee County, Kentucky,* edited by W. S. Webb and W. D. Funkhouser, pp. 147–165. Reports in Archaeology and Anthropology 3(4). University of Kentucky, Lexington.

Knight, V.
1990 Social Organization and the Evolution of Hierarchy in Southeastern Chiefdoms. *Journal of Anthropological Research* 46:1–23.

Linton, R.
1924 The Significance of Certain Traits in North American Maize Culture. *American Anthropologist* 26:345–349.

Lowe, E. H.
1921 *Plants of Mississippi.* Mississippi State Geological Survey Bulletin 17.

de Montigny, Dumont
1753 *Memoires Historiques de la Louisiane.* Paris.

Munson, P.
1984 Weedy Plant Communities on Mud-Flats and Other Disturbed Habitats in the Central Illinois River Valley. In *Experiments and Observations on Aboriginal Wild Plant Food Utilization in Eastern North America,* edited by P. Munson, pp. 379–385. Prehistoric Research Series 6(2). Indiana Historical Society, Bloomington.

Le Page du Pratz, Antoine S.
1758 *Histoire de la Louisiane.* 3 volumes. Paris.

1975 *The History of Louisiana.* A facsimile reproduction of the 1774 English translation published by T. Becket, with an Introduction by J. Tregle. Louisiana State University Press, Baton Rouge.

Richelet, Pierre
1759 *Dictionnaire de la langue Françoise.* Paris.

Seeman, M., and H. Wilson
1984 The Food Potential of *Chenopodium* for the Prehistoric Midwest. In *Experiments and Observations on Aboriginal Wild Plant Food Utilization in Eastern North America,* edited by P. Munson, pp. 299–316. Prehistoric Research Series 6(2). Indiana Historical Society, Bloomington.

Smith, B., and C. W. Cowan
1987 Domesticated *Chenopodium* in Prehistoric Eastern North America: New Accelerator Dates from Eastern Kentucky. *American Antiquity* 52:355–357.

Steyermark, J.
1963 *Flora of Missouri.* Iowa University Press, Iowa City.

Sturtevant, W. C.
1965 Historic Carolina Algonkian Cultivation of *Chenopodium* or *Amaranthus. Southeastern Archaeological Conference Bulletin* 3:64–65.

Swanton, J. R.
1911 *Indian Tribes of the Lower Mississippi Valley and Adjacent Coast of the Gulf of Mexico.* Bureau of American Ethnology Bulletin 43. Smithsonian Institution, Washington, D.C.

1946 *The Indians of the Southeastern United States.* Bureau of American Ethnology, Bulletin 137. Smithsonian Institution, Washington, D.C.

Tregle, J. G.
1975 Introduction. In *The History of Louisiana,* by Le Page du Pratz. Facsimile reproduction of the 1774 English translation. Louisiana State University Press, Baton Rouge.

Usher, G.
1974 *A Dictionary of Plants Used By Man.* Hafner Press, New York.

Wahl, H.
1954 A Preliminary Study of the Genus *Chenopodium* in North America. *Bartonia* 27:1–46.

Wilson, H.
1981 Domesticated *Chenopodium* of the Ozark Bluff Dwellers. *Economic Botany* 35:233–239.

Wilson, H., and T. Walters
n.d. Vascular Flora of the Southeastern United States Chenopodiaceae. Ms. on file, Department of Biology, Texas A&M University, College Station.

Yarnell, R.
1972 Sunflower, Sumpweed, Small Grains, and Crops of Lesser Status. Chapter prepared for the *Handbook of North American Indians.* Ms. on file, Department of Anthropology, National Museum of Natural History, Smithsonian Institution, Washington, D.C.

Yarnell, R., and J. Black
1985 Temporal Trends Indicated by a Survey of Archaic and Woodland Plant Food Remains from Southeastern North America. *Southeastern Archaeology* 4:93–107.

IV. Synthesis

CHAPTER 11

Origins of Agriculture in Eastern North America

Introduction

As a result of research carried out over the past decade, eastern North America now provides one of the most detailed records of the origins of agriculture available. Spanning a full three millennia, the transition from forager to farmer in eastern North America involved the domestication of four North American seed plants during the second millennium B.C., the initial emergence of food production economies based on local crop plants between 250 B.C. and A.D. 200, and the rapid and broad-scale shift to maize-centered agriculture during the three centuries from A.D. 800 to 1100.

The development of agriculture has long been considered a major milestone in human evolution. During the past 10,000 years agricultural economies have also caused significant changes in the earth's ecosystems. This post-Pleistocene transition from foragers to farmers, from a reliance on wild species of plants and animals to food production economies, took place at different rates and times in various regions of the world and involved a rich variety of crop plants (Cowan and Watson 1992). In many regions this developmental transition with its major consequence for human societies and terrestrial ecosystems is far from well documented. As a result of a substantial increase in the amount, quality, and variety of information gained for eastern North America during the past ten years (Ford 1985; Yarnell and Black 1985), this region now provides one of the most detailed records of agricultural origins available.

The recent increase of essential information on agriculture in eastern North America is in large part attributable to the application of four important technological advances to archaeology, and serves to underscore the important role of instrumentation in guiding and stimulating research (Adams 1988). (1) Water flotation technology has produced dramatic improvements in the recovery of small carbonized seeds and other plant parts from archaeological contexts (Pearsall 1989); this technology has allowed the

This chapter originally appeared in *Science* (1989) 246:1566–1571 development of detailed temporally long archaeobotanical sequences for different areas of eastern North America [see, for example, Chapman and Shea 1981; Johannessen 1984; Crites 1987; Asch and Asch 1985). (2) Scanning electron microscopy (SEM) has documented minute morphological changes in seed structure associated with the process of domestication (Smith 1988). (3) Accelerator mass spectrometer (AMS) radiocarbon dating has provided, for the first time, accurate and direct age determinations of very small samples of seeds and other plant parts (Chapman and Crites 1987; Ford 1987; Fritz and Smith 1988). (4) Stable carbon isotope analysis of human bone (Bender et al. 1981; Lynott et al. 1986; Buikstra et al. 1987; Ambrose 1987) has been a direct means of observing temporal and geographical trends in the relative consumption of maize by prehistoric groups in eastern North America.

As might be expected, recent advances in the recovery, identification and quantification, and accurate temporal placement of domesticated plant species in eastern North America have shown that the transformation of forager to farmer was a longer and more complex process than previously thought. In general treatments of the topic, for example, three aspects of the agricultural transition—plant domestication, the development of food production economies, and the shift to monocrop systems—are often either singled out or causally and temporally conflated as marking the origin of agriculture. In the archaeological record of eastern North America, however, these three shifts can be clearly recognized as distinct episodes of change that are linked developmentally yet separated temporally, forming a sequence that spans three millennia.

Between 2000 and 1000 B.C. native North American crop plant species were initially brought under domestication in eastern North America, as indicated by distinctive morphological changes in the structure of reproductive propagules (fruits and seeds) associated with the adaptive syndrome of domestication (Smith 1987a).

Seven to twelve centuries later, between 250 B.C. and A.D. 200, a subsequent initial emergence of food production economies took place, with local crop plants gaining considerable economic importance. This is reflected in increased representation in seed assemblages, as well as by other related developments in technology and settlement patterns, for example (Smith 1986, 1987b [Chapter 9]).

Six to nine centuries later, between A.D. 800 and 1100, a shift in food production economies occurred and a single nonindigenous species (maize) came to dominate the fields and diets of farming societies. This shift is reflected in an increased representation of maize in archaeobotanical assemblages. More directly, changing $^{12}C/^{13}C$ ratios in human bone indicate an increased consumption of a C_4 plant (maize) relative to the intake of indigenous wild and cultivated C_3 food plants (Bender et al. 1981; Lynott et al. 1986; Buikstra et al. 1987; Ambrose 1987).

An Independent Center of Plant Domestication

In 1971, Harlan outlined three localized centers of plant domestication (the Near East, north China, and Mesoamerica), along with three larger, dispersed noncenter areas of domestication (Africa, southeast Asia, South America) (Harlan 1971). Although the status of eastern North America was far from clear at that time, it can now be identified as a fourth independent and localized center of plant domestication (Figure 11.1; Smith 1987 [Chapter 3]).

As recently as 1985, however, the established presence during the Middle Holocene (6000 to 2000 B.C.) of a small, thin-walled (< 2.0 mm) gourd in the region appeared to have confirmed eastern North America as a secondary recipient of domesticated plants and agricultural concepts from Mesoamerica. In the 1980s a series of direct AMS radiocarbon dates were reported for rind fragments of a variety of small, thin-walled gourd assignable to the genus *Cucurbita* that had been recovered from five sites: Koster (two dates 5150 and 4870 B.C.; Asch and Asch 1985), Napoleon Hollow (5050 B.C.; Asch and Asch 1985), Carlston Annis (3780 B.C.; Watson 1985), Bowles (2110 B.C.; Watson 1985), and Hayes (4390 B.C.; Crites 1985b). A direct AMS date has also been obtained on a *Cucurbita* seed from Cloudsplitter (2750 B.C.; Cowan et al. 1981). In addition, 65 measurable *Cucurbita* seeds and ten rind fragments were recovered from well-dated contexts at the Phillips Spring site (2300 B.C.) (Figure 11.1; King 1985).

In the apparent absence of any modern species of wild *Cucurbita* gourd north of Texas, these firmly dated rind fragments and seeds appeared to demonstrate that the tropical domesticate *Cucurbita pepo* had been introduced into the region from Mesoamerica, along with the concept of agriculture, long before the 2000 to 1000 B.C. period of initial domestication of North American plants in the region. Alternatively, it was suggested that these early rind fragments and seeds reflected a more widespread Middle Holocene geographical distribution of an indigenous small, thin-walled gourd (compare *C. texana*) (Cowan et al. 1981; Heiser 1985). Morphological analysis of the available archaeological specimens and recent taxonomic research on *Cucurbita* by Decker (1988; Decker and Wilson 1986) now strongly support this latter explanation. None of the rind fragments or seeds of *Cucurbita* gourd recovered from contexts before 2000 B.C. in eastern North America can be identified morphologically as representing a domesticated form of *C. pepo*. On the basis of rind and seed measurements and morphology, specimens recovered from Cloudsplitter (about 850 to 350 B.C.) and Salts Cave (about 550 B.C.) (Figure 11.1) provide the earliest evidence of large, thick-walled, clearly domesticated varieties of *C. pepo* in eastern North America (Smith 1987a [Chapter 3], 1992 [Chapter 12]; King 1985; Yarnell 1974, 1987; Gardner 1987). These thick-walled domesticated varieties of cucurbit may have been introduced from Mesoamerica at about 1000 B.C. It appears more likely, however, that they were initially domesticated in eastern North America during the period 2000 to 1000 B.C. from the indigenous *C. texana*-like wild progenitor discussed above. Decker (1988) has recognized a taxonomic dichotomy of *Cucurbita* at the subspecies level that supports the idea of two independent centers of domestication for the species. Pumpkins, marrows, and a few ornamental gourds of one subspecies *(C. pepo* ssp. *pepo)* appear to have been first domesticated in Mesoamerica, with acorn squashes, scallop squashes, fordhooks, crooknecks, and most of the ornamental gourds of the second subspecies *(C. pepo* ssp. *ovifera)* having been domesticated in eastern North America. Thus rather than relegating the region to a role of secondary recipient of agriculture from Mesoamerica, *Cucurbita* taxonomy and the archaeological record underscore the identity of eastern North America as an independent center of plant domestication.

Similarly, two reports from Middle Holocene contexts of another potential introduced domesticate, the bottle gourd *(Lagenaria siceraria),* do not cast doubt on the evidence for eastern North America as an independent center of plant domestication. Wild forms of the bottle gourd have never been located or described. As a result, morphological differences between wild and domesticated varieties have yet to be documented (Heiser 1985, 1989). The Phillips Spring (2300 B.C.; King 1985) bottle gourd rind fragments and seeds are among the smallest known for the species and cannot be assumed to represent an introduced domesticate (Heiser 1985). The small, fragmentary, thin-walled bottle gourds recovered from the Windover site (5350 B.C.; Newsom 1988) on the east coast of Florida (Figure 11.1) are similarly dubious as domesticates. They do serve, however, to underscore the possibility that undomesticated bottle gourds were first carried to eastern North America by ocean currents, either from Africa or tropical America, at an early date (Cowan and Watson 1992; Heiser 1985; Newsom 1988).

In addition to the likely domestication of squash *(C. pepo* ssp. *ovifera),* three other native North American plants were brought under domestication in eastern North America during the second millennium B.C.: sumpweed or marshelder *(Iva annua),* sunflower *(Helianthus annuus),*[1] and chenopod *(Chenopodium berlandieri)*. For marshelder and sunflower the morphological change indicating domestication is an increase in achene size (Asch and Asch 1985; Yarnell 1978), whereas a reduction in seed coat (testa) thickness reflects the transition to domesticated status in *Chenopodium* (Fritz and Smith 1988; Smith 1984, 1985a [Chapter 5], 1987a [Chapter 3]).

The temporal trend of increase in the size of marshelder and sunflower achenes in eastern North America are shown in Figure 11.2. With a mean achene length of 4.2 mm (a 31 percent increase over modern undomesticated *Iva* populations), the approximately 2000 B.C. Napoleon Hollow *Iva* assemblage marks the earliest evidence for domestication of

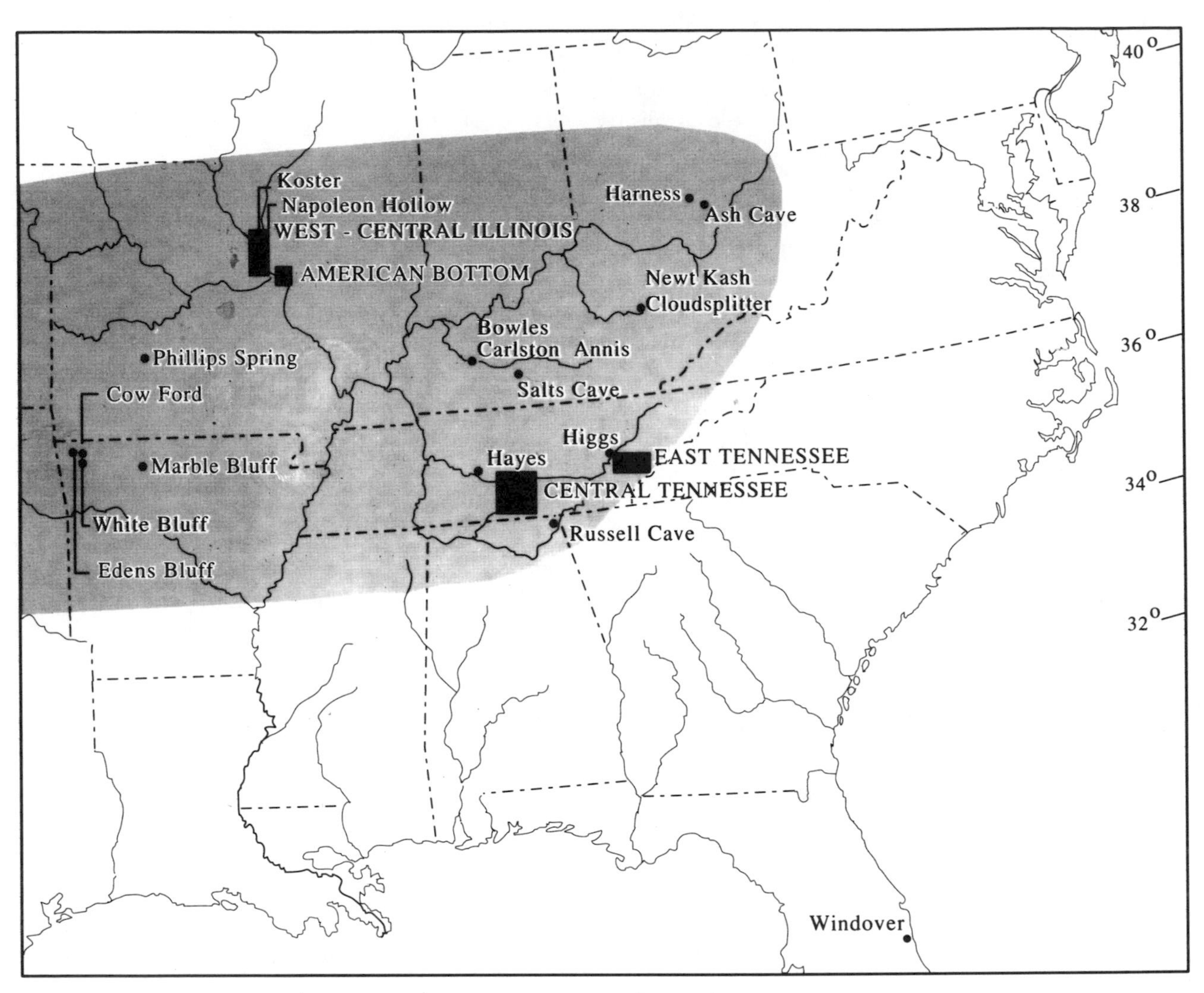

Figure 11.1 Archaeological sites and regions that provide information on agricultural origins in eastern North America. Shaded area indicates the interior mid-latitude zone of domestication of native North American seed crops at 2000 to 1000 B.C. and initial development of early food production economies at 250 B.C. to A.D. 200.

marshelder in the region (Asch and Asch 1985). Similarly, the approximately 900 B.C. Higgs site sunflower achenes, with a mean length of 7.8 mm (as opposed to modern wild populations, which range in mean length from 4.0 to 5.5 mm) provide the earliest evidence for domestication of this species in eastern North America. Representing an early but not initial stage of domestication (Yarnell 1978), the Higgs sunflower achenes suggest that initial domestication of *H. annuus* occurred at about 1500 B.C. (Figure 11.2; Yarnell 1978, 1983; Ford 1985).

This proposed initial domestication of *H. annuus* at about 1500 B.C. corresponds to the earliest evidence for the presence of a thin-testa domesticated form of chenopod *(C. berlandieri* ssp. *jonesianum)* (Smith and Funk 1985) in eastern North America. Whereas modern wild populations of *C. berlandieri* in eastern North America have testa thickness values of 40–80 microns (Smith 1985b [Chapter 6]), chenopod assemblages with thin testas (< 20 microns) have been reported from a number of sites dating from 980 B.C. to A.D. 150 (Ash Cave, Edens Bluff,

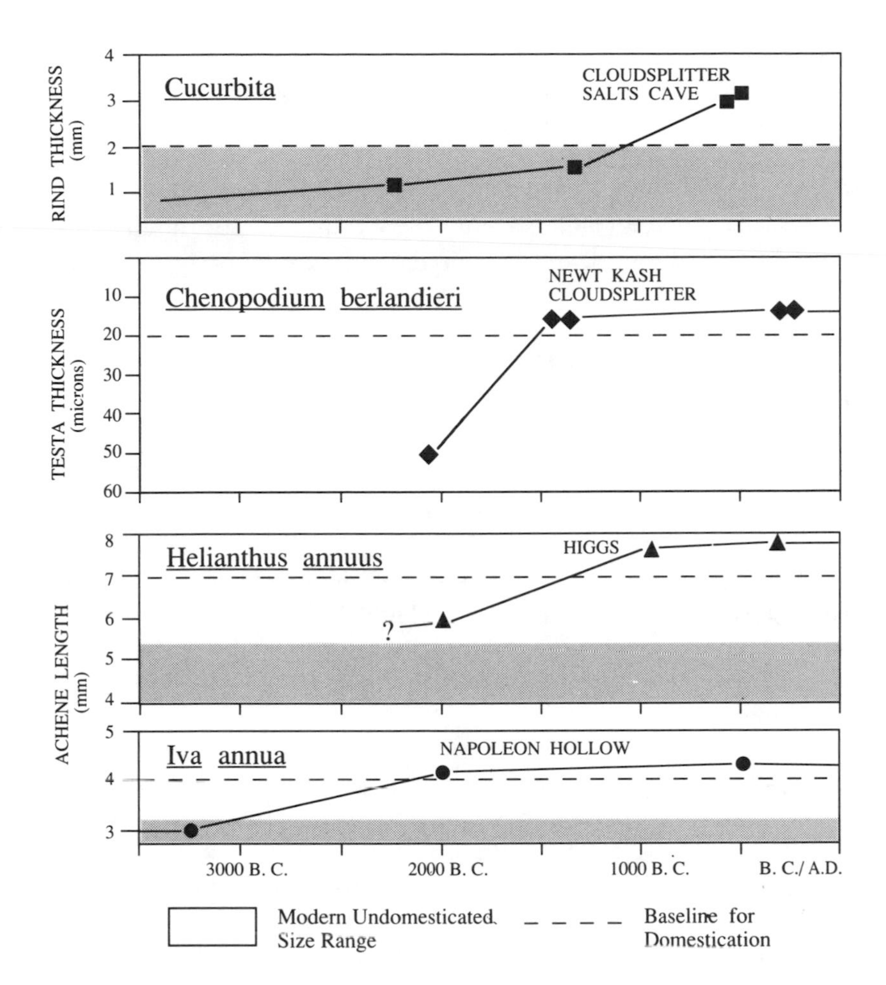

Figure 11.2 The second millennium B.C. domestication of indigenous crop plants in eastern North America based on morphological changes in reproductive propagules. (Adapted from Smith 1987 [Chapter 3, Figure 3.2].)

Marble Bluff, Russell Cave, and White Bluff; Fritz and Smith 1988; Smith 1987a [Chapter 3]). Direct AMS dates on thin-testa chenopod specimens from Newt Kash (1450 B.C.) and Cloudsplitter (1500 B.C.) provide the earliest evidence for domestication of this species in the region (Smith and Cowan 1987). A second type of cultigen *Chenopodium* having an extremely thin, translucent seed coat has been documented at the Cow Ford site by A.D. 330 (Figure 11.1; Fritz and Smith 1988), and specimens recovered from about 700 B.C. contexts at Cloudsplitter suggest that it too may have been brought under domestication during the second millennium B.C.[2]

Both of these second millennium B.C. morphological changes are the result of strong selective pressures within seed beds favoring seedlings that germinate quickly (reduced testa thickness and germination dormancy), and grow rapidly (increased seed size and food reserves) (Smith 1984, 1985a [Chapter 5]; Harlan et al. 1973). As such, they provide clear evidence of a major second millennium B.C. escalation in the level of human intervention in the life cycle of these crop plants—the deliberate storage and planting of seed stock. This major step of planting was preceded by a three millennium long (about 5000 to 2000 B.C.) coevolutionary process, with human and plant populations interacting within river valley settings across a broad interior riverine area of eastern North America (Figure 11.1).

In some segments of mid-latitude (34–40 degrees North) river valleys the widely scattered short-term occupational lenses of antecedent forager bands were replaced, in the period 5000 to 2000 B.C., by deep shell and midden-mound settlements (Smith 1986, 1987a [Chapter 3]). These midden mounds and shell mounds were often located close to oxbow lakes and shoal areas having abundant aquatic resources, as humans narrowed their site preference and responded opportunistically to Middle Holocene river valley stabilization and biotic enrichment. Apparently reoccupied annually through the growing season, such settlements constituted the initial emergence of continually disturbed anthropogenic habitats in eastern North America. Within such habitats, a wide range of human activities would have frequently disturbed and enriched the soil, providing excellent long-term colonizing opportunities (Smith 1987 [Chapter 3]; Anderson 1956; Fowler 1971) for a variety of early successional floodplain plant species, including *C. berlandieri, I. annua, H. annuus* (Heiser 1989), and *Cucurbita* gourd.

It is within such disturbed "domestilocality" (Smith 1987 [Chapter 3]) settings that the long, gradual developmental pathway to plant domestication likely took place. Initially neither requiring nor receiving any human assistance other than inadvertent soil disturbance and enrichment, colonizing weedy stands of these four species, through simple toleration, provided a localized and predictable supplement to "natural" floodplain stands. Although the archaeological record does not show the stages, a gradual co-evolutionary progression of increasing human intervention would have gone from initial toleration through first inadvertent and then active, intentional encouragement of these food plants, to the deliberate storage and planting of seed stock (Smith 1984, 1985a [Chapter 5], 1987a [Chapter 3]). It is this critical and deliberate step of planting that marks the beginning of cultivation and the onset of automatic selection pressures within affected plant populations that produce the morphological changes associated with domestication (Harlan et al. 1973; Heiser 1988). This co-evolutionary progression, which occurred over a span of two to three millennia, transformed stands of colonizing weeds first into inadvertent or incidental gardens and finally, by the second millennium B.C., to intentionally managed and maintained gardens of domesticated crop plants.

Within the annual economic cycle of incipient agricultural groups in eastern North America, these initial domestilocality gardens likely played a significant role in providing a dependable, managed, and storable food supply for late winter to early spring (Watson 1985; Cowan et al. 1981; Cowan 1985). From the limited occurrence of seeds of these species in archaeobotanical assemblages before 500 B.C., however, domesticated crops did not become a substantial food source, and food production did not play a major economic role until 500 B.C. to A.D. 200, a full thousand years after initial domestication.

The Emergence of Food Production Economies

The period from 250 B.C. to A.D. 200 witnessed the initial development, subsequent elaboration, and eventual cultural transformation of the Middle Woodland period Hopewellian societies of eastern North America. The large and impressive geometric earthworks, conical burial mounds, and elaborate mortuary programs of Hopewellian populations have been the focus of archaeological attention for well over a hundred years (Brose and Greber 1979). It is only within the past decade, however, that much information has been recovered concerning Hopewellian food production economies.

For almost thirty years discussions of Hopewellian domesticates centered almost exclusively on the dietary role of maize *(Zea mays),* which had been recovered from contexts that suggested an arrival in eastern North America as early as 500 B.C. Recent direct AMS radiocarbon dates on proposed early corn, however, have substantially revised the timetable of initial introduction. The earliest convincing macrobotanical evidence of the presence of maize in eastern North America are the directly dated carbonized kernel fragments from the Icehouse Bottom site (A.D. 175) and the Harness site (A.D. 220; Chapman and Crites 1987; Ford 1987). In addition, $^{12}C/^{13}C$ values (Bender et al. 1981; Lynott et al. 1986; Buikstra et al. 1987; Ambrose 1987) show no evidence of corn consumption by Hopewellian groups. If *Cucurbita* squash was domesticated independently in eastern North America (Decker 1988), then maize is the first tropical food crop to be introduced into the region. On the basis of stable carbon isotope analysis and AMS dating of early maize, it is now evident that corn did not precipitate either the development of Hopewellian societies or a rapid shift to agricultural economies in eastern North America.[3] Rather it was adopted as a minor, almost invisible addition to well-established food producing economies. These premaize economies were based on the four local domesticates discussed above, as well as on three other cultivated seed crops for which a convincing case for domestication has yet to be made: erect knotweed *(Polygonum erectum)*, maygrass *(Phalaris caroliniana)*, and little barley *(Hordeum pusillum)*.

Early food producing economies based on these seven local crop plants were established within a broad mid-latitude riverine zone (34–40 degrees North) extending from the Appalachian wall west to the prairie margin (Figure 11.1). Rather than being uniform across this broad zone, emergent food production economies exhibited variation between and within different regions in terms of the presence and relative importance of different seed crops (Smith 1984, 1985a [Chapter 5], 1992 [Chapter 12]). Forager economies with little reliance on cultivated plants also likely persisted within some areas of this mid-latitude mosaic of emergent food producing societies. Outside the zone, along the Atlantic and Gulf coastal plains and across the northern latitudes, forager economies based almost exclusively on wild species of animals and plants (with some *Cucurbita* cultivation) persisted until the A.D. 800 to 1100 shift to maize-centered agriculture.

The time frame of initial emergence of premaize food-producing economies in this mid-latitude zone was similarly variable, apparently beginning as early as 500 B.C. in parts of Kentucky and developing over a much broader region between about 250 B.C. and A.D. 200. This transition to food production economies is signaled by a dramatic increase in the representation (and assumed economic importance) of the aforementioned seven crop plants in archaeobotanical seed assemblages (Figure 11.3; Yarnell and Black 1985; Smith 1987c [Chapter 9]). This increase is documented in mid-latitude rock shelters and caves, and in four areas where detailed, temporally long archaeobotanical sequences have been established: east Tennessee (Chapman and Shea 1981), the American Bottom (Johannessen 1984), central Tennessee (Crites 1985, 1987), and west-central Illinois (Asch and Asch 1985) (Figures 11.1 and 11.3). More direct evidence of increasing dietary importance of these local seed crops is provided by the large sample of human paleofecal material recovered from Salts Cave (about 650 to 250 B.C.), which contained substantial numbers of marshelder, maygrass, sunflower, chenopod, and *Cucurbita* seeds. Yarnell estimates that premaize crop plants accounted for 75 percent of the plants consumed by those living at the Salts Cave site and perhaps two-thirds of all foods they consumed (Smith 1992 [Chapter 12]). This paleofecal evidence, along with increased representation of premaize crop plants in Salts Cave deposits by 500 B.C. also indicate that food-producing economies apparently developed in central Kentucky and perhaps some other areas of the

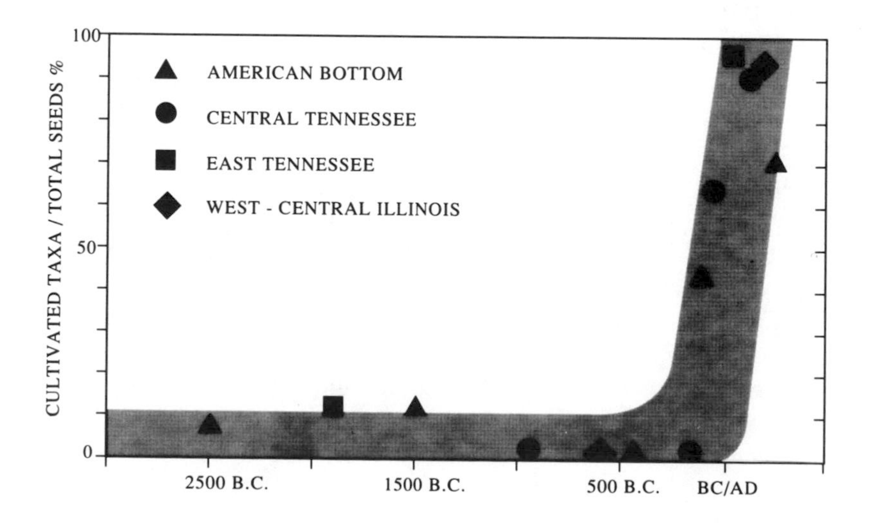

Figure 11.3 The initial emergence of food-producing economies in eastern North America, as reflected by increased representation of cultivated taxa in four regional archaeobotanical sequences. (After Smith 1987 [Chapter 9].)

East somewhat earlier than 250 B.C. (Yarnell and Black 1985; Yarnell 1974, 1987, 1988; Gardner 1987). Other indications of emerging premaize farming economies include the presence of chert hoes and hoe flakes, increases in pollen and macrobotanical indicators of field-clearing activities (Johannessen 1984; Yarnell 1974, 1987, 1988; Gardner 1987; Chapman et al. 1982), evidence of seed storage in a variety of containers and in pit features (Fritz and Smith 1988), and advances in ceramic cooking vessels (Braun 1987) for seed processing.

The indigenous crop plants of eastern North America had considerable economic potential, based both on available harvest yield information and on the nutritional composition of their seeds. Modern commercial production levels for sunflower, along with studies of present-day nondomesticated stands of *I. annua, C. berlandieri,* and *P. erectum* indicate potential harvest yield values in the range of 500 to 1,000 kg per ha (Smith 1987c [Chapters 7, 8, and 9]). This 500 to 1,000 kg per ha harvest yield projection overlaps with harvest estimates for maize in prehistoric eastern North America (400 to 1,400 kg per ha) and *Chenopodium quinoa* in South America (500 to 1,000 kg per ha; Smith 1987c [Chapter 7]). It also compares well with historic (A.D. 1850 to 1900) mean yield values for European wheats grown during the early Neolithic: winter and spring einkorn (835, 645 kg per ha) and winter and spring emmer-spelt (1,045 and 756 kg per ha, respectively; Gregg 1988).

In terms of nutritional composition, premaize crop plants can be divided into those with starchy or oily seeds. Of the five fall maturing crops, two (erect knotweed and chenopod) have high carbohydrate content, while the other three are high in oil or fat (*Cucurbita,* marshelder, and sunflower). Both spring-maturing crops (little barley and maygrass) are high in carbohydrate content (Smith 1987b [Chapter 9], 1992 [Chapter 12]).

On the basis of the most informative regional data sets available (from the American Bottom, central Tennessee, and west-central Illinois) (Figure 11.1), the habitation sites of about A.D. 0–200 Hopewellian farming societies were small, one to three household settlements dispersed along stream and river valley corridors (Smith 1987b [Chapter 9]). Along segments of some river valleys having protein-rich floodplain lakes and marshes, such as the lower Illinois River, small household settlements appear to form loose spatial concentrations that have been called "villages," even in the absence of any indications of an overall community plan. Along with the seeds of crop plants, a wide variety of different species of wild animals and plants are represented in these small household settlements. Variation from household to household in the composition of faunal and floral assemblages indicate both differential species availability within individual household catchment areas and the subsistence autonomy of these basic economic units of society (O'Brien 1987). The broad spectrum and flexible premaize economies of these household units also document the addition of a food production sector, with its storable harvest as a buffer against food shortage, to pre-existing forager subsistence patterns, rather than the wholesale replacement of earlier, largely hunting and gathering economies.

This initial and additive emergence of multicrop food-producing economies in eastern North America at about 250 B.C. to A.D. 200, which involved seven high-yield, high-nutritional profile, spring and fall harvest crop plants, represented a major economic and social transformation. Six to nine centuries later (A.D. 800 to 1100), these indigenous multicrop food-producing economies would in turn be supplemented (and even later largely supplanted) by agricultural systems centering on a single introduced tropical food crop—maize.

The Shift to Maize-Centered Agriculture

Although introduced into at least some areas of eastern North America by A.D. 200 (Chapman and Crites 1987; Ford 1987), probably from the Southwest, maize remained a minor cultigen, or perhaps a high status or ceremonial crop, until after A.D. 800. Although often labeled "Midwestern twelve row," pre-A.D. 800 maize in eastern North America is represented by limited amounts of small kernel and cupule fragments and cannot be easily characterized or compared to either modern or proposed ancestral varieties of corn.

During the six centuries from A.D. 200 to 800, indigenous crops and food-producing economies were becoming increasingly important across the mid-latitude premaize farming zone (Smith 1992 [Chapter 12]). This gradual developmental trend took a rapid and expansive new direction between A.D. 800 and 1100, with agricultural economies based largely on corn developing from north Florida to the northern limits of maize farming. This rapid shift encompassed many of the largely forager societies of the Atlantic and Gulf coastal plains and northern latitudes as well as the mid-latitude premaize farming zone. This shift involved economic transitions from both long-established hunting and gathering subsistence patterns and from indigenous multicrop food production systems.

The evidence for maize agriculture across eastern North America is the marked increase in maize recovered from archaeological contexts after A.D. 800 to 900 and the changing $^{12}C/^{13}C$ values obtained from human bone that indicate an initial substantial consumption of maize during the period from A.D. 900 to 1100 (Figure 11.4).

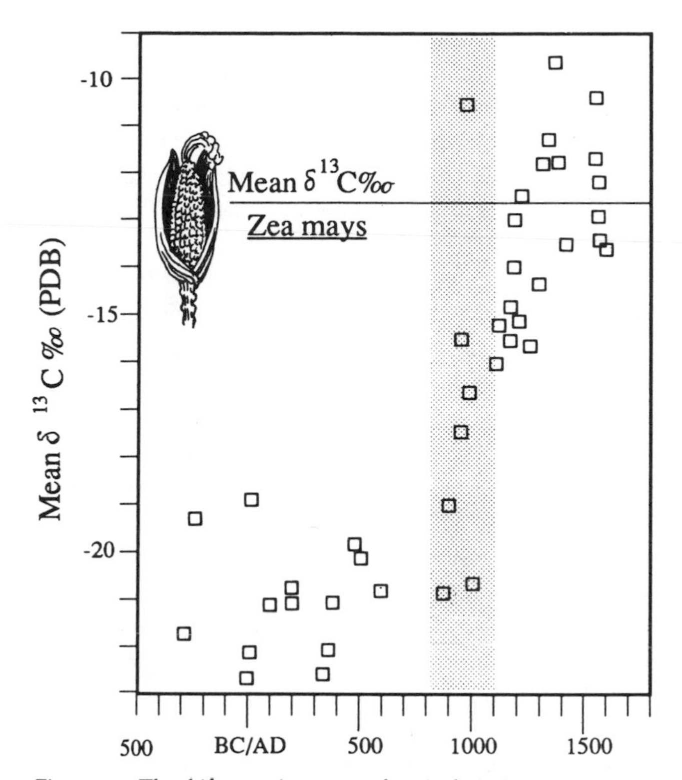

Figure 11.4 The shift to maize-centered agriculture in eastern North America, as reflected by changing mean delta ^{13}C values of human bone collagen samples from sites in the region. (PDB = Pee Dee belemnite; modified from Ambrose 1987.)

The A.D. 800 to 1100 shift to maize-centered agriculture across eastern North America was associated with the emergence of more complex sociopolitical formations. Corn played a central role in the evolution of Iroquoian societies in the Northeast and Ft. Ancient polities along the middle Ohio River Valley, as well as in the diverse array of Mississippian chiefdoms that emerged along the river valleys of the Southeast and Midwest (Smith 1990). These parallel but regionally distinct episodes of major social transformation have attracted a broad range of alternative explanations in which maize agriculture has been variously cast as an adaptive response to growing problems of imbalance between demographics and resources, an opportunistic effort to buffer social and economic uncertainty in the absence of external stress, and as a lever of social inequality for an emerging elite. Similarly, a number of overlapping explanations have been offered for the six-century lag that separated the initial introduction of maize into eastern North America at A.D. 200 and its post A.D. 800 transition from minor cultigen to major crop. These explanations focus both on possible changes in how corn was used and potential changes in the plant itself. Perhaps maize was initially a controlled ceremonial crop, used by the general populace after A.D. 800. Maybe it was initially harvested and eaten in the green state and was not used as a storable food supply until after A.D. 800. Even though it may have had higher harvest rate values relative to local crops (Smith 1987c [Chapter 7]), maize could have remained a minor cultigen prior to A.D. 800 because of its larger initial investment costs in land clearing and field maintenance. The cost-to-yield ratio of maize agriculture may have moved into a more attractive range after A.D. 800, perhaps because of increasing demographic pressure on existing wild and cultivated resources or the development of more productive varieties of maize, or both.

The development of a new eight-row variety of maize that was adapted to short growing seasons and ancestral to the historic period northern flints (Doebley et al. 1986; Wagner 1987) does in fact provide at least part of the explanation for the rapid adoption of maize across the northern latitudes of eastern North America. First appearing in the Northeast, Ohio Valley, and Great Lakes by A.D. 900 to 1000, this distinctive eastern eight-row variety of maize dominated agricultural economies in those regions until European contact.

But the shift to maize-centered agriculture in eastern North America was not simply a matter of the development of new, improved varieties of corn. There is at present no evidence that any higher yield or more storable types of corn played a role in the A.D. 800 to 1100 formation of maize-centered agricultural economies and more complex sociopolitical systems in the Midwest and Southeast (Smith 1992 [Chapter 12]). Moreover, like the second millennium initial domestication of plants in eastern North America and the subsequent emergence of food production economies in the region at about 250 B.C. to A.D. 200 (O'Brien 1987), the shift to maize agriculture was imbedded within a larger and uniquely eastern North American process of social transformation.

Many of the challenging new research questions to be addressed in the coming decades focus on these larger social transformational aspects of the forager to farmer transition in eastern North America and on the intricate patterns of interaction that evolved between human societies and plant populations.[4]

Notes

1. *Helianthus annuus* appears to have arrived in eastern North America as an adventive weed during the Middle Holocene, perhaps not much earlier than 2000 B.C. (see Yarnell 1974, 1983, 1986, 1988; Gardner 1987; Heiser 1989).

2. Similarities in the seeds of these two prehistoric domesticates and those of modern mexican cultivar chenopods *(C. berlandieri* ssp. *nuttalliae* cv. "chia" and "huauzontle") raises the issue of a possible introduction into the East from Mesoamerica at about 1500 B.C. The absence of domesticated *Chenopodium* in the archaeological record of Mesoamerica, however, along with evidence for its independent domestication in Mesoamerica and South America (Wilson and Heiser 1979) support eastern North America as a third independent center of chenopod domestication.

3. See Wills (1988) for a discussion of the similar minor initial impact of maize in the southwestern United States.

4. This article is based in part on a paper prepared for the third American-Soviet Archaeology symposium, Smithsonian Institution, May 1986. Figures done by Marcia Bakry.

Literature Cited

Adams, R. McC.
1988 Contexts of Technological Advance. Darryl Forde Lecture, University of London, November 30.

Ambrose, S.
1987 Chemical and Isotopic Techniques of Diet Reconstruction in Eastern North America. In *Emergent Horticultural Economies of the Eastern Woodlands,* edited by W. Keegan, pp. 87–107. Southern Illinois University at Carbondale, Center for Archaeological Investigations.

Anderson, E.
1956 Man as a Maker of New Plants and New Plant Communities. In *Man's Role in Changing the Face of the Earth,* edited by W. L. Thomas, pp. 763–777. University of Chicago Press, Chicago.

Asch, D., and N. Asch
1985 Prehistoric Plant Cultivation in West-central Illinois. In *Prehistoric Food Production in North America,* edited by R. I. Ford, pp. 149–203. Museum of Anthropology, Anthropological Papers No. 75. University of Michigan, Ann Arbor.

Bender, M., D. Baerreis, and R. Steventon
1981 Further Light on Carbon Isotopes and Hopewell Agriculture. *American Antiquity* 46:346–353.

Braun, D.
1987 Coevolution of Sedentism, Pottery Technology, and Horticulture in the Central Midwest, 200 B.C.–A.D. 600. In *Emergent Horticultural Economies of the Eastern Woodlands,* edited by W. Keegan, pp. 153–181. Southern Illinois University at Carbondale, Center for Archaeological Investigations, Carbondale.

Brose, D., and N. Greber (editors)
1979 *Hopewell Archaeology.* Kent State University Press, Kent, Ohio.

Buikstra, J. E., J. Bullington, D. K. Charles, D. C. Cook, S. R. Frankenberg, L. W. Konigsberg, J. B. Lambert, and Liang Xue

1987 Diet, Demography, and the Development of Horticulture. In *Emergent Horticultural Economies of the Eastern Woodlands,* edited by W. Keegan, pp. 67–85. Southern Illinois University at Carbondale, Center for Archaeological Investigations, Carbondale.

Chapman, J., and G. Crites
1987 Evidence for Early Maize *(Zea mays)* from the Icehouse Bottom Site, Tennessee. *American Antiquity* 52:352–354.

Chapman, J. P., P. A. Delcourt, P. A. Cridlebaugh, A. B. Shea, H. R. Delcourt
1982 Man-Land Interaction: 10,000 Years of American Indian Impact on Native Ecosystems in the Lower Little Tennessee River Valley, Eastern Tennessee. *Southeastern Archaeology* 1:115.

Chapman, J., and A. Shea
1981 The Archaeobotanical Record: Early Archaic Period to Contact in the Lower Little Tennessee River Valley. *Tennessee Anthropologist* 6:61–84.

Cowan, C. W.
1985 Understanding the Evolution of Plant Husbandry in Eastern North America: Lessons from Botany, Ethnography, and Archaeology. In *Prehistoric Food Production in North America,* edited by R. I. Ford, pp. 205–243. Museum of Anthropology, Anthropological Papers No. 75. University of Michigan, Ann Arbor.

Cowan, C. W., and P. J. Watson (editors)
1992 *The Origins of Agriculture: An International Perspective.* Smithsonian Institution Press, Washington, D.C.

Cowan, C. W., H. Edwin Jackson, K. Moore, A. Nickelhoff, and T. Smart
1981 The Cloudsplitter Rockshelter, Menifee County, Kentucky: A Preliminary Report. *Southeastern Archaeological Conference Bulletin* 24:60–75.

Crites, G.
1985a Middle Woodland Paleoethnobotany of the Eastern Highland Rim of Tennessee. M.A. thesis, Department of Anthropology, University of Tennessee, Knoxville.

1985b Middle and Late Holocene Ethnobotany of the Hayes Site. Unpublished report to the Tennessee Valley Authority.

1987 Human-Plant Mutualism and Niche Expression in the Paleoethnobotanical Record. *American Antiquity* 52:725.

Decker, D.
1988 Origin(s), Evolution, and Systematics of *Cucurbita pepo* (Cucurbitaceae). *Economic Botany* 42:4–15.

Decker, D., and H. Wilson
1986 Numerical Analysis of Seed Morphology in *Cucurbita pepo. Systematic Botany* 11:595–607.

Doebley, J. F., M. Goodman, and C. Stuber
1986 Exceptional Genetic Divergence of Northern Flint Corn. *American Journal of Botany* 73:64–69.

Ford, Richard I.
1985 Patterns of Prehistoric Food Production in North American. In *Prehistoric Food Production in North America,* edited by R. I. Ford, pp. 341–364. Museum of Anthropology, Anthropological Papers No. 75. University of Michigan, Ann Arbor.

1987 Dating Early Maize in the Eastern United States. Paper presented at the annual conference of the Society of Ethnobiology, Gainesville, Florida, March 5–8.

Fowler, M.
1971 The Origin of Plant Cultivation in the Central Mississippi Valley: A Hypothesis. In *Prehistoric Agriculture,* edited by S. Struever, pp. 122–128. Natural History Press, New York.

Fritz, G., and B. Smith
1988 Old Collections and New Technology: Documenting the Domestication of *Chenopodium* in Eastern North America. *Midcontinental Journal of Archaeology* 13:3–27.

Gardner, P.
1987 Plant Food Subsistence at Salts Cave, Kentucky: New Evidence. *American Antiquity* 52:358–364.

Gregg, S.
1988 *Foragers and Farmers.* University of Chicago Press, Chicago.

Harlan, J. R.
1971 Agricultural Origins: Centers and Non-Centers. *Science* 174:468–474.

Harlan, J. R., J. M. J. de Wet, and E. G. Price
1973 Comparative Evolution of Cereals. *Evolution* 27:311–325.

Heiser, C.
1985 Some Botanical Considerations of the Early Do-

mesticated Plants North of Mexico. In *Prehistoric Food Production in North America,* edited by R. I. Ford, pp. 57–72. Museum of Anthropology, Anthropological Papers No. 75. University of Michigan, Ann Arbor.

1989 Domestication of Cucurbitaceae: *Cucurbita* and *Lagenaria.* In *Foraging and Farming,* edited by D. Harris and G. Hillman, pp. 472–480. Unwin Hyman, London.

Johannessen, S.
1984 Paleoethnobotany. In *American Bottom Archaeology,* edited by C. Bareis and J. Porter, pp. 197–214. University of Illinois Press, Urbana.

King, F.
1985 Early Cultivated *Cucurbita* in Eastern North America. In *Prehistoric Food Production in North America,* edited by R. I. Ford, pp. 73–99. Museum of Anthropology, Anthropological Papers No. 75. University of Michigan, Ann Arbor.

Lynott, M., T. W. Boutton, J. E. Price, and O. E. Nelson
1986 Stable Carbon Isotope Evidence for Maize Agriculture in Southeast Missouri. *American Antiquity* 51:51–65.

Newsom, L.
1988 Paleoethnobotanical Remains from a Waterlogged Archaic Period Site in Florida. Paper presented at the annual meeting of the Society for American Archaeology, Phoenix, Arizona, April 27 to May 1.

O'Brien, M. J.
1987 Sedentism, Population Growth, and Resource Selection in the Woodland Midwest: A Review of Coevolutionary Developments. *Current Anthropology* 28:177–197.

Pearsall, D.
1989 *Paleoethobotany.* Academic Press, Orlando, Florida.

Smith, Bruce D.
1984 Chenopodium as a Prehistoric Domesticate in Eastern North America: Evidence from Russell Cave, Alabama. *Science* 226:165–167.

1985a The Role of Chenopodium as a Domesticate in Premaize Garden Systems of the Eastern United States. *Southeastern Archaeology* 4:51–72.

1985b *Chenopodium berlandieri* ssp. *jonesianum:* Evidence for a Hopewellian Domesticate from Ash Cave, Ohio. *Southeastern Archaeology* 4:107–133.

1986 The Archaeology of the Southeastern United States: From Dalton to de Soto. In *Advances in World Archaeology,* vol. 5, edited by F. Wendorf and A. Close, pp. 1–92. Academic Press, Orlando, Florida.

1987a The Independent Domestication of Indigenous Seed-Bearing Plants in Eastern North America. In *Emergent Horticultural Economies of the Eastern Woodlands,* edited by W. Keegan, pp. 3–47. Southern Illinois University at Carbondale, Center for Archaeological Investigations, Carbondale.

1987b Hopewellian Farmers of Eastern North America. Plenary session paper, 11th Congress International Union of Prehistoric and Protohistoric Sciences, Mainz, Germany, September 1–6.

1987c The Economic Potential of *Chenopodium berlandieri* in Prehistoric Eastern North America. *Ethnobiology* 7:29–54.

1988 SEM and the Identification of Micro-Morphological Indicators of Domestication in Seed Plants. In *Seed Plants in Scanning Electron Microscopy in Archaeology,* edited by Sandra L. Olsen, pp. 203–213. British Archaeological Reports International Series 452. Oxford.

1992 Prehistoric Plant Husbandry in Eastern North America. In *The Origins of Agriculture in World Perspective,* edited by C. W. Cowan and P. J. Watson. Smithsonian Institution Press, Washington D.C.

Smith, Bruce D. (editor)
1990 *The Mississippian Emergence.* Smithsonian Institution Press, Washington, D.C.

Smith, Bruce D., and C. Cowan
1987 The Age of Domesticated Chenopodium in Prehistoric Eastern North America: New Accelerator Dates from Eastern Kentucky. *American Antiquity* 52:355–357.

Smith, Bruce D., and V. Funk
1985 A Newly Described Subfossil Cultivar of *Chenopodium* (Chenopodiaceae). *Phytologia* 57:445–449.

Wagner, G.
1987 Corn and Cultivated Beans of the Fort Ancient Indians. *Missouri Archaeologist* 47:107–136.

Watson, P. J.
1985 The Impact of Early Horticulture in the Upland

Drainages of the Midwest and Midsouth. In *Prehistoric Food Production in North America,* edited by R. I. Ford, pp. 99–147. Museum of Anthropology, Anthropological Papers No. 75. University of Michigan, Ann Arbor.

Wills, W. H.
1988 Early Agriculture and Sedentism in the American Southwest: Evidence and Interpretations. *Journal of World Prehistory* 2:455–488.

Yarnell, R.
1974 Plant Food and Cultivation of the Salts Cavers. In *Archaeology of the Mammoth Cave Area,* edited by P. J. Watson, pp. 113–122. Academic Press, Orlando, Florida.

1978 Domestication of Sunflower and Sumpweed in Eastern North America. In *The Nature and Status Of Ethnobotany,* edited by R. I. Ford, pp. 289–314. Museum of Anthropology, Anthropological Papers No. 67. University of Michigan, Ann Arbor.

1983 Prehistory of Plant Foods and Husbandry in North America. Paper presented at the annual meeting of the Society for American Archaeology, Pittsburgh, April 11–13.

1987 A Survey of the Prehistoric Crop Plants in Eastern North America. *Missouri Archaeologist* 47:47–60.

1988 The Importance of Native Crops during the Late Archaic and Woodland. Paper presented at the annual meeting of the Southeastern Archaeological Conference, New Orleans, October 20–22.

Yarnell, R., and M. Black
1985 Temporal Trends Indicated by a Survey of Archaic and Woodland Plant Food Remains from Southeastern North America. *Southeastern Archaeology* 4:93–106.

CHAPTER 12

❑ ❑ ❑

Prehistoric Plant Husbandry in Eastern North America

Introduction

From their first radiation into the vast woodlands of eastern North America at the end of the Pleistocene, human populations depended to a substantial degree upon a variety of different plant species as food sources. Spanning more than ten millennia, the ensuing prehistoric period in the Eastern Woodlands witnessed dramatic changes in the nature, strength, and diversity of the relationships between human populations and their plant food sources: wild and cultivated, indigenous and introduced.

As outlined in this summary chapter, current understanding of this long and complex co-evolutionary relationship between prehistoric human and plant populations of the Eastern Woodlands rests both on an archaeobotanical data base that has substantially increased over the past two decades (Yarnell and Black 1985) and associated significant advances in the technology and methodology of archaeobotanical analysis (Watson 1976; Smith 1985a).

To facilitate discussion of evolving plant husbandry systems in the Eastern Woodlands, I have divided what was an unbroken, seamless developmental trajectory of 10,000 years into a temporal sequence of six periods of varying length that bracket ever-increasing levels of human dependence upon, and intervention in, the life cycle of plant species:

1. Early and Middle Holocene foragers prior to 7,000 B.P. (5050 B.C.)
2. Middle Holocene collectors, 7,000 to 4,000 B.P. (5050 to 2050 B.C.)
3. The initial domestication of eastern seed plants, 4,000 to 3,000 B.P. (2050 to 1050 B.C.)

This chapter also will appear in *The Origins of Agriculture: An International Perspective,* edited by C.W. Cowan and P. Watson (1992). Smithsonian Institution Press, Washington, D.C.

4. The development of farming economies, 3,000 to 1,700 B.P. (1050 B.C. to A.D. 250)
5. The expansion of field agriculture, 1,700 to 800 B.P. (A.D. 250 to A.D. 1150)
6. Maize-centered field agriculture after 800 B.P. (A.D. 1150)

While the descriptive headings assigned to each of these temporal periods indicate the most important aspects of the leading edge of the ongoing co-evolutionary process, it is important to emphasize that the nature and timing of this developmental process was not uniform across the Eastern Woodlands. Some areas of the East developed more slowly and in directions different from those emphasized in this chapter.

Early and Middle Holocene Foragers prior to 7,000 B.P. (5050 B.C.)

Prior to 7000 B.P. the bands of hunting and gathering groups that occupied the interior Eastern Woodlands of North America appear to have uniformly exploited plant and animal resources distributed fairly evenly along river valley corridors. Although the small, short-term seasonal camps of these forest foragers have yielded few carbonized plant materials, the impoverished archaeobotanical assemblages (along with faunal materials) suggest a broad-based utilization of both closed-canopy climax forest plant resources (particularly oak mast and hickory nuts) and early successional species more commonly associated with edge areas and more disturbed situations. The seeds, berries, and nuts of twenty plant species are represented in the archaeobotanical assemblages recovered from nine Early Holocene sites (Smith 1986:11; Meltzer and Smith 1986): *Quercus, Carya, Fagus, Corylus, Juglans, Castanea, Celtis, Diospyros, Acalypha, Amaranthus, Ampelopsis, Chenopodium, Galium, Phalaris, Phytolacca, Polygonum, Portulaca, Rhus, Vitis.* These Early Holocene foraging populations filled the role of opportunistic dispersal agents for plant propagules, and their small camps created ephemeral disturbed soil opportunities for some pioneer species. There is no evidence for any greater degree of intervention by human populations in the life cycle of eastern plant species prior to 7,000 B.P.

Middle Holocene Collectors 7,000 to 4,000 B.P. (5050 to 2050 B.C.)

During the Middle Holocene period, the opportunistic response by human populations in the Eastern Woodlands to changes in river valley habitats resulted in both the creation of more permanently disturbed anthropogenic habitat patches, and increased human intervention in the life cycles of a number of plants of economic importance.

The change to zonal atmospheric flow across the Midwest during the Middle Holocene Hypsithermal climatic episode (Knox 1983) resulted in a shift of mid-latitude fluvial systems from the earlier pattern of episodic pulses of sediment removal and river incision, to a phase of river aggradation and stabilization. This change in stream flow characteristics in turn resulted in the expansion of both backwater (backswamp and oxbow lake) and active stream shallow water and shoal area aquatic habitats along segments of some mid-latitude river systems (Smith 1986, 1987a [Chapter 3]). These aquatic habitats supported an increasingly abundant and easily accessible variety of aquatic resources. At the same time, the more geomorphologically active main channel portions of the floodplain environment continued to provide varied and abundant disturbed habitat opportunities for floodplain species of pioneer plants including three of particular interest: *Chenopodium berlandieri*, *Iva annua*, and a wild *Cucurbita* gourd [Chapter 3, 4]. In response to the establishment of biotically richer floodplain segments, the widely scattered short-term occupational lenses of the antecedent Early Holocene and initial Middle Holocene were replaced by deep shell mound and midden mound settlements. Located on suitable topographic highs in the floodplain close to areas of enriched aquatic resources, and reflecting a narrowing of site preference, these midden mounds and shell mounds were annually reoccupied and reused over a long period of time, during the low water, late spring to early winter growing season. This annual reoccupation and reuse of specific floodplain locations through the growing season represented the initial emergence of continually disturbed "anthropogenic" habitats in the interior river valleys of the eastern United States. A wide range of human activities would have resulted in frequent disturbance of the soil of these localities (construction of houses, wind breaks, storage and refuse pits, drying racks, earth ovens, hearths). In addition, floodwater sediments, along with the accumulation of plant and animal debris and human excrement, particularly at site edges, would have both altered the ground surface and improved the soil chemistry of such midden and shell mounds. These continually disturbed and fertilized "anthropogenic" habitat patches provided an excellent "dump heap" colonization opportunity (Anderson 1956; Fowler 1957, 1971; Smith 1987a [Chapter 3]) for a variety of early successional floodplain plant species, with seeds being introduced each year both by floodwaters and by human collectors.

During the 3,000-year period from 7,000 to 4,000 B.P. these anthropogenic habitat localities witnessed increasing human intervention, to varying degrees, in the life cycle of a number of the plant species growing in their disturbed and enriched soils. The degree of human intervention in the life cycle of a plant species, and the extent to which the species in question depends upon humans for its continued survival, can be measured along a continuum extending from wild-status plants through eradicated, tolerated, and encouraged weed categories, to cultivation and domestication (Smith 1985a, fig. 1 [Figure 5.1]). Accurately placing prehistoric human-plant relationships along this continuum is often difficult, however, in the absence of any morphological changes reflecting domesticated status. Such morphological changes represent the single unequivocal indicator of cultivation and domestication that can be observed directly in the archaeological record. As discussed in the following section of this chapter, distinctive changes in the morphology of *Cucurbita*, goosefoot, sumpweed, and sunflower *(Helianthus annuus)* seeds recovered from fourth millennium archaeobotanical assemblages document their domestication by 4,000 to 3,000 B.P., if not earlier.

Although not reflected in the archaeological record of 7,000 to 4,000 B.P., the general sequence of coevolutionary development leading to the fourth millennium domestication of these three seed plants can be reconstructed with a reasonable level of confidence. Introduced from wild floodplain stands into midden mound and shell mound settlements each year either by floodwaters or human collectors, the seeds of *Cucurbita*, sumpweed and goosefoot would have quickly colonized these disturbed habitat patches where they would have thrived as weedy invaders. While the sunflower *(Helianthus annuus)* also initially colonized these "domestilocalities" (Smith 1987a [Chapter 3]) as a weedy invader, it was apparently adventive to the Eastern Woodlands during the Middle Holocene, rather than being indigenous to the region.

Quickly established, and initially neither requiring nor receiving any human assistance other than inadvertent soil disturbance and enrichment, these colonizing weedy stands of *Cucurbita*, goosefoot, sumpweed, and sunflower, through simple toleration, provided a supplement to "natural" floodplain stands. The next level of human intervention in the life cycle of these weedy colonizers—a transition from simple toleration to inadvertent, and then active encouragement—was critical in the co-evolutionary trajectory leading to domestication. When colonizing plant species were separated into "valuable" or "quasi-cultigen" versus "no value" species, with such "no value" plants actively discouraged, then the element of deliberate management or husbandry was introduced, to whatever minimal degree, and domestilocalities were transformed from inadvertent or incidental gardens into managed, or "true" gardens.

From the eradication of competing "weeds," it would have been a small, but very important escalation of the level of human intervention in the life cycle of the quasi-cultigens for their stands to have been expanded by planting within the domestilocality—likely by women (Watson and Kennedy 1991). It is this critical step of planting, even on a very small scale, that, if sustained over the long term, marks both the beginning of cultivation and the onset of automatic selection within affected domestilocality plant populations for interrelated adaptation syndromes associated with domestication (Heiser 1988). From here on, "the initial establishment of a domesticated plant can proceed through automatic selection alone" (Harlan et al. 1973:314). The nature of these automatic selection processes and the morphological changes in plants related to the adaptation syndromes they cause are described in the next section of the chapter, with specific reference to goosefoot, sumpweed, and sunflower.

While fourth millennium morphological changes provide the basis for assigning domesticated status to these three species, *Cucurbita* gourds represented in Middle Holocene archaeobotanical assemblages had until recently also been identified as a domesticate on quite different grounds. Middle Holocene archaeobotanical assemblages from sites in west-central Illinois (Koster, Napoleon Hollow, Kuhlman, Lagoon), Missouri (Phillips Spring), Kentucky (Carlston Annis, Bowles), and Tennessee (Hayes, Bacon Bend) have yielded rind fragments assignable to the genus *Cucurbita* (Figure 12.1, Table 12.1). Eight direct accelerator dates on rind and seed fragments leave no doubt as to the presence of gourdlike *Cucurbita* populations in the middle latitude riverine area of eastern North America between 7,000 and 4,000 B.P.

Because it was thought that no indigenous species of wild *Cucurbita* gourd had ever been present in the East, a number of researchers (Chomko and Crawford 1978; Conard et al. 1984; Asch and Asch 1985c) concluded that the Middle Holocene *Cucurbita* material of the Eastern Woodlands was a variety of the tropical domesticate *C. pepo* that had been introduced, along with the concept of agriculture, from Mesoamerica by 7,000 B.P.

More recently, however, both Decker (1986, 1988) and Smith (1987a [Chapters 3, 4]) have proposed a much more plausible explanation for this early *Cucurbita* rind material—that it reflects a more widespread prehistoric range of distribution of an indigenous wild gourd, that still grows today in the East. All of the Middle Holocene *Cucurbita* rind fragments recovered to date are thin, with reported thickness values falling below 2.0 mm, well within the range of published rind thickness values for present-day indigenous wild gourds of eastern North America (Decker et al. 1992; Cowan and Smith 1992). Rind fragments thicker than 2.0 mm, probably representing domesticated plants, do not, in fact, appear in archaeological contexts in the East until after 3,000 B.P. (at Salts Cave and Cloudsplitter rock shelter in Kentucky).

Along with rind fragments, *Cucurbita* seeds recovered from Middle Holocene contexts should indicate when domesticated forms of *Cucurbita* first appeared in the East. Closely related wild and cultivated varieties of *Cucurbita* can be differentiated on the basis of seed morphology (Decker and Wilson 1986), and temporal changes in seed size and shape should mark the transition of *Cucurbita* from wild to cultivated and domesticated status along a developmental trajectory of incipient plant husbandry. Unfortunately, morphological analysis of Middle Holocene *Cucurbita* seed assemblages from the Eastern Woodlands of North America employing the Decker guidelines have not yet been conducted. In addition, with the exception of a single possible *Cucurbita* seed from the Anderson site dating to 6,990 ± 120 B.P. (AA-1182) (Table 12.1), the 65 seeds securely dated to ca. 4,300 B.P. from the Phillips Spring site in west-central Missouri, along with the 14 specimens apparently of a similar age (about 4,400 B.P.) from the Bacon Bend site in eastern Tennessee, represent the only *Cucurbita* seed material recovered from Middle Holocene contexts in the East. When shape analysis of these early seed assemblages is carried out, however, I think that the results will support the temporal trend already documented in the rind assemblages—a transition to domesticated forms of *Cucurbita* after 3,000 B.P. Presently there is no evidence to support the proposition that a domesticated variety of *C. pepo* was present in the East prior to 3,000 B.P.

It has yet to be determined whether the cultivated fleshier squash varieties of *C. pepo* that first appear in the East after 3,000 B.P. (Salts Cave, Yarnell 1969:51; Cloudsplitter, Cowan 1990) were introduced into the East, or brought under domestication locally from an indigenous wild progenitor. An independent trajectory of domestication for *Cucurbita pepo* in the East, however, is the "much more likely" of the two alternatives (Heiser 1989:474), however, particularly in light of Decker's recent research (1992), and within the context of the first domesticated *Cucurbita* following on the heels of the preceding (4,000–3,500 B.P.) domestication of other indigenous seed plants. In addition, the recent direct dating of several small *Lagenaria* gourds recovered from the Windover site on the east coast of Florida (Brevard County) at 7,300 B.P. (Doran et al. 1990) suggests that wild

Figure 12.1 Archaeological sites and regions discussed in the text that have provided information regarding the development of farming economies in eastern North America.

fruits may have washed ashore and become established during the Middle Holocene or earlier. Thus, a mechanism of human-facilitated diffusion of domesticated forms from Mesoamerica need not be invoked to explain the presence of either the bottle gourd or *Cucurbita* gourd in the East during the Middle Holocene.

Another plant species of considerable economic importance may also have been encouraged by Middle Holocene groups, to their great benefit, without

Table 12.1. *Cucurbita* Remains from Eastern North America Prior to 2,000 B.P.

Site	Temporal/Cultural Context	Rind Fragments (N) Weight	Seeds (N) Size	References
Koster	Horizon 8b	(7) 27 mg		Asch and Asch 1985c
	7100 ± 300 (Accelerator—Rind)			Conard et al. 1984
	7000 ± 80			Conard et al. 1984
	6960 ± 80			Conard et al. 1984
	6820 ± 240 (Accelerator—Rind)			Ford 1985: 345
	Horizon 8a	(1)		Asch and Asch 1985c
	6860 ± 80			
Napoleon Hollow	Lower Helton	(3)		
	7000 ± 250 (Accelerator—Rind)			
	6800 ± 80			
	6730 ± 70			
	6630 ± 100			
	Upper Helton	(1)		
	6130 ± 110			
	5670 ± 90			
	5350 ± 70			
Anderson	6990 ± 120 (Accelerator—Seed)		(1)	Anna Dixon, pers.
	AA—1182			comm. 1987
Hayes	Unit III Levels E 20, 22, 23	(6) 0.4 g		Crites 1985
	5660 ± 190 B.P.			
	5525 ± 290 B.P.			
	Unit IV Level A 25	(16) .01 g		Crites 1985
	5340 ± 120 (Accelerator—Rind)			
	Unit III Level 017	(2)		Crites 1985
	5140 ± 185			
	4390 ± 170			
Carlston Annis	5730 ± 640 (Accelerator—Rind)			Watson 1985: 112
	<4500 ± 60 B.P.			
	>4250 ± 80 B.P.			
	>4040 ± 180 B.P.			
Phillips Spring	4310 ± 70		(65) length x = 10.5	King 1985
	4240 ± 80		width x = 7.0	King 1985
	4222 ± 57			King 1885
	3928 ± 41		(50) fragments	King 1985
Cloudsplitter	4700 ± 250 (Accelerator—seed)		(2) length x = 8.7	Cowan 1990
			width x = 5.3	
Bacon Bend	4390 ± 155	(4)		Chapman 1981:35–36
Lagoon	Feature 102a	(1)		Asch and Asch 1985c
	4300 ± 600			
	Titterington	(1)		Asch and Asch 1985c
	4030 ± 75			
Napoleon Hollow	Titterington	(1)		Asch and Asch 1985c
	4500 ± 500			
	4060 ± 75			
Bowles	4060 ± 220 (Accelerator—Rind)			Watson 1985:113
	>3440 ± 80			
Kuhlman	Titterington	(1)		Asch and Asch 1985c
	4010 ± 130			
Cloudsplitter	3620–3060	(9)		Cowan 1990
Iddens	3655 ± 135	(129+)	(140+)	Chapman 1981:132–34
	3470 ± 75			
	3205 ± 145			
Peter Cave	>3415 ± 105 B.P.		(2)	Watson 1985:113
Riverton	Feature 8a	(1)		Yarnell 1976
	ca. 3200			
Salts Cave	ca. 2500		length: x = 11.3	Yarnell 1969:51
			width: x = 7.3	
Cloudsplitter	2800–2300 B.P.	(29)	(5) length: x = 12.7	Cowan 1985, 1990
			width: x = 7.3	King 1985

the species in question ever being brought under strict cultivation. Munson (1986) has proposed that the dramatic increase in representation of hickory nuts *(Carya)* in archaeobotanical assemblages of the Midwest riverine region after about 7,500 B.P. reflects both improved methods for processing nuts (hide-lined and rock-heated boiling pits for separating nut meats from hulls) and active management of hickory trees to increase the quantity of hickory nuts available for human collectors. By simply girdling (and thereby killing) all nonproductive trees, human foragers could have created large stands of widely spaced hickories with increased long-term nut production and lower squirrel populations. Such stands would have required little maintenance other than annual controlled burning of undergrowth. Alternatively, the increased representation of hickory nuts in Middle Holocene archaeobotanical assemblages may simply reflect the increasing abundance of *Carya* in midwestern forests (Delcourt and Delcourt 1981).

While the degree to which Middle Holocene groups managed and intervened in the life cycles of hickory trees and camp-follower gourds are topics of active and continuing debate, there is unequivocal evidence for the independent domestication of seed plants in the Eastern Woodlands during the fourth millennium B.P.

The Initial Domestication of Eastern Seed Plants 4,000 to 3,000 B.P. (2050 to 1050 B.C.)

The ten-century span from 4,000 to 3,000 B.P. brackets the earliest evidence of morphological changes reflecting domesticated status of the three, and perhaps four, annual seed crops brought under domestication prior to the first appearance of maize in the East at about A.D. 200. For sumpweed *(Iva annua)* and sunflower *(Helianthus annuus)* the morphological change indicating domestication is an increase in achene size, while a reduction in seed coat (testa) thickness reflects the transition to domesticated status in goosefoot *(Chenopodium berlandieri)*. Both of these morphological changes (increase in seed size, reduction in testa thickness) reflect the automatic response by cultivated seed plants to the selective, if inadvertent, pressures inherent in the deliberate planting of seed stock.

Within domestilocality spring seed beds there would have been extremely intense competition between young plants: "The first seed to sprout and the most vigorous seedlings to sprout are more likely to contribute to the next generation than the slow or weak seedlings" (Harlan et al. 1973:318). As a result, strong selective pressure within seed beds would favor those seeds that would both sprout quickly because of a reduced germination dormancy (as reflected in a thinner or absent seed coat or testa), and grow quickly because of greater endosperm food reserves, as reflected by an increase in seed size.

With a mean achene length value of 4.2 mm, the circa 4,000 B.P. Napoleon Hollow *Iva* assemblage, generally acknowledged to represent a domesticated crop (Yarnell 1983; Asch and Asch 1985c:160; Ford 1985:347), reflects a marked (31 percent) size increase over modern wild *Iva* populations, which have mean achene length values of 2.5 to 3.2 mm.

Yarnell (1983) and Ford (1985:348) both identify the circa 2,850 B.P. Higgs site sunflower achenes, which have a mean length value of 7.8 mm (as opposed to modern wild populations that range in mean length from 4.0 to 5.5 mm) as representing the earliest evidence for domestication of this species in the East. In addition, Yarnell (1978:291) suggests that the Higgs site sunflowers were "in early, but not initial stages of domestication." Ford (1985:348) suggests that such initial stages of domestication may have taken place at about 3,500 B.P.

This proposed 3,500 to 3,400 B.P. transition to domesticated status for *Helianthus annuus* corresponds with the earliest available evidence for the presence of a thin-testa domesticated form of goosefoot *(C. berlandieri* ssp. *jonesianum)* (Smith and Funk 1985). In contrast to both sumpweed and sunflower, an increase in seed size is not a reliable indicator of domesticated status in *Chenopodium*. The relative thickness of the seed coat or testa, however, can be employed as a morphological indicator of domestication in *Chenopodium* (Smith 1985a, 1985b [Chapters 5 and 6]) in that it represents a strong selective pressure for reduced germination dormancy. While modern wild populations of *Chenopodium berlandieri* in the Eastern Woodlands have testa thickness values of 40 to 80 microns, assemblages of fruits having mean seed coat thickness values of less than 20 microns have been reported from a number of sites dating from 2,300 B.P. to 1,800 B.P. (Ash Cave, Edens Bluff, Russell Cave, White Bluff; Fritz and Smith 1988), and comparable thin-testa specimens have been recovered from circa 2,500–1,500 B.P. sites in west-central Illinois (John Roy, Asch and Asch 1985a), central Kentucky (Salts Cave, Yarnell 1974) and central Tennessee. In addition, thin-testa fruits of *Chenopodium berlandieri* ssp. *jonesianum* recovered from the Newt Kash and Cloudsplitter rock shelters in eastern Kentucky have recently been accelerator dated to 3,400 ± 150 B.P. and 3,450 ± 150 B.P. respectively, providing the earliest evidence for domesticated *Chenopodium* in the East (Smith and Cowan 1987). Support for these early eastern Kentucky accelerator dates on the thin-testa chenopod cultivar is provided by a direct radiocarbon age determination of 2,926 ± 40 B.P. obtained on a carbonized clump of thin-testa fruits that had been placed in a woven bag before being stored in a rear wall crevice at the Marble Bluff Shelter in northwest Arkansas (Fritz 1986a, 1986b). Pale-colored fruits of a second cultivar variety of *Chenopodium* having an extremely thin testa have also been reported from circa 3,000 B.P. contexts at the Cloudsplitter rock shelter, indicating that the early presence of two distinct crop plants of this species by the fourth millennium B.P. was likely. While the morphological similarity of seeds of these prehistoric thin-testa and pale-colored chenopod cultivars to those of the modern Mexican cultivars *C. berlandieri* ssp. *nuttalliae* cv. "chia" and "huauzontle" raises the issue of a possible prehistoric introduction of Mesoamerican domesticates into the East, it seems reasonable at present to consider them as indigenous eastern domesticates in light of the complete absence of thin-testa and pale-colored domesticated chenopods in the archaeological record of Mesoamerica.

While morphological changes in sumpweed, sunflower, and chenopod indicate that all three had been brought under domestication in the Eastern Woodlands during the period 4,000–3,500 B.P., there is little evidence that this process of domestication occurred within a framework of deliberate human selection, or that these domesticated plant species initially contributed substantially to the diet of fourth millennium B.P. populations. The simple, yet extremely significant step of planting seeds of these three plant species in order to expand their stand size and harvest yield, even on a very small scale, marked the beginning of cultivation and the onset of automatic selection for morphological changes in seeds within affected domestilocality plant populations.

The seeds of domesticates and indigenous cultigens alike are relatively rare in archaeobotanical assemblages dating earlier than 3,000 B.P., making it difficult to establish how closely the transition to domesticated status of sumpweed, sunflower, and chenopod may have been accompanied by an increasingly important economic role. It is quite possible that indigenous cultigens and domesticates initially had significant value to fourth millennium B.P. populations as a dependable, managed, and storable late winter-early spring food supply (Cowan 1985). But based on the continuing limited occurrence of seeds of these species in archaeobotanical assemblages, it also appears likely that indigenous cultigens and domesticates did not become substantial food sources, and food production did not play a major economic role until about 2,500–2,000 B.P., a full 1,000 years after their initial domestication.

The Development of Farming Economies 3,000 to 1,700 B.P. (1050 B.C. to A.D. 250)

The period from 2,500 to 1,700 B.P. (550 B.C.–A.D. 250) witnessed the initial development, subsequent florescence, and eventual cultural transformation of the Middle Woodland Hopewellian cultural systems of the Eastern Woodlands of North America. While the large and impressive geometric earthworks, conical burial mounds, and elaborate mortuary programs of Hopewellian populations have been the focus of archaeological attention for well over a hundred years (Brose and Greber 1979 [Chapter 9]), it is only within the past decade that much information has been recovered concerning Hopewellian plant husbandry systems.

While Hopewellian populations relied on wild species of plants and animals, Middle Woodland archaeobotanical seed assemblages recovered from river valley villages and upland rock shelters and caves over a broad geographical area of the Midwest and Midsouth reflect an increase in the representation (and assumed economic importance) of seeds of seven species of indigenous cultivated plants (Figure 12.1, Table 12.2).

Four of the plant species in question—goosefoot *(Chenopodium berlandieri)* and knotweed *(Polygonum erectum)* (both fall maturing), as well as maygrass *(Phalaris caroliniana)* and little barley *(Hordeum pusillum)* (both spring maturing)—have high carbohydrate content, while the seeds of two other species, marshelder *(Iva annua)* and sunflower *(Helianthus annuus)* are high in oil or fat (Table 12.3).

Although analysis of seed assemblages from this time period has to date yielded morphological evidence only for the domesticated status of four of these seven species (marshelder, goosefoot, *Cucurbita,* and sunflower), knotweed, maygrass, and little barley are classified at present as cultigens or quasi-cultigens (Yarnell 1983) for the period 3,000 to 1,700 B.P., based on geographical range extension (Cowan 1978:282; Asch and Asch 1985c:157; Yarnell 1983), archaeological abundance relative to modern occurrence, and a number of other "plausibility arguments" (Asch and Asch 1982:2; Smith 1985a:55 [Chapter 5]).

Judging from their relative representation in seed assemblages from different areas of the Eastern Woodlands, it is likely that these seven plant species each played roles of differing importance within a developmental mosaic of regionally variable Middle Woodland food production systems (Table 12.2). It is

Table 12.2. The Relative Abundance of Indigenous Starchy-seeded and Oily-seeded Annuals in Archaeobotanical Seed Assemblages Dating to the Period 3,000–1,700 B.P. in Seven Geographical Areas of the Eastern Woodlands

Category	Eastern Missouri		West-Central Illinois		American Bottom		Central Ohio		Central Kentucky		Eastern Kentucky		Central Tennessee		Eastern Tennessee	
	N	%	N	%	N	%	N	%	N	%	N	%	N	%	N	%
Starchy-seeded annuals																
Fall Maturing																
Chenopod	87	17.5	1176	8.6	201	26.4	392	13.6	18444	85.9	2745	52.7	2553	49.0	1431	88.0
Knotweed	29	5.8	4374	31.8	4	0.6	945	32.8	5				475	9.0		
Spring Maturing																
Maygrass	64	12.9	5155	37.5	222	29.1	598	20.7	1150	5.4	519	9.9	1868	36.0		
Little Barley	136	27.4	2040	14.9												
TOTAL	316	63.6	12745	92.8	427	56.1	1935	67.2	19599	91.3	3264	62.6	4896	94.0	1431	88.0
Oily-seeded annuals																
Fall Maturing																
Sumpweed	9	1.8	39	0.3	1	0.1	159	5.5	512	2.4	1154	22.9	3	0.1	26	1.6
Sunflower	1	0.2	35	0.3			11	.4	218	1.0	313	6.0	12	0.2	14	.9
TOTAL	10	2.0	74	0.6	1	0.1	170	5.9	730	3.4	1487	28.9	15	.3	40	2.5
Seed of Other Species	171	34.4	908	6.5	333	43.8	774	26.8	1141	5.3	468	9.0	283	5.4	156	9.5
Total Identified Seeds	497		13727		761		2879		21471		5199		5194		1627	

Sources: Seed count information for the Old Monroe site (Pulliam 1986). Seed count information for the Smiling Dan site (2100–1700 B.P.) (Asch and Asch 1985a, table 15.8). Seed count information for cement Hollow and Hill Lake phases (2100–1700 B.P.) of the American Bottom site (Johannessen 1981, table 29; 1983, table 21); for Central Ohio from the Murphy and O.S.U. Newark sites (ca. 1750 B.P.) (Wymer 1986a); for Central Kentucky from the Salts Cave site (Vestibule JIV, Levels 4–11, 2600–2200 B.P.) (Yarnell 1974, table 16.5; Gardner 1987); for Eastern Kentucky from the Cloudsplitter site (Excavation Units I, 13–18) Cowan 1985, tables 62–63); for Central Tennessee from the McFarland and Owl Hollow sites, late McFarland and early Owl Hollow phases at Normandy Reservoir (1900–1700 B.P.) (Crites 1978, table 2.2; Kline et al. 1982, table 6); and for Eastern Tennessee from the Long Branch Phase sites (2300–1800 B.P.) at Tellico Reservoir (Chapman and Shea 1981, table 4).

Table 12.3. The Nutritive Value of Indigenous Seed Crops of the Prehistoric Eastern Woodlands of North America, Compared to Maize

	% Dry Basis (g/100gm)				
Plant Species	Protein	Fat	Carbohydrate	Fiber	Ash
Chenopodium berlandieri[a]	19.1	1.8	47.6	28.0	3.5
Polygonum erectum[a]	16.9	2.4	65.2	13.3	2.3
Phalaris caroliniana[b]	23.7	6.4	54.3	3.0	2.1
Hordeum pusillum	—	—	—	—	—
Iva annua[c]	32.3	44.5	11.0	1.5	5.8
Helianthus annuus[d]	24.0	47.3	16.1	3.8	4.0
Zea mays (maize)[c]	8.9	3.9	70.2	2.0	1.2

[a]Asch and Asch 1985a
[b]Crites and Terry 1984
[c]Asch and Asch 1978
[d]Watt and Merrill 1963

hoped that as additional archaeobotanical assemblages from different areas of the Eastern Woodlands are analyzed, it will be possible to determine the degree to which observed regional variability in the representation of these seven plants actually indicates different crop mixtures as opposed to simply reflecting differential preservation, recovery, or seasonality of occupation of the site being studied.

Harvesting of sunflower seeds would have been a simple matter of cutting off the small multiple terminal discs for drying and seed removal, while squashes and gourds would have been collected and sectioned, with the flesh being dried in strips and the seeds separated for storage and consumption. Because of the diffuse distribution of seeds along stems and branches, on the other hand, entire knotweed plants were probably cut or pulled for drying and subsequent hand threshing. The terminal "seed clusters" of the other four species (marshelder, goosefoot, maygrass, little barley) could have been beaten or hand stripped into baskets (Asch and Asch 1978; Smith 1987b [Chapters 7 and 8]) or entire plants may have been harvested and flailed after drying.

Once harvested, the initial processing of seeds (other than seed stock) in all likelihood included parching to reduce insect infestation, premature germination, and mold damage (with associated accidental carbonization and loss contributing substantially to the archaeobotanical record). A number of storage contexts have been documented, including grass-lined pits, woven bags, and gourd containers.

Prior to consumption, any number of different technologies, including wooden mortars, stone slabs, and mortar holes in stone slabs were employed to crack protective seed coats, and the seeds could then have been either boiled in ceramic vessels (perhaps with a variety of other plant and animal ingredients), or perhaps ground further into a flour. Analysis of a large sample of human paleofecal material recovered from Salts Cave and dating to about 2,600 to 2,200 B.P., in fact, indicates that once cracked, the small seeds of these indigenous species were sometimes consumed without any further processing (Yarnell 1974). Providing uniquely direct and unequivocal dietary evidence of 1–4 meals (Watson 1974), each of the Salts Cave human paleofecal samples also indicated substantial reliance on the seeds of indigenous cultigens and cucurbits. The four domesticates contributed a full two-thirds of the total food supply of the central Kentucky cavers (chenopod, 25 percent; sunflower, 25 percent; marshelder, 14 percent; cucurbits 3 percent) (Yarnell 1974).

Other, more indirect measures of the increasing dietary importance of cultivated crops during the period 3,000–1,700 B.P. include the presence of chert hoes in Hopewellian artifact assemblages (Watson 1988), an increase in pollen and macrobotanical indicators of land clearing activities (Yarnell 1974:117; Kline et al. 1982:63; Johannessen 1984:201; Delcourt et al. 1986), and changes in settlement patterns and plant storage facilities (Smith 1986).

Based on research done during the past decade, it is now evident that indigenous cultigens increased substantially in dietary importance over a broad area of the eastern United States during the period 2,500 to 1,700 B.P., and that developing food production systems helped to fuel a process of dramatic cultural change and florescence. At the same time, however, it is still not possible to track accurately, from 3,000 to 1,700 B.P. and in different regions, the changing relationship between Woodland period populations and indigenous cultigens in terms of increasing energy investment in land clearing, soil preparation and cultivation, and plant harvesting and processing on the one hand, and total land area under crop and annual harvest yield values, on the other. Even in the absence of this kind of detailed picture of Early and Middle Woodland food production, however, it is important to emphasize that plant husbandry systems focused on indigenous seed-bearing plants having both impressive nutritional profiles and substantial potential harvest yield values (Table 12.3; Smith 1987b, fig. 11 [Chapters 7 and 8]), and had been established over a broad area of the eastern United States prior to the initial introduction of maize into the Eastern Woodlands.

The Expansion of Field Agriculture 1,700 to 800 B.P. (A.D. 250 to 1150)

Although maize pollen has been recovered from earlier temporal contexts in a number of locations (Sears 1982), the earliest convincing macrobotanical evidence recovered to date for the presence of maize in the Eastern Woodlands are carbonized kernel fragments from the Icehouse Bottom site in the Little Tennessee River Valley of eastern Tennessee that have yielded a direct particle accelerator (AMS) radiocarbon date of 1,775 ± 100 B.P. (A.D. 175) (Chapman and Crites 1987). In addition, Ford has recently (1987) reported direct AMS dates of 1,730 ± 85 and 1,720 ± 105 on maize kernels recovered from the Edwin Harness site in Ohio.

While research interest and debate will likely continue to focus in future years on the timing and route(s) of initial entry of maize into the Eastern Woodlands, and its subsequent rate and direction(s) of range extension in the East, these corn-related issues are largely peripheral to the central question of the role of food production in Woodland period cultural development. Judging from its extremely limited representation in archaeobotanical assemblages before A.D. 800, maize played a very minor role in eastern plant husbandry systems during the first 600 years after its initial introduction. But during the six-century span from A.D. 200 to 800 that maize remained a minor crop, at or below the level of archaeological visibility, the Eastern Woodlands witnessed a number of important interrelated developments in plant husbandry, technological innovation, and biocultural evolution.

Plant husbandry systems, centered on indigenous cultigens, continued to provide a significant, if difficult to quantify, percentage of the diet of populations over a broad area of the Midwest during this period (Pulliam 1987; Wymer 1986, 1987), and several new domesticates, in addition to maize, appear in archaeobotanical assemblages for the first time. Although not a food crop, tobacco *(Nicotiana rustica)* is known to have reached west-central Illinois by A.D. 160 (Asch and Asch 1985c) and the American Bottom east of St. Louis by A.D. 500 to 600 (Johannessen 1984). This cultigen was apparently introduced directly from Mexico rather than through the Southwest, and played an important ceremonial role through the remaining portion of the prehistoric sequence.

In addition, the pale-colored, extremely thin-testa variety of *Chenopodium berlandieri,* apparently brought under domestication by 3,000 B.P., occurs in food processing and storage contexts in increased quantities by A.D. 400, often in association with erect knotweed. The carbonized knotweed specimens recovered from parching pit contexts exhibit an A.D. 400–1100 temporal trend of morphological change suggesting that there is at least the possibility that the plant may have been brought under domestication (size increase, smoother pericarp surface, and reduction in pericarp thickness) (Fritz 1987). Although easily recognized when recovered uncarbonized from dry cave contexts, the seeds of this pale-colored chenopod cultivar are subject to substantial morphological distortion when carbonized, making them difficult to identify and resulting in their frequent underrepresentation in archaeobotanical assemblages. Fortunately, the processing (parching?) of large quantities of seeds of this pale-colored chenopod cultivar in shallow pit features sometimes produced, as a result of over-parching, large carbonized seed masses. In addition to documenting the continued role of this pale-colored chenopod as another starchy-seed crop within eastern plant husbandry systems, parching pit seed masses such as those excavated in west-central Illinois (Newbridge, circa A.D. 400), the American Bottom (Sampson's Bluff, Johanning, circa A.D. 800 to 1000), and southeast Missouri (Gypsy Joint, circa A.D. 1300, Smith 1978b; Fritz 1987) also indicate the mixed processing (and cultivation?) of chenopod cultigens (predominantly the pale-colored variety, but with some representation of the thin-testa type) with erect knotweed *(Polygonum erectum)*.

The co-occurrence of these crop plants in cultural contexts 450 km and almost 1,000 years apart suggests the possible development, by A.D. 400, of an indigenous crop complex that endured, at least along the Mississippi River Valley corridor, throughout the late prehistoric period.

The growing socioeconomic importance of plant husbandry systems over a broad area of the Midwest and Midsouth during the period A.D. 200 to 800 is reflected by the continued representation of indigenous cultigens in archaeobotanical assemblages and increasing evidence of deliberate human selection, the tandem increase in abundance of a pale-fruit chenopod and a possibly domesticated knotweed in an apparent crop complex association, and the addition of two tropical domesticates (maize and tobacco). All this takes place against a backdrop of considerable Late Woodland cultural change.

Population levels across the Midwest and Southeast rose markedly during this six-century span, judging from the documented increase, in a number of areas, of small one to four structure farmstead settlements. Such household structure clusters also sometimes loosely coalesced into dispersed communities of up to perhaps 50 inhabitants and covered less than 0.5 to 1.0 ha (Smith 1986). This widespread growth and dispersal of population within river valley landscapes is consistent with a growing reliance on food production.

In at least one region (west-central Illinois) this Late Woodland population increase has been tied to earlier weaning and shorter birth intervals (and higher fertility levels) associated with the availability of soft, palatable, digestible weaning foods (Buikstra et al. 1986). It is suggested that the weaning foods in question were probably the highly nutritious indigenous starchy seeded annuals. Thinner, more efficient ceramic vessels (Braun 1983) facilitated their preparation through boiling.

A variety of technological innovations in field preparation, and crop processing, storage, and preparation also accompanied the seemingly abrupt and widespread appearance of maize in the archaeological record of the Midwest and Southeast at A.D. 800 to 900 (Smith 1986). Large, well made chert hoes begin to be exchanged over substantial distances. Technologically superior calcium carbonate (ground mussel shell or limestone) tempered ceramic assemblages diversify, with a number of new vessel forms probably functioning in the processing ("parching" pans) and cooking (large globular jars) of maize and other cultigens. Below-ground storage pits increase in size.

During the 350-year span following the abrupt movement of maize into archaeological visibility at A.D. 800, eastern North America witnessed the rapid and widespread emergence of "Mississippian" ranked agricultural societies based on maize-dominated field agriculture (Smith 1990).

Although the limited stable carbon isotope analysis done to date (Lynott et al. 1986) suggests that maize did not become a staple food source until after A.D. 1100, there is no question that continued population growth and increasing reliance on maize and other cultigens represented a central theme in the initial Mississippian emergence.

Maize-Centered Field Agriculture after 800 B.P. (A.D. 1150)

By 800 B.P. (A.D. 1150) agricultural economies dominated by corn had been established in river valleys over a broad area of the eastern United States. With the arrival of the common bean *(Phaseolus vulgaris)* in some regions by A.D. 1000 to 1200, the crop triad of maize, beans, and squash was present in the East. During the final four centuries of North American prehistory, however, the eastern United States did not witness a uniform, homogeneous pattern of maize-beans-squash agriculture. On the contrary, currently available information suggests the continuation of a developmental mosaic across the East, with different regions exhibiting variation in the types of maize being grown and its dietary importance relative to indigenous crops and wild animal and plant resources.

The post A.D. 1000 Fort Ancient populations of the Ohio River Valley and its Ohio, Kentucky, and West Virginia tributaries represent the regional manifestation that perhaps most closely matches the common perception of prehistoric agriculture in the East. Although well represented in preceding Middle and Late Woodland archaeobotanical assemblages from the area (Wymer 1987), indigenous seed-bearing crops are thought to have been absent from Fort Ancient food production economies (Wagner 1987).

The common bean, on the other hand, which is not documented in the Central Mississippi Valley until A.D. 1450 (Morse and Morse 1983), and has not been reported at all from the American Bottom (Johannessen 1984), is relatively abundant in Fort Ancient contexts. Wagner's measurement of almost 800 beans from Fort Ancient sites shows them to be small, ranging from 8.6 to 11.7 mm in length and 5.6 to 8.8 mm in width (Wagner 1987, table 6), and to represent several varieties.

Attempting to assess regional variation in the dietary importance of the bean is complicated, however, by its low probability of being preserved in the archaeological record. As a result, its relative abundance or absence in archaeobotanical assemblages may be misleading. In contrast, the measurement of stable carbon isotope levels in human bone allows an accurate assessment of the dietary importance of maize, and studies done to date on Fort Ancient populations suggest that they had a greater reliance on maize than populations in other areas of the East (Wagner 1987). Described by Wagner (1987) as showing little variability and being morphologically similar to the historic early-maturing northern flint corns, typical Fort Ancient corn had a tapered 8-row cob (with some 10-row and a few 12-row and 14-row cobs present) ranging from 8.0 to 12.6 cm in length and 1.3 to 1.6 cm in midpoint diameter, and having an expanded base.

In contrast to the Ohio Valley Fort Ancient pattern of heavy reliance on a single, low variability type of maize, to the apparent total exclusion of indigenous cultigens, populations in other areas of the Eastern Woodlands continued to rely, to an admittedly reduced degree, upon indigenous plants (both cultivated and wild), while at the same time growing corn that exhibited considerably greater genetic diversity (Fritz 1990). Post A.D. 1000 archaeobotanical assemblages from west-central Alabama (Scarry 1986), eastern Tennessee (Chapman and Shea 1981), southeast Missouri (Fritz 1987), northwest Arkansas (Fritz 1986a), the American Bottom (Johannessen 1984), and west-central Illinois (Asch and Asch 1985b) all indicate the continued, if diminished, cultivation of various starchy and oily-seeded indigenous crop plants. In addition, the early eighteenth-century description by Le Page du Pratz of the cultivation of *"La belle dame sauvage,"* a small "grain," by the Natchez, both in corn fields and along the sand banks of the Mississippi, would appear to document the continued husbandry of a domesticated variety of chenopod into the historic period (Smith 1987c [Chapter 10]).

While a low variability, 8-row maize was being grown in the Ohio Valley and into the Northeast and upper Midwest after A.D. 1100, Mississippian populations in other areas of the Eastern Woodlands were growing varieties of maize exhibiting considerably greater variation in size and shape. Caches of well-preserved corn cobs recovered from northwestern Arkansas rock shelters and dating between A.D. 1200 to 1400 contained a high frequency of robust, wide base and wide cupule, 10- and 12-row cobs, along with a few 14-row cobs (Fritz 1986a:184–208). Although 8-row maize was also found to be abundant in Ozark bluff shelter collections, and cobs similar to Fort Ancient maize were occasionally present, most of the 8-row cobs recovered were small with contracting bases and narrow shanks, and were likely "representatives of unfavorable growing seasons, tillers, or possibly at times, of an early season 'little corn' variety" (Fritz 1986a:207). A similar variety of maize to that which dominated the Ozark assemblages was apparently grown throughout much of the Lower and Central Mississippi Valley, as far north as the American Bottom east of St. Louis, Missouri.

In contrast to both the Ozarks and the Ohio Valley, Mississippian period populations in Alabama were apparently growing at least two varieties of maize during the period A.D. 1000 to 1500 (Scarry 1986), one of which exhibited morphological change through time. A predominantly 12-row variety (with some 14-row and occasional 16-row cobs) having both tapered and straight cobs, persisted as a minority type, with a more abundant 8-row variety displaying a number of changes over time (decreasing mean row number, increasing cob diameter and cupule width).

These detailed analyses of large, relatively well preserved and well provenienced cob collections from the Ohio Valley, Ozarks, and Alabama, along with other regional studies, would seem at the present time to suggest that a latitudinal dichotomy in the type(s) of maize grown in the East was established prior to A.D. 1000, well before maize came to dominate fields and food production systems.

An 8-row variety of maize, which was probably adapted to short growing seasons and ancestral to the historic period northern flints, had been developed by A.D. 1000 within existing northern latitude plant husbandry systems, and had been widely adopted across the Northeast, Ohio Valley, and upper Midwest. Most likely because of a climatic adaptation to northern growing seasons, this variety of maize dominated, to the apparently almost total exclusion of other varieties, the post A.D. 1000 agricultural systems of the northern latitudes.

In contrast to the straightforward scenario of unchanging single-variety northern latitude 8-row maize agriculture, the extensive and environmentally diverse, longer growing season areas to the south of the Ohio Valley (and seemingly including a Central Mississippi Valley extension as far north as the American Bottom) witnessed a far more complex and regionally variable mosaic of maize husbandry. It is likely that one or more long growing season varieties of maize had been introduced into the Southeast prior to A.D. 500, and that subsequent, perhaps relatively frequent, arrivals of new varieties, either overland from the Southwest or northern Mexico, or across the Caribbean, contributed to an expanding gene pool and the development of a greater diversity of indigenous cultivar forms (Fritz 1990).

While corn kernels, cupules, and cobs recovered from a large number of prehistoric contexts south of the Ohio Valley have been described, and general developmental trends have been proposed (Blake 1987; Watson 1988), there is still much that is not known concerning the identity, developmental history, relative importance, and geographical and temporal distributions of different varieties of prehistoric eastern maize. These complex issues of prehistoric maize selection in the central and southeastern United States have been masked somewhat by the assignment of general taxonomic labels to corn collections, based largely on cob row number (often projected from kernel or cupule edge angles). Although it may never be possible to identify in archaeological contexts the full range of variation in maize cultivars documented at historic contact (Swanton 1946), the careful and selective analysis of intact cobs from storage contexts (Fritz 1986a; Scarry 1986), as well as kernels and cupules from general refuse deposits, should result in less reliance on general taxonomic gloss, and the recognition of different regional varieties, their range of morphological variation, and their developmental history.

Even though maize came to dominate fields after A.D. 1100, and perhaps contributed more than half of the population's annual caloric intake, judging from the stable carbon isotope studies done to date, the river valley farmers of the Eastern Woodlands continued to cultivate, to varying degrees, a wide range of other plants for a variety of purposes. A number of varieties of the common bean, differentially adopted into the plant husbandry systems of various areas after being introduced into the East around A.D. 1100, were cultivated as food plants, along with fleshy varieties of squash. In addition to fleshy forms of *Cucurbita pepo,* which may represent both indigenous domesticates and tropical cultivars, the green-striped cushaw *(Cucurbita mixta)* was present as far east as northwest Arkansas bluff shelters by circa A.D. 1430 (Fritz 1986a:153). A domesticated variety of pale-seeded amaranth *(Amaranthus hypochondriacus)* was also a late (ca. A.D. 1100) introduction into the Ozarks, but at the present time it does not appear to have been diffused any further east (Fritz 1984).

Along with these introduced tropical cultigens, a wide variety of indigenous food crops continued under cultivation in many areas (sumpweed, sunflower, knotweed, thin-testa and pale-colored varieties of chenopod, maygrass, and little barley). In addition, a number of other quasi-cultigens and weedy camp-followers were components, to varying degrees, of river valley plant husbandry systems after A.D. 1100, including Jerusalem artichoke *(Helianthus tuberosus)*, maypops *(Passiflora incarnata)*, amaranth, purslane, pokeweed, ragweed, chenopod, and carpetweed (Yarnell 1987).

Within this large group of husbanded plants, the emergence of maize as the primary field crop after A.D. 1100, and the associated increased importance of food production, did not represent an abrupt change in course for late prehistoric populations, but rather was an accelerated expansion of a long-established annual cycle of plant husbandry. For at least the preceding millennium, prehistoric farmers had been clearing land, preparing the soil, planting and cultivating crops, and at the end of the growing season, harvesting, processing, and storing them. Underlying this seasonal round of plant husbandry is an even older pattern of dependence upon the rich and diverse wild plant and animal resources of the eastern forests. And this basic, if regionally variable, framework of utilization of wild plants and animals also exhibits considerable continuity throughout the final 2,000 years of prehistory in the eastern United States. The maize-centered food production systems of late prehistoric groups in the East can thus best be viewed as largely compatible extensions of pre-existing, long-evolving subsistence systems. Because of the steep ecological gradient separating the river valleys of the East from intervening upland areas, the resource-rich linear floodplain corridors had attracted human hunter-gatherer groups long before plant husbandry played even a minor supporting role in subsistence systems. One of the most important attractions of these floodplain ecosystems was the localized and dependable aquatic protein sources of both main channel shoal areas and slackwater channel remnant oxbow lakes and backswamps. The need for animal protein certainly did not diminish with the post A.D. 1100 development of maize-centered field agriculture, and as much as half of the protein of at least some river valley maize agriculturalists came from fish and waterfowl (Smith 1975, 1978a). These main channel and backswamp aquatic resource zones were separated and paralleled by linear bands of natural levee soils, coalescing to form broad meander belts in the larger river valleys. Annually replenished by floodwaters, and easily tilled, these sandy, well-drained natural levee soils were highly prized by prehistoric farmers of the Eastern Woodlands, and the fortified mound centers, villages, and small single farmsteads of post A.D. 1100 maize agriculturalists were, with few exceptions, situated on or adjacent to these natural levee soils (Smith 1978a). Providing both easy access to protein-rich aquatic resource zones and excellent soils for simple hoe technology, these river valley natural levee ridges were extensively settled and brought under cultivation during the late prehistoric period. At European contact, early travelers provided accounts of riding for days along such natural levee ridge systems, through a landscape of farmsteads with adjacent "infield" gardens dispersed widely within larger and more extensive "outfield" field systems.

While such seventeenth- and eighteenth-century descriptions of Native American agricultural practices (as summarized by Swanton 1946 and Hudson 1976) provide valuable and tantalizing glimpses of important topics such as field preparation, planting schedules for different varieties of maize, intercropping and multiple cropping, and crop storage and preparation, there is limited evidence, either historical or archaeological, for many of the most basic aspects of late prehistoric, maize-centered agriculture.

It is tempting to assume that the small household "infield" gardens of the contact period represent the continuation of an earlier horticultural system based on indigenous plants, with the larger "outfield" cultivation plots being a late development associated with the emergence of maize agriculture. There is no reason, however, to restrict any of the indigenous crops to garden status, since the modern cultivated (sunflower) and weedy descendants of a number of them (sumpweed, knotweed, chenopod) have produced impressive harvest yield values (Asch and Asch 1978; Smith 1987b, table 5 [Chapters 7 and 8]), and were more than likely grown in "outfields" prior to A.D. 1000.

Evidence of prehistoric hilling or furrowing of fields is relatively rare in the East and with a few notable exceptions (Riley 1987) is restricted to the northern margins of maize agriculture (Gallagher et al. 1985). While the large chert hoes (Brown et al. 1990) employed (along with elk scapula hoes in the Northeast and shell hoes along the Atlantic and Gulf Coast) to prepare and maintain fields through the growing season are frequently recovered, wooden artifacts associated with the planting and preparation of crops, such as digging sticks and wooden mortars, are only rarely preserved in the archaeological record (Smith 1978b). Similarly, while below-ground pits document the storage of agricultural products in some areas, evidence of above-ground storage in gourds, bags, or elevated granaries (the *barbacoas* of the Spanish) is almost nonexistent.

While artifacts and site features, along with ethnohistorical accounts, will provide occasional glimpses of prehistoric plant husbandry practices, it is the plants themselves, preserved in the ground and in museum collections, which, with careful recovery, curation, and analysis, will continue to yield further understanding of the long and complex co-evolutionary relationship between human and plant populations in the Eastern Woodlands of North America.

Acknowledgments

I would like to thank C. Wesley Cowan, Gayle J. Fritz, James B. Griffin, and Patty Jo Watson for taking the time to read early drafts of this chapter. The numerous corrections and suggestions for revision that they offered substantially improved the text.

Literature Cited

Anderson, E.
1956 Man as a Maker of New Plants and Plant Communities. In *Man's Role in Changing the Face of the Earth,* vol. 2, edited by W. L. Thomas, pp. 763–777. University of Chicago Press, Chicago.

Asch, D., and N. Asch
1978 The Economic Potential of *Iva annua* and its Prehistoric Importance in the Lower Illinois Valley. In *The Nature and Status of Ethnobotany,* edited by R. Ford, pp. 300–341. Museum of Anthropology, Anthropological Papers No. 67. University of Michigan, Ann Arbor.

1982 A Chronology for the Development of Prehistoric Agriculture in West-Central Illinois. Paper presented at the 47th annual meeting of the Society for American Archaeology, Minneapolis. Center for American Archaeology, Archaeobotanical Laboratory Report 46. Kampsville, Illinois.

1985a Archeobotany. In *Smiling Dan,* edited by B. D. Stafford and M. B. Sant, pp. 327–399. Center for American Archaeology, Research Series 2. Kampsville, Illinois.

1985b Archeobotany. In *The Hill Creek Homestead,* edited by Michael D. Conner, pp. 115–170. Kampsville Archaeology Center, Research Series No. 1. Kampsville, Illinois.

1985c Prehistoric Plant Cultivation in West-Central Illinois. In *Prehistoric Food Production in North America,* edited by R. I. Ford, pp. 149–203. Museum of Anthropology, Anthropological Paper No. 75. University of Michigan, Ann Arbor.

Blake, Leonard
1987 Corn and Other Plants from Prehistory into History in the Eastern United States. In *Proceedings of the* 1983 Mid-South Archaeological Conference, edited by D. Dye and R. Brister. Mississippi Department of Archives and History, Archaeological Report 18. Jackson, Mississippi.

Braun, D. P.
1983 Pots as Tools. In *Archaeological Hammers and Theories,* edited by J. A. Moore and A. S. Keene, pp. 107–134. Academic Press, New York.

Brose, David, and N'omi Greber
1979 *Hopewell Archaeology.* Kent State University Press, Kent, Ohio.

Brown, J., R. Kerber, and H. Winters
1990 Trade and the Evolution of Exchange Relations at the Beginning of the Mississippi Period. In *The Mississippian Emergence,* edited by B. Smith, pp. 251–280. Smithsonian Institution Press, Washington, D.C.

Buikstra, J., L. W. Koningsberg, and J. Bullington
1986 Fertility and the Development of Agriculture in the Prehistoric Midwest. *American Antiquity* 51:528–546.

Chapman, J.
1981 *The Bacon Bend and Iddins Sites: The Late Archaic Period in the Lower Little Tennessee River Valley.* Department of Anthropology, University of Tennessee, Report of Investigations 31. Knoxville, Tennessee.

Chapman, J., and G. Crites
1987 Evidence for Early Maize *(Zea mays)* from the Icehouse Bottom Site, Tennessee. *American Antiquity* 52:352–354.

Chapman, J., and A. B. Shea
1981 The Archaeobotanical Record: Early Archaic Period to Contact in the Lower Little Tennessee River Valley. *Tennessee Anthropologist* 6:64–84.

Chomko, S. A., and G. W. Crawford
1978 Plant Husbandry in Prehistoric Eastern North America: New Evidence for its Development. *American Antiquity* 43:405–408.

Conard, N., D. L. Asch, N. B. Asch, D. Elmore, H. Gove, M. Rubin, J. A. Brown, M. D. Wiant, K. B. Farnsworth, and T. B. Cook
1984 Accelerator Radiocarbon Dating of Evidence for Prehistoric Horticulture in Illinois. *Nature* 308:443–446.

Cowan, C. W.
1978 The Prehistoric Use and Distribution of Maygrass in Eastern North America: Cultural and Phyto-Geographical Implications. In *The Nature and Status of Ethnobotany,* edited by R. I. Ford, pp. 263–288. Museum of Anthropology, Anthropological Paper No. 67. University of Michigan, Ann Arbor.

1985 *From Foraging to Incipient Food Production: Subsistence Change and Continuity on the Cumberland Plateau of Eastern Kentucky.* Ph.D. dissertation, Department of Anthropology, University of Michigan. University Microfilms, Ann Arbor.

1990 Prehistoric Cucurbits from the Cumberland Plateau of Eastern Kentucky. Paper presented at the Southeast Archaeological Conference, Mobile, Alabama, November 7–10.

Cowan, C.W., and B.D. Smith
1992 New Perspectives on a Free-Living Gourd in Eastern North America. Paper presented at the 15th annual conference of the Society of Ethnobiology, Smithsonian Institution, Washington, D.C.

Crites, G. D.
1978 Plant Food Utilization Patterns during the Middle Woodland Owl Hollow Phase in Tennessee: A Preliminary Report. *Tennessee Anthropologist* 3:79–92.

1985 Middle and Late Holocene Ethnobotany of the Hayes Site (40Mr139): Evidence from Unit 990N918E. Report submitted to the Tennessee Valley Authority. Knoxville.

Crites, G. D., and R. D. Terry
1984 Nutritive Value of Maygrass, *Phalaris caroliniana. Economic Botany* 38:114–120.

Decker, D.
1986 A Biosystematic Study of *Cucurbita pepo.* Ph.D. dissertation, Texas A & M University, College Station.

1988 Origin(s), Evolution, and Systematics of *Cucurbita pepo* (Cucurbitaceae). *Economic Botany* 42:4–15.

Decker-Walters, D., T. Walters, C.W. Cowan, and B.D. Smith
1992 Isozymic Characterization of Wild Populations of *Cucurbita pepo.* Paper presented at the 15th annual conference of the Society of Ethnobiology. Smithsonian Institution, Washington, D.C.

Decker, D., and H. D. Wilson
1986 Numerical Analysis of Seed Morphology in *Cucurbita pepo. Systematic Botany* 11:595–607.

Delcourt, P. A., and H. R. Delcourt
1981 Vegetation Maps for Eastern North America. In *Geobotany II,* edited by R. Romans, pp. 123–165. Plenum Publishing, New York.

Delcourt, P., H. Delcourt, P. Cridlebaugh, and J. Chapman
1986 Holocene Ethnobotanical and Paleoecological Record of Human Impact on Vegetation in the Little Tennessee River Valley, Tennessee. *Quaternary Research* 25:330–349.

Doran, G., D. Dickel, and L. Newsom
1990 A 7,290-Year-Old Bottle Gourd from the Windover Site, Florida. *American Antiquity* 55:354–360.

Ford, Richard I.
1985 Patterns of Prehistoric Food Production in North America. In *Prehistoric Food Production in North America,* edited by R. I. Ford, pp. 341–364. Museum of Anthropology, Anthropological Paper No. 75. University of Michigan, Ann Arbor.

1987 Dating Early Maize in the Eastern United States. Paper presented at the 10th annual conference of the Society of Ethnobiology, Gainesville, Florida, March 5–8.

Fowler, M.
1957 The Origin of Plant Cultivation in the Central Mississippi Valley: A Hypothesis. Paper presented at the annual meeting of the American Anthropological Association, Chicago.

1971 The Origin of Plant Cultivation in the Central Mississippi Valley: A Hypothesis. In *Prehistoric Agriculture,* edited by S. Struever, pp. 122–128. Natural History Press, Garden City, New York.

Fritz, G.
1984 Identification of Cultigen Amaranth and Chenopod from Rock Shelter Sites in Northwest Arkansas. *American Antiquity* 49:558–572.

1986a Prehistoric Ozark Agriculture. The University of Arkansas Rockshelter Collections. Ph.D. dissertation, Department of Anthropology, University of North Carolina, Chapel Hill.

1986b Starchy Grain Crops in the Eastern U.S.: Evidence from the Desiccated Ozark Plant Remains. Paper presented at the 51st annual meeting of the Society for American Archaeology, New Orleans.

1987 The Trajectory of Knotweed Domestication in Prehistoric Eastern North America. Paper presented at the 10th annual conference of the Society of Ethnology, Gainesville, Florida, March 5–8.

1990 Multiple Pathways to Farming in Precontact Eastern North America. In *Journal of World Prehistory,* vol. 4, edited by F. Wendorf and Angela Close, pp. 387–435. Plenum Publishing, New York.

Fritz, G., and B. Smith
1988 Old Collections and New Technology: Documenting the Domestication of Chenopodium in Eastern North America. *Midcontinental Journal of Archaeology* 13:3–27.

Gallagher, J. P., R. Boszhardt, R. Sasso, and K. Stevenson
1985 Oneota Ridged Field Agriculture in Southwestern Wisconsin. *American Antiquity* 50:605–612.

Gardner, P.
1987 Plant Food Subsistence at Salts Cave, Kentucky: New Evidence. *American Antiquity* 52:358–364.

Harlan, J. R., J. M. J. de Wet, and E. G. Price
1973 Comparative Evolution of Cereals. *Evolution* 27:311–325.

Heiser, Charles B.
1988 Aspects of Unconscious Selection and the Evolution of Domesticated Plants. *Euphytica* 37:77–85.

1989 Domestication of Cucurbitaceae: *Cucurbita* and *Lagenaria.* In *Foraging and Farming,* edited by D. Harris and G. Hillman, pp. 471–481. Unwin Hyman, London.

Hudson, C.
1976 *The Southeastern Indians.* University of Tennessee Press, Knoxville, Tennessee.

Johannessen, S.
1981 Plant Remains from the Truck 7 Site. In *Archaeological Investigations of the Middle Woodland Occupation at the Truck 7 and Go-Kart South Sites,* by A. C. Fortier, pp. 116–130. University of Illinois at Urbana-Champaign, Department of Anthropology, FAI-270 Archaeological Mitigation Project Report 30.

1983 Plant Remains from the Cement Hollow Phase. In *The Mund Site,* by A. C. Fortier, F. A. Finney, and R. B. LaCampagne, pp. 94–103. American Bottom Archaeology FAI-270 Site Reports 5. University of Illinois Press, Urbana and Chicago.

1984 Paleoethnobotany. In *American Bottom Archaeology,* edited by C. Bareis and J. Porter, pp. 197–214. University of Illinois Press, Urbana, Illinois.

King, F. B.
1985 Early Cultivated Cucurbits in Eastern North America. In *Prehistoric Food Production in North America,* edited by R. I. Ford, pp. 73–98. Museum of Anthropology, Anthropological Paper No. 75. University of Michigan, Ann Arbor.

Kline, G. W., G. D. Crites, and C. H. Faulkner
1982 *The McFarland Project: Early Middle Woodland Settlement and Subsistence in the Upper Duck*

River Valley in Tennessee. Tennessee Anthropological Association Miscellaneous Paper 8.

Knox, J. C.
1983 Responses of River Systems to Holocene Climates. In *Late Quaternary Environments of the United States,* vol. 2, *The Holocene,* edited by H. E. Wright, Jr., pp. 26–41. University of Minnesota Press, Minneapolis.

Lynott, M., T. Boutton, J. Price, and D. Nelson
1986 Stable Carbon Isotope Evidence for Maize Agriculture in Southeast Missouri and Northeast Arkansas. *American Antiquity* 51:51–65.

Meltzer, D., and B. D. Smith
1986 Paleoindian and Early Archaic Subsistence Strategies in Eastern North America. In *Foraging, Collection, and Harvesting: Archaic Period Subsistence and Settlement in the Eastern Woodlands,* edited by S. W. Neusius, pp. 3–31. Center for Archaeological Investigations, Occasional Paper No. 6. Southern Illinois University, Carbondale.

Morse, D. F., and P. A. Morse
1983 *Archaeology of the Central Mississippi Valley.* Academic Press, Orlando, Florida.

Munson, P. J.
1986 Hickory Silviculture: A Subsistence Revolution in the Prehistory of Eastern North America. Paper presented at a conference on Emergent Horticultural Economies of the Eastern Woodlands, Southern Illinois University, Carbondale, Illinois.

Pulliam, Christopher
1987 Middle and Late Woodland Horticultural Practices in the Western Margin of the Mississippi River Valley. In *Emergent Horticultural Economies of the Eastern Woodlands,* edited by W. Keegan, pp. 185–200. Center for Archaeological Investigations, Occasional Paper No. 7. Southern Illinois University, Carbondale.

Riley, T.
1987 Ridged-Field Agriculture and the Mississippian Economic Pattern. In *Emergent Horticultural Economies of the Eastern Woodlands,* edited by W. Keegan, pp. 295–305. Center for Archaeological Investigations, Occasional Paper No. 7. Southern Illinois University, Carbondale.

Scarry, M.
1986 *Change in Plant Procurement and Production during the Emergence of the Moundville Chiefdom.* Ph.D. dissertation, Department of Anthropology, University of Michigan, Ann Arbor.

Sears, W. H.
1982 *Fort Center.* Ripley P. Bullen Monographs in Anthropology and History No. 4. University of Florida Press, Gainesville, Florida.

Smith, Bruce D.
1975 *Middle Mississippi Exploitation of Animal Populations.* Museum of Anthropology, Anthropological Papers No. 57. University of Michigan, Ann Arbor.

1978a Variation in Mississippian Settlement Patterns. In *Mississippian Settlement Patterns,* edited by B. D. Smith, pp. 479–505. Academic Press, Orlando, Florida.

1978b *Prehistoric Patterns of Human Behavior: A Case Study in the Mississippi Valley.* Academic Press, New York.

1985a The Role of Chenopodium as a Domesticate in Pre-Maize Garden Systems of the Eastern United States. *Southeastern Archaeology* 4:51–72.

1985b *Chenopodium berlandieri* ssp. *jonesianum:* Evidence for a Hopewellian Domesticate from Ash Cave, Ohio. *Southeastern Archaeology* 4:107–133.

1986 The Archaeology of the Southeastern United States: From Dalton to de Soto, 10,500 B.P.–500 B.P. In *Advances in World Archaeology,* vol. 5, edited by F. Wendorf and A. Close, pp. 1–92. Academic Press, Orlando, Florida.

1987a The Independent Domestication of Indigenous Seed-Bearing Plants in Eastern North America. In *Emergent Horticultural Economies of the Eastern Woodlands,* edited by W. Keegan, pp. 1–47. Center for Archaeological Investigations, Occasional Paper No. 7. Southern Illinois University, Carbondale.

1987b The Economic Potential of *Chenopodium berlandieri* in Prehistoric Eastern North America. *Ethnobiology* 7:29–54.

1987c In Search of *Choupichoul,* the Mystical Grain of the Natchez. Keynote address, 10th annual conference of the Society of Ethnobiology, Gainesville, Florida, March 5–8.

Smith, Bruce D. (editor)
1990 *The Mississippian Emergence.* Smithsonian Institution Press, Washington, D.C.

Smith, Bruce D., and C. W. Cowan
1987 The Age of Domesticated *Chenopodium* in Prehistoric Eastern North America: New Accelerator Dates from Eastern Kentucky. *American Antiquity* 52:355–357.

Smith, Bruce D., and V. Funk
1985 A Newly Described Subfossil Cultivar of *Chenopodium* (Chenopodiaceae). *Phytologia* 57:445–449.

Swanton, J. R.
1946 *The Indians of the Southeastern United States.* Bureau of American Ethnology, Bulletin 137. Smithsonian Institution, Washington, D.C.

Wagner, G.
1987 The Corn and Cultivated Beans of the Fort Ancient Indians. *Missouri Archaeologist* 47:107–136.

Watson, Patty Jo
1974 Theoretical and Methodological Difficulties Encountered in Dealing with Paleofecal Material. In *Archaeology of the Mammoth Cave Area,* edited by P. J. Watson, pp. 239–241. Academic Press, Orlando, Florida.

1976 In Pursuit of Prehistoric Subsistence: A Comparative Account of Some Contemporary Flotation Techniques. *Midcontinental Journal of Archaeology* 1:77–100.

1985 The Impact of Early Horticulture in the Upland Drainages of the Midwest and Midsouth. In *Prehistoric Food Production in North America,* edited by R. I. Ford, pp. 99–148. Museum of Anthropology, Anthropological Papers No. 75. University of Michigan, Ann Arbor.

1988 Prehistoric Gardening and Agriculture in the Midwest and Midsouth. In *Interpretations of Culture Change in the Eastern Woodlands during the Late Woodland Period,* edited by R. Yerkes, pp. 39–67. Department of Anthropology, Occasional Papers in Anthropology. Ohio State University, Columbus.

Watson, P., and M. Kennedy
1991 The Development of Horticulture in the Eastern Woodlands of North America: Women's Role. In *Engendering Archaeology*, edited by J. Gero and M. Conkey, pp. 225–275. Basil Blackwell, Oxford, England.

Watt, B., and A. Merrill
1963 *Composition of Foods.* U.S. Department of Agriculture, Agricultural Handbook No. 8.

Wymer, D. A.
1986 The Archaeobotanical Assemblage of the Childers Site: The Late Woodland in Perspective. *West Virginia Archaeologist* 38:24–33.

1987 The Middle Woodland-Late Woodland Interface in Central Ohio: Subsistence Continuity Amid Cultural Change. In *Emergent Horticultural Economies of the Eastern Woodlands,* edited by W. Keegan, pp. 201–216. Center for Archaeological Investigations, Occasional Paper No. 7. Southern Illinois University, Carbondale.

Yarnell, R.
1969 Contents of Human Paleofeces. In *The Prehistory of Salts Cave, Kentucky,* edited by P. J. Watson. Illinois State Museum, Reports of Investigations No. 16. Springfield, Illinois.

1974 Plant Food and Cultivation of the Salts Cavers. In *Archaeology of the Mammoth Cave Area,* edited by P. J. Watson, pp. 113–122. Academic Press, Orlando, Florida.

1976 Early Plant Husbandry in Eastern North America. In *Cultural Change and Continuity,* edited by Charles E. Cleland, pp. 265–273. Academic Press, New York.

1978 Domestication of Sunflower and Sumpweed in Eastern North America. In *The Nature and Status of Ethnobotany,* edited by R. I. Ford, pp. 285–299. Museum of Anthropology, Anthropological Papers No. 67. University of Michigan, Ann Arbor.

1983 Prehistoric Plant Foods and Husbandry in Eastern North America. Paper presented at the 48th annual meeting of the Society for American Archaeology, Pittsburgh, Pennsylvania.

1987 A Survey of the Prehistoric Crop Plants in Eastern North America. *Missouri Archaeologist* 47:47–60.

Yarnell, R., and J. Black
1985 Temporal Trends Indicated by a Survey of Archaic and Woodland Plant Food Remains from Southeastern North America. *Southeastern Archaeology* 4:93–106.

Index